Differential Identities in Rings and Algebras and their Applications

Shakir Ali
Professor, Department of Mathematics
Aligarh Muslim University, India

Mohammad Ashraf
Professor, Department of Mathematics
Aligarh Muslim University, India

Vincenzo De Filippis
Professor, Department of Mathematics
and Computer Science
University of Messina, Messina, Italy

Lahcen Oukhtite
Professor, Department of Mathematics
Faculty of Sciences and Technology
S. M. Ben Abdellah University, Morocco

Nadeem Ur Rehman
Professor, Department of Mathematics
Aligarh Muslim University, India

CRC Press
Taylor & Francis Group
Boca Raton London New York

CRC Press is an imprint of the
Taylor & Francis Group, an **informa** business
A SCIENCE PUBLISHERS BOOK

First edition published 2025
by CRC Press
2385 NW Executive Center Drive, Suite 320, Boca Raton FL 33431

and by CRC Press
4 Park Square, Milton Park, Abingdon, Oxon, OX14 4RN

Library of Congress Cataloging-in-Publication Data (applied for)

ISBN: 978-1-032-81152-9 (hbk)
ISBN: 978-1-032-82454-3 (pbk)
ISBN: 978-1-003-50457-3 (ebk)

DOI: 10.1201/9781003504573

Typeset in Times New Roman
by Prime Publishing Services

Dedicated to Parents,
Families,
and Teachers

Preface

The landscape of modern algebra is as rich and diverse as it is complex, encompassing a wide array of structures and theories that intersect with numerous branches of mathematics. Among these, the study of differential identities in rings and algebras stands out for its profound theoretical significance and extensive applications. This book is a comprehensive exploration of derivations in rings and algebras, an area of ring theory that offers deep insights into the structure and behavior of algebraic systems. Our goal is to present a detailed and accessible account of the various forms and implications of derivations, providing readers with both foundational knowledge and advanced concepts essential for further research in this field. The concept of differential identities serves as a crucial link between algebra and differential equations, offering a unique perspective on the behavior of algebraic structures under differentiation. These identities not only enhance our understanding of the intrinsic properties of rings and algebras but also open up new avenues for solving complex problems in areas such as invariant theory, non-commutative geometry, and mathematical physics. In writing this book, our goal has been to create a resource that is both accessible and thorough, catering to the needs of a diverse audience. We begin with a detailed introduction to the fundamental concepts and techniques, ensuring that readers with a basic background in algebra can follow the material with ease. As the chapters progress, we delve into more advanced topics, exploring the latest research developments and highlighting the interplay between theory and application. Throughout the book, we emphasize the practical significance of differential identities. Each chapter includes examples and references to the most recent findings in the literature designed to reinforce understanding and encourage further exploration. Additionally, we present a series of case studies demonstrating how differential identities are applied in various fields.

In Chapter 1, we lay the groundwork by introducing basic notations and important terminologies. This chapter also includes key results and well-known theorems that will be instrumental in understanding the subsequent chapters. For those interested in the proofs of these theorems, references are provided. This chapter serves as a crucial starting point, offering definitions, results, terminology, and notations that will be consistently used throughout the book. We also give a brief exposition of essential terminologies in ring theory.

Chapter 2 delves into the diverse types of derivations in rings and algebras. We explore how these derivations manifest in different forms and examine their effects on mathematical objects. By doing so, we aim to clarify the fundamental concepts of derivations and underscore their significance in understanding algebraic structures.

In Chapter 3, we investigate the interplay between the structure of certain classes of rings with involution and the behavior of their derivations. Specifically, we characterize the commutativity of these rings by examining derivations that satisfy specific algebraic identities locally on appropriate subsets. This chapter reveals how the intrinsic properties of rings with involution influence their derivations.

Chapter 4 expands the discussion to include generalized derivations, generalized skew-derivations, b-generalized derivations, and their respective properties. The study of these generalized forms represents a fertile area of research within ring theory, offering insights into the complex relationships between algebraic structures and their associated derivations.

Chapter 5 builds a robust framework to address various seemingly unrelated problems, shedding light on issues that previously had efficient solutions but lacked a unified understanding. We focus on two well-known problems: the additivity of a particular class of higher derivable mappings in rings and the characterization of various kinds of higher derivations in settings such as triangular algebras, generalized matrix algebras, von Neumann algebras, and unital algebras.

Finally, Chapter 6 presents characterizations of derivations and their variants when they satisfy certain local properties. Recent literature has established a close and precise connection between the behavior of derivations and the intrinsic structure of the rings in which they are defined. This chapter aims to provide a deeper understanding of these relationships, highlighting the nuanced interactions between derivations and ring structures.

This book is intended for advanced students, researchers, and professionals in mathematics, particularly those specializing in algebra and ring theory. We hope that it will serve as both a valuable reference for seasoned researchers and a stimulating guide for newcomers to the field, offering both a comprehensive overview and detailed insights into the fascinating world of derivations in rings and algebras. This book would not have been possible without the contributions of many individuals. We extend our deepest gratitude to our colleagues and students, whose insightful questions and feedback have greatly enriched this work. We also acknowledge the support of our institutions, which provided the necessary resources and environment for research and writing. By bringing together the theoretical and practical aspects of differential identities, we aim to inspire further investigation and innovation in this fascinating area of algebra. May this book be a source of knowledge and inspiration for all who seek to understand and apply the beautiful intricacies of differential identities in rings and algebras.

Our view in writing this research monograph is also to stimulate new and further research on the topics presented. Several open research problems are listed at the end of some chapters to generate interest in research on these topics. We hope that the reader will find the material presented in an accessible and unified manner. The monograph is intended for research mathematicians in rings & algebras and for advanced graduate students in mathematics.

While we have attempted to make this book as self-contained as possible, it has been difficult because of the limitations on the size of this book. To keep the book to a reasonable length, some proofs (including a few highly technical ones) have been omitted and appropriate references to research papers or books have been included. We have also listed other references related to the material presented in the book. The brief historical notes provided at the beginning of each chapter, should be useful for researchers interested in further investigations. There are many excellent papers which we regrettably could not include in this book in order to keep it within a moderate expansion.

We are grateful to a number of colleagues and researchers for fruitful discussions on the subject of the book. It would be impossible to list all whose influence is felt in our writing. In particular, we are very thankful to Nazia Parveen, Junaid Nisar, Vaishali Varshney and Atif Ahmad Khan, for their final typing in LaTeX, painstaking proof-reading, and suggestions for improvements in this book. The technical help and support provided by Vaishali Varshney & Atif Ahmad Khan and corrections in several proofs they pointed out have been crucial during the preparation of this research monograph. Many others helped in proof-reading of various chapters of the book. These include, Mohammad Salahuddin Khan, Mohammad Afajal Ansari, Shamim Akhter, Pushpendra Sharma, Naira Noor Rafiquee, Mohd. Nazim, Shabir Ahmad Mir, Nazim and Mohammad Azeem. We are thankful to them for their time and efforts. The errors that still may remain in the book, are our fault.

There are others who have also played an important role, directly or indirectly, in the formation of this book. We are thankful to (late) S. Tariq Rizvi, (late) Jae Keol Park, Indah Emilia Wijayanti, Luisa Carini, Intan Muchtadi-Alamsyah, Abdullah Mamouni, Cihat Abdioglu, Elif Segah Oztas, Sharifah Kartini S.

Husain, Kok Bin Wong, Om Prakash, Ahmad Alkenani, Najat Muthana, Nuh Aydin, Mushir A Khan, Mohammad Yahya Abbasi, Ashok Ji Gupta, Atul Gaur, Yaser Tolloei for their influence on our work, words of encouragement, advice and support.

This work would not have been possible without the help and support of our families. We cannot ever fully express our heartfelt thanks to our respective wives, Siddiqa, Farah, Francesca, Fatima and Sana for their unstinting support throughout the preparation of this work. Their continuous support, and that of our children, have been a source of strength for us for which we are truly very grateful.

We are grateful to Vijay Primlani for the support and cooperation in the earlier stages of processing of this research monograph. The helpful cooperation and patience we received from Vijay Primlani and his team in the final processing of the book has been crucial. We greatly appreciate that and thank them for their personal interest and professional support. We are also thankful to other staff at CRC Press, Boca Raton, Florida, USA (A Taylor & Francis Group Company) who have been helpful in our book project.

Aligarh, India	Shakir Ali
Aligarh, India	Mohammad Ashraf
Messina, Italy	Vincenzo De Filippis
Fez, Morocco	Lahchen Oukhtite
Aligarh, India	Nadeem Ur Rehman
June, 2024	

Contents

List of Symbols

$\subseteq$, $\supseteq$	Inclusion
$\subset$, $\supset$	Proper inclusion
$\cap$	Intersection
$\cup$	Union
R	The ring
$\mathbb{N}$	The set of positive integers
$\mathbb{Z}$	The ring of integers
$\mathbb{Z}_n$	The ring of integers modulo n
$\mathbb{Q}$	The field of rational numbers
$\mathbb{R}$	The field of real numbers
$\mathbb{C}$	The field of complex numbers
$\mathbb{H}$	Division algebra of quaternions
δ_{ij}	Kronecker delta
$\|S\|$	Cardinality of the set S
$S \backslash T$	Set difference
id_S	Identity map on S
$\det(M)$	The determinant of the matrix M
$\mathrm{tr}(M)$	The trace of the matrix M
S_n	The Symmetric group on $\{1, ..., n\}$
$R \cong R'$	Isomorphic rings (or algebras, modules, etc.)
$\ker\phi$	Kernel of the homomorphism ϕ
$\mathrm{im}\phi$	Image of the homomorphism ϕ
R/I	The Factor ring
IJ	Product of ideals
$[a, b]$	Commutator of a and b
$a \circ b$	Anticommutator of a and b
$Z(R)$	The center of the ring R
$\mathrm{char}(R)$	The characteristic of R
R°	Opposite ring of R

$M_n(R)$	The ring of all $n \times n$ matrices over R
$T_n(R)$	The algebra of upper triangular $n \times n$ matrices over R
$[r_{ij}]$	The matrix whose (i, j)-position is r_{ij}
$\Pi_{i \in I} R_i$	The direct product of a family of rings
$R_1 \times ... R_n$	The direct product of a finite family of rings
$R[\omega]$	The ring of polynomials over R in one indeterminate
$\mathrm{rad}(R)$	The Jacobson radical of R
$\dim_F V$	Dimension of the vector space V over the field F
$C[a, b]$	Algebra of continuous functions from $[a, b]$ to R
L_a	The left multiplication map $x \longmapsto ax$
R_b	The right multiplication map $x \longmapsto xb$
$C_A(S)$	The centralizer of the set S in A
$\mathrm{ann}_R(M)$	The annihilator of the R-module M
$\mathrm{End}_R(M)$	The ring of endomorphisms of the R-module M
$\mathrm{End}_D(V)$	The ring of linear operators of the vector space V over the division ring D
$\mathrm{End}_F(V)\vert$	The algebra of linear operators of the vector space V over the field F
$M_1 \oplus \cdots \oplus M_n$	The direct sum of a finite family of submodules (or additive subgroups, or subspaces)
$\mathbb{Z} < X >$	The free algebra on X over $\mathbb{Z}$
$Q_r(R)$	The right Martindale ring of quotients of the ring R
$D(R)$	The set of all dense right ideals of the ring R
$\mathrm{Der}(R)$	The set of all derivations on the ring R
$r_R(S)$	The right annihilator of S in R
$\ell_R(S)$	The left annihilator of S in R
$H(R)$	The set of symmetric elements of R
$S(R)$	The set of skew-symmetric elements of R
$Q_{mr}(R)$	The maximal right ring of quotients
$Q_s(R)$	The symmetric ring of quotients of R
$Der(Q_r)$	The set of all derivations of Q_r
$D_{inn}(Q_r)$	The set of all inner derivations of Q_r
$(R, +, \circ)$	The Jordan ring
I_R	The identity map on R
$pDer(\mathcal{A})$	The set of all pre-derivations on $\mathcal{A}$
D^{op}	The opposite division ring
$Q(\mathcal{A})$	The set of all quasi nilpotent elements in $\mathcal{A}$
$Tri(A, M, B)$	The triangular algebra
$M_n(F)$	The ring of all $n \times n$ matrices over a field F
SCP	Strong commutativity preserving

R_C	Central closure
C	Extended centroid
$\in$	is an element of
$\notin$	is not an element of
$\cong$	is isomorphic to
$\ncong$	is not isomorphic to
GPI	Generalized polynomial identity
V_D	Vector space over division ring
m	von Neuman algebra
$f \circ g, fg$	Composition or product of functions *f* and *g*

Chapter 1

Preliminaries

1.1 Introduction

This chapter is devoted to collecting some basic notations and important terminologies. Some key results and well-known theorems are also included, which we shall require for the development of the subject matter in the subsequent chapters. For their proofs, references are mentioned for those who develop an interest in them. In the next section, we shall give a brief exposition of some important terminologies in ring theory. We begin with the basic notions, definitions, results, and notations used in the book. We recommend standard graduate text books such as [104, 141] or [294] for more details on this material.

1.2 Key definitions and notations

The contents presented in this section are largely conventional and aim to offer foundational knowledge. Its purpose is to furnish readers with essential background information. Throughout the text, when referring to a ring (algebra), we are specifically indicating an associative ring (algebra over a commutative ring). This ring may or may not have a unit element, and it is not necessarily commutative. In fact, many of our findings lose significance in a commutative context; they become either trivially true or trivially false. When we need to emphasize the presence of a unit element, we will use the term unital ring or unital algebra. Additionally, we will explore some non-associative rings and algebras, but we will always specify this distinction. For instance, a Lie ring will never be referred to simply as a ring. It is worth noting that a ring or algebra is considered non-associative if multiplication's associativity is not a requirement, but this lack of requirement does not necessarily mean associativity does not hold. For a ring R, we use I_R to denote the identity on R.

Let us outline the definitions of some fundamental types of rings. When we mention a field, we are referring to a commutative division ring. A division ring is essentially a nonzero ring with unity where all nonzero elements have multiplicative inverses. In broader terms, a ring R is labeled as simple if (0) and R are the only ideals of R, and there exist elements $a,\ b \in R$ such that their product equals 0. Ideals without the adjectives left or right represent two-sided ideals. When R is a ring, $I \lhd R$ denotes that I is an ideal of R. A ring R is said to be a domain if for all $a,b \in R$, $ab = 0$ implies $a = 0$ or $b = 0$. Equivalently, R is a domain if it has no left (or right) zero-divisors. In other words, R is a domain if it has the cancellation property: If $a \neq 0$, then each of $ab = ac$ and $ba = ca$ implies $b = c$.

Commutative domains are of extreme importance in rings theory. Noncommutative domains also form a remarkable class of rings, but their role in noncommutative algebra is incomparable to the role of commutative domains in commutative algebra. The most elementary examples of commutative rings are generally domains, however the matrix ring $M_n(F)$ for any field F, which one might consider to be a typical noncommutative ring, is not a domain.

The next proposition introduces a wider class of rings, which can be considered as a noncommutative counterpart of the class of commutative domains.

Proposition 1.2.1 *[140, Lemma 2.17] Let R be a ring. The following assertions are equivalent:*

1. *For all $a,b \in R$, $aRb = (0)$ implies $a = 0$ or $b = 0$.*
2. *For all left ideals I and J of R, $IJ = (0)$ implies $I = (0)$ or $J = (0)$.*
3. *For all right ideals I and J of R, $IJ = (0)$ implies $I = (0)$ or $J = (0)$.*
4. *For all two-sided ideals I and J of R, $IJ = (0)$ implies $I = (0)$ or $J = (0)$.*

Proof: Assume that I and J are left ideals satisfying $IJ = (0)$, then $IRJ = (0)$ because $RJ \subset J$. Hence we see that (1) implies (2). Similarly, (1) implies (3), and each of (2) and (3) obviously implies (4).

Now suppose that (4) holds and consider $a,b \in R$ such that $aRb = (0)$. The two-sided ideals RaR and RbR satisfy $RaR.RbR = (0)$. Hence one of them is 0. Assume that $RaR = (0)$, this implies that Ra and aR are two-sided ideals such that $Ra.R = R.aR = (0)$. Using (4), we get $Ra = aR = (0)$. This proves that Ra is a two-sided ideal of R satisfying $Ra.R = (0)$. Using (4), we get $a = 0$. Hence, (4) implies (1).

♣

Recall that a ring satisfying one (and hence all) of the conditions of Proposition 1.2.1 is called prime. In particular, a commutative ring is prime iff it is a domain. Indeed, for a commutative ring R, $ab = 0$ implies $aRb = (0)$.

Example 1.2.1 (1) *It is easy to check that the ring $M_n(\mathbb{Z})$, $n \geq 2$, is prime.*
(2) *Let* $R = \left\{ \begin{pmatrix} a & b \\ 0 & c \end{pmatrix} | a,b,c \in \mathbb{Z} \right\}$ *and set* $e = \begin{pmatrix} 0 & 1 \\ 0 & 0 \end{pmatrix}$. *It is easy to verify that* $eRe = (0)$ *and hence R is not prime.*
(3) *Let R_1 and R_2 be nonzero arbitrary rings. Then $R_1 \times R_2$ is not a prime ring, as $R_1 \times (0)$ and $0 \times R_2$ are nonzero ideals of $R_1 \times R_2$ whose product is 0.*

Proposition 1.2.2 *[140, Lemma 2.21] Let R be a ring. The following conditions are equivalent:*
1. *For all $a \in R$, $aRa = (0)$ implies $a = 0$.*
2. *For all left ideals I of R, $I^2 = (0)$ implies $I = (0)$.*
3. *For all right ideals I of R, $I^2 = (0)$ implies $I = (0)$.*
4. *For all ideals I of R, $I^2 = (0)$ implies $I = (0)$.*
5. *R has no nonzero nilpotent ideals.*

Proof: The proof that (1)–(4) are equivalent is similar to the proof of Proposition 1.2.1, so we omit it. It is trivial that (5) implies (4). Conversely, if (4) holds and I is an ideal such that $I^n = (0)$, then $(I^{n-1})^2 = (0)$ and hence $I^{n-1} = (0)$. Inductively, we get $I = (0)$.

♣

Recall that a ring satisfying one (and hence all) of the conditions of Proposition 1.2.2 is called semiprime.

Observe that if R is a semiprime ring, then each of the conditions $aR = (0)$ and $Ra = (0)$, as a particular case of (1), yields $a = 0$.

Recall that a ring R is said to be reduced if R has no nonzero nilpotent elements.

Remark 1.2.3 *It is imperative to verify that a commutative ring R is semiprime if and only if R is reduced.*

Let I be a nonzero ideal of a semiprime ring R. If $a \in I$ is such that $aIa = (0)$, then $aRaRaRa = (0)$. Using (1) from Proposition 1.2.2 twice it follows that $a = 0$. This shows that an ideal of a semiprime ring is a semiprime ring.

Example 1.2.2 (1) *Every prime ring is obviously semiprime.*
(2) *Consider the ring* $R = \left\{ \begin{pmatrix} a & b \\ 0 & c \end{pmatrix} | a,b,c \in \mathbb{Z} \right\}$. *If we set* $e = \begin{pmatrix} 0 & 1 \\ 0 & 0 \end{pmatrix}$, *then* $eRe = (0)$ *proving that R is not semiprime.*
(3) *A semiprime ring R is prime if and only if the intersection of any two nonzero ideals of R is nonzero.*
(4) *If N is a maximal nilpotent ideal of a ring R, then the factor ring R/N is semiprime.*

The following example provides some constructions which produce new classes of semiprime rings from the given semiprime rings.

Example 1.2.3 (1) *Let* R_1 *and* R_2 *be semiprime rings. Then their direct product* $R_1 \times R_2$ *is also semiprime (but not prime).*
(2) *Let* R *be a semiprime ring* R*. Then,* $M_n(R)$ *is semiprime.*

A proper ideal P of a ring R is called a *prime ideal* if for any two ideals A and B of R, $AB \subseteq P$ implies that either $A \subseteq P$ or $B \subseteq P$. A ring R is said to be a *prime ring* if its zero ideal is a prime ideal. An ideal P of a ring R is said to be a *semiprime ideal* in R if for every ideal I of R, $I^n \subseteq P$ implies $I \subseteq P$. A ring R is called a *semiprime ring* if all its nonzero ideals have nonzero multiplication, i.e., for an ideal A, the equality $A^2 = (0)$ implies $A = (0)$.

Remark 1.2.4 *If* P *is an ideal in a ring* R*, then the following conditions are equivalent:*

1. P *is a prime ideal of* R*.*
2. *If* $x, y \in R$ *such that* $xRy \subseteq P$*, then* $x \in P$ *or* $y \in P$*.*
3. *If* (x) *and* (y) *are principal ideals of* R *such that* $(x)(y) \subseteq P$*, then* $x \in P$ *or* $y \in P$*.*
4. *If* U *and* V *are right (or left) ideals of* R *such that* $UV \subseteq P$*, then* $U \subseteq P$ *or* $V \subseteq P$*.*

A non-empty set M is said to be an R-module if $(M, +)$ is an abelian group along with operation $\cdot : R \times M \to M$, satisfying the following conditions for all $r, s \in R$ and $x, y \in M$, (*i*) $r \cdot (x+y) = r \cdot x + r \cdot y$, (ii) $(r+s) \cdot x = r \cdot x + s \cdot x$, (iii) $(rs) \cdot x = r \cdot (s \cdot x)$. This structure ensures that the set M is an R-module. A ring R is termed primitive if it possesses a faithful simple module M. Here, simplicity of a module indicates that 0 and M are the only submodules of M, and $RM = (0)$, while faithfulness implies that $aM = (0)$ for every nonzero $a \in M$. The definitions of division, simple, primitive, prime, and semiprime algebras are formulated in a similar manner.

$M_n(R)$ and $T_n(R)$ denote the collection of $n \times n$ matrix rings and the $n \times n$ upper triangular matrix rings over a ring R, respectively. The notation $[r_{ij}]$ stands for the matrix whose (i, j)-position is r_{ij}. The center of R is denoted by $Z(R)$. The letters $\mathbb{Z}$, $\mathbb{Z}_n$, $\mathbb{Q}$, $\mathbb{R}$, and $\mathbb{C}$ are used for the ring of integers, the ring of integers modulo n (n is an integer greater than 1), the field of rational numbers, the field of real numbers, and the field of complex numbers, respectively. For any subset S of R, we will denote by $r_R(S)$, the right annihilator of S in R, that is, $r_R(S) = \{x \in R \mid Sx = (0)\}$ and by $\ell_R(S)$, the left annihilator of S in R, that is, $\ell_R(S) = \{x \in R \mid xS = (0)\}$. If the left and right annihilators coincide, then it is simply called an annihilator of S and is written as $ann_R(S)$. A *ring homomorphism* ϕ between two rings R and S is a function that preserves the ring structure, i.e., $\phi(x+y) = \phi(x) + \phi(y)$ and $\phi(xy) = \phi(x)\phi(y)$ for all $x, y \in R$. An *anti-homomorphism* ψ between two rings R and S is a function such that $\psi(x+y) = \psi(x) + \psi(y)$ and

$\psi(xy) = \psi(y)\psi(x)$ for all $x, y \in R$. An *endomorphism* is a ring homomorphism from a ring R to itself. An *automorphism* is a bijective endomorphism, meaning it is both injective and surjective. Finally, a *ring isomorphism* is a bijective ring homomorphism between two rings R and S, implying that there exists a function $\phi : R \to S$ that is both injective and surjective, and preserves the ring structure, making R and S structurally identical.

For elements x and y in a ring R we shall write $[x, y] = xy - yx$ and $x \circ y = xy + yx$. The element $[x, y]$ is called the Lie product (or the commutator) of elements x and y. Set $[x, y]_1 = [x, y]$, then inductively $[x, y]_k = [[x, y]_{k-1}, y]$ for $k > 1$ is called the kth Lie product. The element $x \circ y$ is called the Jordan product (anti-commutator) of elements x and y. Set $(x \circ y)_1 = (x \circ y)$, then inductively $(x \circ y)_k = ((x \circ y)_{k-1} \circ y)$ for $k > 1$ is called the kth Jordan product. By a commutative Lie ideal, we mean a Lie ideal which generates a commutative subring of R. In the case when $2R = (0)$ $(i.e., 2a = 0 \ for \ every \ a \in R)$, the Lie and the Jordan products coincide, which makes their treatment rather muddled. Therefore, we shall usually restrict our attention to rings which are 2-torsion free. We say that a ring R is n-torsion free, where $n \in \mathbb{N}$, if $na = 0$ implies $a = 0$. In the case when R is a prime ring, then it is more customary to express its n-torsion freeness through the notion of the characteristic, i.e., $char(R)$: R is n-torsion free if and only if $char(R) \neq n$. We also remark that the characteristic of a prime ring is always a prime number or 0.

Let's establish some additional notations. Given a non-empty subset S of a ring R, we denote $\langle S \rangle$ as the subring of R generated by S, and (S) as the ideal of R generated by S. The set $C(S) = \{a \in A \mid [a, S] = (0)\}$ is termed the centralizer of S in R. The center of R is essentially the centralizer of R in R, however, instead of $C(R)$, we typically denote it as $Z(R)$. When $S = \{t\}$, we use $C(t)$, $\langle t \rangle$ and (t) instead of $C(\{t\})$, $\langle \{t\} \rangle$ and $(\{t\})$, respectively. If R and T are additive subgroups of R, we represent by RT, $[R, T]$, and $R \circ T$ the additive subgroup of R generated by all rs, $[r, s]$, and $r \circ s$, respectively, where $r \in R$ and $s \in T$ are arbitrary elements. Similarly, we define $R_1 R_2 \cdots R_n, [R_1, [R_2, [\ldots, R_n] \ldots]]$.

A note of caution: Occasionally, R_n may denote the additive subgroup generated by all $r_1, r_2, \ldots, r_n$, with $r_i \in R_i$, but more commonly, it signifies the Cartesian product of n copies of R. However, the context should clarify in each instance what is meant.

Definition 1 *A mapping $F : R \to R$ is said to be centralizing on a subset S of R if $[F(s), s] \in Z(R)$ for all $s \in S$. In particular, if $[F(s), s] = 0$ for all $s \in S$, then F is commuting on S.*

Example 1.2.4 (1) *Every map having its range in the center of R, is commuting and a fortiori centralizing.*
(2) *Let $f(x) = \phi_0(x)x^n + \phi_1(x)x^{n-1} + \ldots + \phi_{n-1}(x)x + \phi_n(x)$, where $\phi_i : R \longrightarrow Z(R)$. Then f is centralizing.*

(3) *Consider the matrix ring* $R = M_2(\mathbb{Z})$ *and set* $S = \left\{ \begin{pmatrix} a & b \\ 0 & a \end{pmatrix} | a,b \in \mathbb{Z} \right\}$. *Then the mapping* $f : \begin{pmatrix} a & b \\ d & c \end{pmatrix} \longmapsto \begin{pmatrix} 0 & b \\ 0 & 0 \end{pmatrix}$ *is centralizing on* S.
(4) *Let* R *be a ring and* $f : R \longrightarrow R$ *be an arbitrary mapping. Set* $\mathfrak{R} = R \times R$ *and consider the mapping* $F : \mathfrak{R} \longrightarrow \mathfrak{R}$ *defined by* $F(x,y) = (f(x),0)$. *A simple computation shows that* F *is centralizing on* $S = (0) \times R$.

An additive subgroup L of A such that $[L,L] \subseteq L$ is termed a Lie subring of a ring R. If an additive subgroup I of a Lie subring L satisfies $[I,L] \subseteq I$, then I is called a Lie ideal of L. For instance, $[L,L]$ and $Z(L) = C(L) \cap L = \{l \in L | [l,L] = 0\}$, the center of L, are Lie ideals of L. Obviously, a two-sided ideal is a Lie ideal. The following example shows that the converse is not true in general.

Example 1.2.5 *Let* $R = \left\{ \begin{pmatrix} a & b \\ 0 & c \end{pmatrix} | a,b,c \in \mathbb{Z} \right\}$ *and* $U = \left\{ \begin{pmatrix} a & b \\ 0 & a \end{pmatrix} | a,b \in \mathbb{Z} \right\}$. *For* $M = \begin{pmatrix} a & b \\ 0 & c \end{pmatrix} \in R$ *and* $N = \begin{pmatrix} u & v \\ 0 & u \end{pmatrix} \in U$ *we have* $[M,N] = MN - NM = \begin{pmatrix} 0 & av - vc \\ 0 & 0 \end{pmatrix} \in U$ *which proves that* U *is a Lie ideal of* R. *However,* U *is not a two-sided ideal of* R.

Definition 2 *An additive subgroup* J *of* R *is called a Jordan ideal of* R *if* $x \circ r \in J$ *for all* $x \in J$ *and* $r \in R$.

Trivially, a two-sided ideal is a Jordan ideal. The following example proves that the converse is not true in general.

Example 1.2.6 *Let* $R = M_2(\mathbb{Z}/2\mathbb{Z})$ *and* $J = \left\{ \begin{pmatrix} a & b \\ b & a \end{pmatrix} | a,b \in \mathbb{Z}/2\mathbb{Z} \right\}$. *It is easy to check which* J *is a Jordan ideal of* R. *However,* J *is not a two-sided ideal of* R.

Remark 1.2.5 *For any* $x,y,z \in R$, *the following identities hold:*

1. $[xy,z] = x[y,z] + [x,z]y$.
2. $[x,yz] = y[x,z] + [x,y]z$.
3. $[[x,y],z] + [[y,z],x] + [[z,x],y] = 0$ *(this identity is generally known as the Jacobi identity).*
4. $x \circ (yz) = (x \circ y)z - y[x,z] = y(x \circ z) + [x,y]z$.
5. $(xy) \circ z = x(y \circ z) - [x,z]y = (x \circ z)y + x[y,z]$.

We now proceed to expose some well-known results, which will be useful later, and whose proofs can be found in Herstein's famous book "Topics in Ring Theory" [294], or can possibly be easily deduced from what is contained in the first chapter of the book just cited. For a complete description of the main properties of Lie ideals in prime rings, we also refer the reader to [126], [253] and [371].

Definition 3 *The standard identity,* $s_k(x_1,x_2,\ldots,x_k)$ *of degree k is defined by*

$$s_k(x_1,x_2,\ldots,x_k) = \sum_{\sigma\in S_k} (-1)^\sigma x_{\sigma_1}\cdots x_{\sigma_k}$$

where σ *runs over* S_k*, the symmetric group of degree k and where* $(-1)^\sigma$ *is +1 or −1 according as* σ *is even or odd.*

Theorem 1.2.6 *[371, Theorem 4] Let R be a prime ring,* $Z(R)$ *its center and L a commutative Lie ideal of R. Then either* $L\subset Z(R)$ *or* $char(R) = 2$*, R satisfies the standard identity* $s_4(x_1,\ldots,x_4)$ *and* RZ^{-1} *is a simple ring of dimension* 4 *over its center.*

Theorem 1.2.7 *[294, Lemma 1.3] Let R be a prime ring and L a non-commutative Lie ideal of R. Then there exists a non-zero ideal I of R such that* $0 \neq [I,R] \subseteq L$.

As a consequence of the above results, we get:

Theorem 1.2.8 *Let R be a prime ring and L a non-central Lie ideal of R. Then either there exists a non-zero ideal I of R such that* $0 \neq [I,R] \subseteq L$ *or L is commutative,* $char(R) = 2$*, R satisfies the standard identity* $s_4(x_1,\ldots,x_4)$ *and* RZ^{-1} *is a central simple algebra of dimension at most* 4 *over its center.*

The theory of rings characterized by polynomial identities has been extensively covered in various scholarly works such as [320], [489] and [507]. We will examine the key aspects of this theory essential for grasping functional identities. We will omit rigorous proofs, but rather try to give some informal evidence for the truthfulness of the results that will be stated. Let $X = \{x_1,x_2,\ldots\}$ be a countable set, and let $\mathbb{Z}\langle X\rangle$ be the free algebra on X over $\mathbb{Z}$. Let $f = f(x_1,\ldots,x_n) \in \mathbb{Z}\langle X\rangle$ be a polynomial such that at least one of its monomials of highest degree has coefficient 1. Let A be a non-empty subset of a ring R. Then, we say that f is a polynomial identity on A if $f(a_1,\ldots,a_n) = 0$ for all $a_1,\ldots,a_n \in A$, that is, if the polynomial function determined by f vanishes on A^n. In this case, we also say that A satisfies the polynomial identity f. In what follows we will only consider the case where $A = R$, i.e., we will treat rings satisfying polynomial identities. Such rings are called PI-rings. The simplest examples of PI-rings are commutative rings. Indeed, saying that a ring R is commutative is the same as saying that R satisfies the polynomial identity $x_1x_2 - x_2x_1$. Similarly, a ring R is Boolean if and only if it satisfies $x_1^2 = x_1$, and R is a nilpotent ring iff it satisfies $x_1x_2\ldots x_n$ for some positive integer n.

A polynomial $f = f(x_1,\ldots,x_n) \in \mathbb{Z}\langle X\rangle$ is said to be multilinear if every x_i, $1\leq i\leq n$, appears exactly once in each of the monomials of f. Thus, f is of the form

$$f(x_1,\ldots,x_n) = \sum \lim_{\pi\in S_n} n_\pi x_{\pi(1)}x_{\pi(2)}\cdots x_{\pi(n)},$$

where S_n is the symmetric group of order n and n_π are integers. If R satisfies a polynomial identity of degree n, then it also satisfies a multilinear polynomial identity of degree $\leq n$. Let us now recall some results from [373] and [396]. Let T be a ring with 1 and let $e_{ij} \in M_m(T)$ be the matrix unit having 1 as the (i, j)-entry and zero elsewhere. For a sequence $u = (A_1, \ldots, A_n)$ in $M_m(T)$ the value of u is defined as the product $|u| = A_1A_2 \cdots A_n$ and u is nonvanishing if $|u| \neq 0$. For a permutation σ of $\{1, 2, \ldots, n\}$, we write $u^\sigma = (A_{\sigma(1)}, \cdots, A_{\sigma(n)})$. We call u *simple* if it is of the form $u = (a_1e_{i_1j_1}, \ldots, a_ne_{i_nj_n})$, where $a_i \in T$. A simple sequence u is called *even* if for some σ, $|u^\sigma| = be_{ii} \neq 0$, and *odd* if for some σ, $|u^\sigma| = be_{ij} \neq 0$, where $i \neq j$, $b \in T$. We have:

Lemma 1.2.9 [373, Lemma 1] *Let T be a K-algebra with 1 and let $R = M_m(T)$, $m \geq 2$. Suppose that $h(x_1, \ldots, x_n)$ is a multilinear polynomial over K such that $h(u) = 0$ for all odd simple sequences u. Then $h(x_1, \ldots, x_n)$ is centrally valued on R.*

Lemma 1.2.10 [396, Lemma 2] *Let T be a K-algebra with 1 and let $R = M_m(T)$, $m \geq 2$. Suppose that $h(x_1, \ldots, x_n)$ is a multilinear polynomial over K. Let $u = (A_1, \ldots, A_n)$ be a simple sequence from R.*

1. *If u is even, then $h(u)$ is a diagonal matrix.*
2. *If u is odd, then $h(u) = ae_{pq}$ for some $a \in T$ and $p \neq q$.*

Now let $R = M_m(C)$, where C is a field, and let $f(x_1, \cdots, x_n)$ be a multilinear polynomial over C, not centrally valued on R. Then we have the following simple remarks:

Remark 1.2.11 *Assume either $m > 2$ or $m = 2$ and $char(R) \neq 2$. By [396, Lemma 9], there exists some even sequence u from R such that $f(u)$ is not in the center. Since u is even, by Lemma 1.2.10, $f(u)$ is a diagonal matrix.*

Remark 1.2.12 *By Lemma 1.2.9, there exists an odd simple sequence $r = (r_1, \ldots, r_n)$ from R such that $f(r) = f(r_1, \ldots, r_n) \neq 0$. By Lemma 1.2.10, $f(r) = \beta e_{pq}$, where $0 \neq \beta \in C$ and $p \neq q$. Since $f(x_1, \ldots, x_n)$ is a multilinear polynomial and C is a field, we may assume that $\beta = 1$. Now, for distinct i, j, let $\sigma \in S_n$ be such that $\sigma(p) = i$ and $\sigma(q) = j$, and let ψ be the automorphism of R defined by $\psi(\sum_{s,t} \xi_{st}e_{st}) = \sum_{s,t} \xi_{st}e_{\sigma(s)\sigma(t)}$. Then $f(\psi(r)) = f(\psi(r_1), \ldots, \psi(r_n)) = \psi(f(r)) = \beta e_{ij} = e_{ij}$.*

Definition 4 *An additive mapping $* : R \longrightarrow R$ is said to be an involution of the ring R if it obeys the following properties: (i) $(x^*)^* = x$, (ii) $(xy)^* = y^*x^*$ for all $x, y \in R$.*

A ring equipped with an involution is called a $*$-ring or ring with an involution. In other words, an involution is an anti-automorphism of period 2. Let $M_{n\times n}(\mathbb{R})$ be the set of all $n \times n$ matrices over the real field $\mathbb{R}$. Take $A \in M_{n\times n}(\mathbb{R})$ and define $*: M_{n\times n}(\mathbb{R}) \to M_{n\times n}(\mathbb{R})$ such that $A^* = A^T$, the transpose of A. Then $*$ is an involution on $M_{n\times n}(\mathbb{R})$ and hence $M_{n\times n}(\mathbb{R})$ is a ring with an involution. Furthermore, another typical example is the conjugation $z \longmapsto \bar{z}$ on the field of complex numbers $\mathbb{C}$.

Let $(R,*)$ be a ring with an involution. Then $*$ is designated as the first kind if it is the identity on $Z(R)$; otherwise, it is said to be of the second kind. In other words, $*$ is of the first kind if it leaves the center $Z(R)$ elementwise invariant.

Example 1.2.7 (1) *For a field F, a natural example of an involution of the first kind on the matrix algebra $M_n(F)$ is the transpose involution $t : (a_{ij}) \longmapsto (a_{ji})$.*
(2) *Let us consider $R = \left\{\begin{pmatrix} a & b \\ 0 & c\end{pmatrix} | a,b,c \in \mathbb{Z}\right\}$, where $\mathbb{Z}$ is the ring of integers and set $\begin{pmatrix} a & b \\ 0 & c\end{pmatrix}^* = \begin{pmatrix} a & -b \\ 0 & c\end{pmatrix}$. It is straightforward to check that $*$ is an involution of the first kind.*
(3) *Set $R = M_2(\mathbb{C})$ and consider $* : R \longrightarrow R$ defined by $\begin{pmatrix} a & b \\ c & d\end{pmatrix}^* = \begin{pmatrix} \bar{a} & \bar{c} \\ \bar{b} & \bar{d}\end{pmatrix}$ where $\bar{z}$ denotes the conjugate of z. Obviously, $Z(R) = \left\{\begin{pmatrix} a & 0 \\ 0 & a\end{pmatrix} | a \in \mathbb{C}\right\}$. If we put $z = \begin{pmatrix} i & 0 \\ 0 & i\end{pmatrix}$ where $i^2 = -1$, then $z \in Z(R)$ but $z^* = -z$ which shows that the involution $*$ is of the second kind.*

The involution is said to be of the first kind if $Z(R) \subseteq H(R)$, otherwise it is said to be of the second kind. In the latter case $S(R) \cap Z(R) \neq (0)$. If R is a 2-torsion free $*$-ring, then every element $x \in R$ can be uniquely represented in the form $2x = h + k$, where $h \in H(R)$ and $k \in S(R)$.

Suppose $\mathcal{A}$ and $\mathcal{B}$ are two algebras over a ring R and M is a unital $(\mathcal{A},\ \mathcal{B})$-bimodule. Then,

$$\begin{bmatrix} \mathcal{A} & M \\ 0 & \mathcal{B}\end{bmatrix} = \left\{\begin{bmatrix} a & m \\ 0 & b\end{bmatrix} : a \in \mathcal{A},\ m \in M,\ b \in \mathcal{B}\right\},$$

with the usual 2×2 matrix-like addition and the matrix-like multiplication is an algebra. An algebra μ is called a triangular algebra if there exist algebras $\mathcal{A}$ and $\mathcal{B}$ and an $(\mathcal{A},\ \mathcal{B})$-bimodule M such that μ is (algebraically) isomorphic to $\begin{bmatrix} \mathcal{A} & M \\ 0 & \mathcal{B}\end{bmatrix}$ under matrix-like addition and matrix-like multiplication. The algebra μ_n of $n \times n$ upper triangular matrices over the complex field $\mathbb{C}$ may be viewed as triangular algebras when $n > 1$. A C^*-algebra is a Banach algebra $\mathcal{A}$ having an involution $*$ (that is, a conjugate linear map of A into itself satisfying $x^{**} = x$ and $(xy)^* = y^*x^*$, $x,\ y \in \mathcal{A}$) which satisfies $||x^*x|| = ||x||^2$ for all $x \in \mathcal{A}$. A left (right) Leibniz algebra is an algebra A satisfying the left (right) Leibniz identity

$$x(yz) = (xy)z + y(xz) \text{ and } (xy)z = (xz)y + x(yz),$$

for all $x,\ y,\ z \in \mathcal{A}$.

Definition 5 *(Zero-power valued map) [256, Definition 3.4.6] A map $\phi : S \subseteq R \to R$ is called a zero-power valued on S, if for each $s \in S$, there exists a positive integer $n(s) > 1$ such that $\phi^{n(s)}(s) = 0$.*

Example 1.2.8 *Let $R = \mathbb{Z}[x]$ be the ring of polynomials with integral coefficients. Define a map $\phi : R \to R$ by $\phi(f(x)) = f'(x)$. ϕ so defined forms a zero-power valued map.*

Definition 6 *A ring R is defined as right alternative if*

$$xy \cdot y - x \cdot yy = 0,\ \forall\, x,\ y \in R,$$

is an identical relation in ring.

Definition 7 *In a ring R, two idempotents e and f are called orthogonal if $ef = 0$ and $fe = 0$. An idempotent e in a ring R is said to be central if $er = re$ for all $r \in R$.*

Definition 8 *A $\mathbb{Z}$-linear map, also known as an integer-linear map, is a linear map between modules over the ring of integers $\mathbb{Z}$. Specifically, let M and N be $\mathbb{Z}$-modules. A function $f : M \to N$ is $\mathbb{Z}$-linear if it satisfies the following properties for all $x, y \in M$ and $n \in \mathbb{Z}$:*

$$\begin{aligned} f(x+y) &= f(x) + f(y), \\ f(nx) &= nf(x). \end{aligned}$$

Example 1.2.9 *Define a map $f : \mathbb{Z} \to \mathbb{Z}$ by $f(x) = 2x$ for all $x \in \mathbb{Z}$. Then, f is a $\mathbb{Z}$-linear map.*

Example 1.2.10 *Let $M = \left\{ \begin{bmatrix} a & b \\ c & d \end{bmatrix} \mid a,b,c,d \in \mathbb{Z} \right\}$ be a $\mathbb{Z}$-module. Define a map $f : M \to M$ by $f(A) = 6A$ for all $A \in M$. Since f satisfies both properties, f is a $\mathbb{Z}$-linear map.*

1.3 Rings of quotients

Let R be a semiprime ring, ξ be the set of all pairs (U, f), where $U \neq (0)$ is a dense right ideal of R and $f : R \to R$ is a right R-module map of U into R. Define a relation $\sim$ on ξ such that $(U, f) \sim (V, g)$ if $f = g$ on some dense right ideal $W \neq (0)$ of R, where $W \subseteq U \cap V$. It can be easily checked that $\sim$ is an equivalence relation on ξ. Let $Q_{mr}(R)$ be the set of equivalence classes of $\sim$. Denote the equivalence class determined by (U,f) as $\widetilde{f}$. For $\widetilde{f} = cl(U,f)$, $\widetilde{g} = cl(V,g) \in Q_{mr}(R)$, define addition and multiplication on $Q_{mr}(R)$ as $\widetilde{f}+\widetilde{g} = cl(U \cap V, f+g)$ and $\widetilde{f} \times \widetilde{g} = cl(g^{-1}(U), fg)$. Thus, $Q_{mr}(R)$ forms an associative ring with an identity relative to the above defined operations known as the maximal right ring of quotients or right Utumi quotient ring of R. Similarly, we define the maximal left ring of quotients or right Utumi quotient ring of R.

Definition 9 (Dense right (left) ideal) *A right (resp. left) ideal I of R is said to be dense right (resp. left) ideal of R if for any $0 \neq r_1 \in R, r_2 \in R$ there exists $r \in R$ such that $r_1 r \neq 0$ and $r_2 r \in I$ (resp. $rr_1 \neq 0$ and $rr_2 \in I$). The collection of all dense right ideals of R will be denoted by $D(R)$.*

Proposition 1.3.1 *[104, Proposition 2.1.1] Let $I,J,S \in D(R)$ and let $f : I \to R$ be a homomorphism of right R-modules. Then:*

1. *$f^{-1}(J) = \{a \in I \mid f(a) \in J\} \in D(R)$.*
2. *$(a : J) \in D(R)$ for all $a \in R$.*
3. *$I \cap J \in D(R)$.*
4. *$IJ \in D(R)$.*
5. *If K is a right ideal of R and $K \supseteq I$, then $K \in D(R)$.*
6. *$\ell(I) = (0) = r(I)$.*
7. *If K is a right ideal of R and $(a : K) \in D(R)$ for all $a \in I$, then $K \in D(R)$.*
8. *If L is a right ideal of R and $g : L \to R$ is a homomorphism of right S-modules, then g is a homomorphism of right R-modules.*

Corollary 1.3.1 *[104, Corollary 2.1.2] Let J be a right ideal of R. Then $J \in D$ iff $\ell_r((a : J)) = (0)$ for all $a \in R$.*

Let's take a moment to discuss the concept of an essential right ideal. We recall that a right ideal J of R is essential if for every nonzero right ideal K of R, we have $J \cap K \neq (0)$.

Remark 1.3.2 *[104, Remark 2.1.3] Let J be a dense right ideal of R. Then J is an essential right ideal of R.*

Remark 1.3.3 *[104, Remark 2.1.4] Let I be a 2-sided ideal of R. Then the following conditions are equivalent:*

1. *$\ell(I) = (0)$.*
2. *I is a dense right ideal.*
3. *I is an essential right ideal.*
4. *I is essential as a 2-sided ideal (i.e., for any ideal $J \neq (0)$, $I \cap J \neq (0)$).*

Remark 1.3.4 *[104, Remark 2.1.5] Let I be a 2-sided ideal of R. Then:*

1. $\ell(I) = r(I)$.
2. $\ell(I) \cap I = (0)$.
3. $I + \ell(I)$ *is a dense right ideal of R.*

We continue now with our discussion of dense right ideals, highlighting the following useful properties.

Remark 1.3.5 *[104, Remark 2.1.6] Let J be a right ideal of R and let* $f : J \to R$ *be a right R-module homomorphism. Then:*

1. *If* $a \in R$ *and* $r(a) \in D(R)$, *then* $a = 0$.
2. *If* $ker(f) \in D(R)$, *then* $f = 0$.

Let R be a semiprime ring and let $\mathcal{I} = \mathcal{I}(R) = \{I \mid I\ is\ an\ ideal\ of\ R\ and\ \ell_R(I) = 0\}$. We note that $\mathcal{I}$ is closed under products and finite intersections. We also note that any $I \in \mathcal{I}$ is dense and essential as a right (or left) ideal and accordingly we shall call such ideals dense. Consider the set

$$\mathcal{T} = \{(f;J) \mid J \in \mathcal{I},\ f : J_R \to R_R\}$$

and define $(f;J) \simeq (g;K)$ if there exists $L \subseteq J \cap K$ such that $L \in \mathcal{I}$ and $f = g$ on L. We let $\{f;J\}$ denote the equivalence class determined by $(f;J) \in \mathcal{T}$. We now define addition and multiplication of equivalence classes as follows:

$$\{f;J\} + \{g;K\} = \{f + g;KJ\}$$

$$\{f;J\}\{g;K\} = \{fg;KJ\}.$$

We only show that multiplication is well-defined. First of all we note that $KJ \in \mathcal{I}$ whenever $K, J \in \mathcal{I}$. Indeed, let $rKJ = 0$ for some $r \in R$. Then $rK \subseteq \ell_R(J) = 0$ and so $rK = 0$. Hence $r \in \ell_R(K) = 0, r = 0$ and $KJ \in \mathcal{I}$. Further $g(KJ) = g(K)J \subseteq J$, so the composition fg is well-defined on KJ. If $(f_1;J_1) \simeq (f_2;J_2)$ and $(g_1;K_1) \simeq (g_2;K_2)$ we may find $L \in \mathcal{I}$ such that $L \subseteq J_1 \cap J_2$, $f_1 = f_2$ on L and $ME \in \mathcal{I}$ such that $M \subseteq K_1 \cap K_2$, $g_1 = g_2$ on M. Set $N = ML$, and let $x \in N$. Then $N \in \mathcal{I}$ and

$$f_1g_1(x) = f_1(g_1(x)) = f_1(g_2(x)) = f_2(g_2(x))$$

(noting that $g_1(x) = g_2(x) \in L$). Thus, the multiplication is well defined. The reader can readily verify that the ring axioms hold, and so our construction is complete. We shall denote the ring constructed above by $Q_r = Q_r(R)$ and shall call it the two-sided right ring of quotients or right Martindale ring of quotients of R. We are now in a position to characterize the two-sided ring by its properties.

Proposition 1.3.6 *[104, Proposition 2.2.1] Let R be a semiprime ring. Then $Q_r(R)$ satisfies:*

1. *R is a subring of $Q_r(R)$.*
2. *For all $q \in Q_r(R)$, there exists $I \in D(R)$ such that $qI \subseteq R$.*
3. *For all $q \in Q_r(R)$ and $I \in D(R)$, $qI = (0)$ iff $q = 0$.*
4. *For all $I \in D(R)$ and $f : I_R \to R_R$ there exists $q \in Q_r(R)$ such that $f(x) = qx$ for all $x \in I$.*

Furthermore, properties $1-4$ characterize ring $Q_r(R)$ up to isomorphism.

Definition 10 *(Symmetric ring of quotients). Let R be a semiprime ring and $W = \{\ I \mid I\ is\ an\ ideal\ of\ R\ and\ \ell(I) = (0)\}$. We note that W is closed under products and finite intersection. Then the symmetric ring of quotients of R denoted by $Q_s(R)$ is defined as follows:*

$$Q_s(R) = \{q \in Q_r(R) \mid qJ \cup Jq \subseteq R\ \ for\ \ some\ \ J \in W\}.$$

One can easily check that $Q_s(R)$ is a subring of $Q_r(R)$. $Q_s(R)$ may also be characterized by the following four properties:

Proposition 1.3.7 *[104, Proposition 2.2.3] Let R be a semiprime ring. Then $Q_s(R)$ satisfies:*

1. *R is a subring of Q_s.*
2. *For all $q \in Q_s$ there exists $J \in I$ such that $qJ \cup Jq \subseteq R$.*
3. *For all $q \in Q_s$ and $J \in I$, $qJ = 0$ (or $Jq = 0$) iff $q = 0$.*
4. *Given $J \in I$, $f : J_R \to R_R$ and $g : {}_RJ \to {}_RR$ such that $xf(y) = g(x)y$ for all $x, y \in J$, there exists $q \in Q_s$ such that $qx = f(x)$, $xq = g(x)$ for all $x \in J$.*

Furthermore, the above properties $1-4$ characterize ring $Q_s(R)$ up to isomorphism.

Definition 11 *(Extended centroid). The center C of $Q_r(R)$ of R is known as extended centroid of R.*

Definition 12 *(Central closure) Let R be a semiprime ring. Then the subring RC of $Q_r(R)$ is said to be the central closure of R. Further, R is called centrally closed if it coincides with its central closure, i.e., $R = RC$.*

Theorem 1.3.8 *[315, Theorem 13.14]* ***The Jacobson density theorem:*** *Let U be a simple right R-module, $D = End(U_R)$, and $X \subset U$ a finite and D-linearly independent set. If A is a D-linear transformation on U, then there exists $r \in R$ such that $A(x) = x \cdot r$ for all $x \in X$.*

1.4 Differential identities in prime rings

By a differential identity refers to an expression that includes unknowns, ring operations, elements of the given ring and also derivations, considered as unary operations. Here we recall some well known concepts and results which have been studied in detail by Kharchenko [345] and [346] and Lee [372]. We recall that, for a prime ring R and an ideal I of R, an additive map $d: I \to R$ is called a derivation on I if $d(xy) = d(x)y + xd(y)$, for all $x, y \in I$. Furthermore, every derivation defined on an ideal I of a prime ring R can be uniquely extended to a derivation of the right Martindale quotient ring Q_r of R. Hence, without loss of generality, we may assume that all derivations are defined on Q_r. Note that all derivations of Q_r are completely continuous maps for the C-module Q_r. We denote the set of all derivations of Q_r and the set of all inner derivations of Q_r by $Der(Q_r)$ and $D_{\text{int}}(Q_r)$, respectively. For $d \in Der(Q_r)$ and $x \in Q_r$, one usually writes x^d for $d(x)$. Additionally, if $\lambda \in C$, define $x^{d\lambda} = x^d\lambda$. It follows that, in general (for instance when R is semiprime) $Der(Q_r)$ forms a right C-module and $D_{\text{int}}(Q_r)$ can be regarded as a submodule of the C-module $Der(Q_r)$. In particular, by our assumption on the primeness of R, since C is a field we have that $Der(Q_r)$ forms a right vector space over C and $D_{\text{int}}(Q_r)$ can be regarded as a vector subspace of $Der(Q_r)$. Let $d_1, \ldots, d_n$ be derivations. Then, we call $\Delta = d_1 \cdots d_n$ a *derivation word*. A word without derivations is said to be an *empty word*. If Δ is an empty word, then $x^\Delta = x$, for all $x \in Q_r$.

Definition 13 *A differential polynomial means a generalized polynomial with coefficients in Q_r with noncommuting variables x_i which are acted on by derivation words Δ_j. In other words, every differential polynomial can be written in the form $\Phi(x_i^{\Delta_j})$, where $\Phi(z_{ij})$ is a generalized polynomial over Q_r in distinct variables z_{ij} and the Δ_j's are derivation words. Hence, the polynomial $\Phi(x_i^{\Delta_j})$ includes in its presentation not only the variables, ring operations and coefficients from Q_r, but also the derivation operations.*

Definition 14 *A differential polynomial $\Phi(x_i^{\Delta_j})$ is called a differential identity (DI) for a subset T of Q_r if it assumes 0 for assigning values from T to its variables x_i:*

$$\Phi(r_i^{\Delta_j}) = 0 \text{ for all } r_i \in Q_r.$$

The following basic differential identities hold in Q_r:

1. $(xy)^d = x^d y + xy^d$, where $d \in Der(Q_r)$.
2. $(x+y)^d = x^d + y^d$, where $d \in Der(Q_r)$.
3. $x^d = xa - ax$, where d is the inner derivation induced by $a \in Q_r$.
4. $x^{[d,\delta]} = x^{\delta d} - x^{d\delta}$, where $d, \delta \in Der(Q_r)$.

5. $\left(\cdots\left(\left(x^{d}\right)\right)^{d}\cdots\right)^{d} = x^{d^p}$, where $d \in Der(Q_r)$, R has characteristic p, and either p is a prime number or $p = 0$. In the latter case, the previous identity assumes the form $x = x$.

6. $x^{d\alpha+\delta\beta} = x^{d}\alpha + x^{\delta}\beta$, where $d, \delta \in Der(Q_r)$ and $\alpha, \beta \in C$.

Choose a fixed basis M_0 for $D_{int}(Q_r)$ and augment it to a basis M for $Der(Q_r)$. Notice that not only, as it is obvious, M is a C-linearly idependent set, but also the set $M \setminus M_0$ must be C-linear indepenedent modulo $D_{int}(Q_r)$, in the sense that any linear combination $\sum \delta_i \lambda_i$, where $\lambda_i \in C$, for various derivations $\{\delta_i\}$ from $M \setminus M_0$ cannot be an inner derivation for the ring Q_r. Fix a total order in M such that $d_0 > d$, where $d_0 \in M_0$ and $d \in M \setminus M_0$, and extend this order to the set of all derivation words by assuming that, for any pair of derivation words Δ, Δ':

1. If Δ is longer than Δ' then $\Delta > \Delta'$.
2. If Δ and Δ' have the same length then they are ordered lexicographically.

Definition 15 *A* regular word *is a derivation word of the form $\Delta = d_1^{s_1} d_2^{s_2} \cdots d_m^{s_m}$ such that:*

1. *$d_i \in M \setminus M_0$, for $i = 1, \ldots, m$.*
2. *$d_m > d_{m-1} > \cdots > d_1$.*
3. *If $char(R) = p \neq 0$, $s_i < p$, for $i = 1, \ldots, m$.*

Definition 16 *A differential polynomial $\Phi(x_i^{\Delta_j})$ is called* reduced *if all derivations and derivation words appearing in $\Phi(x_i^{\Delta_j})$ satisfy the following conditions:*

1. *$\Phi(z_{ij})$ is a generalized polynomial over Q_r in distinct noncommuting indeterminates z_{ij}.*
2. *Δ_i are distinct regular words in M.*

A differential identity $\Phi(x_i^{\Delta_j}) = 0$ is called reduced *if the associated differential polynomial $\Phi(x_i^{\Delta_j})$ is reduced.*

Proposition 1.4.1 *Any arbitrary differential polynomial $\Phi(x_i^{\Delta_j})$ in a prime ring R can be transformed into a reduced differential polynomial.*

For completeness, we now state Kharchenko's powerful result:

Theorem 1.4.2 *[346, Theorem 2] Let R be a prime ring. If $\Phi(x_i^{\Delta_j}) = 0$ is a reduced differential identity for R, then $\Phi(z_{ij}) = 0$ is a generalized polynomial identity of R, where z_{ij} are distinct indeterminates.*

Example 1.4.1 *The following are the differential identities which are obviously satisfied in the ring:*

1. $(xy)^d = x^d y + xy^d$, *where d is any derivation.*
2. $x^d = xa - ax$, *where d is the inner derivation and a is the corresponding element.*
3. *For $k \geq 1$, $[d(x), x]_k = 0$ for all $x \in R$, where d is a commuting derivation on R.*
4. *For commutative ring R and $k \geq 1$, $[d(x), x]_k = 0$ for all $x \in R$, where d is any derivation on R.*

Chapter 2

Derivations in Rings and Algebras

2.1 Introduction

Rings and algebras are important mathematical structures with certain rules and properties. Derivations are like tools that help us to see how these structures work and how they relate to each other. One of the natural questions which often appears in algebra and analysis is whether a map can be defined by its "local" properties. For example, the question whether a map, which acts like a derivation on the Lie product of some important Lie subalgebra of prime rings, is induced by an ordinary derivation, was a well-known problem posed by Herstein [290]. The first result in this direction was obtained in the unpublished work of Kaplansky (cf.; Herstein [290], p. 529), who considered matrix algebras over a field. With the presence of idempotents, this question has been examined by Martindale III [443] for primitive rings. Herstein's problem [290] was solved in full generality only after the powerful techniques of functional identities were developed (see, for example; [99, 102, 103], where further references can be found). In the year 1993, Bresar [137] studied this problem for prime rings. Further, Beidar & Chebotar [101] solved this problem for Lie ideals of prime rings. The problem of whether a Lie derivation is induced by an ordinary one and related questions were also discussed in analysis viz.; Banning & Mathieu [94], Villena [530], where further details can be looked.

This chapter is all about exploring different types of derivations in rings and algebras and understanding how they affect these mathematical objects. We will take a closer look at how derivations come in different forms and what they do. By doing this, we hope to make it easier to understand the basic ideas behind derivations in rings and how they are important for understanding algebraic structures better. At the very beginning of a calculus class one learns the idea of derivatives, its rules and finds its applications in most of the parts of mathematics and not only there, its wide applications in other sciences like physics, engineering, economics, and information sciences. Properties of a derivative function, especially the famously known Leibniz rule caught attention of many mathematicians, and defined as

$$\frac{d}{dx}(fg) = \left(\frac{d}{dx}f\right)g + f\left(\frac{d}{dx}g\right) \tag{2.1}$$

for any two differentiable functions f and g.

Over the years the idea of Leibniz rule has been used tremendously, but with different notations, the idea remaining the same. Like the notation used by Issac Newton was

$$(\dot{fg}) = \dot{f}g + f\dot{g} \tag{2.2}$$

and that by Lagrange,

$$(fg)' = f'g + fg'. \tag{2.3}$$

Keeping that in mind one should ask whether such an idea can be translated to structures like rings, algebras or groups. A lot of attention has been given to the idea of creating something which behaves like a derivative in the case of rings (algebras) and a lot of work has already been done in this direction. So we can confidently define an additive map d on any ring R. There is no need for R to be commutative nor associative; there are no restrictions on the ring. We name it a derivation on R if

$$d(xy) = d(x)y + yd(x) \text{ holds for all } x, y \in R. \tag{2.4}$$

It can be easily verified that the sum of derivations will fetch us a derivation, and the product zd for any $z \in Z(R)$, d a derivation on R would do the same. The set of all derivations on R, denoted by $Der(R)$ can be thought of as a vector space over the ring R and in addition to that it also forms a Lie ring (Lie algebra) with respect to the operation

$$[d_1, d_2] = d_1d_2 - d_2d_1$$

for $d_1, d_2 \in Der(R)$. Prime and semiprime ideals of Lie rings constructed via derivations of R have been studied by Brown and McCoy [153] and Kawamoto [339]. The principle finding of Brown and McCoy [153] is that if R is prime ring with a charactersitic not equal to 2 (respectively semiprime and 2-torsion free), then $Der(R)$ is a prime (respectively semiprime) Lie ring. Moreover, the noncommutative case was discussed by Jordan and Jordan [335]. In fact, they proved that if R is a noncommutative 2-torsion free prime (resp. semiprime) ring, then $Der(R)$ is a prime Lie ring.

Throughout the discussion, unless otherwise mentioned, R denotes an associative ring having at least two elements, with extended centroid C and symmetric quotient ring Q_s (see Chapter 1 and [142] for details). However, R may not posess unity. The symbol $Z(R)$ stands for the center of R.

2.2 Derivations

The concept of derivation has been studied extensively for a long period of time. Jacobson initially introduced the concept of derivation in associative rings in his classic book [319]. Subsequently, they were explored by Herstein [294] and Posner [487], respectively. Significant advancements in this field have occurred in recent decades, particularly following the development of the theory of generalized/functional identities. In this span, derivations have proven to have numerous applications as they beautifully relate to the structure of rings and unfold many properties of various other structures namely, algebras, alternative algebras, Jordan algebras, Lie algebras, ideals to name a few (see [18] for more information). We begin our discussions with the following notion:

Definition 17 *In a ring R, if we define a map d from R to R satisfying*

$$(i)\ d(x+y) = d(x) + d(y) \text{ and}$$

$$(ii)\ d(xy) = d(x)y + xd(y) \text{ for all } x, y \in R,$$

then such a map d is termed as a derivation on R.

Example 2.2.1 *The most natural example of a nontrivial derivation is the usual differentiation on the ring $R[x]$ of polynomials defined over a field R. That is, consider R to be a commutative ring, define d: $R[x] \longrightarrow R[x]$ by*

$$d(f(x)) = f'(x)$$

where '′' denotes the usual differentiation on $R[x]$. Clearly, d forms a derivation on $R[x]$.

Example 2.2.2 *Let $R = \left\{ \begin{pmatrix} a & b \\ c & d \end{pmatrix} \mid a,b,c,d \in \mathbb{Z} \right\}$ be a ring of 2×2 matrices over the ring of integers. Define a map $d : R \longrightarrow R$ by $d\begin{pmatrix} a & b \\ c & d \end{pmatrix} = \begin{pmatrix} 0 & -b \\ c & 0 \end{pmatrix}$. Then, it can be easily verified that d is a derivation on R.*

Example 2.2.3 *For a fixed element $a \in R$, define an additive map $d : R \longrightarrow R$ by $d(x) = [x,a]$ for all $x \in R$. The function d so defined is additive and satisfies the Leibniz rule, that is,*

$$\begin{aligned} d(xy) &= [xy,a] \\ &= x[y,a] + [x,a]y \\ &= xd(y) + d(x)y. \end{aligned}$$

Thus, d is a derivation that is famously known as an inner derivation of R associated with a fixed element $a \in R$ and generally denoted by I_a.

Remark 2.2.1 *It is obvious that every inner derivation on a ring R is a derivation. However, the converse need not be true in general. The following example justifies the fact.*

Example 2.2.4 *Let* $R = \left\{ \begin{pmatrix} a & b \\ 0 & 0 \end{pmatrix} \mid a, b \in \mathbb{Z} \right\}$ *be a ring of* 2×2 *matrices over the ring of integers. Define a map* $d : R \longrightarrow R$ *by* $d\begin{pmatrix} a & b \\ 0 & 0 \end{pmatrix} = \begin{pmatrix} 0 & -a \\ 0 & 0 \end{pmatrix}$. *Then, it can be verified that d is a derivation but not an inner derivation on R.*

Remark 2.2.2 *Derivations which are not inner are called outer derivations. So, the derivation given in Example 2.2.4 is an outer derivation.*

Let's modify the properties of derivation and see if something interesting emerges. For instance, if we consider the case $a = b$, then property (ii) of Definition 17 becomes,

$$d(x^2) = d(x)x + xd(x) \text{ for all } x \in R, \tag{2.5}$$

which opens up a new class of derivations, known as *Jordan derivations*. Like derivation helps us to know the structure of rings and algebras, Jordan derivations also help to reveal some properties of rings and algebras. One might think Jordan derivations were defined like this, but that's not true. Instead Herstein [289] constructed a Jordan ring $(R, +, \circ)$ then defined an additive map $d : R \longrightarrow R$ satisfying

$$d(x \circ y) = d(x) \circ y + x \circ d(y) \text{ for all } x, y \in R$$

and named such a map as a Jordan derivation (cf. [289]). From the above discussion, it is clear that a Jordan derivation need not be a derivation. This can be justified by the following example.

Example 2.2.5 *Consider any 2-torsion free ring R in which* $a^2 = 0$ *for all* $a \in R$, *but for some* $a, b \in R$ $ab \neq 0$. *If we define* $d : R \to R$ *as the identity map, then clearly d will be a Jordan derivation, but not a derivation.*

Another such example would be an additive map defined on a ring R, as

$$(xy)^* = y^* x + y x^* \text{ for all } x, y \in R.$$

$*$ so defined is clearly a Jordan derivation but not a derivation, such a map is famously known as a reverse derivation. Obviously, if R is commutative, then both derivation and reverse derivation are the same. Each reverse derivation is a Jordan derivation but the converse is not true in general. Consider the ring $R = \left\{ \begin{pmatrix} a & b \\ 0 & 0 \end{pmatrix} : a, b \in S \right\}$, where S is a ring such that $S^2 = (0)$.

Define $d : R \to R$ by $d\begin{pmatrix} a & b \\ 0 & 0 \end{pmatrix} = \begin{pmatrix} 0 & b \\ 0 & 0 \end{pmatrix}$. Then, it is straightforward to verify that d is a Jordan derivation but not a reverse derivation. Now if we go on generalizing the idea of derivations and endomorphisms, we come up with many interesting and important types of derivations. Let's start with an automorphism σ defined on R and let Q_r be its Martindale ring of quotient, defining an additive map $d : R \to R$ which satisfies

$$d(xy) = d(x)y + \sigma(x)d(y) \text{ for all } x, y \in R$$

is what we know as a *skew derivation* (or a *σ-derivation*). We can think of any derivation as a skew derivation considering σ as the identity map. Yet another simple example will be the map $\sigma - I_R$, where I_R is the identity map on R. If a map $d : R \to R$ is defined by

$$d(x) = bx - \sigma(x)b \text{ for all } x \in R$$

for a fixed $b \in Q_r$, then clearly d forms a σ-derivation and such a derivation is termed as an inner skew derivation, and if no such b exists then d is an outer derivation. The existence of skew derivations helps us to relate rings with the theory of GPIs (viz., [104]), like Chuang and Lee [190] studied polynomial identities with skew derivations. They proved that if $\phi(x_i, d(x_i))$ is a generalized polynomial identity for a prime ring R and d is any outer skew derivation of R, then such a ring R also satisfies the generalized polynomial identity $\phi(x_i, y_i)$, where x_i and y_i are distinct indeterminates. Also, they proved in case $\phi(x_i, d(x_i), \sigma(x_i))$ is a generalized polynomial identity for a prime ring R, d an outer skew derivation of R, and σ an outer automorphism of R, then R also satisfies the generalized polynomial identity $\phi(x_i, y_i, z_i)$, where x_i, y_i and z_i are distinct indeterminates (see [104] for details). Till we have used a single endomorphism; if we use two endomorphisms θ and ϕ defined on R and then generalize the definition of a derivation as well as that of a σ-derivation, an additive map $d : R \longrightarrow R$ which satisfies

$$d(xy) = d(x)\theta(y) + \phi(x)d(y) \text{ for all } x, y \in R$$

is known as a (θ, ϕ)-derivation. A mapping $a \mapsto \theta(a)b - b\phi(a)$, where b is a fixed element in R a (θ, ϕ)-derivation. Such a (θ, ϕ)-derivation is said to be inner. A (θ, I_R)-derivation is simply called a θ-derivation. Of course an, I_R-derivation is a derivation. Similarly, we can define a *Jordan* (θ, ϕ)*- derivation* as an additive map $d : R \to R$, satisfying

$$d(x^2) = d(x)\theta(x) + \phi(x)d(x) \text{ for all } x \in R.$$

Clearly, each derivation is a (θ, ϕ)-derivation by taking θ and ϕ both as the identity maps. Kaya [340] and Guven [280] gave a couple of results which linked (θ, ϕ)-derivations with the structure of rings. These types of derivations turn out to be equally important and play an important role in the development of the theory of derivations. Yet another significant class of derivations are left derivations and Jordan left derivations. An additive mapping $d : R \to X$, where R is a ring and X is a left R-module is called a *left derivation* if

$$d(xy) = xd(y) + yd(x) \text{ for all } x, y \in R.$$

Similarly, an additive mapping $d : R \to X$ will be called a *Jordan left derivation* if

$$d(x^2) = 2xd(x) \text{ for all } x \in R.$$

It turns out that the notion of *Jordan left derivations* is in a close connection with commuting mappings which leads us to various fascinating results, see [81] and [235]. In the year 1990, Brešar & Vukman [149] proved that the existence of a nonzero Jordan left derivation on a prime ring R of $char(R) = 2, 3$ forces R to be commutative. Later in the year 1992, Deng [235] improved the above result, proving that if R is a prime ring of $char(R) = 2$, X is a nonzero left R-module that is faithful and prime, and if there exists a nonzero Jordan left derivation $d : R \to X$, then R is commutative.

Let us move to another class of derivations, known as Lie derivations. A *Lie derivation* d on a ring R is an additive map satisfying the condition of $d[x,y] = [d(x),y] + [x,d(y)]$ for all x, $y \in R$. Brešar in [137] proved that every additive Lie derivation on a prime ring R with characteristic not equal to 2 can be decomposed as $\tau + \zeta$, where τ is a derivation from R into its central closure and ζ is an additive map of R into the extended centroid C sending commutators to zero. Johnson [333] proved that every continuous linear Lie derivation from a C^*-algebra $\mathcal{A}$ into a Banach $\mathcal{A}$-bimodule M is standard, that is, it can be decomposed to the form $\tau + h$ where $\tau : \mathcal{A} \to M$ is a derivation and h is a linear map from $\mathcal{A}$ into the center of M vanishing at each commutator. Mathieu and Villena [453] showed that every linear Lie derivation on a C^*-algebra is standard. Qi and Hou [493], proved that the same is true for additive Lie derivations of nest algebras on Banach spaces. Müller in [466] introduced the idea of *pre-morphisms* and in particular of *pre-derivations*, that clubbed the concepts of derivations with Lie products. He showed that, if G is a Lie group endowed with a bi-invariant semi-Riemannian metric and $\mathcal{A}$ its Lie algebra, then the Lie algebra of the group of isometries of G fixing the identity element is a subalgebra of $pDer(\mathcal{A})$ (where $pDer(\mathcal{A})$ is the set of all pre-derivations on $\mathcal{A}$). Thus, the study of the algebra of *pre-derivations* is not only interesting from the algebraic point of view but also from its applications. Let $\mathcal{A}$ be a Lie algebra; we define a linear map $p : \mathcal{A} \to \mathcal{A}$ satisfying

$$p[x,[y,z]] = [px,[y,z]] + [x,[py,z]] + [x,[y,pz]],$$

for all $x, y, z \in \mathcal{A}$ to be a *pre-derivation*. Clearly, $Der(\mathcal{A})$ is contained in $pDer(\mathcal{A})$. The study of derivations of Lie algebras led to the appearance of the notion of their natural generalization, a pre-derivation of a Lie algebra, which is a derivation of a Lie triple system induced by that algebra. In order to generalize the concept of pre-derivations, we define *Leibniz algebras* first and then move on to what is called *Leibniz derivations*. Let V be a vector space and $k \in \mathbb{N}$. A $k-$ *Leibniz algebra structure* on V is a $(k+1)$-linear map $V \times \cdots \times V \to V$: $(x_1,\ldots,x_{k+1}) \mapsto [x_1,\ldots,x_{k+1}]$ satisfying the identity,

$$[[x_1,\ldots,x_{k+1}],y_1,\ldots,y_k] = \sum_{i=1}^{i=n}[x_1,\ldots,x_{i-1},[x_i,y_1,\ldots,y_k],x_{i+1},\ldots,x_{k+1}].$$

We define the multilinear bracket $[x_1,\ldots,x_{k+1}]_{k+1}$, or when the length of the bracket is clear from the context, $[x_1,\ldots,x_{k+1}]$, as the nested expression $[x_1,[x_2,[x_3,\ldots,[x_{k-1},[x_k,x_{k+1}]]\cdots]]]$. Lie algebras and Leibniz algebras, are examples of *1-Leibniz algebras*. The above defined $(k+1)$ linear map defines a *k-Leibniz algebra structure* on $\mathcal{A}$, denoted by $\mathcal{L}_k(\mathcal{A})$. Its subalgebras and ideals are defined in the natural way. A subspace $\mathcal{T}$ of a *Leibniz k-algebra* $\mathcal{L}$ is a subalgebra of $\mathcal{L}$ if it satisfies $[\mathcal{T},\ldots,\mathcal{T}]_{k+1} \subseteq \mathcal{T}$. Similarly, a subspace $\mathcal{I}$ is an ideal, if it satisfies $[\mathcal{I},\mathcal{L},\ldots,\mathcal{L}]_{k+1} + \cdots + [\mathcal{L},\ldots,\mathcal{L},\mathcal{I}]_{k+1} \subseteq \mathcal{I}$.

Now let's define *Leibniz-k derivation* for any $k \in \mathbb{N}$. A Leibniz-derivation of order k for a Lie algebra $\mathcal{A}$ is an endomorphism $P : \mathcal{A} \to \mathcal{A}$ satisfying the identity

$$P[x_1,\ldots,x_{k+1}] = [P(x_1),x_2,\ldots,x_{k+1}] + [x_1,P(x_2),\ldots,x_{k+1}] + \cdots + [x_1,x_2,\ldots,P(x_{k+1})]$$

for every $x_1,x_2,\ldots,x_{k+1} \in \mathcal{A}$. Let $\mathcal{L}Der_k(\mathcal{A})$ be the set of all Leibniz-derivations for $\mathcal{A}$ of order k and let $\mathcal{L}Der(\mathcal{A})$ be the set of all Leibniz-derivations, so we have $\mathcal{L}Der(\mathcal{A}) = \cup_{k\in\mathbb{N}}\mathcal{L}Der_k(\mathcal{A})$. Derivations defined on set $\mathcal{A}$, and pre-derivations defined on $\mathcal{A}$ will be simple examples of Leibniz derivatives of order 1 and 2, respectively. Jacobson [318] proved that if a Lie algebra admits an invertible derivation (by an invertible derivation on a ring R (with unit element) we mean a non zero derivation d which is defined as $d(x) = 0$ or $d(x)$ invertible in R), it must be nilpotent. He also suspected, that the converse might be true, that is, every nilpotent Lie algebra has an invertible derivation but the latter is not true. In the year 2013, Wolfgang [555] proved that a Lie algebra is nilpotent iff it admits an invertible Leibniz-derivation. This was the motivation behind defining these generalizations of derivations.

We are quite familiar with the idea of nilpotency, whether it pertains to nilpotent elements, nilpotent ideals, or, for that matter, nilpotent rings. Now, let's delve into the topic of nilpotent derivations. Following a similar pattern as other definitions, a derivation d defined on a ring R is said to form a nilpotent derivation if there exists a natural number (n) such that

$$d^nR = (0).$$

For $n = 2$ and considering R to be a prime ring of characteristic not 2, Posner [487] has shown that d must be zero. However, for $n > 2$, the situation is less straight forward. Herstein [291] demonstrates that for simple rings with some characteristic restrictions, a nilpotent derivation is equivalent to an inner derivation $x \to [a,x]$, induced by a nilpotent element a. Thus establishes a strong relationship between these types of derivations. Consequently, for a simple ring the index of a nilpotent derivation must be an odd number and precisely equal to $2m-1$, where m is the nilpotent index of the element a. Lets discuss the example of the ring of 2×2 upper triangular matrices over the real field. The derivation induced by the matrix $\begin{pmatrix} 0 & 1 \\ 0 & 0 \end{pmatrix}$ forms a nilpotent derivation of index 2. Further, Kovacs [356] posed the question of whether whether the result regarding odd nilpotency indices holds for all rings. However, it turns out that this result is not universally true, as demonstrated by the example mentioned above. Interestingly, this property extends to prime rings as well. In 1983, Chung and Luh [194] made significant progress by showing that the nilpotency of a derivation on a 2-torsion free semiprime ring is always an odd number. This result expands our understanding to a broader class of rings. For more details of such studies, see Section 2.7 of this chapter.

Let's explore specific structures and examine how certain kinds of derivations behave within them.

2.2.1 Derivations on prime rings

The story of derivations with prime rings dates back to 1957 when Posner [487] proved two remarkable results, the first focusing on the product of derivations and the other on the existence of centralizing derivations with the commutativity of the rings. The result is stated as follows:

Theorem 2.2.3 (Posner's First Theorem): *Let R be a prime ring of characteristic not equal to 2 and d_1, d_2 be two derivations of R such that the iterate $d_1 d_2$ is also a derivation. Then, at least one of d_1, d_2 is zero.*

Proof. Since d_1, d_2 are derivations, we have

$$d_1 d_2(ab) = d_1\{d_2(a)b + ad_2(b)\}$$

$$= d_1 d_2(a)b + d_2(a)d_1(b) + d_1(a)d_2(b) + ad_1 d_2(b), \tag{2.6}$$

for all $a, b \in R$. Also, we have

$$d_1 d_2(ab) = d_1 d_2(a)b + ad_1 d_2(b) \text{ for all } a, b \in R. \tag{2.7}$$

On compairing eqns (2.6) and (2.7), we obtain

$$d_2(a)d_1(b)+d_1(a)d_2(b)=0 \text{ for all } a,b\in R. \tag{2.8}$$

Replace a by $ad_1(c)$ in eqn (2.8) and using the properties of derivations, we have

$$d_2(a)d_1(c)d_1(b)+ad_2d_1(c)d_1(b)+d_1(a)d_1(c)d_2(b)+ad_1^2(c)d_2(b)=0, \tag{2.9}$$

for all $a,b,c\in R$. Now, replacing a by $d_1(c)$ in eqn (2.8), we get

$$d_2(d_1(c))d_1(b)+d_1(d_1(c))d_2(b)=0.$$

This gives,

$$ad_2(d_1(c))d_1(b)+ad_1(d_1(c))d_2(b)=0 \text{ for all } a,b,c\in R. \tag{2.10}$$

On compairing eqns (2.9) and (2.10), we get,

$$d_2(a)d_1(c)d_1(b)+d_1(a)d_1(c)d_2(b)=0 \text{ for all } a,b,c\in R. \tag{2.11}$$

From eqn (2.8), we get,

$$d_2(a)d_1(b)=-d_1(a)d_2(b) \text{ for all } a,b\in R.$$

Therefore,

$$d_2(c)d_1(b)=-d_1(c)d_2(b).$$

Using the above and equation (2.11), we have

$$(d_2(a)d_1(c)-d_1(a)d_2(c))d_1(b)=0.$$

We know that for a derivation d on R, if $ad(x)=0$, for all $x\in R$ then either $a=0$ or $d=0$. Thinking of $d_2(a)d_1(c)-d_1(a)d_2(c)$ as a, we can deduce

$$\text{either } d_2(a)d_1(c)-d_1(a)d_2(c)=0 \text{ or } d_1(b)=0 \text{ for all } a,b,c\in R.$$

If $d_1(b)\neq 0$, it implies,

$$d_2(a)d_1(c)-d_1(a)d_2(c)=0.$$

Also, from eqn (2.8), we have,

$$d_2(a)d_1(c)+d_1(a)d_2(c)=0.$$

Adding the above eqns, we get,

$$2d_2(a)d_1(c)=0 \text{ for all } a,c\in R.$$

Since R is a prime ring with characteristic not equal to 2, it implies R is a 2-torsion free ring. So, we conclude that

$$d_2(a)d_1(c) = 0 \text{ for all } a, c \in R.$$

Thinking of $d_2(a)$ as a and using the fact that $ad(x) = 0$, for all $x \in R$ implies either $a = 0$ or $d = 0$, so we arrive at

$$\text{either } d_2(a) = 0 \text{ or } d_1(c) = 0 \text{ for all } a, c \in R,$$

hence the desired result.

♣

The condition, $char(R) \neq 2$ in the above result is indispensable. To see this consider a ring R of 2×2 matrices with entries from the Galois field $\{0, 1, w, w^2\}$, with inner derivations d_1 and d_2 defined by,

$d_1(X) = \left[\begin{pmatrix} 0 & 1 \\ 0 & 0 \end{pmatrix}, X\right]$ and $d_2(X) = \left[\begin{pmatrix} 0 & w \\ 0 & 0 \end{pmatrix}, X\right]$. The $char(R) = 2$, and we have $d_1 \neq 0$, $d_2 \neq 0$ but $d_1 d_2 = 0$, making the iterate a derivation but none of the derivations are zero. Therefore, we cannot dispose off characteristic "not 2" criteria.

Posner's First Theorem tells us that the composition of two nonzero derivations of a prime ring R can not be a derivation provided that characteristic of R differs from 2. Thereafter, a number of authors have generalized this theorem in several ways (see for example Bergen [124], Chebotar [173], Chuang [185], Hirano et al. [300], Hvala [313], Krempa [357], Lanski [360], Martindale III [451] where further references can be found). Generally speaking, a composition of inner derivations can be a nonzero derivation. For example, if e is a nonzero idempotent in R, i.e., $e^2 = e = 0$, then $(ad(e))^{2k-1} = ad(e)$ for any positive integer k (see in Lanski [361] for more interesting example). If d is any derivation of the prime ring R, then d^p is a derivation of R provided that $char(R) = p$. However, it was not clear in general whether the composition of derivations could be a nonzero derivation if some of them are inner and some of them are outer. Some progress was achieved by applying the result of Kharchenko [345] on the independence of outer derivations. The result on the composition of three derivations was obtained by Lanski [361] in 1992 as follows:

Theorem 2.2.4 *Let R be a prime ring of characteristic different from 2 and $d_1, d_2, d_3 \in Der(R)/\{0\}$ so that $d_1 d_2 d_3 = e \in Der(R)$. Then either $d_1, d_2, d_3 \in Inn(R)$, or else $char(R) = 3$, d_1 is an outer, $d_2 = d_1 z_1$, $d_3 = d_1 z_2$ and $(z_1)^{d_1} = 0$, where $z_i \in C$, so $e = d_1^3 z_1 z_2$.*

In the same paper, Lanski posed the question whether a composition of fewer than $char(R)$ derivations, or any product in case $char(R) = 0$ is a nonzero derivation if some are inner and some are outer. Further, in 1995, Chebotar [173] obtained the necessary condition for the composition of derivations, including both inner and outer ones, to be a derivation (see Beidar et al. [104] for more details). However, Lanski's [361] question remains open till date. Very recently, Chebotar & Lee [175] presented the partial answer to Lanski's question by means of the following example:

Example 2.2.6 *Let F be a field of characteristic different from 2 and let $R = M_2(F[x])$ be the ring of 2×2 matrices over the ring of polynomials in indeterminate x over F. Take $d \in Der(R)$ defined by applying formal differentiation to each entry. Set $d_1 = d_3 = d_5 = ad(e_{12})$, $d_2 = d + ad(e_{21})$, $d_4 = d - ad(e_{21})$. Then d_2 and d_4 are outer derivations and $d_1d_2d_3d_4d_5 = ad(-4e_{12})$.*

Moreover, Posner's First Theorem was rediscovered by Creedon [197] to semiprime algebras. In fact, he proved that if the product of two derivations in an algebra $\mathcal{A}$ is a derivation, then the product maps the algebra into the nil radical, nil($\mathcal{A}$) (the intersection of all prime ideals of $\mathcal{A}$). Thus, if the product of two derivations in a semiprime algebra is a derivation, then the product is zero. Further, Creedon obtained conditions proving that the product of two derivations maps the algebra into the Jacobson radical [197, Proposition 9].

With Posner's first theorem, the debate for the composition of derivations to be a derivation just began. Over a period of time there have been lots of generalizations to it, and restrictions on characteristics of the rings to make the iterate a derivation. Also, various authors focused on whether the product of inner derivations is inner or for that matter the product of a few inner and few outer derivations. Some important results pertaining to the above discussions are as under:

Theorem 2.2.5 *[357, Theorem A] Let Z be a commutative domain, $char(Z) > n > 1$. If $d_1, d_2, \ldots, d_n$ are such derivations of Z that the composition $d_1d_2\cdots d_n$ is a derivation, then $d_i = 0$ for some $1 \leq i \leq n$.*

Theorem 2.2.6 *[357, Theorem C] Let R be a prime ring with non-trivial center Z, $char(R) > n > 1$, and let $d_1, \ldots, d_n$ be such derivations of R such that $d_1 \cdots d_n$ is a derivation. Then, $d_1 \cdots d_n$, is $Z-$linear. Moreover, if $d_v \neq 0$ for any proper subset V of $U = \{1, \ldots, n\}$, then d_i is $Z-$linear for all $i \in U$.*

The proof of Posner's first theorem is easily extendable to ideals of R. Lanski [360] attempted to generalize Posner's first theorem to Lie ideal L of R, assuming that the product of derivations d_1 and d_2 is a Lie derivation on L while many authors have attempted to prove that the product d_1d_2 either takes the Lie ideal L to 0 or takes it in the center of the ring R. In this pursuit in 1988, Lanski [360], attempted to obtain Posner's results using the concept of differential identities which was developed by Kharchenko in [345].

Theorem 2.2.7 *[360, Theorem] If L is a noncommutative Lie ideal of R and d and h are nonzero derivations of R so that dh is a Lie derivation of L into R. Then* $char(R) = 2$ *and either R satisfies* R_4 *or* $h = dc$ *for c in the extended centroid of R.*

The following results proved helpful in understanding whether the product of a few inner and outer derivations is a derivation.

Theorem 2.2.8 *[185, Theorem 2] Let R be a prime ring of characteristic* $p > 0$. *Suppose that* $\delta_1, \delta_2, \ldots, \delta_p$ *are derivations of R such that* $\phi(x) = x^{\delta_1 \delta_2 \cdots \delta_p}$ *for all* $x \in R$ *defining a derivation on R. Then (1) if one of* $\delta_1, \delta_2, \ldots, \delta_p$ *is inner, then* $\phi(x)$ *is also inner, (2) if all* $\delta_1, \delta_2, \ldots, \delta_p$ *are outer, then there exists* $0 \neq \alpha \in C$ *such that* $\phi(x) = \alpha(x)^{\delta_1^p}$ *for all* $x \in R$.

Theorem 2.2.9 *[361, Theorem 6] Let R be a prime ring of characteristic different from 2 and* $d_1, d_2, d \in Der(R)/\{0\}$ *so that* $d_1 d_2 d = E \in Der(R)$. *Then, either* $d_1, d_2, d \in Inn(R)$, *or else* $char(R) = 3$, d_1 *is an outer,* $d_2 = d_1 z_1, d = d_1 z_2$ *and* $(z_1)^{d_1} = 0$, *where* $z_i \in C$, *so* $E = {d_1}^3 z_1 z_2$.

Theorem 2.2.10 *[323, Theorem 1] Let* λ *and* δ *be derivations of a prime ring R, and let* $\lambda \delta^m = 0$ *where m is a positive integer. Then, either* $\lambda = 0$ *or* $\delta^r = 0$, *where* $r < 4m - 1$.

Theorem 2.2.11 *[189, Therorem 1] Let R be a prime GPI-ring with extended centroid C. Suppose that* $dim_C RC \geq 3$ *and C is a finite field with q elements. Then there exist* $2q+3$ *nonzero derivations* d_i *of R,* $i = 1, 2, \ldots, 2q+3$, *such that* $d_1(x) d_2(x) \cdots d_{2q+3}(x) = 0$ *for all* $x \in R$.

Theorem 2.2.12 *[189, Corollary] Let R be a prime ring with infinite extended centroid C and L a non-commutative Lie ideal of R. Suppose that* $d_1, \ldots, d_n$ *are derivations of R satisfying* $d_1(x) d_2(x) \ldots d_n(x) = 0$ *for all* $x \in L$. *Then* $d_i = 0$ *for some i.*

No doubt Posner's first theorem helped to understand derivations and their compositions better but the arrival of Posner's second theorem revolutionized the study of derivations. It started the discussion of a much larger important idea of centralizing and commuting mappings which in turn laid the foundation for the functional identities (FI) theory. Now, functional identities have proved to be of great importance for solving some hundred year old problems. Lets first define certain notions to be used to understand Posner's result. A derivation $d : R \to R$ is *commuting* if $d(x)x = xd(x)$ for all $x \in R$ and *centralizing* if $xd(x) - d(x)x$ is in $Z(R)$, the center of R for all $x \in R$. A derivation is called centralizing on a subset S of R if $[x, d(x)] \in Z(R)$ for all $x \in S$.

Matheiu in [452] proved Posner's theorem in a simple and more general setting under the assumption that the derivation is centralizing on a non zero ideal only instead of the whole ring.

Theorem 2.2.13 (Posner's Second Theorem): *Let R be a prime ring and d be a derivation of R such that, for all $a \in R$, $ad(a) - d(a)a$ is in the center of R. Then, if d is not a zero derivation, R is commutative.*

Proof. Let I be a nonzero ideal of the prime ring R with $char(R) \neq 2$, and let $d{:}R \longrightarrow R$ be a derivation such that $[a,d(a)] \in Z(R)$ for all $a \in I$. Replacing a with $a+b$, $b \in I$, we get,

$$[a,d(b)] + [b,d(a)] \in Z(R), \tag{2.12}$$

in particular, taking $b = a^2$,

$$[a,d(a^2)] + [a^2,d(a)] \in Z(R), \tag{2.13}$$

as d is a derivation, we have

$$[a,d(a^2)] - [a^2,d(a)] = 0. \tag{2.14}$$

Combining eqns (2.13) and (2.14) gives,

$$[a^2,d(a)] \in Z(R).$$

Since $[a,d(a)]$ is also central, we have

$$2[a,d(a)]^2 = [2[a,d(a)]a,d(a)] = [[a^2,d(a)],d(a)] = 0.$$

As R is prime, we conclude $[a,d(a)] = 0$. This shows that in eqn (2.12) we in fact have,

$$[a,d(b)] + [b,d(a)] = 0 \text{ for all } a,b \in I. \tag{2.15}$$

Now, let d_c denote the inner derivation $d_c(x) = [x,c]$, $x \in R$. Then eq (2.15) implies, $d_b \circ d = d_{db}$ on I for all $b \in I$. So the product of these two derivations d_b and d is a derivation, using Posner's first theorem, we conclude that either $d = 0$ or $d_b = 0$ for all $b \in I$, i.e., $xb - bx = 0$ for all $x \in R$, implying $I \subset Z(R)$. Since R is prime, the latter case entails that R is commutative.

♣

This opened up a whole new discussion of whether some other derivations or some other identities involving derivations will help us reveal the properties of rings. Results analogous to that of Posner's second theorem were established, the result motivated mathematicians to look for such centralizing automorphisms as well and one such result is due to Luh.

Theorem 2.2.14 *[432, Theorem] Let R be a prime ring. If R possesses a non-trivial commuting automorphism, then R is a commutative integral domain.*

In 1976, Mayne strengthened this by proving that a prime ring with a non-trivial centralizing automorphism is an integral domain. The result stated;

Theorem 2.2.15 *[455, Theorem] If R is a prime ring with a nontrivial centralizing automorphism, then R is a commutative integral domain.*

The study of derivations on prime rings did not stop here but many authors time and again related this concept with other existing and interesting ideas. In the year 1981 Herstein et al. [126] tried to relate prime rings with derivations to Lie ideals and came up with certain cool results. From taking a nonzero Lie ideal U with $d^2(U) = 0$ in the center of its parent ring R, for any derivation d on R to proving a close relationship between a non central Lie ideal with the center of the ring. In totality they bridged the concept of derivations and that of Lie ideals for prime rings.

Theorem 2.2.16 *[126, Theorem 4] Let R be a prime ring, char$(R) \neq 2$, and let $U \not\subset Z$ be a Lie ideal of R. Suppose that δ and d are derivations of R such that $\delta d(U) = 0$. Then, either $d = 0$ or $\delta = 0$.*

During 1982–84, Mayne (see [456, 457]) proved that in a prime ring, an automorphism or a derivation need only be centralizing and invariant on a nonzero ideal in order to ensure that the ring is commutative. Also, if R is of characteristic not two, then the mapping need only be centralizing and invariant on a nonzero Jordan ideal. Some of her interesting work is stated below;

Theorem 2.2.17 *[456, Theorem] Let R be a prime ring and I be a nonzero ideal of R. If R has a nontrivial automorphism or derivation T such that $uT(u) - T(u)u$ is in the center of R and $T(u) \in I$ for every $u \in I$, then R is commutative.*

Theorem 2.2.18 *[457, Theorem 1] Let R be a prime ring and I be a nonzero ideal in R. If L is a nontrivial automorphism or derivation of R such that $xL(x) - L(x)x$ is in the center of R for every $x \in I$, then the ring R is commutative.*

It is easy to extend the above theorem to the case where the centralized ideal is quadratic Jordan, which generalizes Awtar's theorem for centralizing derivations.

Theorem 2.2.19 *[457, Theorem 2] Let R be a prime ring and I be a nonzero quadratic Jordan ideal of R. If L is a nontrivial automorphism or derivation of R which is centralizing on I, then R is commutative.*

Till now we were talking about two sided ideals, the trend did follow, in 1987, Bell and Martindale [113] showed that instead of a two sided ideal the result holds for a one sided ideal as well;

Theorem 2.2.20 *[113, Theorem 3] Let R be a prime ring and I a nonzero left ideal. If R admits a nonzero derivation which is centralizing on I, then R is commutative.*

Further, Herstein [297] related the commutativity of a derivation with that of the commutativity of the ring giving us a very special result. Precisely, we have the following:

Theorem 2.2.21 *[297, Theorem 2] Let R be a prime ring, $d \neq 0$ a derivation of R such that $d(x)d(y) = d(y)d(x)$ for all $x, y \in R$. Then, if $char(R) \neq 2$, R is a commutative integral domain, and if $char(R) = 2$, R is commutative or is an order in a simple algebra which is 4-dimensional over its center.*

In 1990, Vukman [533] extended Posner's Second Theorem by showing that if d is a derivation of prime ring of characteristic not 2 such that $[[d(x),x],x] = 0$, for all $x \in R$, then $d = 0$ or R is commutative. In fact, in the spirits of Posner's theorem, he merely showed that d is commuting. In this direction, Bell and Daif [110] proved some interesting results for derivations satisfying $d(xy) = d(yx)$ for all $x, y \in R$ but for some specific subsets of the ring. They showed that;

Theorem 2.2.22 *[110, Theorem 3] Let R be a prime ring and I a nonzero two-sided ideal of R. If R admits a nonzero derivation d such that $d(xy) = d(yx)$ for all $x, y \in I$, then R is commutative.*

In this case being a two-sided ideal is important, as can be seen in the example below. Consider a 2×2 ring R with entries from a field F, and take $I = \left\{ \begin{pmatrix} a & b \\ 0 & 0 \end{pmatrix} \mid a, b \in F \right\}$. Let's define a derivation on R as $d(A) = A\begin{pmatrix} 0 & 1 \\ 0 & 0 \end{pmatrix} - \begin{pmatrix} 0 & 1 \\ 0 & 0 \end{pmatrix}A$, for every $A \in R$. Clearly d satisfies $d(xy) = d(yx)$, for all $x, y \in I$. But the ring is noncommutative, this is because I is a one sided ideal. Later on, in 1999, Bell [108] extended the above mentioned result as follows:

Theorem 2.2.23 *[108, Theorem 2.3] Let R be a prime ring with $char(R) \neq 2$. If R admits a non-zero derivation d such that $d([x,y]) \in Z(R)$ for all $x, y \in R$, then R is commutative.*

Proof. We have given that $d([x,y]) \in Z(R)$ for all $x, y \in R$. The above expression can also be written as

$$[x, d(y)] + [d(x), y] \in Z(R) \text{ for all } x, y \in R. \tag{2.16}$$

Replacing y by zy, $z \in Z(R)$, we get $z[x,d(y)] + d(z)[x,y] + z[d(x),y] \in Z(R)$, i.e., $z([x,d(y)] + [d(x),y]) + d(z)[x,y] \in Z(R)$. Recalling (2.16), we see that

$$d(z)[x,y] \in Z(R) \text{ for all } x, y \in R \text{ and } z \in Z(R). \tag{2.17}$$

If $d(Z(R)) \neq (0)$, this implies $[x,y] \in Z(R)$ for all $x,y \in R$, so that R is commutative. We now suppose that $d(Z(R)) = (0)$, in which case

$$d^2(xy - yx) = 0 \text{ for all } x,y \in R. \tag{2.18}$$

Expanding this condition, we get

$$[x,d^2(y)] + [d^2(x),y] + 2[d(x),d(y)] = 0 \text{ for all } x,y \in R. \tag{2.19}$$

Letting y be a commutator c and applying (2.18), we get $[d^2(x),c] = 0$ for all $x \in R$ and all commutators c. It follows from [108, Lemma 1.3] that $d^2(R) \subseteq Z(R)$, so by (2.19), we obtain $[d(zx),d(y)] = 0$ for all $x,y \in R$. Therefore, R is commutative by [297, Theorem 2].

♣

In 2006, Argac [27] generalized the above result for non zero ideals. In fact, he proved that:

Theorem 2.2.24 *[27, Theorem 3.12] Let R be a prime ring and I be a nonzero ideal of R. If d is a derivation of R such that $d([x,y]) \in Z(R)$ for all $x,y \in I$, then $d^2(I) \subseteq Z(R)$ or R is commutative. Furthermore, if $char(R) \neq 2$ and $d \neq 0$, then R is commutative.*

These ideas connecting special forms of derivations with commutativity was taken forward by many authors who showed some parallel results for some additive mappings. Some other fascinating results are;

Theorem 2.2.25 *[533, Theorem 3] Let R be a noncommutative prime ring of characteristic different from two and three. Suppose R contains the identity element 1. Let $d : R \to R$ be an additive mapping such that $d(x^3) = 3xd(x)x$ holds for all $x \in R$. In this case $d = 0$.*

Theorem 2.2.26 *[135, Theorem 1] Let R be a prime ring of characteristic not 2. Suppose that an additive mapping $f : R \to R$ satisfies the relation $[[f(x),x],x] = 0$ for all $x \in R$. In this case, f is commuting on R.*

Theorem 2.2.27 *[136, Theorem B] Let R be a prime ring and I be a nonzero left ideal of R. Suppose derivations d and g of R satisfy $d(u)u - ug(u) \in Z(R)$ for all $u \in I$. If $d \neq 0$, then R is commutative.*

From the above discussion, it is clear that certain additive maps like derivations, automorphisms and endomorphisms cannot be centralizing on certain subsets of a noncommutative prime ring. This motivated Brešar to determine the structure of any additive mapping which is centralizing on a prime ring and he proved the following:

Theorem 2.2.28 *[136, Theorem A] Let R be a prime ring. Suppose an additive mapping F of R into itself is centralizing on R. If either R has a characteristic* *different from 2 or F is commuting on R, then f is of the form* $f(x) = \lambda x + \zeta(x)$, $x \in R$, *where* λ *is an element from the extended centroid C of R and* ζ *is an additive mapping of R into C.*

Motivated by Posner's second theorem, Khan and Khan [344] proved that a derivation d on a noncommutative prime ring R with characteristic not 2 or 3 such that the map $x \longrightarrow [[[[d(x),x],x],x],x]$ becomes centralizing makes the derivation d the zero map.

2.2.2 Derivations on semiprime rings

In an analogous way like prime rings, various algebraists were interested in knowing whether the existence of a special type of map guarantees for a ring to have a central ideal or to be commutative. Many authors tried to generalize the already known results, as semiprime rings are just a bigger class of rings containing prime rings. Some of them achieved such generalizations but a few of the results had to be modified to fit into the semiprime setting. As observed in the previous section, maps like derivations, automorphisms, and endomorphisms do play a vital role in revealing the structure of the rings, so a natural question does arise whether this trend continues for semiprime rings as well. The following results answered this query.

Theorem 2.2.29 *[113, Theorem 3] Let R be a semiprime ring and I a nonzero left ideal. If R admits a derivation d which is nonzero on I and centralizing on I, then R contains a nonzero central ideal.*

Theorem 2.2.30 *[113, Corollary 1] Let R be semiprime and I a nonzero left ideal; and suppose that R admits an endomorphism T which is one-to-one on I, centralizing on I, and not the identity on I. If* $T(I) \subset I$, *then R contains a nonzero central ideal.*

In the year 1992, Daif and Bell [202] proved that the existence of a specific identity like that of Hersteins' unfolds the structure of a semiprime ring. The result is stated as follows:

Theorem 2.2.31 *[111, Theorem 1] If R is a semiprime ring admitting a derivation d such that there exists a nonzero ideal I of R such that either* $xy + d(xy) = yx + d(yx)$ *for all* $x, y \in I$, *or* $xy - d(xy) = yx - d(yx)$ *for all* $x, y \in I$, *then I is a central ideal.*

Theorem 2.2.32 *[111, Theorem 2] Let R be a semiprime ring admitting a derivation d for which either* $xy + d(xy) = yx + d(yx)$ *for all* $x, y \in R$ *or* $xy - d(xy) = yx - d(yx)$ *for all* $x, y \in R$. *Then, R is commutative.*

Consider $R = R_1 \oplus R_2$, where R_1 is an integral domain, R_2 is a prime ring which is not commutative, and d is the direct sum of derivations on the summands R_1 and

R_2. It means we cannot conclude from the above result that the ring is commutative. Now let's talk about the rings admitting a special kind of commutativity-preserving maps. If R is a ring and $S \subset R$, a map f: $S \longrightarrow S$ is called *strong commutativity preserving (SCP)* on R if $[x,y] = [f(x), f(y)]$ for all $x,y \in S$. Bell and Daif [109] were able to establish the commutativity property for semiprime rings admitting a derivation which is SCP on a nonzero right ideal.

Theorem 2.2.33 *[109, Theorem 1] Let R be a semiprime ring and I a nonzero right ideal. If R admits a derivation d which is scp on I, then $I \subset Z$.*

Theorem 2.2.34 *[109, Corollary 1] If R is a semiprime ring admitting a derivation d which is strong commutativity preserving on R, then R is commutative.*

Example 2.2.7 *Let R be a 3-dimensional algebra over a field of characteristic 2, with basis $\{u_0, u_1, u_2\}$ and multiplication defined by*

$$u_i u_j = \begin{cases} u_0 & \text{if } (i,j) = (1,2) \\ 0 & \text{otherwise.} \end{cases}$$

Next, let d be the linear transformation on R defined by $d(u_0) = 0, d(u_1) = u_1, d(u_2) = u_2$. It is easily verified that d is a derivation which is SCP on R. This shows that the hypothesis of semiprimeness cannot be omitted from the above corollary. The derivation d in this case is not an inner derivation. Indeed, it is easy to show that any ring R admitting an inner derivation which is SCP on R must be commutative.

Example 2.2.8 *Let $R = R_1 \oplus R_2$, where R_1 is a non-commutative prime ring with derivation d_1 and R_2 is a commutative domain. Define $d: R \longrightarrow R$ by $d((r_1, r_2)) = (d_1(r_1), 0)$. Then, R is a semiprime ring and d is a derivation which is SCP on the ideal I consisting of elements of form $(0, r_2)$. Thus, under the hypothesis of Theorem 2.2.34, we cannot prove that R must be commutative.*

Inspired by the work on SCP maps and that of Bell and Daif [109], Deng and Ashraf [237] initiated the study of a more general concept than SCP mappings by considering the situation when mappings F and G of a ring R satisfy $[F(x), G(y)] = [x,y]$ for all x,y in some subset of R. In fact, they proved the followings theorems:

Theorem 2.2.35 *[237, Theorem 1] Let R be a semiprime ring and I a nonzero ideal of R. If R admits a mapping f and a derivation d such that $[f(x), d(y)] = [x,y]$ for all $x,y \in I$, then R contains a nonzero central ideal.*

Similarly in 1997, Hongan [301] generalized the results of Bell and Daif by taking the identities in the center instead of mapping it to the zero element. Precisely, they proved the following:

Theorem 2.2.36 *[301, Theorem 1] Let R be a 2-torsion free semiprime ring, and let I be a nonzero ideal of R. Then the following conditions are equivalent:*

1. *R admits a derivation d such that $d[x,y]-[x,y]\in Z(R)$ for all $x,y\in I$.*
2. *R admits a derivation d such that $d[x,y]+[x,y]\in Z(R)$ for all $x,y\in I$.*
3. *R admits a derivation d such that $d[x,y]+[x,y]\in Z(R)$ or $d[x,y]-[x,y]\in Z(R)$ for all $x,y\in I$.*
4. *$I\subset Z(R)$.*

From the above result, we cannot exclude the condition of being 2-torsion free, which is elaborated by the following example;

Example 2.2.9 *Let $R=\begin{pmatrix}\mathbb{Z}_2 & \mathbb{Z}_2\\ \mathbb{Z}_2 & \mathbb{Z}_2\end{pmatrix}$, $a=\begin{pmatrix}1 & 0\\ 0 & 0\end{pmatrix}$ and d be an inner derivation induced by a, that is, $d(z)=[a,z]$ for all $z\in R$. Then R is a non-commutative prime ring with $char(R)=2$, and $d[z,y]\pm[z,y]\in Z(R)$ for all $z\in R$.*

Motivated by his work on prime rings, Vukman proved the following result;

Theorem 2.2.37 *[542, Theorem 2] For integers m,n with $m\geq 0, n\geq 0$, and $m+n\neq 0$, let R be a $(m+n+2)!$-torsion free semiprime ring with an identity element. Suppose there exists an additive mapping $d:R\to R$ such that $d(x^{m+n+1})=(m+n+1)x^m d(x)x^n$ is fulfilled for all $x\in R$. In this case, d is a derivation, which maps R into its center. In case R is a noncommutative prime ring, we have $d=0$.*

Theorem 2.2.38 *[201, Theorem 2.1] Let R be a semiprime ring and d a derivation of R with $d^3\neq 0$. If $[d(x),d(y)]=0$ for all $x,y\in R$, then R contains a nonzero central ideal.*

Daif [201] tried to extend the theorem of Herstein in the situation when the ring was semiprime and the condition $[d(x),d(y)]=0$ was satisfied on just an ideal of the ring.

Theorem 2.2.39 *[201, Theorem 2.2] Let R be a two-torsion-free semiprime ring and I a nonzero two-sided ideal of R. If R admits a derivation d which is nonzero on I and $[d(x),d(y)]=0$ for all $x,y\in I$, then R contains a nonzero central ideal.*

What happens if the two sided ideals are replaced by one sided ideals? Is the above result still valid? To understand this let's consider a ring R to be the set of 2×2 matrices over a field F and let $I=\begin{pmatrix}1 & 0\\ 0 & 0\end{pmatrix}R$. Next, let d be an inner derivation given by $d(x)=x\begin{pmatrix}0 & 1\\ 0 & 0\end{pmatrix}-\begin{pmatrix}0 & 1\\ 0 & 0\end{pmatrix}x$ for all $x\in R$. For any two elements x and $y\in I$, we have, $[d(x),d(y)]=0$, but the conclusion of the theorem is not true.

In 2006, Argac [27] gave the generalization of a well-known result [487, Lemma 3] and a partial extension of [136, Theorem 4.1] as follows:

Theorem 2.2.40 *[27, Theorem 3.1] Let R be a semiprime ring. Let d and g be derivations of R such that at least one of them is nonzero. If* $d(x)x = xg(x)$ *for all* $x \in R$*, then R has a nonzero central ideal.*

Proof. Assume that g is nonzero. A linearization of $d(x)x = xg(x)$ gives

$$d(x)y + d(y)x = xg(y) + yg(x) \text{ for all } x, y \in R. \tag{2.20}$$

Replacing x with xy in (2.20) we get,

$$d(x)y^2 + xd(y)y + d(y)xy = xyg(y) + yg(x)y + yxg(y). \tag{2.21}$$

Right multiplying (2.20) by y, we get,

$$d(x)y^2 + d(y)xy = xg(y)y + yg(x)y \text{ for all } x, y \in R. \tag{2.22}$$

Combining (2.21) and (2.22), we get,

$$xd(y)y = xyg(y) + yxg(y) - xg(y)y \text{ for all } x, y \in R. \tag{2.23}$$

Since the left-hand side of the above relation is equal to $xyg(y)$ by the hypothesis, we get,

$$[x, y]g(y) = x[y, g(y)] \text{ for all } x, y \in R. \tag{2.24}$$

Substituting rx for x in (2.24), where $r \in R$, we get,

$$[r, y]xg(y) + r[x, y]g(y) = rx[y, g(y)] \text{ for all } r, x, y \in R. \tag{2.25}$$

Left multiplying (2.25) by r, we get

$$r[x, y]g(y) = rx[y, g(y)] \text{ for all } r, x, y \in R. \tag{2.26}$$

Using (2.26) in (2.25), we get,

$$[r, y]xg(y) = 0 \text{ for all } r, x, y \in R. \tag{2.27}$$

Replacing r by $g(y)$ in (2.27) yields $[g(y), y]xg(y) = 0$. In particular, $[g(y), y]xyg(y) = 0$ and also $[g(y), y]xg(y)y = 0$. Hence, combining these relationships we conclude that $[g(y), y]R[g(y), y] = (0)$ for all $y \in R$. By the semiprimeness of R, we get $[g(y), y] = 0$ for all $y \in R$. Therefore, R has a nonzero central ideal by [113, Theorem 3]. Hence, the proof of the theorem is complete.

♣

In the same paper, the author has also proved Theorem 2.2.41.

Theorem 2.2.41 *[27, Theorem 3.10] Let R be a 2-torsion free semiprime ring and I a nonzero ideal of R. Let d be a derivation of R. If one of the following conditions holds:*

1. $[d(x),d(y)] = d([x,y])$ *for all* $x,y \in I$
2. $[d(x),d(y)] = d([y,x])$ *for all* $x,y \in I$
3. *for all* $x,y \in I$, *either* $[d(x),d(y)] = d([x,y])$ *or* $[d(x),d(y)] = d([y,x])$,

then d is commuting on I. Furthermore, if $d(I) \neq (0)$, *then R has a nonzero central ideal.*

The next result is the generalization of the result proved by Brešar and Hvala [144]. In 2011, Fošner and Vukman [275] proved that:

Theorem 2.2.42 *[275, Theorem 2] Let R be a 2-torsion free semiprime ring. Suppose that an additive mapping* $f : R \to R$ *satisfies the relation* $[f(x),x^2] = 0$ *for all* $x \in R$. *In this case f is commuting on R.*

Theorem 2.2.43 *[275, Theorem 6] Let R be a 2-torsion free semiprime ring and next let* $D,G : R \to R$ *be derivations. Suppose that either*

$$[[D^2(x)+G(x),x],x] = 0 \text{ or } [[D^2(x)+G(x),x^2] = 0$$

holds for all $x \in R$. *Then D and G map R into* $Z(R)$.

Further, in the year 2012, Ali and Shuliang [20] proved the Theorem 2.2.41 in the setting of a Jordan Product. In fact, they proved that

Theorem 2.2.44 *[20, Theorem 3.1] Let R be a 2-torsion free semiprime ring and I a nonzero ideal of R. Let d be a derivation of R. If one of the following conditions holds:*

1. $d(x) \circ d(y) = x \circ y$ *for all* $x,y \in I$,
2. $d(x) \circ d(y) = -(x \circ y)$ *for all* $x,y \in I$,
3. *for all* $x,y \in I$, *either* $d(x) \circ d(y) = x \circ y$ *or* $d(x) \circ d(y) = -(x \circ y)$,

then d is commuting on I. Further, if $d(I) \neq 0$, *then R has a nonzero central ideal.*

They also extended Herstein's Theorem for the Jordan Product as follows:

Theorem 2.2.45 *[20, Theorem 3.2] Let R be a 2-torsion free semiprime ring and I a nonzero ideal of R. If R admits a derivation d such that* $d(x) \circ d(y) = 0$ *for all* $x,y \in I$, *then d is commuting on I. Further, if* $d(I) \neq 0$, *then R has a nonzero central ideal.*

Fošner et al. [269] proved the following results:

Theorem 2.2.46 *[269, Theorem 1] Let n be a fixed positive integer, let R be a $(2n)!$-torsion free semiprime ring, α be an automorphism or an antiautomorphism of R. Also, let $d_1, d_2 : R \to R$ be derivations. Suppose that*

$$(d_1^2(x) + d_2(x)) \circ \alpha(x)^n = 0$$

holds for all $x \in R$. Then, $d_1 = d_2 = 0$.

Theorem 2.2.47 *[269, Theorem 2] Let R be a 2-torsion-free semiprime ring, let α be an automorphism or an antiautomorphism of R, and let $d_1, d_2 : R \to R$ be derivations. Suppose that $f : R \to R$ is a mapping defined by*

$$f(x) = (d_1^2(x) + d_2(x)) \circ \alpha(x), \ x \in R.$$

If $f(x) \circ \beta(x) = 0$ holds for all $x \in R$ and some automorphism or antiautomorphism β of R, then $d_1 = d_2 = 0$.

In the year 2016, Dhara et al. [247] gave various identities by generalizing some of the earlier results for orthogonal central idempotents. The result states;

Theorem 2.2.48 *[247, Theorem 3.2] Let R be a 2-torsion free noncommutative semiprime ring, Q_{mr} the left Utumi quotient ring of R, d a nonzero derivation of R and $0 \neq a \in R$. If $ad(R) \subseteq Z(R)$, then there exist orthogonal central idempotents $e_1, e_2, e_3 \in Q_{mr}$ with $e_1 + e_2 + e_3 = 1$ such that $d(e_1 Q_{mr}) = 0$, $e_2 a = 0$, and $e_3 Q_{mr}$ is commutative.*

Theorem 2.2.49 *[247, Theorem 3.3] Let n be a positive integer, R be a noncommutative 2-torsion free semiprime ring, Q_{mr} the left Utumi quotient ring of R and $0 \neq a \in R$. Let d, g be two nonzero derivations of R such that $a(d(x)x^n - x^n g(x)) = 0$ for all $x \in R$. Then, there exist orthogonal central idempotents e_1, e_2, $e_3 \in Q_{mr}$ with $e_1 + e_2 + e_3 = 1$ such that $d(e_1 Q_{mr}) = g(e_1 Q_{mr})$, $e_2 a = 0$, and $e_3 Q_{mr}$ being commutative.*

Theorem 2.2.50 *[247, Theorem 3.5] Let $\mathcal{A}$ be a Banach algebra, d, g a pair of continuous Jordan derivations of $\mathcal{A}$, $0 \neq a \in \mathcal{A}$ and n a fixed positive integer. If $a(d(x)x^n - x^n g(x)) \in rad(\mathcal{A})$ for all $x \in \mathcal{A}$, then $d(\mathcal{A}) \subseteq rad(\mathcal{A})$ and $g(\mathcal{A}) \subseteq rad(\mathcal{A})$.*

2.2.3 Jordan derivations on prime and semiprime rings

An additive mapping θ of a ring R into a 2-torsion free ring R' is called a Jordan homomorphism if $\theta(xy + yx) = \theta(x)\theta(y) + \theta(y)\theta(x)$ for all $x, y \in R$. A well known result due to Herstein [288] states that every Jordan homomorphism of a ring R onto a prime ring R' of characteristic different from 2 and 3 is either homomorphism or an anti homomorphism. Later, Smilley [523] provided a brief proof of this result and also at the same time removed the requirement that the characteristic of R' is different from 3. Suppose R' contains ideals U' and V' with null intersection. Let $\phi : R \to U'$ be a homomorphism and $\Psi : R \to V'$ be an anti homomorphism. A mapping $\theta = \phi + \Psi$ is a so-called direct sum of mappings θ and Ψ. Obviously, θ is a Jordan homomorphism.

An additive mapping θ of a ring R into a ring R' which satisfies $\theta(xyx) = \theta(x)\theta(y)\theta(x)$ for all $x, y \in R$ will be called a Jordan triple homomorphism. An easy computation shows that every Jordan homomorphism is also a Jordan triple homomorphism (see for example [294, Lemma 3.1]). In [292] Herstein proved that a Jordan triple homomorphism θ of a ring R into a prime ring R' of characteristic different from 2 and 3 is of the form $\theta = \pm\phi$ where ϕ is a homomorphism or an anti homomorphism of R onto R'. Brešar [132, Theorem 3.3] generalized this result by removing the requirement that the characteristic be different from 3. He obtained a more general result where θ is a Jordan triple homomorphism of a ring R onto a 2-torsion free semiprime ring R'. In fact, Brešar obtained the following result which generalizes the principal theorem of [292].

Theorem 2.2.51 *[132, Theorem 3.3] Let θ be a Jordan triple homomorphism of a ring R onto a prime ring R'. Then $\theta = \pm\phi$ where ϕ is a homomorphism or an antihomomorphism of R onto R'.*

In the 1950's, Herstein [294] initiated a study of the relationship between Jordan and the Lie structures of associative rings. We refer the reader to ([289, 294, 295]), where one can find further references and more detailed explanations concerning the motivation and the background of these concepts. A Jordan derivation d of a ring R is an additive mapping $d : R \longrightarrow R$ such that $d(x^2) = d(x)x + xd(x)$, for all $x \in R$. Every derivation is obviously a Jordan derivation and the converse is not true in general. The following example demonstrates that there exists a Jordan derivation which is not a derivation.

Example 2.2.10 *Let R be a 2-torsion free ring and $a \in R$ such that $xax = 0$ for all $x \in R$, but $xay \neq 0$ for some $x \neq y \in R$. Define a map $d : R \to R$ such that $d(x) = ax$ for all $x \in R$. Then, it can be verified that d is a Jordan derivation but not a derivation.*

For a prime ring with characteristic 2, he managed to prove the same but with a specific condition on its structure:

Theorem 2.2.52 *[289, Theorem 4.1] Let R be a prime ring of characteristic 2. Then if R is not a commutative integral domain, any Jordan derivation of R is an ordinary derivation.*

Later on, Awtar [89] generalized the result of Herstein on Lie ideals and proved the following theorem:

Theorem 2.2.53 *[89, Theorem] Let R be a prime ring of characteristic differing from two and U be a square closed Lie ideal of R. If $d : R \to R$ is an additive mapping such that $d(u^2) = d(u)u + ud(u)$ for all $u \in U$, then $d(uv) = d(u)v + ud(v)$ for all $u, v \in U$.*

In 1988, Brešar and Vukman [148] gave an alternative proof of Herstein's result [294]. In addition to that Brešar [131] generalized the result of Herstein even further to 2-torsion-free semiprime rings;

Theorem 2.2.54 *[131, Theorem 1] Let R be a 2-torsion free semiprime ring and let $d{:}R \longrightarrow R$ be a Jordan derivation. In this case, d is a derivation.*

Jacobson in his classical book "Structure of Rings" [319] has given a passing reference of (s_1, s_2)-derivation which was later more commonly referred as (σ, τ)-derivation or (α, β)-derivation by some authors and (θ, ϕ)-derivation by others. Let θ and ϕ be endomorphisms of R. An additive mapping $d : R \longrightarrow R$ is called a (θ, ϕ)-derivation (resp. Jordan (θ, ϕ)-derivation) on R if $d(xy) = d(x)\theta(x) + \phi(x)d(y)(d(x^2) = d(x)\theta(x) + \phi(x)d(x))$ holds for all $x, y \in R$. The mappings $a \mapsto \theta(x)y - y\phi(x)$, where y is a fixed element in R, is a (θ, ϕ)-derivation. Such a (θ, ϕ)-derivation is said to be inner. A (θ, I_R) derivation, where I_R is the identity map on R is simply called a θ-derivation. Of course, I_R-derivation is a derivation.

Leroy and Matczuk [397] generalized Herstein's [294] result for Jordan θ-derivations, where θ is an automorphism [397, Theorem 2.6]. Further, in the year 1991, Brešar and Vukman [150] extended Herstein's [294] result for Jordan (θ, ϕ)-derivations and proved the following:

Theorem 2.2.55 *[150, Theorem] Let R be any ring and R' be a non-commutative ring. Let θ and ϕ be homomorphisms of R into R'. Let X be a 2-torsion free R'-bimodule. Suppose that either ϕ is onto and $xR'a = (0)$ with $x \in X$; $a \in R'$ implies that $x = 0$ or $a = 0$ or that θ is onto and $aR'x = (0)$ with $x \in X$, $a \in R'$ implies that $x = 0$ or $a = 0$. In this case, every Jordan (θ, ϕ)-derivation $d : R \to X$ is a (θ, ϕ) -derivation.*

Remark 2.2.56 *If a ring R and a nonzero R-bimodule X are such that $Ra = (0)$ with $x \in X$, $a \in R$ implies that $x = 0$ or $a = 0$, then R is prime. Indeed, suppose that $aRb = (0)$ for some $a, b \in R$. Given any nonzero $x \in X$ we have $(xRa)Rb = (0)$. Hence, it simply follows that either $a = 0$ or $b = 0$. Then R is prime. Suppose further that X is 2-torsion free. If $2a = 0$ for some $a \in R$ then $2xRa = (0)$ for all $x \in X$, which yields $a = 0$. Thus R is 2-torsion free.*

If d is a Jordan derivation of a 2-torsion free, commutative, semiprime ring, then an easy linearization shows that d must be a derivation. The simple approach fails for a Jordan (θ,ϕ)-derivation, although the result remains true. Very recently Lanski [369] proved the following results for Jordan (θ,ϕ)-derivations:

Theorem 2.2.57 *[369, Theorem 1] Let R be a 2-torsion free, commutative, semiprime ring. If d is a Jordan (θ,ϕ)-derivation of R, then d is a (θ,ϕ)-derivation of R.*

Theorem 2.2.58 *[369, Theorem 2] Let R be a 2-torsion free semiprime ring and d a Jordan (θ,ϕ) derivation of R with θ or ϕ an automorphism of R. Then d is a (θ,ϕ)-derivation of R.*

Lanski [369] presented an example of a prime ring of characteristic 2 (commutative) that admits Jordan (θ,ϕ)-derivations which are not (θ,ϕ)-derivations (see [369] for other examples).

Example 2.2.11 *Let $R = F[x,y]$ for F a field with char(F)=2. Note that any square in R is contained in $R = F[x^2,y^2]$. If $V = xR + yR + xyR$, then as an F-vector space $R = R \oplus V$, so $q \in R$ has a unique representation $q = s + v$ for $s \in R$ and $v \in V$. Defining $\theta,\phi,d : R \to R$ follows: $d(s+v) = v$ and $\theta = \phi = I_R$. Clearly d is additive, and by our definition, for any $g \in R$, $g^2 \in R$ so $d(g^2) = 0$ and $\theta(g)d(g) + d(g)\phi(g) = 2\phi(g)d(g) = 0$. Thus d is a Jordan (θ,ϕ)-derivation of R. Further, $d(xy) = xy$ but $\theta(x)d(y) + d(x)\phi(y) = xy + xy = 0$, so d is not a (θ,ϕ)-derivation of R. This example works as well as if we take $\phi = \theta$, given by $\phi(g(x,y)) = g(x,y)$ or take either ϕ to be described, but with $d((s,v(x,y)) = v(x,y)$.*

The following example highlights the existence of automorphisms θ or ϕ. In fact, there exist semiprime rings R having a Jordan (θ,ϕ)-derivation d, where neither θ nor ϕ is an automorphism of R, yet d is not a (θ,ϕ) derivation of R.

Example 2.2.12 *Let $F\{x,y\}$ be a free algebra over a field F and consider its ideal $I = (x,y)$, which consists of elements with zero constant terms. For $b \in I$, we can express it uniquely as, $b = f_1x + f_2y + g(x,y)$ where $f_i \in F$ and $g(x,y) \in I^2$. Thus $b^2 = a^2 + f_2^2y^2 + f_1f_2(xy+yx) + h(x,y)$ with $h(x,y) \in I^3$. We define a set of $T = \{c_1x^2 + c_2y^2 + c3(xy+yx) + g(x,y) \mid g(x,y) \in I^3\}$. Each element has a unique representation $b = c_1x + c_2y + c_3xy + t$ for each $b \in I$ and some $t \in T$. Note that all $b^2 \in T$. For any nonzero semiprime ring R set $R' = A \oplus I$. Define $\phi, d : R \to R$ by $\phi((a,b)) = (a,0)$ and $d(a, c_1x + c_2y + c_3xy + t) = (0, c_1x + c_2y + c_3xy)$. Clearly d is additive and is a homomorphism but not an automorphism, and $\phi(R)d(R) = 0 = d(R)\phi(R)$. Our observation above about squares shows that d is a Jordan (θ,ϕ)-derivation for $\theta = \phi$ or for $\theta((a,b)) = (h(a),0)$ for any endomorphism h of A. However, d is not a (θ,ϕ)-derivation since $(0,xy) = d((0,x)(0,y))$, whereas $\theta((0,x))d((0,y)) + d((0,x))\phi((0,y)) = 0$.*

In 2001, Ashraf et al. [74] extended Awtar's [89] result for Jordan (θ,ϕ)-derivation as follows:

Theorem 2.2.59 *[74, Theorem] Let R be a prime ring of characteristic different from two and U be a non-commutative Lie ideal of R such that $u^2 \in U$. Suppose that θ, ϕ are automorphisms of R. If $d : R \to R$ is an additive mapping satisfying $d(u^2) = d(u)\theta(u) + \phi(u)d(u)$ for all $u \in U$, then $d(uv) = d(u)\theta(u) + \phi(u)d(v)$ for all $u, v \in U$.*

So one might think, those Jordan derivations which are not derivations are very difficult to find. Indeed that turns out to be true. In 1996, Johnson [333] thought about those (Banach) algebras $\mathcal{A}$ which had no proper Jordan derivations from $\mathcal{A}$ into any arbitrary (Banach) $\mathcal{A}$-bimodule $\mathbb{M}$. While it turned out that this was true for some important classes of algebras (in particular, for the algebra of all $n \times n$ complex matrices). On the other hand, there were simple counter examples on some algebras and their (special) bimodules. The fundamental example in this direction is given on the algebra $\mathbb{T}_2$ of 2×2 upper triangular matrices over R (a commutative ring with unity). We can make R a $\mathbb{T}_2$-bimodule by defining $x\lambda = x_{22}\lambda$ and $\lambda x = \lambda x_{11}$ for all $\lambda \in R$, $x \in \mathbb{T}_2$. A map $d : \mathbb{T}_2 \to R$ defined by $d(x) = x_{12}$ is a proper Jordan derivation, which, Johnson proposed to be an antiderivation, i.e., satisfies

$$d(xy) = d(y)x + yd(x),$$

for all $x, y \in \mathbb{T}_2$. In this direction Benkovic [118] gave a very interesting result which helps to decompose a Jordan derivation into a derivation and an antiderivation; the result is stated as follows:

Theorem 2.2.60 *[118, Theorem 1.1] Let $\mathbb{M}$ be a $\mathbb{T}_n$-bimodule and let $\Delta : \mathbb{T}_n \to \mathbb{M}$ be a Jordan derivation. Then there exists a derivation $d : \mathbb{T}_n \to \mathbb{M}$ and an antiderivation $\delta : \mathbb{T}_n \to \mathbb{M}$ such that $\Delta = d + \delta$ and $\delta(\mathbb{D}_n) = 0$. Moreover, d and δ are uniquely determined, where $\mathbb{D}_n$ is the subalgebra of all diagonal matrices.*

As a corollary to this result, Benkovic [118] proved that a proper Jordan derivation for $n \times n$ upper triangular matrices does not exist.

Corollary 2.2.1 *[118, Corollary 1.2] Let $2 \leq n \leq m$ be fixed positive integers. There are no proper Jordan derivations from $\mathbb{T}_n$ into $\mathbb{M}_n$. In particular, there are no proper Jordan derivations from $\mathbb{T}_n$ into itself.*

In the previous section we discussed how Wang [545] described Lie derivations on triangular algebras. He did the same for Jordan derivations and proved the following:

Theorem 2.2.61 *[545, Corollary 3.1] Let $\mathfrak{A} = Tri(\mathcal{A}, M, \mathcal{B})$ be a triangular algebra. Let d be a Jordan derivation on $\mathfrak{A}$. Suppose that $\mathfrak{A}$ is 2-torsion free. Then d is a derivation. Otherwise, there exists a triangular algebra $\mathfrak{A}^0$ such that $\mathfrak{A}$ is a subalgebra of $\mathfrak{A}^0$ having the same unity and d; it can be written as $d = \delta + \tau$, where $\delta : \mathfrak{A} \to \mathfrak{A}^0$ is a derivation and $\tau : \mathfrak{A} \to Z(\mathfrak{A}^0)$ is a linear map such that $\tau(x \circ y) = 0$ for all $x, y \in \mathfrak{A}$.*

Motivated by Brešar's results in [132], Vukman [540] proved the following theorem:

Theorem 2.2.62 *[540, Theorem 2] Let R be a 2-torsion free semiprime ring and let $d : R \to R$ be an additive mapping. Suppose that either*

$$d(xyx) = d(xy)x + xyd(x)$$

or

$$d(xyx) = d(x)yx + xd(yx)$$

holds for all pairs $x, y \in R$. In both cases, d is a derivation.

Theorem 2.2.63 *[540, Theorem 6] Let X be a real or complex Banach space. Also, let $\mathcal{A}(X)$ be a standard operator algebra on X and let $d : \mathcal{A}(X) \to L(X)$ be a linear derivation. In this case d is of the form $d(A) = [A, B]$ for all $A \in \mathcal{A}(X)$ and some fixed $B \in L(X)$.*

Motivated by the above theorem, Fošner and Peršin [272], showed the following:

Theorem 2.2.64 *[272, Theorem 3] Let R be a prime ring with $char(R) = 0$ or $4 < char(R)$ and let $d : R \to R$ be an additive mapping satisfying either the relation*

$$d(x^3) = d(x^2)x + x^2d(x)$$

or the relation

$$d(x^3) = d(x)x^2 + xd(x^2)$$

for all $x \in R$. In both cases, d is a derivation.

Recently in the year 2020, Fošner and Jing [271] proved that every Jordan derivation on a triangular ring is a derivation and were also able to establish that, under some conditions, every Jordan derivation on a 2-torsion free ring is a derivation.

2.3 Left derivations

The concepts of a left derivation, (an additive mapping $d_1 : R \to R$ satisfying $d_1(xy) = xd_1(y) + yd_1(x)$, for all $x, y \in R$) and that of a Jordan left derivation (additive mapping $d_2 : R \to R$ such that $d_2(x^2) = 2xd_2(x),\ x \in R$) were introduced by Brešar and Vukman in [149]. In the past three decades, there a significant amount of work has been done around Jordan left derivations and related mappings, as these are in a close connection with the so-called commuting mappings. The main motivation comes from Posner's second theorem. Way back in 1955, a fundamental result was obtained by Singer and Wermer [521] while studying the range of derivations on Banach algebras, which is famously called the Singer-Wermer theorem. It states that any continuous derivation on a commutative Banach algebra maps the algebra into the Jacobson radical and in the same paper they conjectured that the assumption of continuity was unnecessary. This conjecture was later proved in 1988 by Thomas [529]. So it was inferred that every derivation on a commutative semisimple Banach algebra is zero. Various authors time and again have been able to prove many non-commutative forms of the Singer-Wermer theorem (see, e.g., [146], [349] and [381]). So the question that under what conditions all derivations on a Banach algebra are zero attracted a lot of attention. In this direction the range of left derivations was reviewed. In the year 1998, Jung [336] proved that every spectrally bounded left derivation maps algebra into its Jacobson radical. Vukman [539] showed that every Jordan left derivation on a semisimple Banach algebra is identically zero. In the same paper, he conjectured that every Jordan left derivation on a Banach algebra maps the algebra into its radical (see [10, 48] for more recent results).

Thinking of Posner's results for left derivations, Brešar and Vukman [149] came up with some interesting results in 1990 where they guaranteed the commutativensss of the ring once a Jordan left derivation was found. The result states the following:

Theorem 2.3.1 *[149, Theorem 1.2] Let R be a ring and X be a 2-torsion free and 3-torsion free left $R-$module. Suppose that $aRx = 0$ with $a \in R$, $x \in X$ implies that either $a = 0$ or $x = 0$. If there exists a nonzero Jordan left derivation $d : R \to X$, then R is commutative.*

Theorem 2.3.2 *[149, Corollary 1.3] Let R be a prime ring of characteristic different from 2 and 3. If R admits a nonzero Jordan left derivation $d : R \longrightarrow R$, then R is commutative.*

In the same paper they proposed the condition for X to be 3-torsion free can be omitted. This was generalized by Ashraf et al. [81] for a 2-torsion free prime ring;

Theorem 2.3.3 *[81, Main Theorem] Let R be a 2-torsion free prime ring. Suppose J is a nonzero Jordan ideal and a subring of R. If $d : R \to R$ is an additive mapping such that $d(u^2) = 2ud(u)$ for all $u \in J$, then either $J \subseteq Z(R)$ or $d(J) = (0)$.*

In the year 1992, Deng [235] proved the commutativity of a prime ring using a left $R-$ module, precisely;

Theorem 2.3.4 *[235, Theorem 1] Let R be a prime ring with characteristic $\neq 2$, and X a nonzero left R-module. Suppose that X is faithful and prime. If there exists a nonzero Jordan left derivation $d : R \longrightarrow X$, then R is commutative.*

In 1992, Vukman [534] proved the above mentioned result for semiprime rings. He was able to accomplish the following;

Theorem 2.3.5 *[534, Theorem 1] Let R be a 2-torsion free semiprime ring and $d : R \longrightarrow R$ be a Jordan left derivation. If there exists a positive integer n such that $d(x)^n = 0$ for all $x \in R$, then $d = 0$.*

In 1998, Jung [336] investigated the above study for semiprime Banach algebra and was able to establish the following;

Theorem 2.3.6 *[336, Theorem] Every left derivation on a semiprime Banach algebra $\mathcal{A}$ is a derivation which maps $\mathcal{A}$ into the intersection of the center of $\mathcal{A}$ and the Jacobson radical of $\mathcal{A}$.*

Theorem 2.3.7 *[336, Corollary 3.7] Let d be a left derivation on a semisimple Banach algebra. Then $d = 0$.*

It is evident that every left derivation forms a Jordan left derivation but the converse need not be true in general. In this direction in 2000 Ashraf and Rehman [77] obtained a general result which established that under appropriate restrictions on a Lie ideal U of a 2-torsion free prime ring, every Jordan left derivation on U turns out to be a left derivation on U.

Theorem 2.3.8 *[77, Corollary] Let R be a 2-torsion free prime ring and $d : R \rightarrow R$ be a Jordan left derivation. Then, d is a left derivation.*

Further, Vukman in 2008 proved the following results;

Theorem 2.3.9 *[539, Theorem 1] Let R be a noncommutative 2-torsion free prime ring and let $d : R \longrightarrow R$ be a left Jordan derivation. In this case $d = 0$.*

Theorem 2.3.10 *[539, Theorem 2] Let R be a 2-torsion free semiprime ring and let $d : R \rightarrow R$ be a left Jordan derivation. In this case d is a derivation which maps R into $Z(R)$.*

Also, Vukman proved Theorem 2.3.10 in the case of semisimple Banach algebra. He proved the following theorem:

Theorem 2.3.11 *[539, Theorem 4] Let $\mathcal{A}$ be a semisimple Banach algebra and let $d : \mathcal{A} \rightarrow \mathcal{A}$ be a linear left Jordan derivation. In this case $d = 0$.*

The above theorem is used in the proof of the next result which was inspired by the work of Kurepa [358].

Theorem 2.3.12 *[539, Theorem 5] Let $\mathcal{A}$ be a semisimple Banach algebra with an identity element and let $f, g : \mathcal{A} \rightarrow \mathcal{A}$ be linear mappings. Suppose that the relation $f(x) = x^2 g(x^{-1})$ holds for all invertible elements $x \in \mathcal{A}$. In this case the following statements hold:*

1. *$f = g$;*
2. *$f(x) = xf(e)$ for all $x \in R$, where e denotes the identity element.*

Recently in 2017, Hosseini [303] gave certain conditions under which the left derivations become zero. A few of these interesting results are as follows:

Theorem 2.3.13 *[303, Theorem 2.13] Let $\mathcal{A}$ be a semisimple Banach algebra and let $d : \mathcal{A} \to \mathcal{A}$ be a Jordan left derivation. Then d is identically zero.*

Theorem 2.3.14 *[303, Theorem 2.14] Let $\mathcal{A}$ be a Banach algebra, $d : \mathcal{A} \to \mathcal{A}$ be a Jordan left derivation, and let P be a primitive ideal of $\mathcal{A}$. If $d(P) \subset P$, then $d(\mathcal{A}) \subseteq P$.*

Theorem 2.3.15 *[303, Theorem 2.17] Let R be a 2-torsion free semiprime ring, and let $d : R \to R$ be a Jordan left derivation. Then, $d(x)[y,z] = 0$ for all $x, y, z \in R$.*

Theorem 2.3.16 *[303, Corollary 2.18] Let R be a 2-torsion free semiprime ring such that $ann\{[y,z] \mid y,z \in R\} = (0)$. Then, every Jordan left derivation $d : R \to R$ is identically zero.*

Theorem 2.3.17 *[303, Theorem 2.20] Let R be a non-commutative 2-torsion free simple ring, and let $d : R \to R$ be a Jordan left derivation. In this case $d = 0$.*

Brešar and Vukman [149] proved that every Jordan left derivation on a non-commutative 2-torsion free and 3-torsion free prime ring is identically zero. But the condition for a 3-torsion free prime ring can be omitted. It was achieved by Hosseini [303];

Theorem 2.3.18 *[303, Theorem 2.21] Let R be a non-commutative 2-torsion free prime ring, and let $d : R \to R$ be a Jordan left derivation. In this case, d is zero.*

Theorem 2.3.19 *[303, Corollary 2.25] Let R be a (commutative or non-commutative) 2-torsion free prime ring, I be a non-zero Jordan ideal of R, and let $d : R \to R$ be a Jordan left derivation such that $ad(a)a = 0$ for all $a \in I$. Then, d is identically zero.*

Theorem 2.3.20 *[304, Theorem 2.1] Let I be a semiprime ideal of an algebra $\mathcal{A}$, and let $d : \mathcal{A} \to \mathcal{A}$ be a Jordan left derivation such that $d(I) \subset I$. If $dim\{d(a) + I \mid a \in \mathcal{A}\} \leq 1$, then $d(\mathcal{A}) \subset I$.*

2.4 (α, β)-derivations

The union of the idea of derivations with that of endomorphisms lets us dive in a new and interesting type of derivation. An additive map $d : R \to R$ defined by the identity

$$d(xy) = d(x)y + \alpha(x)d(y) \text{ for all } x, y \in R,$$

where α is an endomorphism of R, what we know as a *Skew derivation* (or an *α-derivation*). Clearly, every derivation is a skew derivation. Introducing another endomorphism β of R and defining an additive map $d : R \longrightarrow R$ by

$$d(xy) = d(x)\alpha(y) + \beta(x)d(y) \text{ for all } x, y \in R$$

gives us an (α, β)-derivation. Similarly, we define a *Jordan* (α, β)*-derivation* as an additive map $d : R \rightarrow R$, satisfying

$$d(x^2) = d(x)\alpha(x) + \beta(x)d(x) \text{ for all } x \in R.$$

In [297], Herstein showed that an element a of R satisfying $ad(x) = d(x)a$ for all $x \in R$, for any derivation d should be central. Later on in [162], Chang extended this result by assuming that $[a, \delta(x)] = 0$ for all $x \in R$, where δ is an (α, β)-derivation of R such that $\delta\alpha = \alpha\delta$, $\delta\beta = \beta\delta$. Argac et al. generalized these results and proved the following;

Theorem 2.4.1 *[32, Theorem 1] Let I be a nonzero ideal of R and $d : R \rightarrow R$ a nonzero (α, β)-derivation such that $d\alpha = \alpha d$, $d\beta = \beta d$. If $a \in R$ and $[a, d(I)]_{\lambda,\mu} = 0$, then $a \in C_{\lambda,\mu}$, where $C_{\alpha,\beta} = \{c \in R | c\alpha(r) = \beta(r)c \text{ for all } r \in R\}$.*

Theorem 2.4.2 *[32, Theorem 2] Let I be a nonzero ideal of R and $d_1 : R \rightarrow R$, a nonzero (σ, τ) derivation and $d_2 : R \rightarrow R$, a (α, β) derivation such that $d_2\alpha = \alpha d_2$, $d_2\beta = \beta d_2$. If $d_1 d_2(I) = 0$ then $d_1 = 0$ or $d_2 = 0$.*

Corollary 2.4.1 *[32, Corollary 3] Let I be a nonzero ideal of R and d: $R \longrightarrow R$, a nonzero (α, β)-derivation such that $d\alpha = \alpha d$, $d\beta = \beta d$. If $[d(R), I]_{\alpha,\beta} = 0$, then R is commutative.*

Generalizations to centralizing theorems:

Theorem 2.4.3 *[90, Theorem 1] Let α, β be endomorphisms on R, let d be a nonzero (α, β)-derivation, I an ideal of R and $a \in R$. If $[d(I), a]_{\alpha,\beta} = 0$, then $a \in Z(R)$.*

Theorem 2.4.4 *[90, Theorem 2] Let d be a nonzero (α, β)-derivation with endomorphisms α and β, and I an ideal of R. If $[d(I), d(I)]_{\alpha,\beta} = 0$, then R is commutative.*

Theorem 2.4.5 *[90, Theorem 3] Let α, β be endomorphisms on R, let d be a nonzero (α, β)-derivation. If $[d(R), d(R)]_{\alpha,\beta} \subseteq C_{\alpha,\beta}$ then R is commutative.*

In 1997, Chang provided a description of derivations, automorphisms d, g, h, as well as (α, β)-derivations of a prime ring R. These derivations satisfy the equation $d(x) = ag(x) + h(x)b$ for all $x \in I$, where a and b are fixed noncentral elements of R and I is a nonzero ideal of R. Chang's work led to a fascinating result connecting (α, β)-derivations with elements from the symmetric Martindale quotient ring.

Theorem 2.4.6 *[163, Theorem 1] Let R be a prime ring, I a nonzero ideal of R, Q_s the symmetric Martindale quotient ring of R and C the extended centroid of R. Further, let δ, g and h be (α,β)-derivations of I into R and $a,b \in Q_s \setminus C$. Suppose that either $g \neq 0$ or $h \neq 0$. Then the following conditions are equivalent:*

1. $\delta(x) = ag(x) + h(x)b$ *for all* $x \in I$.
2. *There exists an invertible element* $s \in Q_s$ *such that,*

$$\beta(x) = s^{-1}\alpha(x)s, \delta(x) = [asbs^{-1}, \alpha(x)]s,$$

$$g(x) = s[b, \beta(x)] = [sbs^{-1}, \alpha(x)]s \text{ and } h(x) = [a, \alpha(x)]s$$

for all $x \in I$.

Moreover, they showed that under some conditions two (α,β)-derivations are identical.

Corollary 2.4.2 *[163, Corollary] Let R be a commutative domain and let $0 \neq \delta_1$ be an (α_1,β_1)-derivation, $0 \neq \delta_2$ be an (α_2,β_2)-derivation of R. If $\delta_1 = \delta_2$, then we have either $\alpha_1 = \alpha_2\beta_1$ or $\alpha_1 = \beta_2$ or $\alpha_2 = \beta_1$.*

Theorem 2.4.7 *[163, Theorem 4] Let R be a prime ring, I a nonzero ideal of R, and δ a nonzero (α,β)-derivation of R. If $[\delta(I),\delta(I)]_{\alpha,\beta} = 0$, then,*

1. *If $char(R) \neq 2$, then R is commutative. In this case, if $\alpha \neq \beta$, then $\alpha^2 = \beta^2$, $\alpha\beta = \beta\alpha$ and $\delta(x) = \lambda(\alpha(x) - \beta(x))$, for some $\lambda \in C$ such that $\alpha(\lambda) + \beta(\lambda) = 0$. Also $\delta^2 = 0$.*
2. *If $char(R) = 2$, then R is an s_4-ring.*

Generalizing Posner's theorem to (α,β)-derivations:

Theorem 2.4.8 *[80, Theorem 1] Let R be a 2-torsion free prime ring. Suppose there exists an (α,β)-derivation $d : R \to R$ such that $[d(x),x]_{\alpha,\beta} = 0$, for all $x \in R$. Then, either $d = 0$ or R is commutative.*

Results from Herstein [297], Bell and Daif [110] paved the way for proving similar looking identities for (α,β)-derivations. Motivated by these results, Ashraf and Rehman [80] proved the following:

Theorem 2.4.9 *[80, Theorem 2] Let R be a 2-torsion free prime ring, and I a nonzero ideal of R. If R admits a (α,β)-derivation d such that $[d(x),d(y)] = 0$ for all $x,y \in I$ and d commutes with both α,β, then either $d = 0$ or R is commutative.*

Theorem 2.4.10 *[80, Theorem 3] Let R be a 2-torsion free prime ring, and I a nonzero ideal of R. If R admits a nonzero (α,β)-derivation d such that $d(xy) = d(yx)$, for all $x,y \in I$ and d commutes with β, then R is commutative.*

Theorem 2.4.11 *[80, Theorem 4] Let R be a 2-torsion free prime ring and α,β be automorphisms of R. Suppose d_1 and d_2 are two (α,β)-derivations of R such* *that* $d_1\alpha = \alpha d_1$, $d_1\beta = \beta d_1$, $d_2\alpha = \alpha d_2$ *and* $d_2\beta = \beta d_2$. *If* $d_1d_2(R) = 0$, *then either* $d_1 = 0$ *or* $d_2 = 0$.

Motivated by the above results Kaya et al. [341] were able to generalize these results as follows:

Theorem 2.4.12 *[341, Theorem 2] Let* d_1 *be a* (σ, τ) *derivation and* d_2 *an* (α, β)*-derivation on* R *such that* $d_2\alpha = \alpha d_2$, $d_2\beta = \beta d_2$. *If* $d_1d_2(R) = 0$ *then,* $d_1 = 0$ *or* $d_2 = 0$.

Theorem 2.4.13 *[341, Corollary 3] Let* $0 \neq d_1\colon R \longrightarrow R$ *be a* (σ, τ)*-derivation and* $0 \neq d_2\colon R \longrightarrow R$ *an* (α, β)*-derivation such that* $d_2\alpha = \alpha d_2$ *and* $d_2\beta = \beta d_2$. *If* $[d_1(R), d_2(R)] = (0)$, *then* R *is commutative.*

An additive mapping $\delta : R \to R$ is called a left (θ, ϕ)-derivation (resp., Jordan left (θ, ϕ)-derivation) of R if $\delta(xy) = \theta(x)\delta(y) + \phi(y)\delta(x)$ (resp., $\delta(x^2) = \theta(x)\delta(x) + \phi(x)\delta(x)$) holds for all $x, y \in R$. In 2004, Zaidi et al. [566] worked with left (θ, ϕ)-derivations and Jordan left (θ, ϕ)-derivations and were able to prove that for a Jordan ideal J which is also a subring of a 2-torsion-free prime ring R if $\delta(x^2) = 2\theta(x)\delta(x)$ for all $x \in J$, for any automorphism θ of R then either $J \subseteq Z(R)$ or $\delta(J) = (0)$. Further, they also talked about (θ, θ)-derivations and deduced that left (θ, θ)-derivations of a prime ring R act either as a homomorphism or as an antihomomorphism on the ring R. Some more interesting results given in [566] are stated as follows:

Theorem 2.4.14 *[566, Theorem 3.1] Let* R *be a 2-torsion-free prime ring and let* J *be a Jordan ideal and a subring of* R. *If* θ *is an automorphism of* R *and* $\delta : R \to R$ *is an additive mapping satisfying* $\delta(u^2) = 2\theta(u)\delta(u)$ *for all* $u \in J$, *then either* $J \subseteq Z(R)$ *or* $\delta(J) = (0)$.

In the above result if we consider J to be a subring only then the result does not hold. To visualize this consider a ring R such that the square of each element of R is zero, but the product of some elements in R is nonzero. Also, suppose that $\mathcal{R} = \left\{ \begin{pmatrix} x & y \\ 0 & 0 \end{pmatrix} \mid x, y \in R \right\}$, consider $J = \left\{ \begin{pmatrix} 0 & y \\ 0 & 0 \end{pmatrix} \mid y \in R \right\}$. Define maps $\delta\colon \mathcal{R} \longrightarrow \mathcal{R}$ and $\theta\colon \mathcal{R} \longrightarrow \mathcal{R}$ by

$\delta\begin{pmatrix} x & y \\ 0 & 0 \end{pmatrix} = \begin{pmatrix} 0 & y \\ 0 & 0 \end{pmatrix}$ and $\theta\begin{pmatrix} x & y \\ 0 & 0 \end{pmatrix} = \begin{pmatrix} x & -y \\ 0 & 0 \end{pmatrix}$. It can be verified that δ forms a Jordan left (θ, θ)-derivation, but neither $J \subseteq Z(\mathcal{R})$ nor $\delta(J) = (0)$. So the subring J has to be an ideal as well.

In 1989, Bell and Kappe [112] were able to show that if d, a derivation of a prime ring R acts as a homomorphism or as an antihomomorphism on a nonzero right ideal I of R, then the derivation d must be zero. Later on in [83] this result was extended for (α, β)-derivations as under:

Theorem 2.4.15 *[83, Theorem] Let R be a prime ring, I a nonzero right ideal of R, and let α, β be automorphisms of R. Suppose that $\delta : R \to R$ is an (α, β)-derivation of R.*

1. *If δ acts as a homomorphism on I, then $\delta = 0$ on R.*
2. *If δ acts as an antihomomorphism on I, then $\delta = 0$ on R.*

Zaidi et al. [566] were able to extend it to a left (θ, θ)-derivation of a prime ring R, with appropriate restrictions, the result states the following:

Theorem 2.4.16 *[566, Theorem 4.1] Let R be a 2-torsion-free prime ring and J a nonzero Jordan ideal and a subring of R. Suppose θ is an automorphism of R and $\delta : R \to R$ is a left (θ, θ)-derivation of R. (i) If δ acts as a homomorphism on J, then $\delta = 0$ on R. (ii) If δ acts as an antihomomorphism on J, then $\delta = 0$ on R.*

Taking Lanski's [361] result forward, Vukman proved the following:

Theorem 2.4.17 *[537, Theorem 1] Let R be a noncommutative prime ring of characteristic different from two and let d and $g \neq 0$ be (α, β)-derivations of R such that g commutes with α and β. If $[d(x), g(x)] = 0$ holds for all $x \in R$, then $d = \lambda g$ where λ is an element from $C(R)$.*

In case we have α- derivations d and g on a prime ring R of characteristic different from two satisfying the relation $d(x)g(x) = 0$ for all $x \in R$ then, either $d = 0$ or $g = 0$ (see Theorem 3 in [536]). Our next result is in the spirit of the result we have just mentioned.

Theorem 2.4.18 *[537, Theorem 3] Let R be a prime ring of characteristic different from two and let $d, g : R \to R$ be (α, β)-derivations such that $d(x)g(x) = 0$ holds for all $x \in R$. Suppose that either d or g commutes with α and β. In both cases either $d = 0$ or $g = 0$.*

2.5 Bi-derivations

After the concept of derivation was defined in [487] by Jacobson, lots of researchers studied derivations in ring theory in different ways and various forms. In 1980, Maksa introduced and studied the concept of symmetric bi-derivations in [436]. In [437] where he also gave the representation theorem of bi-derivations in Hilbert spaces. Lots of work has been done in this direction. Let's first equip ourselves with the terminology associated with bi-derivations.

Definition 18 *A mapping $D : R \times R \to R$ is called bi-additive if it is additive in each argument, i.e., $D(x+w, y) = D(x, y) + D(w, y)$ and $D(x, y+w) = D(x, y) + D(x, w)$ for all $x, y, w \in R$.*

Definition 19 *A mapping $D : R \times R \to R$ is called symmetric if $D(x,y) = D(y,x)$ for all $x, y \in R$.*

For example, let R be a ring. The mapping $D : R \times R \to R$ defined by $D(x,y) = x+y$ for all $x, y \in R$ is a symmetric mapping.

Definition 20 *A mapping $D : R \times R \to R$ is called symmetric bi-additive if it is both bi-additive and symmetric.*

Something that helped the transition of various results of ordinary derivations to that of symmetric bi-derivations is the idea of a trace of a symmetric bi-derivation. The way it is defined is as follows:

Definition 21 *A mapping $\delta : R \to R$ defined by $\delta(x) = D(x,x)$ is called a trace of D where $D : R \times R \to R$ is a symmetric mapping.*

Clubbing all these concepts we say a symmetric bi-additve mapping $D : R \times R \to R$ is called a symmetric bi-derivation of R if it is a derivation in each argument; that is, for every $x \in R$, the maps $y \to D(x,y)$ and $y \to D(y,x)$ are derivations of R into R or if it satisfies the relations $D(xy,z) = D(x,z)y + xD(y,z)$ for all $x,y,z \in R$. It is evident that in this case the relation $D(x,yz) = D(x,y)z + yD(x,z)$ for all $x,y,z \in R$ also holds. If $D : R \times R \to R$ is a symmetric mapping, which is also bi-additive then d, the trace of D satisfies the relation

$$\delta(x+y) = \delta(x) + \delta(y) + 2D(x,y) \; for \; all \; x, \; y \in R.$$

Since we have

$$D(0,y) = D(0+0,y) = D(0,y) + D(0,y)$$

for all $y \in R$, we obtain $D(0,y) = 0$ for all $y \in R$. Hence, we get $D(-x,y) = -D(x,y)$, similarly $D(x,-y) = -D(x,y)$ for all $x, y \in R$. This tells us that d is an even function.

Example 2.5.1 *For a commutative ring R, let $M := \left\{ \begin{pmatrix} a & b \\ 0 & 0 \end{pmatrix} \mid a,b \in R \right\}$. It is obvious that M is a non-commutative ring under matrix addition and matrix multiplication. We define a map $D : M \times M \to M$ by $\left(\begin{pmatrix} a_1 & b_1 \\ 0 & 0 \end{pmatrix}, \begin{pmatrix} a_2 & b_2 \\ 0 & 0 \end{pmatrix} \right) \mapsto \begin{pmatrix} 0 & a_1 a_2 \\ 0 & 0 \end{pmatrix}$. It is easy to see that D forms a symmetric bi-derivation.*

Example 2.5.2 *Let R be a commutative ring. Then a map $D : R \times R \to R$ defined by*

$$(x_1, x_2) \mapsto d(x_1)d(x_2) \; for \; x_1, \; x_2 \in R,$$

where d is a derivation on R. Then, D is a symmetric bi-derivation on R.

Example 2.5.3 *A typical example of a bi-derivation are mappings of the form $(x,y) \mapsto c[x,y]$ where c is an element of the center of R.*

In [437], functional equations for the traces of symmetric bi-derivations was introduced. Also the non-negative solution for functional equations were characterized. When studying additive commuting maps, the concept of bi-derivation naturally develops. On linearizing $[f(x),x]=0$ for all $x \in R$, we get $[f(x),y]=[x,f(y)]$, $x,y \in R$ and hence the map $(x,y) \mapsto [f(x),y]$ is a bi-derivation, where f is a commuting map. In [142] it was shown that every bi-derivation D on a non-commutative prime ring R is of the form $D(x,y)=\lambda[x,y]$ for some $\lambda \in C$. In 1995, the same author has given a more general version of this result(see, [144]). Indeed, he showed the following:

Theorem 2.5.1 *[144, Theorem 1] Let I be an ideal of R, Q_r be the Martindale ring of quotient of R and $D: I \times I \to Q_r$ be a bi-derivation. If R is non-commutative, then there exists $\lambda \in C$ such that $D(x,y)=\lambda[x,y]$ for all $x,y \in I$.*

As a consequence of the above result, a slight generalization of the main result of [136] has been obtained as follows:

Theorem 2.5.2 *[136, Theorem 3.2] Let $f: I \to Q_r$ be an additive modulo C. If f is commuting then there exists $\lambda \in C$ and a map $\zeta: I \to C$ such that $f(x)=\lambda x+\zeta(x)$ for all $x \in I$.*

Some results concerning symmetric bi-derivations in prime and semiprime rings can be found in [511, 531, 532]. In [531], the author has generalized the very famous and classical Posner's result for bi-derivations. The following result is in the spirit of Posner's second result [487] which states that the existence of a non-zero centralizing derivation on a prime ring forces the ring to be commutative. He showed the following:

Theorem 2.5.3 *[531, Theorem 2] Let R be a non-commutative prime ring of characteristic not 2 and 3. Let $D: R \times R \to R$ and d be a symmetric bi-derivation and trace of D, respectively. Suppose that d is centralizing on R. In this case, $D=0$.*

In [487], Posner proved that in case of a prime ring R with characteristic not two and D_1, D_2 non-zero derivations on R, the mapping $x \mapsto D_1(D_2(x))$ cannot form a derivation. The result below was motivated by Posner's result mentioned above.

Theorem 2.5.4 *[531, Theorem 5] Let R be a prime ring of characteristic not 2 and 3. Let $D_1: R \times R \to R$ and $D_2: R \times R \to R$ be symmetric bi-derivations. Suppose further that there exists a symmetric bi-additive mapping $B: R \times R \to R$, such that $\delta_1(\delta_2(x))=f(x)$ holds for all $x \in R$, where δ_1 and δ_2 are the traces of D_1 and D_2, respectively, and f is the trace of B. Then, either $D_1=0$ or $D_2=0$.*

The above result was also proved for semiprime rings for $D_1 = D_2$ which is as follows:

Theorem 2.5.5 *[531, Theorem 6] Let R be a semiprime ring which is 2-torsion and 3-torsion free. Let* $D : R \times R \to R$ *and* $B : R \times R \to R$ *be a symmetric bi-derivation and a symmetric bi-additive mapping, respectively. Suppose that* $\delta(\delta(x)) = f(x)$ *holds for all* $x \in R$*, where* δ *is the trace of D and f is the trace of B. In this case we have* $D = 0$.

One could have worked on the necessity of the characteristic of R not equal to three in Theorem 2.5.3. In fact, in 1996, Hongan [302] has proved the result without assuming that R is of characteristic different from three. He generalized the above mentioned result to non-zero two sided ideals of a prime ring with a characteristic not two.

Theorem 2.5.6 *[302, Theorem 2] Let R be a non-commutative prime ring of* $char(R) \neq 2$ *and I a nonzero two-sided ideal of R. Let* $D : R \times R \to R$ *be a symmetric bi-derivation and* δ *the trace of D. If* δ *is centralizing on I, then* $D = 0$.

The following result is an addition to the theory of centralizing maps. In [532], Vukman has generalized Theorem 2.5.3 as follows:

Theorem 2.5.7 *[532, Theorem 2] Let R be a non-commutative prime ring of characteristic different from 2 and 3. Suppose there exists a symmetric bi-derivation* $D : R \times R \to R$*, such that the mapping* $x \mapsto [f(x), x]$*, where f stands for the trace of D, is centralizing on R. Then* $D = 0$.

Based on the above theorem, Vukman proposed the conjecture for an arbitrary n, where $n \in \mathbb{N}$, which he felt, would be very hard to prove. The conjecture is as follows:

Conjecture 1 *[532] Let R be a non-commutative prime ring of characteristic different from 2 and 3, and let* $D : R \times R \to R$ *and* $f : R \to R$ *be a symmetric bi-derivation and its trace, respectively. Suppose that for some* $n \geq 1$*, we have* $f_n(x) \in Z(R)$ *for all* $x \in R$*, where* $f_{k+1}(x) = [f_k(x), x]$ *for* $k = 1, 2, 3\ldots$ *and* $f_1(x)$ *stands for* $f(x)$*. Then* $D = 0$.

In 1997, Deng [236] proved the above mentioned Vukman's conjecture for an arbitrary n. The result is as follows:

Theorem 2.5.8 *[236, Theorem 2] Let n be a fixed positive integer and R be a prime ring with* $char(R) = 0$ *or* $char(R) > n+2$*. Let* $f_{k+1} = [f_k(x), x]$ *for* $k > 1$*, and* $f_1(x) = f(x)$*, be the traces of a symmetric bi-derivation D of R. If* $f_n(x) \in Z(R)$ *for all* $x \in R$*, then either* $D = 0$ *or R is commutative.*

Proof. Linearizing $f_n(x) \in Z(R)$, we get,

$$[[\dots[f(x)+f(y)+2D(x,y),x+y],\dots,x+y],x+y] \in Z(R);$$

and using [236, Proposition], we get,

$$[\dots[[f(x),y],x],\dots,x]+[\dots[[f(x),x],y],\dots,x]+\dots+ \\ [\dots[f(x),x],\dots,y]+2[\dots[[D(x,y),x],x],\dots,x] \in Z(R),$$

equivalently,

$$(-1)^{n-2}I_x^{n-2}([f_1(x),y])+(-1)^{n-3}I_x^{n-3}([f_2(x),y])+\dots \\ +[f_{n-1}(x),y]+2(-1)^{n-1}I_x^{n-1}(D(x,y)) \in Z(R). \quad (2.28)$$

Noting that,

$$(-1)^{n-2}I_x^{n-2}([f_1(x),x^2])=(-1)^{n-3}I_x^{n-3}([f_2(x),x^2])=\dots \\ =[f_{n-1}(x),x^2]=(-1)^{n-1}I_x^{n-1}(D(x,x^2))=2f_n(x)x,$$

and replacing y by x^2 in (2.28), we get $2(n+1)f_n(x)x \in Z(R)$. Since $f_n(x) \in Z(R)$, it follows that $f_n(x)=0$. The linearization of $f_n(x)=0$ gives,

$$(-1)^{n-2}I_x^{n-2}([f_1(x),y])+(-1)^{n-3}I_x^{n-3}([f_2(x),y])+\dots \\ +[f_{n-1}(x),y]+2(-1)^{n-1}I_x^{n-1}(D(x,y))=0. \quad (2.29)$$

We know $I_x^{n-k}([f_{k-1}(x),xy])=xI_x^{n-k}([f_{k-1}(x),y])+I_x^{n-k}(f_k(x)y)$ for $k=2,3,\dots,n$ and $I_x^{n-1}(D(x,xy)) = xI_x^{n-1}(D(x,y)) + I_x^{n-1}(f_1(x)y)$. Substituting xy for y in (2.29), gives,

$$(-1)^{n-2}I_x^{n-2}(f_2(x)y)+(-1)^{n-3}I_x^{n-3}(f_3(x)y)+\dots \\ +(-1)I_x(f_{n-1}(x)y)+2(-1)^{n-1}I_x^{n-1}(f_1(x)y)=0.$$

Taking $y=f_{n-2}(x)$, applying $I_x^k(ab)=\sum_{j=0}^k \binom{k}{j} I_x^{k-j}(a)I_k^j(b)$ and noting $I_x^i(f_j(x))=0$ for $i+j>n$, we conclude that

$$2(-1)^{n-1}\binom{n-1}{1}I_x^{n-2}(f_1(x))I_x(f_{n-2}(x))+(-1)^{n-2}\binom{n-2}{1}I_x^{n-3}(f_2(x)) \\ I_x(f_{n-2}(x))+\dots+(-1)(f_{n-1}(x))I_x(f_{n-2}(x))=0.$$

But $(-1)^k I_x^{k-1}(f_{n-k}(x))I_x(f_{n-2}(x)) = (f_{n-1}(x))^2$, so $(n+2)(n-1)(f_{n-1}(x))^2 = 0$, and from the hypotheses on the characteristic, we get $(f_{n-1}(x))^2 = 0$. Suppose that $D \neq 0$. By [[236], Theorem 1], $f_{n-1}(x) = 0$, and by induction, $f_2(x) = [f(x), x] = 0$. Using Vukman's result [[531], Theorem 1], R is commutative, thus the proof of the theorem is complete.

♣

Let σ be an automorphism of R. Following Brešar [136], a bi-additive map $\Delta : R \times R \to R$ will be called a σ-bi-derivation of R if for every $x \in R$, the maps $y \to \Delta(x,y)$ and $y \to \Delta(y,x)$ are σ-derivations of R into R. In the same paper, the structure of an arbitrary σ-bi-derivation of a prime ring R has been characterized. Basically he proved the following:

Theorem 2.5.9 *Let R be a non-commutative prime ring, σ be an automorphism of R and $\Delta : R \times R \to R$ be a nonzero σ-bi-derivation of R. Then, σ is X-inner and there exists an invertible element $b \in Q_s$ such that $\sigma(x) = bxb^{-1}$ and $\Delta(x,y) = b[x,y]$ for all $x, y \in R$.*

Ashraf [36], studied bi-derivations and characterized the commutativity of a ring satisfying certain identities involving symmetric bi-derivations. Also, he showed the following:

Theorem 2.5.10 *[36, Theorem 2.1] Let R be a 2-torsion free and 3-torsion free semiprime ring. Suppose there exists a symmetric bi-derivation $D : R \times R \to R$ such that the mapping $x \mapsto [f(x), x]$ is centralizing on R, where f denotes the trace of D. Then, f is commuting on R.*

In [350], commutativity of a prime ring involving a symmetric bi-derivation has been discussed. In [19], the authors have considered a relationship between Lie ideal L of a ring and bi-additive symmetric map D fulfilling certain identities and proved that L must be central. Furthermore, it is shown that if R permits a symmetric bi-derivation D such that the trace d of D is n-centralizing on L, then d is n-commuting on L. This was driven by Posner's well known theorem and a work by Deng and Bell. Very recently, in 2021, Leerawat and Khun-in [388], looked into some results on the concept of symmetric bi-derivations on rings, which expand some of Vukman's ideas. They also looked into various circumstances involving symmetric bi-derivations, derivations and endomorphisms in a prime ring that change it into a commutative ring.

2.6 n-derivations

In order to broaden the scope derivations, Maksa [436] introduced the notion of symmetric bi-derivations on rings, which Vukman examined in greater detail in [531, 532]. The study of tri-derivation was initiated in [481] by Öztürk, in which he proved various results. Several results have been obtained by various authors in this direction (see [481, 565] and references therein). In light of the concepts of bi-derivation and tri-derivation, Park [482] introduced the concept of permuting n-derivations as follows:

Definition 22 *Let $n \geq 2$ be a fixed positive integer and $R^n = R \times R \times \cdots \times R$. A map $\Delta : R^n \to R$ is said to be symmetric (or permuting) if the equation $\Delta(x_1, x_2, \ldots, x_n) = \Delta(x_{\pi(1)}, x_{\pi(2)}, \ldots, x_{\pi(n)})$ holds for all $x_i \in R$, $1 \leq i \leq n$ and for every permutation $\{\pi(1), \pi(2), \ldots, \pi(n)\}$.*

Definition 23 *Let $n \geq 2$ be a fixed positive integer. An n-additive map $\Delta : R^n \to R$ (i.e., additive in each component) is called a $n-$derivation if the relations*

$$\Delta(x_1x_1^{'}, x_2, \ldots, x_n) = \Delta(x_1, x_2, \ldots, x_n)x_1^{'} + x_1\Delta(x_1^{'}, x_2, \ldots, x_n)$$

$$\Delta(x_1, x_2x_2^{'}, \ldots, x_n) = \Delta(x_1, x_2, \ldots, x_n)x_2^{'} + x_2\Delta(x_1, x_2^{'}, \ldots, x_n)$$

$$\vdots$$

$$\Delta(x_1, x_2, \ldots, x_nx_n{'}) = \Delta(x_1, x_2, \ldots, x_n)x_n^{'} + x_n\Delta(x_1, x_2, \ldots, x_n^{'})$$

are valid for all $x_i, x_i^{'} \in R$, $1 \leq i \leq n$

i.e., n-derivation, behaves like an ordinary derivation in each component. One can easily see that a 1-derivation is a derivation and a 2-derivation will give us a bi-derivation. To the above definition of n-derivation if we add the condition that Δ is symmetric, it will fetch symmetric n-derivations, due to the symmetric nature of Δ, the relations, will of course be equivalent to one another, and the validity of one will imply the rest of the identities.

Example 2.6.1 *Let $R := \left\{ \begin{pmatrix} a & b \\ 0 & 0 \end{pmatrix} \mid a, b \in \mathbb{C} \right\}$. Define a map $\Delta : R^n \to R$ by*

$$\left(\begin{pmatrix} a_1 & b_1 \\ 0 & 0 \end{pmatrix}, \begin{pmatrix} a_2 & b_2 \\ 0 & 0 \end{pmatrix}, \ldots, \begin{pmatrix} a_n & b_n \\ 0 & 0 \end{pmatrix} \right) \mapsto \begin{pmatrix} 0 & a_1a_2\cdots a_n \\ 0 & 0 \end{pmatrix}.$$

Δ so defined gives us a symmetric n-derivation.

Definition 24 *A map $\delta : R \to R$ is the trace of a symmetric n-additive map Δ and is defined by $\delta(x) = \Delta(x, x, \ldots, x)$ for all $x \in R$.*

Moreover, it can be easily seen that $\Delta(x_1, x_2, \ldots, -x_i, \ldots, x_n) = -\Delta(x_1, x_2, \ldots, x_i, \ldots, x_n)$ for all $x_i \in R$, $i = 1, 2, \ldots, n$. If $\Delta : R^n \to R$ is permuting and n-additive, then the trace δ of Δ satisfies the relation $\delta(x+y) = \delta(x) + \delta(y) + \sum_{k=1}^{n-1} {}^nC_k \, h_k(x;y)$ for all $x, y \in R$, where ${}^nC_k = \binom{n}{k}$ and

$$h_k(x;y) = \Delta(\underbrace{x, \ldots, x}_{(n-k)-\text{times}}, \underbrace{y, \ldots, y}_{k-\text{times}}).$$

The trace of a symmetric n-derivation plays an important role as it helps to bridge the gap between a n-derivation and an ordinary derivation. It becomes useful while generalizing the results already proved for derivations or bi-derivations to that of n-derivations. In this direction Park in [482] proved various interesting results for symmetric n-derivations like generalization to the famous Posner's result and many more. Some of his initial work on n-derivations which paved the way for future developments in this area are stated as under:

Lemma 2.6.1 *[482, Lemma 2.4] Let n be a fixed positive integer and let R be a n!-torsion free ring. Suppose that* $y_1, y_2, \ldots, y_n \in R$ *satisfy* $\lambda y_1 + \lambda^2 y_2 + \cdots + \lambda^n y_n = 0$ *for* $\lambda = 1, 2, \ldots, n$. *Then,* $y_i = 0$ *for all i.*

Lemma 2.6.2 *[482, Lemma 2.4] Let n be a fixed positive integer and let R be a n!-torsion free ring. Suppose that* $y_1, y_2, \ldots, y_n \in R$ *satisfy* $\lambda y_1 + \lambda^2 y_2 + \cdots + \lambda^n y_n \in Z$ *for* $\lambda = 1, 2, \ldots, n$. *Then,* $y_i \in Z(R)$ *for all i.*

Theorem 2.6.3 *[482, Theorem 2.6] Let* $n \geq 2$ *be a fixed positive integer and let R be a* $n!$*- torsion free prime ring. Suppose there exists a nonzero symmetric n-derivation* $\Delta : R^n \to R$ *such that the trace* δ *of* Δ *is centralizing on R. Then, R is commutative.*

In [62], Ashraf and Jamal provided interesting identities that reveal information about the structure of the ring, similar to what Daif and Bell [202] did for semiprime rings. The authors were able to relate that to the trace δ of an n-derivation and proved that a ring R is commutative if there exists a permuting n-additive map $\Delta : R^n \to R$ such that $xy + \delta(xy) = yx + \delta(yx)$, $xy - \delta(xy) = yx - \delta(yx)$, $xy - yx = \delta(x) \pm \delta(y)$ and $xy + yx = \delta(x) \pm \delta(y)$ holds for all $x, y \in R$. In addition to this they proved that if R is a prime ring with a suitable torsion restriction then R is commutative whenever there exist non-zero permuting n-derivations Δ_1 and Δ_2 from $R^n \to R$ such that $\Delta_1(\delta_2(x), x, \ldots, x) = 0$ for all $x \in R$, where δ_2 is the trace of Δ_2. Finally, they showed that in a prime ring R with suitable torsion restriction, if $\Delta_1, \Delta_2 : R^n \to R$ are non-zero permuting n-derivations with traces δ_1, δ_2, respectively, and $B : R^n \to R$ is a permuting n-additive map with trace f such that $\delta_1 \delta_2(x) = f(x)$ holds for all $x \in R$, then R is commutative. Park [482] proved Posner's result for n-derivations in the case of prime rings. This was taken forward by Ashraf et al. [73] with a specific condition.

Theorem 2.6.4 *[73, Theorem 2.2] For any fixed integer* $n \geq 2$, *let R be a n!-torsion free semiprime ring. Suppose that R admits a nonzero permuting n-derivation* $\Delta : R^n \to R$ *with trace* $\delta : R \to R$. *If* δ *satisfies* $[\delta(x), \delta(y)] = 0$ *for all* $x, y \in R$, *then* δ *is commuting on R.*

The condition of semiprimeness is important, as it is supported by the following example;

Example 2.6.2 *Let* $R := \left\{ \begin{pmatrix} a & b \\ 0 & 0 \end{pmatrix} \mid a, b \in \mathbb{Z} \right\}$, *R so defined is a noncommutative ring under matrix addition and matrix multiplication. We define* $\Delta : R^n \to R$ *as in Example 2.6.1, then* Δ *is a symmetric n-derivation with trace*

$$\delta \begin{pmatrix} a & b \\ 0 & 0 \end{pmatrix} = \begin{pmatrix} 0 & a^n \\ 0 & 0 \end{pmatrix}.$$

Though δ *satisfies* $[\delta(x), \delta(y)] = 0$ *for all* $x, y \in R$ *but it is not commuting, making the semiprimeness condition indispensable in this case.*

In the same paper, they were able to establish analogous results for any ring, albeit satisfying different identities. One of such result is given below.

Theorem 2.6.5 *[73, Theorem 2.3] For any fixed integer* $n \geq 2$, *let R be a n!-torsion free ring. If R admits a nonzero permuting n-derivation* $\Delta : R^n \to R$ *with trace* $\delta : R \to R$ *satisfying any one of the following conditions:*

(1) $[\delta(x), \delta(y)] = [x, y]$ *for all* $x, y \in R$,

(2) $[\delta(x), \delta(y)] = [y, x]$ *for all* $x, y \in R$.

Then, R is commutative.

Ashraf et al. [63] used properties of an ideal and the trace of a permuting n-derivation to achieve results for the existence of central ideals which eventually helps to make the ring commutative. A few of their results are as follows:

Theorem 2.6.6 *[63, Theorem 2.2] For a fixed integer* $n \geq 2$, *let R be a n!-torsion free semiprime ring and I be a nonzero ideal of R. Let* $\Delta : R^n \to R$ *be a permuting n-derivation such that* $\Delta(I, \ldots, I) \neq (0)$ *and* $[\delta(x), x] = 0$ *for all* $x \in R$, *where* δ *is the trace of* Δ. *Then, R contains a nonzero central ideal.*

Theorem 2.6.7 *[63, Theorem 2.7] For a fixed integer* $n \geq 2$, *let R be a (2n-1)!-torsion free semiprime ring and I be a nonzero left ideal of R. Suppose that there exists a permuting n-derivation* $\Delta : R^n \to R$ *with trace* δ. *Then, R is commutative if* $\delta(x^2) = \pm x^2$ *for all* $x \in R$.

Inspired by the result of Brešar [136, Theorem 4.1], Ali et al. [12] proved analogous results related to permuting n-derivations in the setting of semiprime rings. In fact, they investigated the structure of semiprime rings and described the forms of maps (traces of n-derivations) satisfying certain functional identities. In fact, they showed the following:

Theorem 2.6.8 *[12, Theorem 2.5] Let* $n \geq 2$ *be a fixed integer and R be a n!-torsion free semiprime ring and I be a nonzero ideal of R. If R admits two nonzero symmetric n-derivations* $\Delta_1 : R^n \to R$ *with trace* $\delta :\to R$ *and* $\Delta_2 : R^n \to R$ *with trace* $\xi : R \to R$ *satisfying* $\delta(x)y \pm x\xi(y) \in Z(R)$ *for all* $x, y \in I$, *then R has a nonzero central ideal.*

The idea of a permuting n-multiplier was initially suggested by Ashraf et al. [68]. They defined it as follows:

Definition 25 *A permuting n-additive map $\Lambda : R^n \to R$ is called a permuting left n-multiplier (resp. permuting right n-multiplier) if*

$$\Lambda(x_1,x_2,\ldots,x_tx_t',\ldots,x_n) = \Lambda(x_1,x_2,\ldots,x_t,\ldots,x_n)x_t'$$

$$(\textit{resp. } \Lambda(x_1,x_2,\ldots,x_tx_t',\ldots,x_n) = x_t\Lambda(x_1,x_2,\ldots,x_t',\ldots,x_n))$$

hold for all $x_t, x_t' \in R$, $t = 1,2,\ldots,n$. If Λ is both a permuting left n-multiplier and a permuting right n-multiplier, it is referred to as a permuting n-multiplier.

Recently, Ashraf et al. [68] and Ali et al. [12] proved the following interesting results related to permuting n-multipliers.

Theorem 2.6.9 *[12, Theorem 3.2] Let R be a n!-torsion free prime ring and I be a nonzero left ideal of R. Suppose that R admits two symmetric n-derivations $\Delta_1 : R^n \to R$ and $\Delta_2 : R^n \to R$ with δ and ξ as traces of Δ and $\mathcal{G}$, respectively. If $\delta(x)x + x\xi(x) = 0$ for all $x \in I$, then either R is commutative or $\mathcal{G}$ acts as a left n-multiplier on I. Furthermore, in the last case, either $\Delta = 0$ or $I[I,I] = (0)$.*

Theorem 2.6.10 *[12, Theorem 3.6] Let R be a n!-torsion free prime ring, I be a nonzero ideal of R and $q \geq 1$, be a fixed integer. Consider $\Delta_1, \Delta_2, \ldots, \Delta_q : R^n \to R$ to be n-derivations on R such that $\delta_1(x_1)\delta_2(x_2)\cdots\delta_q(x_q) = 0$ for all $x_1, x_2, \ldots, x_q \in I$ where δ_i's are traces of Δ_i's respectively. Then one of the following holds:*

1. *$\delta_1(x_1) = 0$ for all $x_1 \in I$,*
2. *All Δ_p act as left n-multipliers on R for $p = 2,3,\ldots,q$.*

Very recently, Ali et al. [11] extended Posner's second theorem [487] for the trace of symmetric n-derivations which involves prime ideals. Basically, they proved the following:

Theorem 2.6.11 *[11, Theorem 1.1] For a fixed integer $n \geq 2$, let R be a ring and P be a prime ideal of R such that R/P is n!-torsion free and $\Delta : R^n \to R$ be a nonzero symmetric n-derivation on R with trace $\delta : R \to R$. If $[\delta(x), x] \in P$ for all $x \in R$, then we have one of the following assertions:*

1. *$\delta(R) \subseteq P$,*
2. *R/P is a commutative integral domain.*

Inspired by the idea of centralizing and commuting mappings, Idrissi and Oukhtite [314] introduced the concepts of I-centralizing and I-commuting mappings, where I is an ideal of a ring R. Basically, they investigated the relationship between the structure of the ring R/P and generalized derivations of R that are P-centralizing, where P is a prime ideal of R.

Definition 26 *[314] Let I be an ideal of a ring R.*

1. *A map $f : R \to R$ is I-centralizing if $\overline{[f(x),x]} \in Z(R/I)$ for all $x \in R$ or equivalently $[[f(x),x],y] \in I$ for all $x,y \in R$.*
2. *A map $f : R \to R$ is I-commuting if $[f(x),x] \in I$ for all $x \in R$.*

In the next theorem, Ali et al. [11] utilized the above-mentioned approach and explored algebraic identities associated with the traces of symmetric n-derivations acting on prime ideals P of R, but without imposing the assumption of primeness on the ring under consideration. The result is stated as under:

Theorem 2.6.12 *For a fixed integer $n \geq 2$, let R be any ring and P be a prime ideal of R such that R/P is $(n+1)!$-torsion free and Δ be a nonzero symmetric n-derivation of R with trace $\delta : R \to R$. If δ is P-centralizing, then one of the following holds:*

1. *$\delta(R) \subseteq P$,*
2. *R/P is a commutative integral domain.*

2.7 Nilpotent and nil derivations

2.7.1 Nilpotent derivations

A derivation d on a ring R is called nilpotent if $d^n R = 0$ for some positive integer n. For $n = 2$ and R a prime ring of characteristic not equal 2, a basic theorem of Posner [487] implies that d is actually zero. For $n > 2$, the situation is less clear cut. Herstein [291] showed that for simple rings with some characteristic restrictions, a nilpotent derivation is identical to an inner derivation $x \longrightarrow [a,x]$ induced by a nilpotent element a. As a consequence, the index of a nilpotent derivation (for simple rings) must be an odd number and it is equal to $2m-1$, where m is the nilpotent index of the element a above. Kovacs [356] thus asks the possibility of generalizing these to prime rings. Chung and Luh [192] showed that the index of a nilpotent derivation is indeed an odd number even for semiprime rings without 2-torsion. They proved the following:

Theorem 2.7.1 *[192, Theorem] Let R be a 2-torsion free semiprime ring and d be a derivation of R. If $d^{2n}R = (0)$, then $d^{2n-1}R = (0)$.*

The restriction of semiprimeness and that of being 2-torsion free are both necessary, as can be seen from the following examples;

Example 2.7.1 *The ring of all 2×2 upper triangular matrices over the real field is not semiprime but an inner derivation determined by $\begin{pmatrix} 0 & 1 \\ 0 & 0 \end{pmatrix}$ has nilpotency two. Thus, the assumption of semiprimeness is necessary.*

Example 2.7.2 *In the simple ring of* 2×2 *matrices over* $\mathbb{Z}_2$*, an inner derivation determined by* $\begin{pmatrix} 0 & 1 \\ 0 & 0 \end{pmatrix}$ *has nilpotency two. Hence, the restriction on characteristics is necessary.*

Taking this idea forward Chung and Luh [193] generalized this result for any semiprime ring without any torsion restrictions and were able to prove the following:

Theorem 2.7.2 *[193, Theorem 2] Let* R *be a semiprime ring and* d *a nilpotent derivation of* R*. Then, the nilpotency of* d *is either a power of 2 or an odd number.*

The hypothesis that R is semiprime is essential. It can be seen from the following example; let R be the ring of 6×6 upper triangular matrices over $GF(2)$ and d the inner derivation determined by the matrix

$$A = E_{12} + E_{23} + E_{34} + E_{45} + E_{56},$$

where E_{ij} denotes the unit matrix having 1 at the (i,j)-position and zero elsewhere. We can easily see that $A^6 = 0$, $A^5 \neq 0$ and the nilpotency of d is 6 which is neither a power of 2 nor an odd number.

In 1992, Lanski [362] proved some interesting results by relating nilpotency of a derivation on a subset to that of its nilpotency on the whole ring. He was able to prove that a derivation nilpotent on any noncommutative Lie ideal of a ring R will be a nilpotent derivation with the same nilpotency. Further, he studied the results for one sided ideals and was able to establish the following result:

Theorem 2.7.3 *[362, Theorem 2] Let* T *be a nonzero right ideal of* R *and* d *a derivation of* R *so that* $d^n(T) = 0$*. If either* $char(R) = 0$ *or* $char(R) > n$*, then* $d = ad(A)$ *for* $A \in Q$*, and for some* $c \in C$*,* $(A-c)^n = 0$*, so in particular,* $d^t = 0$ *for* $t = min\ positive(2n-1, char(R))$*.*

Posner [487], gave the idea of the product of derivations to be a derivation in case of prime rings. Taking this idea forward, Jensen [323] proved that a specific product of powers of derivations gives us either a zero derivation or a nilpotent one. He proved the following results;

Theorem 2.7.4 *[323, Theorem 1] Let* d_1 *and* d_2 *be derivations of a prime ring* R*, and let* $d_1 d_2^m = 0$*, where* m *is a positive integer. Then, either* $d_1 = 0$ *or* $d_2^r = 0$*, where* $r \leq 4m-1$*.*

Theorem 2.7.5 *[323, Theorem 2] Let* d_1 *and* d_2 *be derivations of a prime ring* R*, and let* $d_1^n d_2 = 0$*, where* n *is a positive integer. Then, either* $d_2^2 = 0$ *or* $d_1^t = 0$*, where* $t \leq 12n-9$*.*

Theorem 2.7.6 *[323, Theorem 3] Assume* d_1 *and* d_2 *are derivations of a prime ring* R*, and assume* $d_1^n d_2^m = 0$*, where* n *and* m *are positive integers. If* d_1 *and* d_2 *commute, then at least one of them is nilpotent.*

Till now we have been dealing with nilpotent derivations, let's shift our focus to a broader concept which contains the idea of nilpotent derivations called the nil derivations.

2.7.2 Nil derivations

A mapping $f : R \to R$ is said to be nil on an ideal I of a ring R if for each $x \in I$, there exists a number n which depends on x such that $f^n(x) = 0$. There can be a number of values for n such that $f^n(x) = 0$ holds, but we are interested in the least value of n and known as the index of nilpotency of f with respect to x, which is denoted by $nil(f,x)$. f is said to be 2-sided nil if for each $x \in R$, there exists a number n such that $x(f^n(y)) = 0 = f^n(x)$ for all $y \in R$. One can observe that nilpotent derivations are always nil. However, the converse need not be true.

Example 2.7.3 *Let d be a derivation on $R[x]$, the ring of polynomials over a ring R satisfying $d(x^n) = nx^{n-1}$. Clearly, d forms a nil derivation which is not nilpotent.*

Felzenszwalb and Lanski [261] studied nil derivations with the aim of seeing whether nonzero nil derivations exist or not. They obtained the following:

Theorem 2.7.7 *[261, Theorem 3] Let R be a prime ring with a nontrivial idempotent so that R contains no nonzero nil right ideal. If d is a nil derivation on I, then d = 0.*

Similar results were proven for primitive rings and semi-simple rings. Further, they were interested in exploring whether dropping the condition of no nil right ideals will affect the above result, indeed it did alter the result as follows:

Theorem 2.7.8 *[261, Theorem 4] Let R be a prime ring with nontrivial idempotent. If d is a nil derivation on I, then d = 0 if either R is a simple ring, or if $d(x)^N = 0$ for all $x \in I$ and N a fixed integer.*

In the same paper, they dropped every other restriction such as having a nontrivial idempotent and ring primeness and achieved the following result:

Theorem 2.7.9 *[261, Theorem 7] Let R be a ring containing no nonzero nil right ideal. If d is a nil derivation on I, then d(I) = (0).*

Kharchenko [345] was able to prove an interesting result: all algebraic derivations are X-inner. Various authors tried to generalize and refine this result (see [195], [451]). A natural question can be asked whether nil derivations are inner. This was answered by Chung [191] in the year 1985. The result the following:

Theorem 2.7.10 *[191, Theorem] Let d be a nil derivation on a prime ring of characteristic 0. Then, the following are equivalent:*

1. *There is a nonzero ideal I of R such that d is E-inner for E = E(I) and d is induced by a 2-sided nil element in E(I).*
2. *There is a nonzero ideal I of R such that exp d is E-inner for E = E(I) and exp d is induced by an unipotent element f in E(I) such that f has a nonzero fixed point.*
3. *There is an element $a \neq 0$ in R such that $da = 0$ and that $ad(x)a = 0$ for all $x \in R$.*

A derivation d is said to be $E - inner$ if there is an $f \in E(I)$ (here $E(I) = End(I_R, I_R)$) such that $d(x)(w) = (adf)(x)(w) = [fx - xf](w)$ for all $x \in R$ and $w \in I$. Also $exp\ d$ refers to the automorphism generated by nilpotent derivation d, $exp\ d : R \to R$ given by the relation $exp\ d = \sum_{n=0}^{\infty} \frac{1}{n!} d^n$ (cf.; [191] and [261]).

2.8 Derivations with invertible values

The idea of derivations with invertible values was first introduced by Bergen et al. [127] in the year 1983 in an attempt to know the properties of a ring which possesses an invertible derivation. By an invertible derivation on a ring R (with unit element) we mean a non zero derivation d which is defined as $d(x) = 0$ or $d(x)$ is invertible in R. They were successful in proving that such a ring admitting an invertible derivation had a special structure.

Theorem 2.8.1 *[127, Theorem 1] Let R be a ring with 1 and $d \neq 0$, a derivation of R such that for each $x \in R$, $d(x) = 0$ or $d(x)$ is invertible in R. Then, R is either*

1. *A division ring D, or*
2. *D_2, a 2×2 matrix over D or*
3. *$D[x]/(x^2)$, where $char(D) = 2$, $d(D) = 0$ and $d(x) = 1 + ax$ for some $a \in Z(D)$, the center of D.*

Furthermore, if $2R \neq 0$ then $R = D_2$ is possible iff D does not contain all quadratic extensions of $Z(D)$, the center of D; equivalently iff some element in $Z(D)$ is not a square in D.

Later, shifting the idea of invertible derivations from the entire ring to its one sided ideals, they were able to establish an analogous result as the previous one. The result is stated as follows:

Theorem 2.8.2 *[127, Theorem 2] Let R be a ring with 1 and suppose that $d \neq 0$ is a derivation of R such that $d(L) \neq (0)$ for some left ideal of R and $d(x) = 0$ or $d(x)$ is invertible for every $x \in L$. Then $R = D$, $R = D_2$, or $R = D[x]/(x^2)$, where $2R = 0$ for some division ring D.*

Taking this idea forward, Bergen and Carini in [125] generalized the above result by taking a non-central Lie ideal and to their surprise a similar story unfolded. The outcome of their thinking was as under:

Theorem 2.8.3 *[125, Theorem 1] Let R be a ring with 1, $U \not\subseteq Z(R)$ a Lie ideal of R, and d a derivation of R such that $d(U) \neq (0)$ and $d(u) = 0$ or d(u) is invertible, for every $u \in U$. Then, R is either*

1. *A division ring D, or*
2. *D_2, a 2×2 matrix over D,*

Unless 2R or 3R is zero, d is not an inner, and R is not semiprime. In this case, $R = M + d(M)$, where M is the unique maximal ideal of R and $M^3 = (0)$.

Further, they focused on a specific question: what if $R = D_2$ and they proposed the following theorem.

Theorem 2.8.4 *[125, Theorem 2] Suppose $R = D_2$. Then,*

1. *if D is not commutative and $2R \neq (0)$, every derivation of R such that d(u) = 0 or d(u) is invertible, for all u in a non-central Lie ideal, must be an inner derivation.*
2. *there exists an inner derivation d such that $d(U) \neq (0)$ and $d(u) = 0$ or $d(u)$ is invertible, for all u contained in a non-central Lie ideal U, if and only if D does not contain all quadratic extensions of Z(D) or D is a field of characteristic 2.*

In 1986 Giambruno et al. [279] introduced the idea of involution in the already established idea of invertible derivations and came up with certain interesting results.

Theorem 2.8.5 *[279, Theorem] Let R be a 2-torsion free semiprime ring with involution. Let d be a derivation of R such that $d(R) \neq (0)$ and the non-zero elements of $d(R)$ are invertible in R. Then R is either:*

1. *a division ring D, or*
2. *D_2, a ring of 2×2 matrices over D, or*
3. *$D \oplus D^{op}$, the direct sum of a division ring and its opposite relative to the exchange involution, or*
4. *$D_2 \oplus D_2^{op}$, with an exchange involution, or*
5. *F_4, the ring of 4×4 matrices over a field F with a symplectic involution.*

Once involution was introduced with invertible derivations, there was no looking back. In [405], Lin proved something similar for arbitrary semiprime rings. However, he used the skew symmetric part K of the ring.

Theorem 2.8.6 *[405, Theorem 1] Let R be a semiprime ring with involution $*$. Let d be a derivation of R such that $d(K) \neq (0)$ and the nonzero elements of $d(K)$ are invertible in R. Then R is either*

1. *a division ring D, or*
2. *D_2, a ring of 2×2 matrices over a division ring, or*
3. *$D \oplus D^{op}$, the direct sum of a division ring and its opposite, relative to the exchange involution $(x,y)^* = (y,x)$, or*
4. *$D_2 \oplus D^{op}$ with the exchange involution, or*
5. *F_3, the 3×3 matrices over a field, or*
6. *F_4, the 4×4 matrices over a field.*

Example 2.8.1 *Let $R = F_4$, where F is a field in which -1 is not a square, and let $*$ be the transpose involution in F_4. Let d be the inner derivation on F_4 induced by $\begin{pmatrix} J & 0 \\ 0 & J \end{pmatrix}$, where $J = \begin{pmatrix} 0 & -1 \\ 1 & 0 \end{pmatrix}$. Then, it can be seen that $d(K) \neq (0)$ and the nonzero elements of $d(K)$ are invertible.*

In 2014, Kaygorodov and Popov [342] described all the alternative algebras that admitted derivations with invertible values and proposed a similar problem for Jordan algebras. They stated the following:

Theorem 2.8.7 *[342, Proposition] How does one classify all the Jordan (and, more generally, structurable) algebras that admit derivations with invertible values?*

2.9 Lie derivations

The study of Lie and Jordan mappings goes back to Ancochea (cf.; [22] and [23]), Herstein [288], Hua ([310], [306]), Jacobson and Rickart [321], Kaplansky [338], and Smiley [523]. Since then for almost 30 years, the study of Lie isomorphisms and Lie derivations was carried on mainly by Martindale III and his students [305], [317], [442], [443], [444], [445], [448], [449], [450], [506]. A Lie derivation d on a ring R is an additive map satisfying the condition

$$d[x,y] = [d(x),y] + [x,d(y)] \tag{2.30}$$

for all $x, y \in R$. In 1964, Martindale [443] described the various forms of Lie derivations of primitive rings of characteristic not 2 with nontrivial idempotents which lead to a further understanding of such derivations. The result is as follows:

Theorem 2.9.1 *[443, Theorem 2] Let L be a Lie derivation of a primitive ring R into itself, where R contains a nontrivial idempotent and the characteristic of R is not 2. Then, L is of the form $d + f$, where d is an ordinary derivation of R into a primitive ring $\bar{R}$ containing R and f is an additive mapping of R into the center of $\bar{R}$ that maps commutators into zero.*

Later on he considered several cases of prime rings with involution [317], [348], [525], [524], and von Neumann algebras [461] under the similar assumptions. Similarly in 1993, Brešar [137] determined the structure of Lie derivations of prime rings which do not satisfy the standard polynomial identity s_4 and gave the following result:

Theorem 2.9.2 *[137, Theorem 3] Let R be a prime ring of characteristic not 2. Let d be a Lie derivation of R. If R does not satisfy s_4 then, d is of the form $d + \tau$, where d is a derivation of R into its central closure and τ is an additive mapping of R into C sending commutators to zero.*

This result combined with other results initiated the theory of functional identities on rings. For further understanding the concept of functional identities on rings, one can see the book [141]. Mathieu and Banning [94] tried to generalize what Bresar did to prime rings on semiprime rings and established the following result:

Theorem 2.9.3 *[94, Theorem 3.2] Let d be a Lie derivation on a 2-torsion free semiprime ring R. Then $d = \sigma + \tau$ where σ and τ are additive mappings from R to $Q_s(R)$ and to eC, respectively, such that $e\sigma$ is a derivation and τ sends commutators to zero. Moreover, σ and τ are uniquely determined by these properties.*

In the year 1995, Swain [525] tried to extend Lie derivations defined on the skew symmetric elements of a prime ring with involution to derivations on the associative ring generated by the skew elements. He was able to extend Lie derivations to ordinary derivations in both the cases when the ring satisfied a GPI and when it did not the following result was stated:

Theorem 2.9.4 *[525, Theorem 1.1] Let R be a prime ring with an involution of the first kind, and characteristic $\neq 2, 3$. Let K denote the skew elements of R, and let C denote the extended centroid of R. Assume that $(RC : C) \neq 1, 4, 16$. Then, any Lie derivation of K into itself can be extended to an ordinary derivation of $< K >$.*

In the same paper, Swain gave other interesting results as well that led to the extension of Lie derivations to ordinary derivations.

Theorem 2.9.5 *[525, Theorem 4.1] Let $A \cong M_3(C)$ or $M_5(C)$ where C is a field with characteristic $\neq 2, 3$, and let A have an involution of the first kind. Then, any Lie derivation of K into itself can be extended to an ordinary derivation of A.*

Swain [524], with the help of Blau's result [129] concerning triadditive mappings on skew elements, was able to generalize the result to non-GPI rings with a second-kind involution. For the other case of GPI rings with involution of second kind, they believe the idea is tractable, and propose using the socle for the results on prime rings without involution. The result is stated as follows:

Theorem 2.9.6 *[524, Theorem 1.1] Let R be a non-GPI prime ring with involution and characteristic $\neq 2,3$. Let K denote the skew elements of R, and C denote the extended centroid of R. Suppose d is a Lie derivation of K into itself. Then, $d = \sigma + \tau$ where τ is an additive map into the skew elements of the extended centroid of R which is zero on $[K,K]$, and σ is a Lie derivation which can be extended to an ordinary derivation of $< K >$ into RC.*

Lie derivations, as well as other Lie maps have been extensively studied for a long time now (see [102, 119, 141, 147, 453, 530]). In this direction, Wang recently in [545] gave the description of Lie (Jordan) derivations in case of arbitrary triangular algebras. By a triangular algebra we mean the following;

Definition 27 *Let $\mathcal{A}$ and $\mathcal{B}$ be algebras over a ring R with unity. Let M be a faithful $(\mathcal{A},\mathcal{B})$-bimodule. $\mathfrak{A} = Tri(\mathcal{A},M,\mathcal{B}) := \left\{ \begin{pmatrix} a & m \\ 0 & b \end{pmatrix} : a \in \mathcal{A}, b \in \mathcal{B}, m \in M \right\}$, under the usual matrix operations is said to be a triangular algebra.*

Upper triangular matrix algebras serve as typical examples of triangular algebras. Triangular algebras, in turn, are classical examples of algebras which are not semiprime. Wang's result for triangular algebra with Lie derivations can be summarized as follows:

Theorem 2.9.7 *[545, Theorem 3.1] Let $\mathfrak{A} = Tri(\mathcal{A},M,\mathcal{B})$ be a triangular algebra. Let d be a Lie derivation on $\mathfrak{A}$. Then, there exists a triangular algebra $\mathfrak{A}^0$ such that $\mathfrak{A}$ is a subalgebra of $\mathfrak{A}^0$ having the same unity. Moreover, d can be expressed as*

$$d = \sigma + \tau,$$

where $\sigma : \mathfrak{A} \to \mathfrak{A}$ is a derivation and $\tau : \mathfrak{A} \to Z(\mathfrak{A})$ is a linear map such that $\tau([x,y]) = 0$ for all $x, y \in \mathfrak{A}$.

One of the corollaries of this result gives a description of Jordan derivations on triangular algebras which will be discussed in the next section.

2.10 Pre-derivations

The idea of pre-derivations was first introduced by M*ü*ller [466] in the year 1989 while studying certain bi-invariant pseudo-Riemannian metrics. In the same paper, he initially started with the ideas of pre-morphisms, pre-endomorphisms and pre-isomorphisms. On similar lines, he defined pre-derivations that resonated with the rest of the notions, a club of pre-morphisms and derivations.

Definition 28 *Let $\mathcal{A}$ be a Lie algebra. A pre-derivation of $\mathcal{A}$ is a linear mapping $P:\mathcal{A}\to\mathcal{A}$ such that*

$$P[x,[y,z]] = [Px,[y,z]] + [x,[Py,z]] + [x,[y,Pz]] \tag{2.31}$$

for all $x,y,z\in\mathcal{A}$.

Let us denote the set of all pre-derivations on $\mathcal{A}$ to be $Pder(\mathcal{A})$. It is easy to verify that this set forms a subalgebra of the Lie algebra $gl(\mathcal{A})$ and also contains the algebra $Der(\mathcal{A})$, the set of all derivations on $\mathcal{A}$. Müller proved that, if G is a Lie group endowed with a bi-invariant semi-Riemannian metric and $\mathcal{A}$ its Lie algebra, then the Lie algebra of the group of isometries of G fixing the identity element is a subalgebra of $Pder(\mathcal{A})$. This makes the study of the algebra of pre-derivations not only interesting from the algebraic point of view but also from its applications. In the same paper, Müller tried to establish when pre-derivations can be ordinary derivations. In the process he was successful in discovering the following:

Theorem 2.10.1 *[466, Corollary] Every pre-derivation of a finite-dimensional semi-simple Lie algebra $\mathcal{A}$ over K is a derivation, hence also an inner derivation.*

In 1997, Bajo [93] explored the connection between Lie algebras equipped with a pre-derivation and their nilpotency. This investigation was inspired by Jacobson's well known result [318], which asserts that a Lie algebra with a non-singular derivation must be nilpotent. This characterization of Lie algebra structures using a pre-derivation marked a significant development for subsequent research in this field. The result states the following:

Theorem 2.10.2 *[93, Theorem 1] If $\mathcal{A}$ is a real or complex Lie algebra admitting a non-singular pre-derivation, then $\mathcal{A}$ is nilpotent.*

In 1955, Jacobson [318] addressed a fascinating question: whether a Lie algebra admits a non-singular derivation. He proved that any Lie algebra over a field of characteristic zero, admitting a non-singular derivation, must be nilpotent. He also investigated its converse, i.e., whether any nilpotent Lie algebra admits a non-singular derivation. The answer was a negative. Later on, Dixmier and Lister [254] later a nilpotent Lie algebra that exclusively possess nilpotent derivations. This particular algebra was called the characteristically nilpotent Lie algebra which received extensive study in subsequent work [395]. As for pre-derivations, Bajo provided a partial answer. He demonstrated that characteristically nilpotent algebra introduced by Favre in [260] exhibits only singular pre-derivations. The result is stated as follows:

Theorem 2.10.3 *[93, Proposition 1] Let $\mathcal{A}$ be the real Lie algebra spanned by elements $\{x_1, x_2, \ldots, x_7\}$ with non-trivial brackets*

$$[x_1, x_i] = x_{i+1}, 2 \leq i \leq 6,$$

$$[x_2, x_3] = x_6,$$

$$[x_2, x_4] = [x_2, x_5] = -[x_3, x_4] = x_7.$$

Every pre-derivation of $\mathcal{A}$ is nilpotent and hence singular.

Analogous to the definition of characteristically nilpotent Lie algebras, we call a nilpotent Lie algebra strongly nilpotent if all its pre-derivations are nilpotent. Burde [154] in 2002 worked extensively on nilpotent Lie algebras and was able to relate the dimension of $Pder(\mathcal{A})$ to that of $Der(\mathcal{A})$ and proved the following.

Theorem 2.10.4 *[154, Proposition] Let $\mathcal{A}$ be a nilpotent of degree $p > 1$ and dimension $n \geq 3$. Then, $dimPder(\mathcal{A}) > dimDer(\mathcal{A})$.*

Burde [154] provided an example of a characteristically nilpotent algebra which is not strongly nilpotent.

Example 2.10.1 *Consider the following 7-dimensional Lie algebra, defined by*

$$[e_1, e_i] = e_{i+1}, 2 \leq i \leq 6$$

$$[e_2, e_3] = e_6 + e_7$$

$$[e_2, e_4] = e_7.$$

It can be easily verified that all the derivations are nilpotent (see [435]). On the other hand, $P = diag(1,3,3,5,5,7,7)$ is a non-singular pre-derivation. One has dim $Pder(\mathcal{A}) = 16$ and dim $Der(\mathcal{A}) = 11$.

He gave yet another example of a Lie algebra possessing only nilpotent pre-derivations.

Theorem 2.10.5 *[154, Proposition] Let $\mathcal{A}$ be the n-dimensional Lie algebra with basis $(e_1, \ldots, e_n)$, $n \geq 7$ and defining brackets*

$$[e_1, e_i] = e_{i+1}, 2 \leq i \leq n-1,$$

$$[e_2, e_3] = e_{n-1},$$

$$[e_2, e_4] = e_n,$$

$$[e_2, e_5] = -e_n,$$

$$[e_3, e_4] = e_n.$$

Then, $\mathcal{A}$ is strongly nilpotent.

Thus, the study of the algebra of pre-derivations is not only interesting from the algebraic point of view but also from its applications.

2.11 Homoderivations

In the year 2000, El Sofy [256] gave an interesting concept which clubs both the concepts of homomorphism and that of derivation. Such a map is called a homoderivation. El Sofy showed that these maps are distinct from both homomorphisms and derivations.

Definition 29 *An additive map on a ring R satisfying the condition*

$$h(xy) = h(x)h(y) + h(x)y + xh(y) \text{ for all } x, y \in R,$$

is termed as a homoderivation.

For example, $h(x) = f(x) - x$ for any endomorphism f on R forms a homoderivation. A homoderivation h is also a derivation if $h(x)h(y) = 0$ for all $x, y \in R$. In this case $h(x)Rh(y) = (0)$ for all $x, y \in R$. In the case of a prime ring R, the only additive mapping which is both a derivation and a homoderivation is the zero mapping.

Example 2.11.1 *Let* $R = \left\{ \begin{pmatrix} 0 & 0 & 0 \\ a & 0 & b \\ c & 0 & 0 \end{pmatrix} : a, b, c \in \mathbb{R} \right\}$. *Define* $h : R \to R$ *as* $h\begin{pmatrix} 0 & 0 & 0 \\ a & 0 & b \\ c & 0 & 0 \end{pmatrix} = \begin{pmatrix} 0 & 0 & 0 \\ a & 0 & 0 \\ c & 0 & 0 \end{pmatrix}$. *Then, h so defined forms a homoderivation on R.*

Bell and Daif [110] in 1998, while extending Herstein's results [297] proved that for a derivation d, whenever $d([x, y]) = 0$ for all x, y in a two sided ideal of R, the ring R is commutative. Extending this to homoderivations Melaibari et al. [458] showed the following:

Theorem 2.11.1 *[458, Theorem 3.2] Let R be a prime ring and I a nonzero ideal of R. If R admits a nonzero homoderivation h that satisfies $h([x, y]) = 0$ for all $x, y \in I$, then R is commutative.*

Theorem 2.11.2 *[458, Theorem 4.5] Let R be a prime ring with characteristic different from 2. If $0 \neq h$ is a zero-power valued homoderivation on R such that $h([x, y]) \in Z(R)$ for all $x, y \in R$, then R is commutative.*

In [106], the focus was shifted to nilpotent homoderivations in order to prove Herstein's results [297] for homoderivations. The composition of homoderivations was also studied in the same paper.

Theorem 2.11.3 *[106, Theorem A] Let R be a prime ring and $s \geq 1$ be a fixed integer. Next, let h_1 and h_2 be two homoderivations on R such that $[h_1^s(x), h_2^s(y)] = 0$ for all $x, y \in R$. Then, either R is a commutative integral domain or $h_1^{3s-1} = 0$ or $h_2^{3s-1} = 0$.*

Theorem 2.11.4 *[106, Theorem C] Let R be a prime ring. Next, let h_1 and h_2 be homoderivations of R such that $h_1 \circ h_2 = h_2 \circ h_1$ and $h_1^{s_1} \circ h_2^{s_2} = 0$ for positive integers s_1 and s_2. Then, at least one of the homoderivations must be nilpotent.*

The above result does not hold for more than two homoderivations. Consider the ring $R = M_3(D)$, where D is a division ring. Now, consider three maps h_1, h_2, h_3 on R defined by,

$$h_1(X) = AXA^{-1} - X,$$

$$h_2(X) = BXB^{-1} - X,$$

$$h_3(X) = CXC^{-1} - X,$$

where $A = \begin{pmatrix} 2 & 0 & 0 \\ 0 & 1 & 0 \\ 0 & 0 & 1 \end{pmatrix}$, $B = \begin{pmatrix} 1 & 0 & 0 \\ 0 & 2 & 0 \\ 0 & 0 & 1 \end{pmatrix}$, $C = \begin{pmatrix} 1 & 0 & 0 \\ 0 & 1 & 0 \\ 0 & 0 & 2 \end{pmatrix}$. One can observe that h_1, h_2, h_3 so defined are homoderivations on R. Also, these homoderivations are commuting with each other such that $h_1 \circ h_2 \circ h_3 = 0$. But, none of them is a nilpotent homoderivation. In 2023, Sögütcü [352] characterized semiprime rings with homoderivations and generalized some of the existing results that included derivations on prime rings and semiprime rings. A few of his results are as follows:

Theorem 2.11.5 *[352, Theorem 3.1] Let R be a semiprime ring and I a nonzero ideal of R. Then, R contains a nonzero central ideal if R admits a nonzero homoderivation h on I such that $h(I) \subseteq Z$.*

Theorem 2.11.6 *[352, Theorem 3.2] Let R be a semiprime ring and I a nonzero ideal of R. Then, R contains a nonzero central ideal, if R admits a nonzero homoderivation h on I such that,*

1. $h([I,I]) = (0)$ *or*
2. $[h(I),I] \subseteq Z$ *or*
3. $[h(I),h(I)] = (0)$, $Ih^2(I) \neq (0)$, *and* $h(I) \subseteq I$.

Very recently, Belkadi et. al. [107] introduced the concepts of Jordan homoderivations and n-Jordan homoderivations as follows:

Definition 30 *Let R be a ring. An additive mapping $h : R \to R$ is called a Jordan homoderivation if $h(x^2) = h^2(x) + h(x)x + xh(x)$ holds for all $x \in R$.*

Definition 31 *Let R be a ring, let $n \geq 1$ be a fixed integer and let $h : R \to R$ be an additive mapping. Then, h is said to be a n-Jordan homoderivation if it satisfies the relationship,*

$$h(x^n) = (h(x) + x)^n - x^n \tag{2.32}$$

for all $x \in R$.

We call equation (2.32) the nth-power property. For $n = 2,3$, the nth-power property makes h a 2-Jordan homoderivation and a 3-Jordan homoderivation, respectively. In particular, we called 2-Jordan homoderivation a Jordan homoderivation for simplicity. In order to get a Jordan homoderivation which is not a homoderivation, consider a ring R having a non-trivial central idempotent e with $x^2 = 0$ for all $x \in R$, $xy \neq 0$ for some non-zero elements x and y in R, then the map of the form $x \mapsto ex$ for all $x \in R$ is a Jordan homoderivation. Every homoderivation is a Jordan homoderivation (resp. n-Jordan homoderivation). However, the converse in general is not true. In [107], the authors were able to establish that every homoderivation forms a n-Jordan homoderivation. Also, for a 2-torsion free commutative ring any Jordan homoderivation was proved to be a homoderivation. Further, they showed that under suitable torsion conditions a n-Jordan homoderivation forms a homoderivation. The result is stated as follows:

Theorem 2.11.7 *[107, Theorem 3.1] Let $n \geq 2$ be a fixed integer, let R be a unital ring with identity e and let $h : R \to R$ be a n-Jordan homoderivation with $h(e) = 0$ and let δ be the determinant of the $(n-1) \times (n-1)$ matrix. Then the following hold:*

1. *If R is commutative and δ-torsion free, then h is a homoderivation.*
2. *If R is prime, δ-torsion free and h is such that $h + Id_R$ is an onto mapping that is not an anti-homomorphism, then h is a homoderivation.*
3. *If R is semi-prime, δ-torsion free and h is such that $h + Id_R$ is an onto mapping, then there exists an essential ideal I of R such that the restriction of h to I is a direct sum $h_1 \oplus g_2$, where h_1 is a homoderivation of I into R and g_2 is an anti-homomorphism of I into R.*

2.12 Open problems

Derivations, which are additive mappings satisfying the Leibniz rule, play a crucial role in understanding the structure and behavior of these algebraic systems. They provide deep insights into the intrinsic properties of rings and algebras, helping mathematicians uncover the underlying frameworks that govern their operations and interactions. Despite significant advancements, numerous open problems persist, particularly concerning the characterization and classification of derivations, their extensions, and their impact on the structure of rings and algebras. This section delves into these unresolved issues, highlighting key questions and challenges that continue to intrigue researchers. By exploring these open problems, we aim to shed light on the complex landscape of derivations in rings and algebras and inspire further investigation into this fascinating area of study.

Open Problem 2.12.1 *[344, Open Problem 5.2] Let $\mathcal{A}$ be a Banach algebra. If $f,g:\mathcal{A}\to\mathcal{A}$ are continuous linear derivations and $f(x^m)x^n+x^n g(x^m)\in Q(\mathcal{A})$, for all $x\in\mathcal{A}$ and some positive integers m,n, then f and g map $\mathcal{A}$ into $rad(\mathcal{A})$, where $Q(\mathcal{A})$ stands for the set of all quasinilpotent elements in $\mathcal{A}$ and $rad(\mathcal{A})$ denotes the radical of a Banach algebra $\mathcal{A}$.*

Open Problem 2.12.2 *[542, Conjecture 5] Let R be a semiprime ring with suitable torsion restrictions. Suppose there exists an additive mapping $d:R\to R$ satisfying the relation*

$$(m+n)d(x^2)=2nd(x)x+2mxd(x),$$

for all $x\in R$ and some integers $m\geq 0, n\geq 0$, $m+n\neq 0$. If $m\neq n$, the mapping d is a derivation which maps R into $Z(R)$.

Open Problem 2.12.3 *[539, Conjecture] Let $\mathcal{A}$ be a Banach algebra and let $d:\mathcal{A}\to\mathcal{A}$ be a linear left Jordan derivation. In this case d maps $\mathcal{A}$ into its radical.*

Open Problem 2.12.4 *[537, Conjecture] Let R be a prime ring of characteristic different from two and let d and g be (α,β)-derivations of R into itself with suitable additional assumptions. Suppose that the relation $d(x)g(x)+g(x)d(x)=0$ holds for all $x\in R$. In this case either $d=0$ or $g=0$.*

Open Problem 2.12.5 *[548, Conjecture 2] Let R be a semiprime ring with a suitable torsion restriction and α,β be automorphisms of R. Suppose $d:R\to R$ is an additive mapping such that*

$$2d(x^n)=d(x^{n-1})\alpha(x)+\beta(x^{n-1})d(x)+d(x)\alpha(x^{n-1})+\beta(x)d(x^{n-1}),$$

holds for all $x\in R$ and some fixed integer $n\geq 2$. Then, d is an (α,β)-derivation.

Open Problem 2.12.6 *[548, Conjecture 5] Let R be a semiprime ring with a suitable torsion restriction and α,β be automorphisms of R. Suppose $d:R\to R$ is an additive mapping such that*

$$d(x^n)=\sum_{j=1}^{n}\alpha(x^{n-j})d(x)\beta(x^{j-1}),$$

holds for all $x\in R$ and some fixed integer $n\geq 2$. Then, d is an (α,β)-derivation.

Open Problem 2.12.7 *[68, Conjecture 1] Let R be a semi-prime ring with suitable torsion restrictions and Δ be a non-zero permuting n-derivation. Suppose that for some integer $m\geq 1$, $\delta_m(x)\in Z(R)$ for all $x\in R$ where $\delta_{k+1}(x)=[\delta_k(x),x]$ for $k>1$ and $\delta_1(x)=\delta(x)$ stands for the trace of Δ. Then, $[\delta(x),x]=0$ for all $x\in R$.*

Open Problem 2.12.8 *[530, Conjecture] For every Lie derivation L on a complex Banach algebra $\mathcal{A}$ and every primitive ideal P of $\mathcal{A}$, we have*

$$[L(P), \mathcal{A}] \subset P.$$

Open Problem 2.12.9 *[107, Conjecture 3.7] Let R be a (semi)-prime ring with suitable torsion restrictions. Then any Jordan homoderivation of R is a homoderivation.*

Open Problem 2.12.10 *[107, Conjecture 3.8] Let $n \geq 2$ be a fixed integer, let R be a (semi)-prime ring with suitable torsion restrictions and let h be a n-Jordan homoderivation. Then there exists an essential ideal I of R such that the restriction of h to I is a direct sum $h_1 \oplus g_2$, where h_1 is a homoderivation of I into R and g_2 is an anti-homomorphism of I into R.*

Chapter 3

Derivations in Rings with Involution

3.1 Introduction

Rings and algebras with involution have been the object of many studies since von Neumann remarked on the role played by the classical adjoint in the algebra of linear operators on a Hilbert space. Especially, the theory of rings with involution has been developed to investigate Lie algebras, Jordan algebras, and rings of operators. It was known, that there is a connection between semi-simple algebras with involution and the classical semi-simple Lie groups. The book of involutions published by Knus et al. [351] in 1998, gives a complete description of the new investigations concerning this topic.

The purpose of this section is to study the relationship between the structure of some classes of rings with involution and the behavior of their derivations. More precisely, we characterize the commutativity of certain rings with involution by considering derivations satisfying some specific algebraic identities locally on appropriate subsets. Let us first start with a class of rings with which the reader is presumably already familiar, especially in the commutative context.

Definition 32 *Let $(R,*)$ be a ring with involution. Then R is called $*$-prime if $aRb=(0)$ and $aRb^*=(0)$ implies that $a=0$ or $b=0$.*

Obviously, a prime ring with involution $(R,*)$ is also $*$-prime, but the converse need not be true in general. Indeed, if R° denotes the opposite ring of a prime ring R, then $R\times R^\circ$ equipped with the exchange involution $*_{ex}$, defined by $*_{ex}(x,y)=(y,x)$, is $*_{ex}$-prime but not prime. This means that a prime ring can be injected into a $*$-prime ring and from this point of view $*$-prime rings constitute a more general class of prime rings.

Definition 33 *If R is equipped with an involution $*$, then a Lie ideal U which satisfies $U^* \subseteq U$ is called a $*$-Lie ideal.*

Definition 34 *If U is a Lie (resp. $*$-Lie) ideal of R, then U is called a square closed Lie (resp. $*$-Lie) ideal if $u^2 \in U$ for all $u \in U$.*

We remark that for a square closed Lie ideal U, since $(u+v)^2 \in U$ and $[u,v] \in U$, then $2uv \in U$ for all $u,v \in U$. Therefore, for all $r \in R$ we get $2r[u,v] = 2[u,rv] - 2[u,r]v \in U$ and $2[u,v]r = 2[u,vr] - 2v[u,r] \in U$, so that $2R[U,U] \subseteq U$ and $2[U,U]R \subseteq U$. This simple fact will be used frequently.

3.2 Some special elements in rings with involution

In this section we introduce two particular kinds of elements in rings with involution, namely symmetric (or hermitian) elements and skew-symmetric (or skew-hermitian) elements. Let $(R,*)$ be a ring with involution. Let $H(R) = \{x \in R | x^* = x\}$ be the set of hermitian elements and $S(R) = \{x \in R | x^* = -x\}$ the set of skew-hermitian elements of R. $Sa_*(R)$ will denote the set $H(R) \cup S(R)$, that is, $Sa_*(R) = \{x \in R / x^* = \pm x\}$. If R is 2-torsion free, then $H(R) \cap S(R) = 0$ and $2R \subset H(R) + S(R)$ since $2x = (x+x^*) + (x-x^*)$ for all $x \in R$. Let us put $H' = H(R) - \{0\}$. Hermitian and skew-hermitian elements play an important role in studying the structure of rings with involution. In fact, many authors established the structure of a ring with involution by the imposition of a special behavior on elements of $H(R)$ or on elements of $S(R)$. In [359], the author determines the structure of 2-torsion free rings with involution whose nonzero symmetric elements do not annihilate one another. Precisely, he studied the following results.

Lemma 3.2.1 *[359, Lemma 2] If H' has no nilpotent elements then every nil right (left) ideal N of R satisfies*
(1) $N^2 = (0)$;
(2) $N \subset S(R)$;
(3) if $x \in N$, then $x^2 = 0$.

Theorem 3.2.2 *[359, Theorem 3] Suppose H' has no zero divisors. Then R has a unique maximal nilpotent ideal N satisfying:*
(1) $N \subset S(R)$;
(2) $N^3 = (0)$;
(3) If $x \in N$, then $x^2 = 0$;
(4) N contains all nil one sided-ideals of R;
(5) R/N is a 2-torsion free ring with involution containing no nil ideals;

(6) *If* $H(R/N)$ *are the symmetric elements of* R/N *then* $H'(R/N)$ *has non zero divisors.*

Theorem 3.2.3 *[359, Lemma 5] If* R *is semiprime and* H' *has no zero divisors, then either* $H(R) \subset Z$*, the center of* R*, or* R *has no nilpotent elements.*

Theorem 3.2.4 *[359, Theorem 6] Let* R *be semiprime and suppose* H' *has no zero divisors. Then one of the following holds:*
(1) R *is a domain;*
(2) $H' \subset Z$*, the center of* R*, and* $R(H')^{-1}$*, the localization of* R *at* H'*, is the complete* 2×2 *matrix ring over a field;*
(3) R *is a subring of* $A \oplus A^{\circ}$*, where* $A \cong R/P$ *is a domain,* A° *is its opposite ring,* $P \cup P^*$ *is the set of zero divisors of* R*, and* $*$ *in* R *is induced by interchanging co-ordinates in* $A \oplus A^{\circ}$*.*

Corollary 3.2.1 *[359, Corollary 7] If* R *is semiprime and* $\overline{H}$*, the subring generated by* H*, is a domain, then* R *is either commutative, a domain, or* $H(R) \subset Z(R)$ *and* $R(H')^{-1}$ *is a complete* 2×2 *matrix ring over a field.*

Theorem 3.2.5 *[359, Theorem 8] Suppose* R *is semiprime with* 1 *and that every element of* H' *is a unit in* R*. Then one of the following holds:*
(1) $H(R) \subset Z(R)$ *and* R *is the complete* 2×2 *matrix ring over a field;*
(2) R *is a division ring;*
(3) $R = D \oplus D^{\circ}$ *where* D *is a division ring,* D° *its opposite ring, and* $*$ *in* R *is induced by interchanging co-ordinates in* $D \oplus D^{\circ}$*.*

In [161] the authors studied rings with involution for which the symmetric elements are nilpotent or regular. They obtained the following results.

Theorem 3.2.6 *[161, Theorem] Let* $(R,*)$ *be any ring with involution in which every symmetric element is nilpotent or regular (or equivalently nilpotent or right regular). If* R *is a non-nil noetherian ring then* R *is an order in an artinian ring with unity* Q *in which each symmetric element is nilpotent or invertible.*

Corollary 3.2.2 *[161, Corollary] If* R *is as in the mentioned theorem, then the nilpotent radical* N *of* R *contains any nilpotent symmetric element of the form* xx^* *or* $x+x^*$ *and the factor ring* R/N *is one of the following types: (i) a skew-domain; (ii) a subdirect product of two isomorphic skew-domains; (iii) an order in the* 2×2*-matrices over afield.*

Proposition 3.2.7 *[161, Proposition 1] Let* R *be any ring with involution having all its symmetric elements nilpotent. Assume, further, at least one of the following conditions:*
(1) R *is noetherian;*
(2) *the symmetric elements of* R *have a bounded index.*
Then R *is a nil ring.*

Proposition 3.2.8 *[161, Proposition 2] Let R be any noetherian ring with involution in which each symmetric element is nilpotent or regular. If R is not nil (or, better, not nilpotent), then R is an order in an artinian ring Q with an involution extending the base involution.*

Proposition 3.2.9 *[161, Proposition 3] The nilpotent symmetric elements of R have a bounded index.*

Proposition 3.2.10 *[161, Proposition 7] If R is a semiprime noetherian ring in which each symmetric element is nilpotent or regular, then R is of one of the following types:*
(1) *R is a skew-domain;*
(2) *R is a subdirect product of a skew-domain by itself;*
(3) *R is an order in the 2×2-matrices over a field.*

3.3 Derivations acting on Lie ideals

Many authors have established a connection between the structure of a ring R and the behavior of specific additive mappings defined on R. Especially, the structure of a ring is very tightly determined by the imposition of a special behavior on one of its derivations. The first famous result in this area is Posner's Second Theorem. This remarkable theorem by Posner has been influential and it has played a key role in the development of various notions, especially the concept of functional identities. Motivated by this result, many authors have studied the commutativity of rings and Banach algebras, admitting suitably constrained additive mappings acting on appropriate subsets of the rings.

Various authors have considered the following problem: Suppose R is a ring with unity element 1 and d is a nonzero derivation of R such that for every element $x \in R$, either $d(x) = 0$ or $d(x)$ is invertible in R; must R then have a very special structure? This question has been solved in [127] by proving that except for a special case which occurs when $2R = 0$, R must be a division ring D or the ring $M_2(D)$ of 2×2 matrices over a division ring. In [127], a local version of the above problem has been considered by treating the case of a ring R with 1 provided with a nonzero derivation d such that $d(x) = 0$ or $d(x)$ is invertible in R for each element x in a nonzero left ideal L of R with $d(L) \neq 0$.

Hermitian and skew-hermitian elements play an important role in studying the structure of rings with involution admitting some specific additive mappings. In fact, many authors established the structure of a ring with involution by imposing a special behavior on one of its derivations locally on the set of hermitian elements $H(R)$ or on the set of skew-hermitian elements $S(R)$. In this direction, the above problem has also been studied for rings with involution by considering the action of the derivation d on $H(R)$. In this area, Giambruno et al. [279] explore the structure of a 2-torsion free semiprime ring with involution $(R,*)$ equipped with a derivation d, non-zero on $H(R)$, such that for all $x \in H(R)$ either $d(x) = 0$ or $d(x)$ is invertible. More exactly, it is proved that R must be one of the following: (1) a division ring, (2) a 2×2 matrix over a division ring, (3) the direct sum of a division ring and its opposite with exchange involution, (4) the direct sum of 2×2 matrices over a division ring and its opposite with exchange involution, (5) 4×4 matrices over a field with symplectic involution.

In [296], Herstein proved that if R is a prime ring of characteristic not 2 which has a nonzero derivation d such that $[d(x), d(y)] = 0$ for all $x, y \in R$, then R is commutative. Motivated by this result, Bell [110], studied derivations d satisfying $d([x,y]) = 0$ for all $x, y \in R$. In [473] and [477], Oukhtite and Salhi generalized these results to $*$-prime rings. In particular, they proved that if R is a 2-torsion free $*$-prime ring equipped with a nonzero derivation which is centralizing on R, then R is necessarily commutative. To start, we collect some lemmas which will be frequently used in the sequel.

Lemma 3.3.1 *Let R be a 2-torsion free $*$-prime ring and I a nonzero $*$-ideal of R. If a, b in R are such that $aIb = (0) = aIb^*$, then $a = 0$ or $b = 0$.*

Lemma 3.3.2 *[475, Lemma 2.3] Let R be a 2-torsion free $*$-prime ring and U a nonzero $*$-Lie ideal of R. If $[U,U] = (0)$, then $U \subseteq Z(R)$.*

Lemma 3.3.3 *Let R be a 2-torsion free $*$-prime ring and U a nonzero $*$-Lie ideal of R. If $[U,U] \neq (0)$, then there exists a $*$-ideal M of R such that $[M,R] \subseteq U$ with $[M,R] \not\subseteq Z(R)$.*

Lemma 3.3.4 *[478, Lemma 1] Let R be a 2-torsion free $*$-prime ring and $U \not\subseteq Z(R)$ a $*$-Lie ideal of R. If $a, b \in R$ are such that $aUb = a^*Ub = (0)$, then $a = 0$ or $b = 0$.*

Proof. Suppose $b \neq 0$, then conclude that $Ub \neq (0)$. Otherwise, for any $x \in U$ we get $[x,r]b = 0$ for all $r \in R$. Replace r by rt, where $t \in R$, to get

$$0 = [x, rt]b = [x, r]tb$$

so that $[x,r]Rb = 0$ for all $r \in R$. As $U^* = U$, it follows that $[x,r]Rb = 0 = [x,r]^*Rb$ and the $*$-primeness of R yields $[r,x] = 0$ so that $U \subseteq Z(R)$ which contradicts $[U,U] \neq (0)$. Consequently, there exists $u \in U$ such that $ub \neq 0$. On the other hand, by Lemma 3.3.3, there exists a nonzero $*$-ideal M such that $[M,R] \subseteq U$ and $[M,R] \not\subseteq Z(R)$. Let $u \in U$, $m \in M$ and $r \in R$. Then, we have $[mau, r] \in [M,R] \subset U$

and hence it follows that,

$$\begin{aligned} 0 &= a[mau,r]b \\ &= a[ma,r]ub + ama[u,r]b \\ &= a[ma,r]ub \\ &= amarub - armaub \\ &= amarub \end{aligned}$$

Accordingly,

$$amaRub = (0) \text{ for all } m \in M.$$

Similarly, from $0 = a^*[ma^*u, r]b$ we find that $a^*ma^*Rub = 0$ for all $m \in M$. As $M^* = M$, then

$$amaRub = (0) = (ama)^*Rub.$$

Since R is $*$-prime and $ub \neq 0$, so $ama = 0$. Therefore $aMa = (0)$. On the other hand, we have

$$\begin{aligned} 0 &= a[ma^*u,r]b \\ &= a[ma^*,r]ub + ama^*[u,r]b \\ &= a[ma^*,r]ub \\ &= ama^*rub - arma^*ub \\ &= ama^*rub. \end{aligned}$$

Hence $ama^*Rub = (0)$ for all $m \in M$. Since M is a $*$-ideal, we conclude that

$$ama^*Rub = 0 = (ama^*)^*Rub.$$

Since $ub \neq 0$, the $*$-primeness of R yields $ama^* = 0$ thereby $aMa^* = (0)$. Consequently, $aMa = (0) = aMa^*$ and $a = 0$ by Lemma 3.3.1.

♣

Lemma 3.3.5 *[476, Lemma 2.2] Let R be a 2-torsion free $*$-prime ring and U a nonzero $*$-Lie ideal of R. If d is a derivation of R which commutes with $*$ and satisfies $d(U) = (0)$, then either $d = 0$ or $U \subseteq Z(R)$.*

Remark 3.3.6 *One can easily verify that Lemma 3.3.5 is still valid if the condition that d commutes with $*$ is replaced by $d \circ * = - * \circ d$.*

Centralizing maps play a crucial role in studying the structure of certain classes of rings. Several authors have deeply explored additive mapping centralizing on particular subsets of the considered ring or on specific elements in the case of rings with involution. The following theorem due to Brešar gives a classification of additive mappings centralizing on a prime ring.

Theorem 3.3.7 *[136, Theorem A] Let R be a prime ring. Suppose an additive mapping f of R into itself is centralizing on R. If either R has a characteristic different from 2 or f is commuting on R, then f is of the form* $f(x) = \lambda x + \zeta(x)$, *for all* $x \in R$, *where* λ *is an element from the extended centroid C of R and* ζ *is an additive mapping of R into C.*

Furthermore, Brešar initiated the study of a more general concept then centralizing mappings by considering the situation of additive mappings f and g of a ring R satisfying $f(x)x - xg(x) \in Z(R)$ for all x in some subset S of R. This also covers the concept of a skew-centralizing mapping: A mapping f of a ring R is said to be skew-centralizing on a subset S of R is $f(x)x + xf(x) \in Z(R)$ for all $x \in S$. In the case when f and g are derivations, Brešar established a commutativity criteria for prime rings with involution as follows:

Theorem 3.3.8 *[136, Theorem B] Let R be a prime ring and U be a nonzero left ideal of R. Suppose derivations d and g of R satisfy* $d(x)x - xg(x) \in Z(R)$ *for all* $x \in U$. *If* $d \neq 0$, *then R is commutative.*

Corollary 3.3.1 *[136, Corollary 4.3] Let R be a prime ring and U be a nonzero left ideal of R. If there exists a nonzero derivation of R which is centralizing or skew-centralizing on U, then R is commutative.*

Corollary 3.3.2 *[136, Corollary 4.4] Let R be a noncommutative prime ring and U be a nonzero left ideal of R. Suppose there exist* $a, b \in R$ *and a derivation d of R such that the mapping* $x \longmapsto d(x) + ax + xb$ *is centralizing on U. Then, d is an inner derivation given by* $d(x) = [x, a]$ *for all* $x \in U$.

Proof. It suffices to remark that the relation $[d(x) + ax + xb, x] \in Z(R)$ can be reformulated as $\big(d(x) - [x,a]\big)x - x\big(d(x) - [b,x]\big) \in Z(R)$.

♣

If we put $d = 0$ in the preceding corollary, the element a should be central.

Corollary 3.3.3 *[136, Corollary 4.4] Let R be a prime ring and U be a nonzero left ideal of R. If* $a, b \in R$ *are such that the mapping* $x \longmapsto ax + xb$ *is centralizing on U, then* $a \in Z(R)$.

The next theorem deals with derivations centralizing on a Lie ideal of a $*$-prime ring.

Theorem 3.3.9 *[478, Theorem 1] Let R be a 2-torsion free $*$-prime ring and U a square closed $*$-Lie ideal of R. If d is a derivation of R satisfying* $[d(u), u] \in Z(R)$ *for all* $u \in U$, *then* $U \subseteq Z(R)$ *or* $d = 0$.

Proof. Suppose that $U \not\subseteq Z(R)$. As $[d(x), x] \in Z(R)$ for all $x \in U$. By linearizing of the above relation, we get

$$[d(x), y] + [d(y), x] \in Z(R) \text{ for al } x, y \in U.$$

Since $\mathrm{char}(R) \neq 2$, the fact that $[d(x),x^2]+[d(x^2),x] \in Z(R)$ yields $x[d(x),x] \in Z(R)$ for all $x \in U$. Hence

$$[r,x][d(x),x] = 0 \quad \text{for all} \quad x \in U, r \in R,$$

and therefore $[d(x),x]^2 = 0$ for all $x \in U$. Since $[d(x),x] \in Z(R)$, we have

$$[d(x),x]R[d(x),x][d(x),x]^* = (0) \quad \text{for all} \quad x \in U$$

and the $*$-primeness of R yields $[d(x),x] = 0$ or $[d(x),x][d(x),x]^* = 0$. Assume that $[d(x),x][d(x),x]^* = 0$, then $[d(x),x]R[d(x),x]^* = (0)$; and the fact that $[d(x),x]^2 = 0$ gives

$$[d(x),x]R[d(x),x]^* = [d(x),x]R[d(x),x] = (0).$$

Since R is *-prime, we obtain

$$[d(x),x] = 0 \quad \text{for all} \quad x \in U.$$

Let us consider the map $\delta : R \longmapsto R$ defined by $\delta(x) = d(x) + * \circ d \circ x^*$. One can easily verify that δ is a derivation of R which commutes with $*$ and satisfies

$$[\delta(x),x] = 0 \quad \text{for all} \quad x \in U.$$

Linearizing this equality, we obtain

$$[\delta(x),y] + [\delta(y),x] = 0 \quad \text{for all} \quad x,y \in U.$$

Writing $2xz$ instead of y and using $\mathrm{char}(R) \neq 2$, we find that

$$\delta(x)[x,z] = 0 \quad \text{for all} \quad x,z \in U.$$

Replacing z by $2zy$ in this equality, we conclude that $\delta(x)z[x,y] = 0$, so that

$$\delta(x)U[x,y] = (0) \quad \text{for all} \quad x,y \in U. \tag{3.1}$$

By virtue of Lemma 3.3.1, it then follows that

$$\delta(x) = 0 \quad \text{or} \quad [x,U] = (0), \quad \text{for all} \quad x \in U \cap Sa_*(R).$$

Let $u \in U$. Since $u - u^* \in U \cap Sa_*(R)$, it follows that

$$\delta(u-u^*) = 0 \quad \text{or} \quad [u-u^*,U] = (0).$$

If $\delta(u-u^*) = 0$, then $\delta(u) \in Sa_*(R)$ and (3.1) yields $\delta(u) = 0$ or $[u,U] = (0)$. If $[u-u^*,U] = (0)$, then $[u,y] = [u^*,y]$ for all $y \in U$ and (3.1) assures that

$$\delta(u)U[u,y] = (0) = \delta(u)U[u,y]^*, \quad \text{for all} \quad y \in U.$$

Applying Lemma 3.3.1, we find that $\delta(u) = 0$ or $[u,U] = (0)$. Hence, U is a union of two additive subgroups G_1 and G_2, where

$$G_1 = \{u \in U \text{ such that } \delta(u) = 0\} \text{ and } G_2 = \{u \in U \text{ such that } [u,U] = (0)\}.$$

Since a group cannot be a union of two of its proper subgroups, we are forced to set $U = G_1$ or $U = G_2$. Since U is non-central. So, Lemma 3.3.2 assures that $U = G_1$ and therefore $\delta(U) = (0)$. Invoking Lemma 3.3.5, we get $\delta = 0$ and therefore $d \circ * = - * \circ d$. As $[d(x),x] = 0$ for all $x \in U$, in view of the above remark a similar reasoning leads to $d = 0$.

♣

Corollary 3.3.4 *[477, Theorem 1.1] Let R be a 2-torsion free $*$-prime ring and d a nonzero derivation of R. If d is centralizing on R, then R is commutative.*

Theorem 3.3.10 *[478, Theorem 2] Let U be a square closed $*$-Lie ideal of a 2-torsion free $*$-prime ring R and d a derivation of R which commutes with $*$ on U. If $[d(x),d(y)] = d([y,x])$ for all $x,y \in U$, then $d = 0$ or $U \subseteq Z(R)$.*

Proof. Suppose that $U \not\subseteq Z(R)$. We have

$$[d(x),d(y)] = d([y,x]) \text{ for all } x,y \in U. \tag{3.2}$$

Substituting $2xy$ for y in (3.2) and using $\text{char}(R) \neq 2$, we get

$$d(x)[y,x] = [d(x),x]d(y) + d(x)[d(x),y] \text{ for all } x,y \in U. \tag{3.3}$$

Replacing y by $2[y,z]x$ and using (3.3), we find that

$$[d(x),x][y,z]d(x) + d(x)[y,z][d(x),x] = 0 \text{ for all } x,y,z \in U.$$

Replace y by $2[y,z]d(x)$ in (3.3) to get

$$d(x)[y,z][d(x),x] - [d(x),x][y,z]d^2(x) = 0 \text{ for all } x,y,z \in U.$$

Comparing the last two equations, we obtain

$$[d(x),x][y,z](d(x) + d^2(x)) = 0 \text{ for all } x,y,z \in U. \tag{3.4}$$

Writing $2[u,v](d(x) + d^2(x))y$ instead of y in (3.4), where $u,v \in U$, we obtain

$$[d(x),x][u,v]z(d(x) + d^2(x))y(d(x) + d^2(x)) = 0,$$

so that

$$[d(x),x][u,v]z(d(x) + d^2(x))U(d(x) + d^2(x)) = (0) \text{ for all } x,u,v,z \in U. \tag{3.5}$$

If $x \in U \cap Sa_*(R)$, then Lemma 3.3.1 combined with (3.5) assures that

$$d(x) + d^2(x) = 0 \text{ or } [d(x),x][u,v]z(d(x) + d^2(x)) = 0 \text{ for all } u,v,z \in U.$$

Suppose that $[d(x),x][u,v]z(d(x)+d^2(x))=0$. Then

$$[d(x),x][u,v]U(d(x)+d^2(x))=(0). \tag{3.6}$$

Since d commutes with $*$ and $x\in Sa_*(R)$, in view of (3.6), Lemma 3.3.1 gives

$$d(x)+d^2(x)=0 \text{ or } [d(x),x][u,v]=0 \text{ for all } u,v\in U. \tag{3.7}$$

If $[d(x),x][u,v]=0$, then replacing u by $2uw$ in (3.7) where $w\in U$, we obtain

$$[d(x),x]U[u,v]=(0). \tag{3.8}$$

As $U^*=U$ and $[U,U]\neq(0)$, by (3.8), Lemma 3.3.2 yields $[d(x),x]=0$. Thus, in any event,

$$[d(x),x]=0 \text{ or } d(x)+d^2(x)=0 \text{ for all } x\in U\cap Sa_*(R).$$

Let $x\in U$. Since $x+x^*\in U\cap Sa_*(R)$, either

$$d(x+x^*)+d^2(x+x^*)=0 \text{ or } [d(x+x^*),x+x^*]=0.$$

If $d(x+x^*)+d^2(x+x^*)=0$, then $d(x)+d^2(x)\in Sa_*(R)$ and equation (3.5) yields $d(x)+d^2(x)=0$ or $[d(x),x][u,v]U(d(x)+d^2(x))=(0)$.

If $[d(x),x][u,v]U(d(x)+d^2(x))=(0)$, once again using $d(x)+d^2(x)\in Sa_*(R)$, we find that $d(x)+d^2(x)=0$, or $[d(x),x][u,v]$ for all $u,v\in U$, in which case $[d(x),x]=0$. Now suppose that $[d(x+x^*),x+x^*]=0$. As $x-x^*\in U\cap Sa_*(R)$ we have to distinguish two cases:

1) If $d(x-x^*)+d^2(x-x^*)=0$, then $d(x)+d^2(x)\in Sa_*(R)$. Reasoning as above we get $d(x)+d^2(x)=0$ or $[d(x),x]=0$.
2) If $[d(x-x^*),x-x^*]=0$, then $[d(x),x]\in Sa_*(R)$. Replace u by $2yu$ in (3.5), with $y\in U$, to get $[d(x),x]y[u,v]z(d(x)+d^2(x))U(d(x)+d^2(x))=0$, so that

$$[d(x),x]U[u,v]z(d(x)+d^2(x))U(d(x)+d^2(x))=0 \text{ for all } x,u,v,z\in U. \tag{3.9}$$

Since $[d(x),x]\in Sa_*(R)$ and from (3.9) it follows that

$$[d(x),x]=0 \text{ or } [u,v]U(d(x)+d^2(x))U(d(x)+d^2(x))=(0) \text{ for all } u,v\in U.$$

Suppose $[u,v]U(d(x)+d^2(x))U(d(x)+d^2(x))=(0)$. As $U^*=U$ and $[U,U]\neq(0)$, then

$$(d(x)+d^2(x))U(d(x)+d^2(x))=(0). \tag{3.10}$$

In (3.4), write $2[u,v](d(x)+d^2(x))r$ instead of z, where $u,v\in U$ and $r\in R$, to obtain

$$[d(x),x][u,v]y(d(x)+d^2(x))r(d(x)+d^2(x))=0, \text{ for all } u,v,y\in U, r\in R. \tag{3.11}$$

Replacing r by $r(d(x)+d^2(x))^*z$ in (3.11), where $z \in U$, we find that

$$[d(x),x][u,v]y(d(x)+d^2(x))r(d(x)+d^2(x))^*z(d(x)+d^2(x)) = 0,$$

which leads us to

$$[d(x),x][u,v]y(d(x)+d^2(x))U(d(x)+d^2(x))^*U(d(x)+d^2(x)) = (0). \quad (3.12)$$

Since $(d(x)+d^2(x))^*U(d(x)+d^2(x))$ is invariant under $*$, by virtue of (3.12), Lemma 3.3.1 yields

$$(d(x)+d^2(x))^*U(d(x)+d^2(x)) = (0) \text{ or } [d(x),x][u,v]y(d(x)+d^2(x)) = 0.$$

If $(d(x)+d^2(x))^*U(d(x)+d^2(x)) = (0)$, then (3.10) implies that $d(x)+d^2(x) = 0$.

Now assume that

$$[d(x),x][u,v]y(d(x)+d^2(x)) = 0 \text{ for all } u,v,y \in U. \quad (3.13)$$

Replace v by $2wv$ in (3.13), where $w \in U$, and use (3.13) to get

$$[d(x),x]w[u,v]y(d(x)+d^2(x)) = 0,$$

so that $[d(x),x]U[u,v]y(d(x)+d^2(x)) = (0)$ for all $u,v,y \in U$. As $[d(x),x] \in Sa_*(R)$, then $[u,v]U(d(x)+d^2(x)) = (0)$, in which case $d(x)+d^2(x) = 0$, or $[d(x),x] = 0$. In conclusion, for all $x \in U$ we have either $[d(x),x] = 0$ or $d(x)+d^2(x) = 0$.

Now let $x \in U$ such that $d(x)+d^2(x) = 0$. In (3.2), put $y = 2[y,z]d(x)$ to get

$$d([y,z])[d(x),x] - [[y,z],x]d(x) + [d(x),[y,z]]d(x) = [y,z][d(x),x]. \quad (3.14)$$

If in (3.2) we put $y = 2[y,z]x$, we get

$$[[y,z],x]d(x) = [d(x),[y,z]]d(x) + d([y,z])[d(x),x] = 0. \quad (3.15)$$

From (3.14) and (3.15) it follows that $[y,z][d(x),x] = 0$ for all $y,z \in U$, hence $[y,z]U[d(x),x] = (0)$ for all $y,z \in U$. Applying Lemma 3.3.1, this leads to $[d(x),x] = 0$, for all $x \in U$. By virtue of Theorem 3.3.9, this yields $d = 0$.

♣

Note that if d is a derivation of R which acts as an anti-homomorphism on U, then d satisfies the condition $[d(x),d(y)] = d([y,x])$ for all $x,y \in U$. Thus we have the following corollary.

Corollary 3.3.5 *[476, Theorem 1.1] Let d be a derivation of a 2-torsion free $*$-prime ring R which acts as an anti-homomorphism on a nonzero square closed $*$-Lie ideal U of R. If d commutes with $*$, then either $d = 0$ or $U \subseteq Z(R)$.*

Theorem 3.3.11 *[478, Theorem 3] Let U be a square closed $*$-Lie ideal of a 2-torsion free $*$-prime ring R and d a derivation of R. If either $d([x,y]) = 0$ for all $x,y \in U$, or $[d(x),d(y)] = 0$ for all $x,y \in U$ and d commutes with $*$ on U, then $d = 0$ or $U \subseteq Z(R)$.*

Corollary 3.3.6 *[473, Theorem 3.3] Let d be a nonzero derivation of a 2-torsion free $*$-prime ring R. If $d([x,y]) = 0$ for all $x,y \in R$, then R is commutative.*

3.4 Posner's second theorem for Jordan ideals

In this section, we extend Posner's second theorem to Jordan ideals of rings with involution. Moreover, some related results are also discussed. We use without explicit mention the fact that if J is a nonzero Jordan ideal of a ring R, then $2[R,R]J \subseteq J$ and $2J[R,R] \subseteq J$ (cf.; [566, Lemma 2.4]).

In order to prove our main theorem, we need the following lemmas.

Lemma 3.4.1 *[471, Lemma 2] Let R be a 2-torsion free $*$-prime ring and J a nonzero $*$-Jordan ideal of R. If $aJb = a^*Jb = (0)$, then $a = 0$ or $b = 0$.*

Proof. Assume that $a \neq 0$. Since $2[R,R]J \subseteq J$, then $2a[r,s]jb = 0$ for all $r,s \in R, j \in J$. This implies that

$$a[r,s]jb = 0 \quad \text{for all } r,s \in R, j \in J. \tag{3.16}$$

Replacing s by sa in (3.16), because of $ajb = 0$, we find that $asarjb = 0$ and thus

$$aRarjb = 0 \quad \text{for all } r \in R, j \in J. \tag{3.17}$$

On the other hand, from $a^*Jb = (0)$ it follows that $a^*[r,sa]jb = 0$, which leads to $a^*sarjb = 0$ for all $r,s \in R$ and therefore

$$a^*Rarjb = (0) \quad \text{for all } r \in R, j \in J. \tag{3.18}$$

From equations (3.17) and (3.18), because of $a \neq 0$, the $*$-primeness of R yields $arjb = 0$ for all $r \in R, j \in J$. Accordingly, we have

$$aRjb = (0) \quad \text{for all } j \in J. \tag{3.19}$$

Writing sa^* instead of s in (3.16), because of $a^*Jb = (0)$, we get $asa^*rjb = 0$ so that

$$aRa^*rjb = (0) \quad \text{for all } r \in R, j \in J. \tag{3.20}$$

In view of $a^*Jb = (0)$, we find that $a^*[r,sa^*]jb = 0$ and thus $a^*sa^*rjb = 0$ for all $r,s \in R, j \in J$. Hence

$$a^*Ra^*rjb = (0) \quad \text{for all } r \in R, j \in J. \tag{3.21}$$

Using (3.20) and (3.21), because of $a \neq 0$, the $*$-primeness of R yields $a^* r jb = 0$ therefore

$$a^* Rjb = (0) \quad \text{for all} \quad j \in J. \tag{3.22}$$

Again, because of equations (3.19) and (3.22), $*$-primeness of R assures that $jb = 0$ for all $j \in J$. Whence it follows that

$$Jb = (0). \tag{3.23}$$

From $(j \circ r)b = 0$ and in view of relation (3.23), we get $jrb = 0$ for all $r \in R, j \in J$ and thus $jRb = (0)$ for all $j \in J$. Since J is invariant under $*$, it follows that $j^*Rb = (0)$ for all $j \in J$. Using the $*$-primeness of R, because of $J \neq (0)$, we conclude that $b = 0$.

♣

Lemma 3.4.2 *[471, Lemma 3] Let R be a 2-torsion free $*$-prime ring and J a nonzero $*$-Jordan ideal of R. If $[J,J] = (0)$, then $J \subseteq Z(R)$.*

Lemma 3.4.3 *[471, Lemma 4] Let R be a 2-torsion free $*$-prime ring and J a nonzero $*$-Jordan ideal of R. If d is a derivation of R such that $d(J) = (0)$, then $d = 0$ or $J \subseteq Z(R)$.*

Lemma 3.4.4 *[472, Lemma 3] Let R be a 2-torsion free $*$-prime ring and J a nonzero $*$-Jordan ideal of R. If $J \subseteq Z(R)$, then R is commutative.*

The main result of the present section is the following theorem.

Theorem 3.4.5 *[472, Theorem 1] Let $(R,*)$ be a 2-torsion free ring with involution. Let J be a nonzero $*$-Jordan ideal of R and d be a nonzero derivation centralizing on J. If R is $*$-prime, then R is commutative.*

Proof. Linearizing $[d(x),x] \in Z(R)$ for all $x \in J$, we obtain

$$[d(x),y] + [d(y),x] \in Z(R) \quad \text{for all} \quad x \in J. \tag{3.24}$$

Replacing y by $2x^2$ in (3.24), as $\text{char}(R) \neq 2$, we find that $x[d(x),x] \in Z(R)$. Therefore, $[d(x),x[d(x),x]] = 0$ which leads to $[d(x),x]^2 = 0$ for all $x \in J$. Since $[d(x),x] \in Z(R)$, then $[d(x),x]R[d(x),x][d(x),x]^* = (0)$ for all $x \in J$. R being $*$-prime, then $[d(x),x] = 0$ or $[d(x),x][d(x),x]^* = 0$. Assume that $[d(x),x][d(x),x]^* = 0$, then $[d(x),x]R[d(x),x]^* = (0)$. Since $[d(x),x]R[d(x),x] = (0)$, we conclude that $[d(x),x] = 0$ for all $x \in J$. Once again linearizing this equation, we obtain

$$[d(x),y] + [d(y),x] = 0 \quad \text{for all} \quad x,y \in J.$$

Writing $2x[r,s]$ instead of y, where $r,s \in R$, we find that

$$x[d(x),[r,s]] + d(x)[[r,s],x] + x[d([r,s]),x] = 0 \quad \text{for all} \quad x \in J, r,s \in R.$$

Replacing s by $2uv$, where $u,v \in J$, since $[r,2uv] = 2u[r,v]+2[r,u]v \in J$, we get

$$d(x)[[r,uv],x] = 0 \quad \text{for all } u,v,x \in J, r \in R.$$

Substituting xt for r in this equation, we find that

$$d(x)x[[t,uv],x] + d(x)[x,uv][t,x] = 0 \quad \text{for all } u,v,x \in J, t \in R.$$

Since $d(x)x = xd(x)$, then $d(x)x[[t,uv],x] = xd(x)[[t,uv],x] = 0$ and the last two equations assure that

$$d(x)[x,uv][t,x] = 0 \quad \text{for all } u,v,x \in J, t \in R.$$

Writing rt instead of t, we get $d(x)[x,uv]r[t,x] = 0$ and thus

$$d(x)[x,uv]R[t,x] = (0) \quad \text{for all } u,v,x \in J, t \in R. \tag{3.25}$$

Suppose $x_0 \in J \cap Sa_*(R)$; from (3.25) it follows that $d(x_0)[x_0,uv]R[t,x_0]^* = (0)$, so either $x_0 \in Z(R)$ or $d(x_0)[x_0,uv] = 0$ for all $u,v \in J$.

Suppose

$$d(x_0)[x_0,uv] = 0 \quad \text{for all } u,v \in J. \tag{3.26}$$

Replacing v by $2v[r,s]$ in (3.26), where $r,s \in R$, we find

$$d(x_0)uv[x_0,[r,s]] = 0 \quad \text{for all } u,v \in J, r,s \in R,$$

and therefore

$$d(x_0)uJ[x_0,[r,s]] = (0) \quad \text{for all } u \in J, r,s \in R. \tag{3.27}$$

Since $x_0 \in Sa_*(R)$, the relation (3.27) assures that

$$d(x_0)uJ([x_0,[r,s]])^* = (0) \quad \text{for all } u \in J, r,s \in R. \tag{3.28}$$

In view of equations (3.27) and (3.28), Lemma 3.4.1 forces $[x_0,[r,s]] = 0$ for all $r,s \in R$ or $d(x_0)u = 0$ for all $u \in J$, in which case $d(x_0)J = (0)$. If $d(x_0)J = (0)$ then $d(x_0)Jd(x_0) = (0) = d(x_0)J(d(x_0))^*$ and once again using Lemma 3.4.1 we arrive at $d(x_0) = 0$. Now assume

$$[x_0,[r,s]] = 0 \quad \text{for all } r,s \in R. \tag{3.29}$$

Substituting x_0r for r in (3.29) we obtain

$$[x_0,s][x_0,r] = 0 \quad \text{for all } r,s \in R. \tag{3.30}$$

Replacing s by st in (3.30), where $t \in R$, we find that $[x_0,s]t[x_0,r] = 0$ and thus

$$[x_0,s]R[x_0,r] = (0) \quad \text{for all} \ \ r,s \in R. \tag{3.31}$$

Once again using the fact that $x_0 \in Sa_*(R)$, from (3.31) it follows that

$$[x_0,s]R[x_0,r]^* = (0) \quad \text{for all} \ \ r,s \in R. \tag{3.32}$$

Since R is $*$-prime, then (3.31) together with (3.32) assure that $x_0 \in Z(R)$.

In conclusion,

$$d(x_0) = 0 \ \ \text{or} \ \ x_0 \in Z(R) \ \ \text{for all} \ \ x_0 \in J \cap Sa_*(R). \tag{3.33}$$

Let $x \in J$; as $x^* - x \in J \cap Sa_*(R)$, in view of (3.33) we obtain $d(x^* - x) = 0$ or $x^* - x \in Z(R)$.

(i) Assume that $d(x^* - x) = 0$. Similarly, the fact that $x^* + x \in J \cap Sa_*(R)$ implies that $d(x^* + x) = 0$ or $x^* + x \in Z(R)$.

If $d(x^* + x) = 0$ then $2d(x) = 0$ and 2-torsion freeness forces $d(x) = 0$.

If $x^* + x \in Z(R)$, then $[t,x] = -[t,x^*]$ for all $t \in R$. Hence, (3.25) implies that

$$d(x)[x,uv]R[t,x^*] = (0) \quad \text{for all} \ \ u,v \in J, t \in R,$$

and therefore

$$d(x)[x,uv]R[t,x]^* = (0) \quad \text{for all} \ \ u,v \in J, t \in R. \tag{3.34}$$

Using (3.25) together with (3.34), because of the $*$-primeness of R, we get $x \in Z(R)$ or $d(x)[x,uv] = 0$ for all $u,v \in J$. Suppose

$$d(x)[x,uv] = 0 \quad \text{for all} \ \ u,v \in J. \tag{3.35}$$

Replacing v by $2v[r,s]$ in (3.35), where $r,s \in R$, and reasoning as in (3.26), the fact that $[x,[r,s]] = -[x^*,[r,s]]$ yields $d(x) = 0$ or $x \in Z(R)$.

(ii) Now assume that $x^* - x \in Z(R)$. Thus $[t,x] = [t,x^*]$ for all $t \in R$ and in view of (3.25) we get

$$d(x)[x,uv]R[t,x]^* = (0) \quad \text{for all} \ \ u,v \in J, t \in R. \tag{3.36}$$

Since R is $*$-prime, from equations (3.36) and (3.35) we obtain $x \in Z(R)$ or $d(x)[x,uv] = 0$ for all $u,v \in J$. Since $[t,x] = [t,x^*]$ for all $t \in R$, arguing as above we find that $d(x) = 0$ or $x \in Z(R)$. Accordingly, in all cases we have $d(x) = 0$ or $x \in Z(R)$ for all $x \in J$. It follows that J is a union of two additive subgroups G_1 and G_2, where

$$G_1 = \{x \in J \text{ such that } d(x) = 0\} \ \ \text{and} \ \ G_2 = \{x \in J \text{ such that } x \in Z(R)\}.$$

Since a group cannot be a union of two of its proper subgroups, we are forced to have $J = G_1$ or $J = G_2$. If $J = G_1$, then $d(J) = (0)$ and Lemma 3.4.3 yields $J \subseteq Z(R)$. If $J = G_2$, then $J \subseteq Z(R)$. Hence, in both the cases we find that $J \subseteq Z(R)$ and Lemma 3.4.4 assures that R is commutative.

♣

The following example proves that the $*$-primeness hypothesis in Theorem 3.4.5 is not superfluous.

Example 3.4.1 *[472] Suppose that* $R = \left\{ \begin{pmatrix} x & 0 \\ y & z \end{pmatrix} \mid x, y, z \in \mathbb{Z} \right\}$*, where* $\mathbb{Z}$ *is the ring of integers. Let us consider* $d\begin{pmatrix} x & 0 \\ y & z \end{pmatrix} = \begin{pmatrix} 0 & 0 \\ y & 0 \end{pmatrix}$ *and* $\begin{pmatrix} x & 0 \\ y & z \end{pmatrix}^* = \begin{pmatrix} z & 0 \\ -y & x \end{pmatrix}$*. It is easy to check that R is a non-$*$-prime ring and d is a nonzero derivation of R. Moreover, if we set* $J = \left\{ \begin{pmatrix} 0 & 0 \\ y & 0 \end{pmatrix} \mid y \in \mathbb{Z} \right\}$*, then J is a nonzero $*$-Jordan ideal and d is centralizing on J. However, R is not commutative.*

Corollary 3.4.1 *[452, Theorem 2] Let R be a 2-torsion free prime ring and I a nonzero ideal of R. If R admits a nonzero derivation centralizing on I, then R is commutative.*

Proof. Assume that d is a nonzero derivation of R centralizing on I. Let $\mathcal{D}$ be the additive mapping defined on $\mathcal{R} = R \times R^0$ by $\mathcal{D}(x,y) = (d(x),0)$. Clearly, $\mathcal{D}$ is a nonzero derivation of $\mathcal{R}$. Moreover, if we set $W = I \times I$, then W is a $*_{ex}$-Jordan ideal of $\mathcal{R}$. As d is centralizing on I, it is easy to check that $\mathcal{D}$ is centralizing on W. Since $\mathcal{R}$ is a $*_{ex}$-prime ring, in view of Theorem 3.4.5 we deduce that $\mathcal{R}$ is commutative and a fortiori R is commutative.

♣

Awtar in [88, Theorem 3] proved that if a 2-torsion free prime ring R admits a nonzero derivation d centralizing on a nonzero Jordan ideal J, then $J \subseteq Z(R)$. However, the conclusion is less precise and incomplete. Indeed, in light of Lemma 3.4.4, if $J \neq (0)$ then the condition $J \subseteq Z(R)$ forces R to be commutative. The application of Theorem 3.4.5 yields the following result which can be viewed as a short proof of [88, Theorem 3].

Theorem 3.4.6 *[472, Theorem 2] Let R be a prime ring of characteristic not equal to 2. Next, let d be a nonzero derivation of R and J be a nonzero Jordan ideal of R such that* $[x, d(x)] \in Z(R)$ *for all* $x \in J$*. Then, R is commutative.*

Proof. Let us consider the nonzero derivation $\mathcal{D}$ defined on $\mathcal{R} = R \times R^0$ by $\mathcal{D}(x,y) = (d(x),0)$. If we set $\mathcal{J} = J \times J$, then $\mathcal{J}$ is a $*_{ex}$-Jordan ideal of $\mathcal{R}$. Moreover, the fact that d is centralizing on J implies that $\mathcal{D}$ is centralizing on $\mathcal{J}$. Since

$\mathcal{R}$ is a $*_{ex}$-prime ring, application of Theorem 3.4.5 assures that $\mathcal{R}$ is commutative. Accordingly, R is commutative.

♣

3.5 Strong commutativity preserving derivations

In this section, we introduce two important classes of additive mappings, commutativity preserving and strong commutativity preserving mappings. Moreover, we provide a connection between the commutativity of some rings with involution and the behavior of these kind of additive mappings by considering their action on some specific subsets of the considered ring.

Definition 35 *We say that a map $f : R \longrightarrow R$ preserves commutativity on a subset S of R if $[f(x), f(y)] = 0$ whenever $[x,y] = 0$ for $x, y \in R$.*

The problem of describing bijective additive (or linear) commutativity preserving maps is initiated by Watkins [551] in the case when the ring R is a matrix algebra. Later, maps preserving the commutativity of symmetric matrices (and operators) were investigated by various authors (see, for example [498]). The first ring-theoretic result, in this direction, was established by Brešar [137] who characterized the bijective commutativity preserving linear maps between centrally closed prime algebras. Analogous results obtained for maps between symmetric elements of prime rings with involution were treated in [105], where further references can be found. These results have been extended in various directions (cf.; [516] for details). In [109], Bell and Daif introduced a certain kind of commutativity preserving map as follows.

Definition 36 *Let R be a ring and S be a subset of R. A map $f : S \longrightarrow R$ is called strong commutativity preserving (SCP) on S if $[f(x), f(y)] = [x,y]$ for all $x, y \in S$.*

More precisely, they proved that a prime ring R must be commutative if it possesses a nonzero derivation or a non-identity endomorphism which is SCP on a right ideal of R. Later, Brešar and Miers [143] described the form of an additive map $f : R \longrightarrow R$ which is SCP on a semiprime ring R. More specifically, they showed that f must be of the form $f(x) = \lambda x + \mu(x)$, where $\lambda \in C, \lambda^2 = 1$ and $\mu : R \longrightarrow C$ is an additive map where C is the extended centroid of R. Later, this result has been extended to Lie ideals of prime rings. In this section, we continue to study SCP maps in prime rings with involution.

The study of additive maps in rings with involution was initiated by Brešar et al. [142] to describe maps which are centralizing on the skew-symmetric elements in prime rings. Later many related generalizations were established by several authors. Motivated by these results, Beidar and Martindale [103] studied the theory of functional identities in prime rings with involution and then simultaneously generalized the previous results. According to Brešar and Chebotar

[99], a functional identity can be informally described as an identical relation involving some elements in a ring, together with some maps of the considered ring. The general goal of the functional identity theory is to determine the maps appearing in the identity, or when this is not possible, to determine the structure of the ring admitting the identity in question. Hence to characterize SCP maps can be viewed as solving a special functional identity. Using the theory of functional identities, Lin and Liu [415] describes the structure of SCP maps of the symmetric elements in prime rings with involution.

Theorem 3.5.1 *[415, Theorem 1.1] Let R be a prime ring of characteristic not 2, with center $Z(R)$ and involution $*$. Let S be the set of symmetric elements of R. Suppose that $f : S \longrightarrow R$ is an additive map such that $[f(x), f(y)] = [x,y]$ for all $x,y \in S$. Then unless R is an order in a 4-dimensional central simple algebra, there exists an additive map $\mu : S \longrightarrow Z(R)$ such that $f(x) = x + \mu(x)$ for all $x \in S$ or $f(x) = -x + \mu(x)$ for all $x \in S$.*

The proof of Theorem 3.5.1 is mainly based on the following proposition.

Proposition 3.5.2 *[415, Proposition 4.1] Let R be a prime ring of characteristic not 2, with involution $*$, with extended centroid C and let S be the set of symmetric elements of R. In case $*$ is of the first kind, we assume that R is not an order in a 4 or 16-dimensional central simple algebra. In case $*$ is of the second kind, we assume that R is not an order in a 4-dimensional central simple algebra. Suppose that an additive map $f : S \longrightarrow Q_{mr}$ satisfies*

$$[f(x),[y,z]] + [f(y),[z,x]] + [f(z),[x,y]] = 0 \text{ for all } x,y,z \in S.$$

Then there exist $\lambda \in C$ and an additive map $\mu : S \longrightarrow C$ such that $f(x) = \lambda x + \mu(x)$ for all $x \in S$, where Q_{mr} denotes the maximal right ring of quotients of R.

Application of Theorem 3.5.1 yields the following result:

Corollary 3.5.1 *[415, Corollary 1.2] Let R be a prime ring of characteristic not 2, with involution $*$ and let S be the set of symmetric elements of R. Suppose that $d : R \longrightarrow R$ is a nonzero derivation such that $[d(x), d(y)] = [x,y]$ for all $x,y \in S$. Then, R is a commutative domain or an order in a 4-dimensional central simple algebra.*

The following examples illustrate the exceptional case in Theorem 3.5.1.

Example 3.5.1 *[415, Example 1] Let $R = M_2(\mathbb{F})$ under the ordinary transpose involution, where $\mathbb{F}$ is a field of characteristic not 2. Then, S is $\mathbb{F}$-linear spanned by $\{e_{11}, e_{22}, e_{12} + e_{21}\}$. Suppose that $f : S \longrightarrow R$ is a $\mathbb{F}$-linear map defined by $f(e_{11}) = e_{11} + e_{12} + e_{21}, f(e_{22}) = -e_{12} - e_{21}$, and $f(e_{12} + e_{21}) = e_{22}$. Then f is SCP on S. However, f cannot be expressed as the form given in Theorem 3.5.1.*

Example 3.5.2 *[415, Example 2] Let $R = M_2(\mathbb{F})$ under the involution $*$ defined by $\begin{pmatrix} a & b \\ c & d \end{pmatrix}^* = \begin{pmatrix} d & -b \\ -c & a \end{pmatrix}$, where $\mathbb{F}$ is a field of characteristic not 2. Then $S = \mathbb{F}.(e_{11} + e_{22})$. Clearly, the $\mathbb{F}$-linear $f : S \longrightarrow R$ defined by $f(e_{11} + e_{22}) = e_{12}$ is SCP on S, but f can not be expressed as the form given in Theorem 3.5.1.*

The next aim of this section is to explore commutativity of a $*$-prime ring provided with a derivation which is either strong commutativity preserving or preserves commutativity on a nonzero $*$-Jordan ideal of R. Furthermore, examples are given to demonstrate that our results cannot be extended to semiprime rings. We recall some basic properties of a $*$-prime ring which will be used frequently throughout the text.

(I) Let J be a nonzero Jordan ideal of a 2-torsion free $*$-prime ring R such that $aJ = (0)$ (resp. $Ja = (0)$), then $a = 0$.

(II) Let R be a 2-torsion free $*$-prime ring and J a nonzero $*$-Jordan ideal of R. If $aJa = (0)$, then $a = 0$.

Let R be a 2-torsion free $*$-prime ring and J a nonzero $*$-Jordan ideal. We leave the proof of the following easy upto the reader.

Fact 1. If R admits a derivation d such that $d^2(J) = (0)$, then $d = 0$.

Lemma 3.5.3 *[440, Lemma 3] Let R be a 2-torsion free $*$-prime ring and J a nonzero $*$-Jordan ideal of R. If d is a derivation such that $d(x^2) = 0$ for all $x \in J$, then $d = 0$.*

Lemma 3.5.4 *[439, Lemma 2.2] Let R be a 2-torsion free $*$-prime ring and J a nonzero $*$-Jordan ideal. If R admits a nonzero derivation d such that $[d(x), y^2] = 0$ for all $x, y \in J$, then R is commutative.*

Proof. First suppose that $Z(R) \cap J = (0)$. From $[d(4xy^2), y^2] = 0$ it follows that

$$[x, y^2]d(y^2) = 0 \quad \text{for all } x, y \in J. \tag{3.37}$$

Substituting $2[r,s]x$ for x in (3.37), where $r, s \in R$, we obtain $[[r,s], y^2]xd(y^2) = (0)$ which leads to

$$[[r,s], y^2]Jd(y^2) = (0) \quad \text{for all } y \in J, \quad r, s \in R. \tag{3.38}$$

For $y \in Sa_*(R) \cap J$, equation (3.38) together with Lemma 3.4.1 force $[[r,s], y^2] = 0$, in which case $y^2 \in Z(R)$, or $d(y^2) = 0$. Since $J \cap Z(R) = (0)$, in both the cases we arrive at

$$d(y^2) = 0 \quad \text{for all } y \in J \cap Sa_*(R).$$

Let $y \in J$; using the fact that $d((y + y^*)^2) = 0$ and $d((y - y^*)^2) = 0$, we obtain $d(y^2) = -d((y^*)^2)$. Replacing y by y^* in (3.38), we get $([[r,s], y^2])^* Jd(y^2) = 0$

which combined with (3.38) yields either $d(y^2)=0$ or $y^2 \in Z(R)$ so that $y^2=0$. This implies that $d(y^2)=0$ for all $y \in J$ and hence by Lemma 3.5.3, $d=0$ which contradicts our hypothesis and thus $Z(R)\cap J \neq (0)$. Now let $0 \neq u \in Z(R)\cap J$; from $[d(4xu^2),y^2]=0$ it follows that $[x,y^2]d(u^2)=0$ for all $x,y \in J$ and thus either $[x,y^2]=0$ or $d(u^2)=0$. If $[x,y^2]=0$, then using similar arguments as used in ([438], proof of Theorem 3) we conclude that R is commutative. Assume that $d(u^2)=0$; the fact that $[d(4u^2r),y^2]=0$ yields $u^2[d(r),y^2]=0$. Similarly in view of $[d(4(u^*)^2r),y^2]=0$, we find that $(u^2)^*R[d(r),y^2]=0$ therefore

$$[d(r),y^2]=0 \quad \text{for all } y \in J, \ r \in R.$$

Replacing r by $d(r)s$ we get $d^2(r)[s,y^2]=0$. Again, replace s by st in the last equation, to get $d^2(r)s[t,y^2]=0$ so that

$$d^2(r)R[t,y^2]=(0) \quad \text{for all } y \in J, \ r,t \in R.$$

In view of $*$-primeness, the above equation assures that either $d^2(r)=0$ or $[t,y^2]=0$. If $d^2(r)=0$ for all $r \in R$, then $d=0$ by [440, Lemma 3] contradiction, thus $[t,y^2]=0$ for all $t \in R$ and $y \in J$, then [438, Theorem 3] implies that R is commutative.

♣

Theorem 3.5.5 *[439, Theorem 2.3] Let R be a 2-torsion free $*$-prime ring and J a nonzero $*$-Jordan ideal. If R admits a nonzero derivation d, commuting with $*$, such that either d is strong commutativity preserving on J or $[d(x),d(y)]=0$ for all $x,y \in J$, then R is commutative.*

Proof. (i) Assume that d is strong commutativity preserving on J. In this case the condition $d \neq 0$ is not necessary. Indeed, if $d=0$, then $[J,J]=(0)$ and Lemma 3.4.2 forces $J \subseteq Z(R)$. Thus R is commutative by Lemma 3.4.4.

We are given that $[d(x),d(y)]=[x,y]$ for all $x,y \in J$. Replacing y by $4y^2z$, where $z \in J$, we get

$$d(y^2)[d(x),z]+[d(x),y^2]d(z)=0 \quad \text{for all } x,y,z \in J. \tag{3.39}$$

Again, replacing z by $2[a,uv]z$ in (3.39), where $u,v \in J$ and $a \in R$, and using (3.39) we obtain

$$d(y^2)[a,uv][d(x),z]+[d(x),y^2][a,uv]d(z)=0 \quad \text{for all } u,v,x,y,z \in J, \ a \in R. \tag{3.40}$$

Putting $4z^2t$ for z in (3.40), where $t \in R$ we find that

$$d(y^2)[a,uv]z^2[d(x),t]+[d(x),y^2][a,uv]z^2d(t)=0 \tag{3.41}$$

for all $u,v,x,y,z \in J$ and $a,t \in R$. Substituting $td(x)$ for t in (3.41), we have $[d(x),y^2][a,uv]z^2td^2(x)=0$ therefore

$$[d(x),y^2][a,uv]z^2Rd^2(x)=(0) \quad \text{for all } u,v,x,y,z \in J, \ a \in R. \tag{3.42}$$

Let $x_0 \in J \cap Sa_*(R)$; from (3.42) it follows that $[d(x_0), y^2][a, uv]z^2 R(d^2(x_0))^* = 0$, so either $d^2(x_0) = 0$ or $[d(x_0), y^2][a, uv]z^2 = 0$.

Suppose that

$$[d(x_0), y^2][a, uv]z^2 = 0 \quad \text{for all } u, v, y, z \in J, \ a \in R. \tag{3.43}$$

Replacing a by aj^2 in (3.43), where $j \in J$, we find that $[d(x_0), y^2]a[j^2, uv]z^2 = 0$ therefore

$$[d(x_0), y^2]R[j^2, uv]z^2 = (0) \quad \text{for all } j, u, v, y, z \in J. \tag{3.44}$$

Since $x_0 \in Sa_*(R)$, then (3.44) assures that

$$([d(x_0), y^2])^* R[j^2, uv]z^2 = (0) \quad \text{for all } j, u, v, y, z \in J. \tag{3.45}$$

In view of eqns (3.44) and (3.45), the $*$-primeness of R forces $[d(x_0), y^2] = 0$ for all $y \in J$ or $[j^2, uv]z^2 = 0$ for all $j, u, v, z \in J$. Suppose that $[j^2, uv]z^2 = 0$ thus

$$u[j^2, v]z^2 + [j^2, u]vz^2 = 0 \quad \text{for all } z, u, v, j \in J. \tag{3.46}$$

Replacing u by $2[r,s]u$ in (3.46) we arrive at $[j^2, [r,s]]uvz^2 = 0$ and thus $[j^2, [r,s]]uJz^2 = (0)$ which leads to $[j^2, [r,s]] = 0$ or $z^2 = 0$. Hence, in both the cases we find that $z^2 \in Z(R)$ for all $z \in J$ and [438, Theorem 3] assures that R is commutative, thereby $[d(x_0), y^2] = 0$ for all $y \in J$. In conclusion,

$$d^2(x_0) = 0 \text{ or } ([d(x_0), y^2] = 0 \text{ for all } y \in J) \text{ for all } x_0 \in J \cap Sa_*(R).$$

Let $x \in J$; as $x - x^* \in J \cap Sa_*(R)$, from the above relation it follows that $d^2(x) = d^2(x^*) = d^2(x)^*$ or $[d(x), y^2] = [d(x^*), y^2]$ for all $y \in J$.

Assume that $d^2(x) = d^2(x)^*$. Hence equation (3.42) can be rewritten as

$$[d(x), y^2][a, uv]z^2 R d^2(x)^* = (0) \quad \text{for all } u, v, y, z \in J, \ a \in R.$$

Combining the latter relation with (3.42), we get $d^2(x) = 0$ or $[d(x), y^2][a, uv]z^2 = 0$ which makes it possible to conclude, using similar arguments as above, that $[d(x), y^2] = 0$ or $d^2(x) = 0$.

Now suppose $[d(x), y^2] = [d(x^*), y^2]$. In relation (3.42) let x be x^*; then we have

$$[d(x), y^2][a, uv]z^2 R d^2(x^*) = (0) \quad \text{for all } u, v, y, z \in J, \ a \in R,$$

which gives, because d commutes with $*$,

$$[d(x), y^2][a, uv]z^2 R d^2(x)^* = (0) \quad \text{for all } u, v, y, z \in J, \ a \in R.$$

Whence it follows, according to (3.42), that $d^2(x) = 0$ or $[d(x), y^2] = 0$. Thus in both cases we find that

$$d^2(x) = 0 \text{ or } [d(x), y^2] = 0 \text{ for all } x, y \in J.$$

Now let $J_1 = \{x \in J \mid d^2(x) = 0\}$ and $J_2 = \{x \in J \mid [d(x), y^2] = 0 \text{ for all } y \in J\}$. Clearly, J_1 and J_2 are additive subgroups of J whose union is J. Hence, by Brauer's trick, either $J = J_1$ or $J = J_2$. If $J = J_1$, then $d^2(J) = 0$ and hence, by Fact1, $d = 0$ thus $J = J_2$, so $[d(x), y^2] = 0$ for all $x, y \in J$, whence it follows, according to Lemma 3.5.4, that R is commutative.

(ii) Assume that

$$[d(x), d(y)] = 0 \quad \text{for all } x, y \in J.$$

Substituting $4y^2z$ for y, where $z \in J$, we get

$$d(y^2)[d(x), z] + [d(x), y^2]d(z) = 0 \quad \text{for all } x, y, z \in J.$$

Since this equation is the same as equation (3.39), reasoning as in the first case, we conclude that R is commutative.

♣

In [297], Herstein proved that if R is a prime ring of characteristic not 2 equipped with a nonzero derivation d such that $[d(x), d(y)] = 0$ for all $x, y \in R$, then R is commutative. As an application of the above theorem, we get the following theorem which generalizes Herstein's result for Jordan ideals.

Theorem 3.5.6 *[439, Theorem 2.4] Let R be a 2-torsion free prime ring and J a nonzero Jordan ideal. If R admits a nonzero derivation d such that $[d(x), d(y)] = 0$ for all $x, y \in J$, then R is commutative.*

Proof. Let $\mathcal{D}$ be the additive mapping defined on $\mathcal{R} = R \times R^0$ by $\mathcal{D}(x, y) = (d(x), d(y))$. Clearly, $\mathcal{D}$ is a nonzero derivation of $\mathcal{R}$. Moreover, if we set $\mathcal{J} = J \times J$, then $\mathcal{J}$ is a $*_{ex}$-Jordan ideal of $\mathcal{R}$ such that $[\mathcal{D}(x, y), \mathcal{D}(u, v)] = (0, 0)$ for all $(x, y),\ (u, v) \in \mathcal{J}$. Since $\mathcal{D}$ commutes with $*_{ex}$ and $\mathcal{R}$ is $*_{ex}$-prime, then Theorem 3.5.5 assures that $\mathcal{R}$ is commutative and thus so is R.

♣

An application of similar arguments yields the following:

Theorem 3.5.7 *Let R be a 2-torsion free prime ring and J a nonzero Jordan ideal. If R admits a derivation d strong commutativity preserving on J, then R is commutative.*

In [478], it is proved that if a 2-torsion free $*$-prime ring admits a nonzero derivation d such that $[d(x), d(y)] = d([x, y])$ for all x, y in a nonzero square closed Lie ideal U, then $d = 0$ or $U \subset Z(R)$. Motivated by this result, the following theorem explores the commutativity of a $*$-prime ring provided with a nonzero derivation d satisfying the above condition on a nonzero $*$-Jordan ideal.

Theorem 3.5.8 *[439, Theorem 2.6] Let R be a 2-torsion free $*$-prime ring and J a nonzero $*$-Jordan ideal. If R admits a nonzero derivation d which commutes with $*$ such that $[d(x), d(y)] = d([x, y])$ for all $x, y \in J$, then R is commutative.*

As an application of Theorem 3.5.8, we get the following theorem for which the proof goes through in the same way as the proof of Theorem 3.5.6.

Theorem 3.5.9 *Let R be a 2-torsion free prime ring and J a nonzero Jordan ideal of R. If R admits a nonzero derivation d such that $[d(x),d(y)] = d([x,y])$ for all $x,y \in J$, then R is commutative.*

The next part of this section is devoted to finding out if commutativity still holds when the commutator in the conditions of the preceding results is replaced by an anti-commutator.

Theorem 3.5.10 *[439, Theorem 3.1] Let R be a 2-torsion free $*$-prime ring and J be a nonzero $*$-Jordan ideal. Then R admits no nonzero derivation d which commutes with $*$ such that $d(x)\circ d(y) = 0$ for all $x,y \in J$.*

Proof. Assume that there exists a nonzero derivation d which commutes with $*$ and satisfies

$$d(x)\circ d(y) = 0 \quad \text{for all } x,y \in J. \tag{3.47}$$

Replacing y by $4y^2z$ in (3.47), where $z \in J$, we get

$$-d(y^2)[d(x),z] + [d(x),y^2]d(z) = 0. \tag{3.48}$$

Substituting $2[a,uv]z$ for z in (3.48) with $u,v \in J$, $a \in R$ and using (3.48) again, we obtain

$$-d(y^2)[a,uv][d(x),z] + [d(x),y^2][a,uv]d(z) = 0. \tag{3.49}$$

Replacing z by $4z^2t$ in (3.49) with $t \in R$, and using (3.49) again, we get

$$-d(y^2)[a,uv]z^2[d(x),t] + [d(x),y^2][a,uv]z^2d(t) = 0. \tag{3.50}$$

Writing tw instead of t in (3.50) with $w \in R$, we find that

$$-d(y^2)[a,uv]z^2t[d(x),w] + [d(x),y^2][a,uv]z^2td(w) = 0. \tag{3.51}$$

Taking $w = d(x)$ in (3.51), we get $[d(x),y^2][a,uv]z^2td^2(x) = 0$ and thus

$$[d(x),y^2][a,uv]z^2Rd^2(x) = 0$$

which is the same as equation (3.42). Reasoning as in the proof of Theorem 3.5.5 we conclude that R is commutative and equation (3.47) becomes $2d(x)d(y) = 0$ for all $x,y \in J$ hence $d(x)Rd(y) = (0)$ for all $x,y \in J$ so that $d(J) = (0)$ and $d = 0$, are contradictions.

♣

Using the same arguments as used in the proof of Theorem 3.5.6, an application of Theorem 3.5.10 yields the following result.

Theorem 3.5.11 *Let R be a 2-torsion free prime ring and J a nonzero Jordan ideal. Then R admits no nonzero derivation d such that $d(x) \circ d(y) = 0$ for all $x, y \in J$.*

Theorem 3.5.12 *[439, Theorem 3.3] Let R be a 2-torsion free $*$-prime ring and J a nonzero $*$-Jordan ideal. Then R admits no nonzero derivation d such that $d(x) \circ d(y) = d(x \circ y)$ for all $x, y \in J$.*

Proof. Suppose that there exists a derivation d such that

$$d(x) \circ d(y) = d(x \circ y) \text{ for all } x, y \in J. \tag{3.52}$$

Replacing y by $2y[r,s]$ in (3.52), we find that

$$-d(y)[d(x) - x, [r,s]] + ((d(x) - x) \circ y)d([r,s]) - y[d(x), d([r,s])] + yd([x, [r,s]]) = 0. \tag{3.53}$$

Substituting $2y[a,uv]$ for y in (3.53), where $u, v \in J$ and $a \in R$, we get

$$-d(y)[a,uv][d(x) - x, [r,s]] + [d(x) - x, y][a,uv]d([r,s]) = 0. \tag{3.54}$$

Replacing y by $4ty^2$ in (3.54), where $t \in R$, and using (3.54) again, we find that

$$-d(t)y^2[a,uv][d(x) - x, [r,s]] + [d(x) - x, t]y^2[a,uv]d([r,s]) = 0. \tag{3.55}$$

Writing wt instead of t in (3.55), where $w \in R$, and using (3.55) we get

$$-d(w)ty^2[a,uv][d(x) - x, [r,s]] + [d(x) - x, w]ty^2[a,uv]d([r,s]) = 0. \tag{3.56}$$

Taking $w = d(x) - x$ in (3.56), we obtain $d(d(x) - x)ty^2[a,uv][d(x) - x, [r,s]] = 0$, that is,

$$d(d(x) - x)Ry^2[a,uv][d(x) - x, [r,s]] = 0. \tag{3.57}$$

As equation (3.57) is the same as equation in the proof of Theorem 3.5.8, it follows that R is commutative and thus equation (3.52) becomes

$$2d(x)d(y) - 2d(x)y - 2xd(y) = 0.$$

Replacing y by $2yz$, where $z \in J$, we get $d(x)yd(z) - xyd(z) = 0$. Writing $2xt$ instead of x, with $t \in J$, we obtain $d(t)xyd(z) = 0$ which leads to $d(J) = (0)$ and therefore $d = 0$.

♣

Using the same arguments as used in the proof of Theorem 3.5.6, an application of Theorem 3.5.12 yields the following result.

Theorem 3.5.13 *Let R be a 2-torsion free prime ring and J a nonzero Jordan ideal. Then R admits no nonzero derivation d such that $d(x) \circ d(y) = d(x \circ y)$ for all $x, y \in J$.*

The next example proves that the above results cannot be extended to semiprime rings.

Example 3.5.3 *[439, Example 3.5] Let* $(R_1,*)$ *be a noncommutative semiprime ring, with involution, which admits a nonzero derivation* d *and let* $R = R_1 \times R_1$. *Consider* $J = 0 \times R_1$ *and define a derivation* δ *on* R *by setting* $\delta(x,y) = (d(x),0)$. *Obviously,* J *is a nonzero* τ*-Jordan ideal of* R*, where* τ *is the involution defined on* R *by* $\tau(x,y) = (x^*,y^*)$. *Furthermore,*

$$[\delta(u),\delta(v)] = 0,\ [\delta(u),\delta(v)] = \delta([u,v]),\ \delta(u)\circ\delta(v) = 0,\ \delta(u)\circ\delta(v) = \delta(u\circ v)$$

for all $u,v \in J$*; but* R *is noncommutative. Hence Theorems 3.5.5, 3.5.8, 3.5.10 and 3.5.12 can not be extended to a semiprime ring.*

3.6 Derivations with special algebraic identities

Daif and Bell [202] established commutativity of a semiprime ring R satisfying $d([x,y]) = [x,y]$ for all x,y in a nonzero ideal of R, and d a derivation of R. In the year 1997, Hongan [301] established commutativity of 2-torsion free semiprime ring R which admits a derivation d satisfying $d([x,y]) + [x,y] \in Z(R)$ for all $x,y \in I$ or $d([x,y]) - [x,y] \in Z(R)$ for all $x,y \in I$, where I is an ideal of R. The purpose of the present section is to study derivations of a $*$-prime ring R satisfying any one of the properties: $(i)\ d[x,y] = 0$, $(ii)\ d([x,y]) - [x,y] \in Z(R)$, $(iii)\ d([x,y]) + [x,y] \in Z(R)$, $(iv)\ d(x\circ y) = 0$, $(v)\ d(x\circ y) - x\circ y \in Z(R)$, $(vi)\ d(x\circ y) + x\circ y \in Z(R)$ for all $x,y \in J$, a nonzero Jordan ideal of R. We begin with the following lemmas which are essential for developing the proof of our main results.

Lemma 3.6.1 *[301, Corollary 1] Let* R *be a 2-torsion free semiprime ring. If* R *admits a derivation* d *such that* $d([x,y]) + [x,y] \in Z(R)$ *or* $d([x,y]) - [x,y] \in Z(R)$ *for all* $x,y \in R$*, then* R *is commutative.*

Lemma 3.6.2 *[479, Lemma 4] Let* R *be a 2-torsion free* $*$*-prime ring. If* R *admits a nonzero derivation* d *such that* $d([x,y]) \in Z(R)$ *for all* $x,y \in R$*, then* R *is commutative.*

Proof. Suppose that

$$d([x,y]) \in Z(R) \quad \text{for all } x,y \in R. \tag{3.58}$$

We claim that $Z(R) \neq (0)$. Indeed, otherwise equation (3.58) reduces to

$$d([x,y]) = 0 \quad \text{for all } x,y \in R.$$

Hence, R is commutative by [477, Theorem 1.3], and thus $R = (0)$, a contradiction. Now replacing x by xz in (3.58), where $0 \neq z \in Z(R)$, we get $[x,y]d(z) \in Z(R)$. This yields

$$[[x,y],r]Rd(z) = (0) \quad \text{for all } x,y,r \in R.$$

In the light of $*$-primeness, either $[[x,y],r] = 0$ for all $x,y,r \in R$ and hence R is commutative or $d(z) = 0$. Assume that $d(z) = 0$ for all $z \in Z(R)$. Then

$$d^2([x,y]) = 0 \quad \text{for all } x,y \in R.$$

Replacing y by xy in the above expression, we obtain

$$d^2(x)[x,y] + 2d(x)d([x,y]) = 0 \quad \text{for all } x,y \in R.$$

Substituting yx for y, we get

$$d(x)[x,y]d(x) = 0 \quad \text{for all } x,y \in R.$$

Writing $yd(x)r$ instead of y, we obtain

$$d(x)y[x,d(x)]rd(x) = 0 \quad \text{for all } r,x,y \in R.$$

Right multiplying the last equation with $y[x,d(x)]$ yields

$$d(x)y[x,d(x)]Rd(x)y[x,d(x)] = (0) \quad \text{for all } x,y \in R.$$

Since R is semiprime, the above equation forces

$$d(x)y[x,d(x)] = 0 \quad \text{for all } x,y \in R. \tag{3.59}$$

Left multiplying the last equation with x implies that

$$xd(x)y[x,d(x)] = 0 \quad \text{for all } x,y \in R. \tag{3.60}$$

Replacing y by xy in (3.59), we obtain

$$d(x)xy[x,d(x)] = 0 \quad \text{for all } x,y \in R.$$

Subtracting this equation from (3.60), we find that

$$[d(x),x]R[d(x),x] = (0) \quad \text{for all } x \in R.$$

Once again using semiprimeness, we conclude that $[d(x),x] = 0$ for all $x \in R$ and the commutativity of R follows from Theorem 3.4.5.

♣

Lemma 3.6.3 *[479, Lemma 5] Let R be a 2-torsion free $*$-prime ring and J be a non-zero $*$-Jordan ideal of R such that $[J,J] \subseteq Z(R)$. Then, R is commutative.*

Theorem 3.6.4 *[479, Theorem 1] Let R be a 2-torsion free $*$-prime ring and J be a nonzero $*$-Jordan ideal of R. If R admits a nonzero derivation d such that $d([x,y]) \in Z(R)$ for all $x,y \in J$, then R is commutative.*

Proof. We have

$$d([x,y]) \in Z(R) \quad \text{for all } x,y \in J. \tag{3.61}$$

Let us consider $J_1 = \{x \in J \mid d(x) \in J\}$. It is easy to see that J_1 is a nonzero $*$-Jordan ideal of R. If $J \cap Z(R) = (0)$, then because of

$$d([x,2y^2]) = [d(x),2y^2] + [x,2d(y) \circ y] \in Z(R) \cap J \quad \text{for all } x,y \in J_1$$

we get

$$d([x,2y^2]) = 0 \quad \text{for all } x,y \in J_1.$$

Replacing x by $4xy^2$ we arrive at

$$[x,y^2]d(y^2) = 0 \quad \text{for all } x,y \in J_1. \tag{3.62}$$

Substituting $2[r,s]x$ for x, where $r,s \in R$, we obtain

$$[[r,s],y^2]J_1 d(y^2) = (0) \quad \text{for all } y \in J_1, \; r,s \in R. \tag{3.63}$$

For $y \in Sa_*(R) \cap J_1$ we obtain $2y^2 \in Z(R) \cap J_1$ or $d(y^2) = 0$. But $2y^2 \in Z(R) \cap J_1$ forces $y^2 = 0$ so that in both the cases we arrive at $d(y^2) = 0$. Let $y \in J_1$. Since $y - y^*, \; y + y^* \in Sa_*(R) \cap J_1$, $d(y+y^*)^2 = d(y-y^*)^2 = 0$. Therefore $d((y^*)^2) = -d(y^2)$. Replacing y by y^* in (3.63) we obtain

$$([[r,s],y^2])^* J_1 d(y^2) = (0) \quad \text{for all } y \in J_1, \; r,s \in R.$$

Combining the last equation with (3.63) and reasoning as above we deduce that

$$d(y^2) = 0 \quad \text{for all } y \in J_1.$$

Hence by Lemma 3.5.3, $d = 0$ is a contradiction. Consequently, $Z(R) \cap J \neq (0)$. Let $0 \neq z \in Z(R) \cap J$, replacing $2xz$ instead of x in (3.61) we obtain $[x,y]d(z) \in Z(R)$ so that $[[x,y],r]d(z) = 0$ for all $x,y \in J, r \in R$ and therefore

$$[[x,y],r]Rd(z) = (0) \quad \text{for all } x,y \in J, \quad r \in R. \tag{3.64}$$

As $[[x,y],r]^* Rd(z) = (0)$ by the above expression, hence $d(z) = 0$ or $[J,J] \subseteq Z(R)$. If $d(z) = 0$ for all $z \in Z(R)$, then replacing x by $2rz$ in (3.61) and using (3.61), we obtain

$$d([r,y]) \in Z(R) \quad \text{for all } y \in J, \quad r \in R.$$

Substituting $2zs$ for y we find that $d([r,s]) \in Z(R)$ for all $r,s \in R$ and hence R is commutative by Lemma 3.6.2. If $[J,J] \subseteq Z(R)$, then R is commutative by Lemma 3.6.3.

♣

If $d([x,y]) \in Z(R)$ for all $x,y \in J$, then in that case we find the following theorem which improves the result of [110] for the case when R is 2-torsion free.

Theorem 3.6.5 *[479, Theorem 2] Let R be a prime ring of characteristic different from two and J be a nonzero Jordan ideal of R. If R admits a nonzero derivation d such that $d([x,y]) \in Z(R)$ for all $x,y \in J$, then R is commutative.*

Proof. Assume that d is a nonzero derivation of R such that $d([x,y]) \in Z(R)$ for all $x,y \in J$. Let $\mathcal{D}$ be the additive mapping defined on $\mathcal{R} = R \times R^0$ by $\mathcal{D}(x,y) = (d(x),0)$. Clearly, $\mathcal{D}$ is a nonzero derivation of $\mathcal{R}$. Moreover, if we set $\mathcal{J} = J \times J$, then $\mathcal{J}$ is a $*_{ex}$-Jordan ideal of $\mathcal{R}$ and $\mathcal{D}([x,y]) \in Z(\mathcal{R})$ for all $x,y \in \mathcal{J}$. Since $\mathcal{R}$ is a $*_{ex}$-prime ring, in view of Theorem 3.6.4 we deduce that $\mathcal{R}$ is commutative and a fortiori R is commutative.

♣

Corollary 3.6.1 *[438, Theorem 1] Let R be a 2-torsion free $*$-prime ring and J be a nonzero $*$-Jordan ideal of R. If R admits a nonzero derivation d such that $d([x,y]) = 0$ for all $x,y \in J$, then R is commutative.*

Theorem 3.6.6 *[479, Theorem 3] Let R be a 2-torsion free $*$-prime ring and J a nonzero $*$-Jordan ideal of R. Then the following statements are equivalent:*
(1) R is commutative.
(2) R admits a derivation d such that $d([x,y]) - [x,y] \in Z(R)$ for all $x,y \in J$.
(3) R admits a derivation d such that $d([x,y]) + [x,y] \in Z(R)$ for all $x,y \in J$.

Proof. Obviously $(1) \Longrightarrow (3)$. Moreover, $(3) \Longrightarrow (2)$ indeed if $d([x,y]) + [x,y] \in Z(R)$ then the derivation $(-d)$ satisfies $(-d)([x,y]) - [x,y] \in Z(R)$.

$(2) \Longrightarrow (1)$. Suppose that R satisfies (2). If $d = 0$ then $[J,J] \subseteq Z(R)$ and R is commutative by Lemma 3.6.3. Assume that d is a nonzero derivation such that

$$d([x,y]) - [x,y] \in Z(R) \quad \text{for all } x,y \in J. \tag{3.65}$$

Define J_1 as in Theorem 3.6.4. If $J \cap Z(R) = (0)$, then because

$$d([x,2y^2]) - [x,2y^2] = [d(x),2y^2] + [x,2d(y) \circ y] - [x,2y^2] \in Z(R) \cap J$$

for all $x,y \in J_1$, we get

$$d([x,2y^2]) - [x,2y^2] = 0 \quad \text{for all } x,y \in J_1. \tag{3.66}$$

Replacing x by $4xy^2$ in (3.66) we get,

$$[x,y^2]d(y^2) = 0 \quad \text{for all } x,y \in J_1. \tag{3.67}$$

Since equation (3.67) is the same as equation (3.62), arguing as in the proof of Theorem 3.6.4 we are forced to $Z(R) \cap J \neq (0)$.

Let $0 \neq z \in Z(R) \cap J$; writing $2xz$ instead of x in (3.65) and employing (3.65) we obtain $[x,y]d(z) \in Z(R)$ and thus

$$[[x,y],r]Rd(z) = (0) \quad \text{for all } x,y \in J \text{ and } r \in R. \tag{3.68}$$

Since equation (3.68) is the same as equation (3.64), arguing as in the proof of Theorem 3.6.4 we arrive at R is commutative or $d(z)=0$.

Assume that $d(z)=0$. Replacing x by $2zr = z\circ r$ in (3.65), where $r\in R$, we obtain

$$d([r,y])-[r,y]\in Z(R) \quad \text{for all } y\in J,\ r\in R.$$

Replacing y by $2zs$, with $s\in R$, we get

$$d([r,s])-[r,s]\in Z(R) \quad \text{for all } r,s\in R.$$

Hence, R is commutative by Lemma 3.6.1.

♣

Theorem 3.6.7 *[479, Theorem 4] Let R be a 2-torsion free ∗-prime ring and J a nonzero ∗-Jordan ideal of R. Then R is commutative iff R admits a derivation d such that for all* $x,y\in J$, *either* $d([x,y])-[x,y]\in Z(R)$ *or* $d([x,y])+[x,y]\in Z(R)$.

Proof. Obviously, every commutative ring satisfies $d([x,y])-[x,y]\in Z(R)$ and $d([x,y])+[x,y]\in Z(R)$. Conversely, for each fixed $x\in J$ we put

$$J_x=\{y\in J \mid d([x,y])-[x,y]\in Z(R)\} \text{ and } J'_x=\{y\in J \mid d([x,y])+[x,y]\in Z(R)\}.$$

Then obviously J_x and J'_x are additive subgroups of J whose union is J. But a group cannot be a theoretic union set of two of its proper subgroups. Hence either $J=J_x$ or $J=J'_x$. Further using similar arguments as above, we find that $J=\{x\in J \mid J_x=J\}$ or $J=\{x\in J \mid J'_x=J\}$. Therefore, R is commutative by Theorem 3.6.6.

♣

Using similar arguments as used in proving Theorem 3.6.5 and employing Theorem 3.6.6, we get the following theorem which improves Theorem 1 of [202].

Theorem 3.6.8 *Let R be a prime ring of characteristic different from two and J a nonzero Jordan ideal of R. If R admits a derivation d such that either* $d([x,y])-[x,y]\in Z(R)$ *for all* $x,y\in J$, *or* $d([x,y])+[x,y]\in Z(R)$ *for all* $x,y\in J$, *then R is commutative.*

The following is an immediate consequence of Theorem 3.6.6.

Corollary 3.6.2 *Let R be a 2-torsion free ∗-prime ring and J be a nonzero ∗-Jordan ideal of R. If R admits a nonzero derivation d such that* $d([x,y])=[x,y]$ *for all* $x,y\in J$, *then R is commutative.*

The last part of this section is devoted to knowing whether the commutativity of the ring still holds if the commutator in the conditions of the preceding results is replaced by the anti-commutator.

Theorem 3.6.9 *[479, Theorem 6] Let R be a 2-torsion free $*$-prime ring and J be a nonzero $*$-Jordan ideal of R. If R admits a nonzero derivation d such that $d(x \circ y) \in Z(R)$ for all $x, y \in J$, then R is commutative.*

Proof. Suppose that

$$d(x \circ y) \in Z(R) \quad \text{for all } x, y \in J. \tag{3.69}$$

If $J \cap Z(R) = 0$ then, because of $d(x \circ y) \in Z(R) \cap J$, we get,

$$d(x \circ y) = 0 \quad \text{for all } x, y \in J. \tag{3.70}$$

Hence $d = 0$ by Lemma 3.5.3, a contradiction. Accordingly, $J \cap Z(R) \neq (0)$. Let $0 \neq z \in J \cap Z(R)$. Replacing x by $2zx = z \circ x$ in (3.69) we arrive at

$$d(x \circ y)z + (x \circ y)d(z) \in Z(R) \quad \text{for all } x, y \in J. \tag{3.71}$$

Comparing (3.69) with (3.71), we obtain $(x \circ y)d(z) \in Z(R)$ so that

$$[(x \circ y), r]Rd(z) = (0) \quad \text{for all } x, y \in J, \ r \in R. \tag{3.72}$$

In particular

$$[x^2, r]Rd(z) = ([x^2, r])^* Rd(z) = 0 \quad \text{for all } x \in J, \ r \in R. \tag{3.73}$$

Consequently, either $[x^2, r] = 0$ or $d(z) = 0$ for all $z \in J \cap Z(R)$. If $d(J \cap Z(R)) = (0)$, then equation (3.69) assures that

$$d^2(x \circ y) = 0 \quad \text{for all } x, y \in J.$$

Substituting $2x^2$ for x and $2y^2$ for y and invoking (3.69)we arrive at $d(x^2)Rd(y^2) = (0)$ in such a way that $d(x^2)Rd(x^2) = (0)$ for all $x \in J$ and the semiprimeness gives $d(x^2) = 0$ for all $x \in J$. Hence again by Lemma 3.5.3, $d = 0$ is a contradiction. Thus $[x^2, r] = 0$ for all $x \in J, r \in R$. Now applying a similar method as used in the proof of Lemma 3.6.3, we find that R is commutative.

♣

Further using similar arguments as used to prove Theorem 3.6.5, applying Theorem 3.6.9, we get the following result.

Theorem 3.6.10 *[479, Theorem 7] Let R be a prime ring of characteristic different from two and J be a nonzero Jordan ideal of R. If R admits a nonzero derivation d such that $d(x \circ y) \in Z(R)$ for all $x, y \in J$, then R is commutative.*

Theorem 3.6.11 *[479, Theorem 8] Let R be a 2-torsion free $*$-prime ring and J be a nonzero $*$-Jordan ideal of R. Then the following conditions are equivalent:*
(1) R is commutative.
(2) R admits a derivation d such that $d(x \circ y) - x \circ y \in Z(R)$ for all $x, y \in J$.
(3) R admits a derivation d such that $d(x \circ y) + x \circ y \in Z(R)$ for all $x, y \in J$.

Proof. Clearly $(1) \Longrightarrow (3) \Longrightarrow (2)$.
$(2) \Longrightarrow (1)$. Assume that R satisfies (2). If $d = 0$, then $x^2 \in Z(R)$ for all $x \in J$ and arguing as in the proof of Lemma 3.6.3 we infer that R is commutative. Suppose d is a nonzero derivation such that

$$d(x \circ y) - x \circ y \in Z(R) \quad \text{for all } x, y \in J. \tag{3.74}$$

If $Z(R) \cap J = 0$, then by (3.74) we obtain

$$d(x \circ y) = x \circ y \quad \text{for all } x, y \in J. \tag{3.75}$$

Writing x^2 instead of x and $4yx^2$ instead of y in (3.75) we get

$$(x^2 \circ y)d(x^2) = 0 \quad \text{for all } x, y \in J. \tag{3.76}$$

Substituting $2[r,s]y$ for y in (3.76), where $r, s \in R$, we obtain $[x^2,[r,s]]yd(x^2) = 0$ and therefore

$$[x^2,[r,s]]Jd(x^2) = (0) \quad \text{for all } x \in J, \ r, \ s \in R. \tag{3.77}$$

Since equation (3.77) is the same as equation (3.63), arguing as in the proof of Theorem 3.6.4 we find that $d = 0$, a contradiction. Consequently, $J \cap Z(R) \neq (0)$. Let $0 \neq z \in J \cap Z(R)$. Replacing x by $2xz = x \circ z$ in (3.74) we find that $(x \circ y)d(z) \in Z(R)$ and hence

$$[x^2, r]Rd(z) = [x^2, r]^* Rd(z) = (0) \quad \text{for all } x \in J, \ r \in R. \tag{3.78}$$

R being $*$-prime, from (3.78) it follows that either $d(Z(R) \cap J) = (0)$ or $[x^2, r] = 0$ for all $x \in J$, $r \in R$. Assume that $d(Z(R) \cap J) = (0)$. Thus (3.74) implies

$$d^2(x \circ y) - d(x \circ y) = 0 \quad \text{for all } x, y \in J. \tag{3.79}$$

In particular, for $0 \neq y \in J \cap Z(R)$ equation (3.79) leads to

$$(d^2(x) - d(x))y = 0 \quad \text{for all } x \in J, \tag{3.80}$$

and hence $(d^2(x) - d(x))Ry = (d^2(x) - d(x))Ry^* = (0)$ for all $x \in J$. R being $*$-prime implies that $d^2(x) - d(x) = 0$ for all $x \in J$. Writing $4xu^2$ instead of x, where $u \in J$, we obtain $d(x)d(u^2) = 0$ for all $x, u \in J$. Replacing x by $4u^2x$ we find that $d(u^2)Jd(u^2) = (0)$ for all $u \in J$. Now application of (II) assures that $d(u^2) = 0$ for all $u \in J$, and hence by Lemma 3.5.3, $d = 0$, a contradiction. Consequently, $[x^2, r] = 0$ for all $x \in J$ $r \in R$ and using similar arguments as given in the proof of Lemma 3.6.3, we get the required result.

♣

Using a similar procedure as used to prove Theorem 3.6.7, we get the following.

Theorem 3.6.12 *Let R be a 2-torsion free $*$-prime ring and J a nonzero $*$-Jordan ideal of R. Then R is commutative iff R admits a derivation d such that for all $x,y \in J$, either $d(x\circ y)-x\circ y \in Z(R)$ or $d(x\circ y)+x\circ y \in Z(R)$.*

If R is prime with a characteristic not equal to two, then one can prove the following.

Theorem 3.6.13 *Let R be a prime ring of characteristic not equal to two and J be a nonzero Jordan ideal of R. If R admits a derivation d such that either $d(x\circ y)-x\circ y \in Z(R)$ for all $x,y \in J$, or $d(x\circ y)+x\circ y \in Z(R)$ for all $x,y \in J$, then R is commutative.*

Corollary 3.6.3 *Let R be a 2-torsion free $*$-prime ring and J be a nonzero $*$-Jordan ideal of R. Suppose that R admits a derivation d. Then the following conditions are equivalent:*
(1) $d(xy)-xy \in Z(R)$ for all $x,y \in J$.
(2) $d(xy)+xy \in Z(R)$ for all $x,y \in J$.
(3) $d(xy)-yx \in Z(R)$ for all $x,y \in J$.
(4) $d(xy)+yx \in Z(R)$ for all $x,y \in J$.
(5) R is commutative.

The following example demonstrates that Theorems 3.6.4, 3.6.6, 3.6.7, 3.6.9 and 3.6.11 cannot be extended to semiprime rings.

Example 3.6.1 *[479, Example 1] Let (R,σ) be a noncommutative prime ring with an involution. Let us consider $\mathcal{R}= R\times \mathbb{Q}[X]$. It is obvious that $\mathcal{R}$ is semiprime. Moreover, if we define $(r,P(X))^* = (\sigma(r),P(X))$, then $*$ is an involution of $\mathcal{R}$ for which $J=\{0\}\times \mathbb{Q}[X]$ is a nonzero $*$-Jordan ideal. Furthermore, $d(r,P(X)) = (0,P'(X))$ is a nonzero derivation of $\mathcal{R}$ such that $d([u,v]) \in Z(\mathcal{R})$, $d([u,v])-[u,v] \in Z(\mathcal{R})$, $d(u\circ v) \in Z(\mathcal{R})$, $d(u\circ v)-u\circ v \in Z(\mathcal{R})$, for all $u,v \in J$. However, $\mathcal{R}$ is not commutative.*

The following example shows that in Theorems 3.6.4 and 3.6.9 the hypothesis of $*$-Jordan ideal is crucial.

Example 3.6.2 *[479, Example 2] Let R be a noncommutative prime ring which admits a nonzero derivation d and let $\mathcal{R}=R\times R^0$. If we set $J=R\times 0$, then J is a nonzero Jordan ideal of the $*_{ex}$-prime ring $\mathcal{R}$. Furthermore, if we define $\delta(x,y) = (0,d(y))$, then δ is a derivation of $\mathcal{R}$ which satisfies $\delta[u,v] \in Z(\mathcal{R})$, $\delta(u\circ v) \in Z(\mathcal{R})$ for all $u,v \in J$. However, $\mathcal{R}$ is noncommutative.*

Theorem 3.6.14 *[438, Theorem 1] Let R be a 2-torsion free $*$-prime ring and J a nonzero $*$-Jordan ideal of R. If R admits a nonzero derivation d such that $d([x,y]) = 0$ for all $x,y \in J$, then R is commutative.*

Proof. Assume that $d([x,y]) = 0$ for all $x,y \in J$. Let us consider $\delta : R \longrightarrow R$ defined by $\delta(x) = d(x)+(*\circ d\circ *)(x)$. Obviously, δ is a derivation which commutes with $*$ and $\delta([x,y]) = 0$ for all $x,y \in J$. Hence

$$[\delta(x),y] = [\delta(y),x] \text{ for all } x,y \in J. \tag{3.81}$$

Replacing y by $2[y,u]v$ in (3.81), because $\delta([y,u]) = 0$, we find that

$$[\delta(x),[y,u]]v = [[y,u],x]\delta(v) \text{ for all } u,v,x,y \in J. \tag{3.82}$$

Substituting $2v[r,s]$ for v in (3.82), where $r,s \in R$, and using (3.82) we obtain $[[y,u],x]v\delta([r,s]) = 0$ for all $u,v,x,y \in J$ and $r,s \in R$ in such a way that

$$[[y,u],x]J\delta([r,s]) = (0) \text{ for all } u,x,y \in J, r,s \in R. \tag{3.83}$$

Since δ commutes with $*$, in view of (3.83), it then follows that

$$[[y,u],x]J(\delta([r,s]))^* = (0) \text{ for all } u,x,y \in J, r,s \in R. \tag{3.84}$$

Using Lemma 3.4.1, eqns (3.83) and (3.84) ensures that $\delta([r,s]) = 0$ for all $r,s \in R$ or $[[y,u],x] = 0$ for all $u,x,y \in J$. If $\delta([r,s]) = 0$ for all $r,s \in R$, then ([473], Theorem 3.3) assures the commutativity of R. Now assume that

$$[[y,u],x] = 0 \text{ for all } u,x,y \in J. \tag{3.85}$$

Substituting $yu + uy$ for y in (3.85) we find that

$$[y,u][u,x] + [u,x][y,u] = 0 \text{ for all } u,x,y \in J. \tag{3.86}$$

Since $[[y,u],[u,x]] = 0$ by (3.85), then (3.86) with a 2-torsion freeness yield

$$[y,u][u,x] = 0 \text{ for all } u,x,y \in J. \tag{3.87}$$

Replacing y by $2[r,s]y$ in (3.87), where $r,s \in R$, we obtain $[[r,s],u]y[u,x] = 0$ for all $u,x,y \in J$ and $r,s \in R$ such that

$$[[r,s],u]J[u,x] = (0) \text{ for all } u,x \in J, r,s \in R. \tag{3.88}$$

Let $u \in J \cap Sa_*(R)$, then from (3.88) it follows that

$$[[r,s],u]J[u,x]^* = (0) \text{ for all } x \in J, r,s \in R. \tag{3.89}$$

Using (3.88) and (3.89), Lemma 3.4.1 forces $[u,J] = (0)$ or $[[r,s],u] = 0$ for all $r,s \in R$.

Assume that

$$[[r,s],u] = 0 \text{ for all } r,s \in R. \tag{3.90}$$

Substituting ru for r in (3.90) and employing (3.90) we obtain

$$[r,u][u,s] = 0 \text{ for all } r,s \in R. \tag{3.91}$$

Replacing s by tr in (3.91), where $t \in R$, we find that $[r,u]t[u,r]=0$, hence,

$$[u,r]R[u,r]=(0) \quad \text{for all } r \in R. \tag{3.92}$$

Since a $*$-prime ring is semiprime, from (3.92) it follows that $[r,u]=0$ so that $u \in Z(R)$ and a fortiori $[u,J]=(0)$. Therefore, in both the cases we find that

$$[u,J]=(0) \quad \text{for all } u \in J \cap Sa_*(R).$$

Let $x \in J$; as $x-x^* \in J \cap Sa_*(R)$ and $x+x^* \in J \cap Sa_*(R)$, then

$$[x-x^*,j]=0 \text{ and } [x+x^*,j]=0 \quad \text{for all } j \in J.$$

Since R is 2-torsion free, this equation yields $[x,j]=0$ for all $x,j \in J$ and Lemma 3.4.2 assures that $J \subseteq Z(R)$. Applying Lemma 3.4.4, we conclude that R is commutative.

♣

Corollary 3.6.4 *[477, Theorem 1.3] Let R be a 2-torsion free $*$-prime ring and I a nonzero $*$-ideal of R. If R admits a nonzero derivation d such that $d([x,y])=0$ for all $x,y \in I$, then R is commutative.*

Corollary 3.6.5 *[473, Theorem 3.3] Let R be a 2-torsion free $*$-prime ring. If R admits a nonzero derivation d such that $d([x,y])=0$ for all $x,y \in R$, then R is commutative.*

Bell and Daif [110] showed that if a 2-torsion free prime ring R admits a nonzero derivation d satisfying $d([x,y])=0$ for all $x,y \in R$, then R is commutative. As an application of Theorem 3.6.14, the following theorem extends the result of [110] to Jordan ideals.

Theorem 3.6.15 *Let R be a 2-torsion free prime ring and J a nonzero Jordan ideal of R. If R admits a nonzero derivation d satisfying $d([x,y])=0$ for all $x,y \in J$, then R is commutative.*

Proof. Assume that d is a nonzero derivation of R such that $d([x,y])=0$ for all $x,y \in J$. Let δ be the additive mapping defined on $\mathcal{R}=R \times R^0$ by $\delta(x,y)=(d(x),0)$. Clearly, δ is a nonzero derivation of $\mathcal{R}$. Moreover, if we set $\mathcal{J}=J \times J$, then $\mathcal{J}$ is a $*_{ex}$-Jordan ideal of $\mathcal{R}$ and $\delta([x,y])=0$ for all $x,y \in \mathcal{J}$. Since $\mathcal{R}$ is a $*_{ex}$-prime ring, in view of Theorem 3.6.14 we deduce that $\mathcal{R}$ is commutative and a fortiori R is commutative.

♣

Corollary 3.6.6 *Let R be a 2-torsion free prime ring and I a nonzero ideal of R. If R admits a nonzero derivation d such that $d([x,y])=0$ for all $x,y \in I$, then R is commutative.*

Corollary 3.6.7 *Let R be a 2-torsion free prime ring. If R admits a nonzero derivation d such that $d([x,y]) = 0$ for all $x,y \in R$, then R is commutative.*

Theorem 3.6.16 *[438, Theorem 3] Let R be a 2-torsion free $*$-prime ring and J a nonzero $*$-Jordan ideal of R. If R admits a nonzero derivation d such that $d([x,y]) = [x,y]$ for all $x,y \in J$, then R is commutative.*

Proof. Assume that

$$d([x,y]) = [x,y] \quad \text{for all } x,y \in J. \tag{3.93}$$

Suppose that $Z(R) \cap J = (0)$; replacing x by $4xy^2$ in (3.93) we get,

$$[x,y]d(y^2) = (0) \quad \text{for all } x,y \in J. \tag{3.94}$$

Writing $2[r,s]x$ instead of x in (3.94) we find that $[[r,s],y]xd(y^2) = 0$ and thus

$$[[r,s],y]Jd(y^2) = 0 \quad \text{for all } y \in J \text{ and } r,s \in R. \tag{3.95}$$

For $y \in J \cap Sa_*(R)$, in view of (3.95), Lemma 3.4.1 yields $[[r,s],y] = 0$ or $d(y^2) = 0$,

if

$$[[r,s],y] = 0 \quad \text{for all } r,s \in R. \tag{3.96}$$

Since equation (3.96) is the same as equation (3.90), reasoning as in the proof of Theorem 3.6.14, we get $y \in Z(R)$. Hence $y = 0$ and therefore $d(y^2) = 0$ for all $y \in J \cap Sa_*(R)$. Let $y \in J$, since $y + y^*, y - y^* \in J \cap Sa_*(R)$, then $d((y+y^*)^2) = 0$ and $d((y-y^*)^2) = 0$, which forces $d(y^2) = -d((y^*)^2)$. Replacing y by y^* in (3.95) we get,

$$([[r,s],y])^*Jd(y^2) = (0) \quad \text{for all } y \in J \text{ and } r,s \in R. \tag{3.97}$$

Using (3.95) together with (3.97), Lemma 3.4.1 forces $d(y^2) = 0$ for all $y \in J$. According to Lemma 3.5.3, the last equality implies that R is commutative. But $J \cap Z(R) = (0)$ forces $J = (0)$; this contradicts our hypothesis and consequently $J \cap Z(R) \neq (0)$.

Suppose that $d(u) = 0$ for all $u \in J \cap Z(R)$. Substituting $4u^2r$ for x in (3.93) where $r \in R$ we get $u^2(d([r,y]) - [r,y]) = 0$, thus,

$$u^2R(d([r,y]) - [r,y]) = (0) \quad \text{for all } y \in J \text{ and } r \in R. \tag{3.98}$$

Since $u^* \in J \cap Z(R)$, similarly we get,

$$(u^2)^*R(d([r,y]) - [r,y]) = (0) \quad \text{for all } y \in J \text{ and } r \in R. \tag{3.99}$$

Combining (3.98) and (3.99), the $*$-primeness forces

$$d([r,y]) = [r,y] \quad \text{for all } y \in J \text{ and } r \in R. \tag{3.100}$$

Replacing r by ry in (3.100), we obtain

$$[r,y]d(y) = (0) \quad \text{for all } y \in J \text{ and } r \in R. \tag{3.101}$$

Substituting rs for r in (3.101) we find that $[r,y]sd(y) = 0$ such that

$$[r,y]Rd(y) = (0) \quad \text{for all } y \in J \text{ and } r \in R. \tag{3.102}$$

For $y \in J \cap Sa_*(R)$, equation (3.102) yields $d(y) = 0$ or $y \in Z(R)$.

Let $y \in J$; since $y - y^* \in J \cap Sa_*(R)$, then $y - y^* \in Z(R)$ or $d(y - y^*) = 0$.

If $y - y^* \in Z(R)$, then $[r,y] = [r,y^*]$ for all $r \in R$. Hence, (3.102) becomes $[r,y^*]Rd(y) = (0)$ such that

$$[r,y]^*Rd(y) = (0) \quad \text{for all } y \in J \text{ and } r \in R. \tag{3.103}$$

Using (3.102) with (3.103), we conclude that either $y \in Z(R)$ or $d(y) = 0$.

If $d(y - y^*) = 0$, then $d(y) = d(y^*)$ and replacing y by y^* in (3.102) we obtain $[r,y]^*Rd(y) = (0)$ which combined with (3.102) leads to $d(y) = 0$ or $y \in Z(R)$. Accordingly, in all the cases we find that $d(y) = 0$ or $y \in Z(R)$ for all $y \in J$. Let

$$J_1 = \{x \in J | x \in Z(R)\} \quad \text{and} \quad J_2 = \{x \in J | d(x) = 0\}.$$

Clearly, each of J_1 and J_2 is an additive subgroup of J such that $J = J_1 \cup J_2$. But, a group can not be the set-theoretic union of two proper belonging to it. Hence, $J = J_1$ and as in the proof of Theorem 3.6.14 R is commutative or $J = J_2$ and, because $0 \neq d$, thus [471, Lemma 4] assures that $J \subseteq Z(R)$ thereby R is commutative.

Now suppose there exists $u \in J \cap Z(R)$ such that $d(u) \neq 0$. Replacing x by $4x^2u$ in (3.93), we obtain $[x^2,y]d(u) = 0$ which, in the light of $d(u) \in Z(R)$, forces

$$[x^2,y]Rd(u) = (0) \quad \text{for all } x,y \in J. \tag{3.104}$$

Since J is a $*$-ideal, then (3.104) leads to

$$[x^2,y]^*Rd(u) = (0) \quad \text{for all } x,y \in J. \tag{3.105}$$

Combining (3.104) and (3.105) we conclude that $[x^2,y] = 0$ for all $x,y \in J$. Replacing y by $4ry^2$ we obtain $[x^2,r]y^2 = 0$ for all $x,y \in J$ and $r \in R$. Substituting rt for r we find that $[x^2,r]ty^2 = 0$ and thus $[x^2,r]Ry^2 = (0)$ and $[x^2,r]^*Ry^2 = (0)$ for all $x,y \in J$ and $r \in R$. Since R is $*$-prime, this equation yields $x^2 \in Z(R)$ for all $x \in J$ or $y^2 = 0$ for all $y \in J$. Hence, in both the cases we have $x^2 \in Z(R)$ for all $x \in J$. Therefore $xy + yx \in Z(R)$ and replacing x by $2x^2$, 2-torsion freeness forces $x^2y \in Z(R)$ for all $x,y \in J$. Accordingly, we have $x^2[y,r] = 0$ for all $x,y \in J$ and $r \in R$. Since $x^2 \in Z(R)$, we get,

$$x^2R[y,r] = (0) \quad \text{and} \quad x^2R[y,r]^* = (0) \quad \text{for all } x,y \in J \text{ and } r \in R.$$

Once again using the $*$-primeness, we have $J \subseteq Z(R)$, in which case R is commutative, or $x^2 = 0$ for all $x \in J$. But the last case forces $d(x^2) = 0$ for all $x \in J$ and in view of ([440], Lemma 3) concludes that R is commutative.

♣

3.7 A particular class of derivations

This section is devoted to study some special class of derivations acting as a homomorphism or an anti-homomorphism on an appropriate subset of the considered ring.

Theorem 3.7.1 *[471, Theorem 1] Let R be a 2-torsion free $*$-prime ring, d a derivation which commutes with $*$ and J a nonzero $*$-Jordan ideal and a subring of R. If d acts as a homomorphism or as an anti-homomorphism on J, then $d = 0$ or $J \subseteq Z(R)$.*

Proof. Assume that $d(xy) = d(x)d(y)$ for all $x, y \in J$. This implies

$$d(x)y + xd(y) = d(x)d(y) \quad \text{for all } x, y \in J. \tag{3.106}$$

Replacing y by yz in (3.106) and using (3.106), we obtain $(d(x) - x)yd(z) = 0$ for all $x, y, z \in J$. Thus

$$(d(x) - x)Jd(z) = (0) \quad \text{for all } x, z \in J. \tag{3.107}$$

Since d commutes with $*$ and $J^* = J$, then (3.107) yields

$$(d(x) - x)Jd(z)^* = (0) \quad \text{for all } x, z \in J. \tag{3.108}$$

Applying Lemma 3.4.1, from (3.107) and (3.108) it follows that $d(z) = 0$ for all $z \in J$ or $d(x) = x$ for all $x \in J$. If $d(x) = x$ for all $x \in J$; then from $d(xy) = xy$ we find, due to 2-torsion freeness, $xy = 0$ for all $x, y \in J$. Since $x(r \circ y) = 0$, we get $xry = 0$ for all $x, y \in J, r \in R$, whence it follows that

$$xRy = (0) = xRy^* \quad \text{for all } x, y \in J. \tag{3.109}$$

Since R is $*$-prime, equation (3.109) contradicts the fact that $(0) \neq J$. Hence, $d(z) = 0$ for all $z \in J$ so that $d(J) = (0)$ and, by Lemma 3.4.3, $d = 0$ or $J \subseteq Z(R)$.

Let us now assume that d acts as an anti-homomorphism on J, then,

$$d(y)d(x) = d(x)y + xd(y) \quad \text{for all } x, y \in J. \tag{3.110}$$

Replacing x by xy in (3.110), we get,

$$d(y)xd(y) = xyd(y) \quad \text{for all } x, y \in J. \tag{3.111}$$

Substituting zx for x in (3.111) and using (3.111) we get $[d(y),z]xd(y) = 0$ in such that

$$[d(y),z]Jd(y) = (0) \quad \text{for all } y,z \in J. \tag{3.112}$$

Since d commutes with $*$, because of Lemma 3.4.1, equation (3.112) implies that

$$\text{for all } y \in J \cap Sa_*(R) \text{ either } d(y) = 0 \text{ or } [d(y),z] \text{ for all } z \in J.$$

Let $y \in J$, since $y^* - y \in J \cap Sa_*(R)$, then $d(y^* - y) = 0$ or $[d(y^* - y), J] = (0)$.

If $d(y^* - y) = 0$, as d commutes with $*$, then $d(y) \in Sa_*(R)$ and equation (3.112) implies that $d(y) = 0$ or $[d(y),J] = 0$. If $[d(y^* - y),J] = 0$, then $[d(y^*),z] = [d(y),z]$ for all $z \in J$. Substituting y^* for y in (3.112), we get,

$$[d(y),z]Jd(y^*) = (0) \quad \text{for all } z \in J. \tag{3.113}$$

Since d commutes with $*$, then (3.113) becomes

$$[d(y),z]J(d(y))^* = (0) \quad \text{for all } z \in J. \tag{3.114}$$

In view of equations (3.112) and (3.114), Lemma 3.4.1 yields $d(y) = 0$ or $[d(y),J] = (0)$. In conclusion, we have $d(y) = 0$ or $[d(y),J] = (0)$ for all $y \in J$.

Let us consider $J_1 = \{y \in J \mid d(y) = 0\}$ and $J_2 = \{y \in J \mid [d(y),J] = 0\}$; it is clear that J_1 and J_2 are additive subgroups of J such that $J = J_1 \cup J_2$. But a group can not be a union of two of its proper subgroups so that $J = J_1$ or $J = J_2$. If $J = J_1$, then $d(J) = 0$ and Lemma 3.4.3 forces $d = 0$ or $J \subseteq Z(R)$. Suppose $J = J_2$, then

$$[d(x),y] = 0 \quad \text{for all } x,y \in J. \tag{3.115}$$

Replacing x in (3.115) by xy, we get

$$x[d(y),y] + [x,y]d(y) = 0 \quad \text{for all } x,y \in J. \tag{3.116}$$

Substituting zx for x in (3.116) we obtain $[z,y]xd(y) = 0$, thus

$$[z,y]Jd(y) = (0) \quad \text{for all } y,z \in J.$$

Reasoning as above, the above equation leads to $d(y) = 0$ or $[y,J] = (0)$ for all $y \in J$. Consider $U_1 = \{y \in J \mid d(y) = 0\}$ and $U_2 = \{y \in J \mid [y,J] = (0)\}$; clearly U_1 and U_2 are additive subgroups of J such that $J = U_1 \cup U_2$. Therefore$J = U_1$ or $J = U_2$. If $J = U_1$, then $d(J) = (0)$ and Lemma 3.4.3 forces $d = 0$ or $J \subseteq Z(R)$. If $J = U_2$, then $[J,J] = (0)$ and Lemma 3.4.2 yields $J \subseteq Z(R)$.

♣

Using the fact that a $*$-prime ring which admits a nonzero central $*$-ideal must be commutative, Theorem 3.7.1 yields the following result.

Theorem 3.7.2 *[471, Theorem 2] Let R be a 2-torsion free $*$-prime ring, d a nonzero derivation commuting with $*$ and I a nonzero $*$-ideal of R. If either d acts as a homomorphism or as an anti-homomorphism on I, then R is commutative.*

Lemma 3.7.3 *[470, Lemma 2.1] Let R be a prime ring with an involution of the second kind. Then, $*$ is centralizing iff R is commutative.*

Theorem 3.7.4 *[470, Theorem 3.1] Let R be a 2-torsion free prime ring with involution $*$ of the second kind. If d is a derivation of R, then the following assertions are equivalent:*
(1) $[d(x),d(x^*)]-[x,x^*]\in Z(R)$ for all $x\in R$;
(2) $[d(x),d(x^*)]+[x,x^*]\in Z(R)$ for all $x\in R$;
(3) R is commutative.

Moreover, if $d\neq 0$, then $[d(x),d(x^*)]\in Z(R)$ for all $x\in R$, implies that R is commutative.

Proof. It is obvious that (3) implies both of (1) and (2). So we need to prove that (1) $\Rightarrow$ (3) and (2) $\Rightarrow$ (3).

If $d=0$, then $[x,x^*]\in Z(R)$, so our theorem follows from Lemma 3.7.3. Accordingly, one can assume that $d\neq 0$.

Since (1) $\Rightarrow$ (3) we suppose

$$[d(x),d(x^*)]-[x,x^*]\in Z(R)\quad \text{for all } x\in R.$$

Linearizing this equation, we discover,

$$[d(x),d(y^*)]+[d(y),d(x^*)]-[x,y^*]-[y,x^*]\in Z(R)\quad \text{for all } x,y\in R. \tag{3.117}$$

Replacing y by yh where $h\in Z(R)\cap H(R)$ and using the last equation, we get,

$$[[d(x),y^*]+[y,d(x^*)],r]d(h)=0\quad \text{for all } r,x,y\in R, \tag{3.118}$$

and thus

$$[[d(x),y^*]+[y,d(x^*)],r]Rd(h)=0\quad \text{for all } r,x,y\in R.$$

Since R is prime, then either $d(h)=0$ or $[[d(x),y^*]+[y,d(x^*)],r]=0$.

If $d(h)=0$ for all $h\in Z(R)\cap H(R)$, then

$$d(z)=0\quad \text{for all } z\in Z(R).$$

Substituting yz for y, where $z\in Z(R)$, we have

$$z^*[d(x),d(y^*)]+[d(y),d(x^*)]z-[x,y^*]z^*+[y,x^*]z\in Z(R).\quad \text{for all } x,y\in R.$$

Invoking (3.117), the last equation yields

$$[[d(x),d(y)]-[x,y],r](z^*-z)=0\quad \text{for all } r,x,y\in R,$$

and thus

$$[[d(x),d(y)]-[x,y],r]R(z^*-z)=(0) \quad \text{for all } r,x,y\in R.$$

Since the involution is of the second kind, then taking $y=x^2$ we get

$$[[d(x^2),d(x)],x]=0. \quad \text{for all } x\in R$$

and [355, Theorem 1.1] concludes, R is commutative. If $[[d(x),y^*]+[y,d(x^*)],r]=0$ for all $r,x,y\in R$, replacing y by yz where $z\in Z(R)$ we get,

$$[[d(x),y^*],r]z^*+[[y,d(x^*)],r]z=0 \quad \text{for all } r,x,y\in R.$$

Then we obtain

$$[[d(x),y],r](z^*-z)=0 \quad \text{for all } r,x,y\in R$$

so that

$$[[d(x),y],r]R(z^*-z)=(0) \quad \text{for all } r,x,y\in R.$$

Since the involution is of the second kind, the last equation becomes

$$[[d(x),y],r]=0 \quad \text{for all } r,x,y\in R.$$

Replacing y by ry we obtain

$$[d(x),r][y,r]=0 \quad \text{for all } r,x,y\in R.$$

Now replacing y by yt where $t\in R$, we get,

$$[d(x),r]y[t,r]=0 \quad \text{for all } r,t,x,y\in R$$

and the primeness of R forces

$$[d(x),x]=0 \quad \text{for all } x\in R.$$

By Posner's Theorem, we conclude that R is commutative.

Since (2) $\Rightarrow$ (3) we suppose

$$[d(x),d(x^*)]+[x,x^*]\in Z(R) \quad \text{for all } x\in R.$$

Linearizing this equation, we get

$$[d(x),d(y^*)]+[d(y),d(x^*)]+[x,y^*]+[y,x^*]\in Z(R) \quad \text{for all } x,y\in R.$$

Substituting yh for y where $h\in Z(R)\cap H(R)$ and using the last equation, we obtain

$$[[d(x),y^*]+[y,d(x^*)],r]d(h)=0 \quad \text{for all } r,x,y\in R.$$

Since this equation is the same as (3.118), arguing as above, we conclude that R is commutative.

If $d \neq 0$ and $[d(x), d(x^*)] \in Z(R)$ for all $x \in R$, replacing x by $x+y$ we obtain

$$[d(x), d(y^*)] + [d(y), d(x^*)] \in Z(R) \quad \text{for all } x, y \in R.$$

Substituting yh for y where $h \in Z(R) \cap H(R)$, we have

$$[[d(x), y^*] + [y, d(x^*)], r]d(h) = 0 \quad \text{for all } r, x, y \in R.$$

That is identical to equation (3.118), so we may argue as before that R is commutative.

♣

Corollary 3.7.1 *[14, Theorem 2.6] Let R be a 2-torsion free prime ring with an involution $*$ of the second kind. If R admits a derivation d such that $d(x)d(x^*) \pm xx^* = 0$ for all $x \in R$, then R is commutative.*

Corollary 3.7.2 *Let R be a 2-torsion free prime ring with involution $*$ of the second kind. Let d be a derivation of R, then the following assertions are equivalent:*
(1) $[d(x), d(y)] - [x, y] \in Z(R)$ for all $x, y \in R$;
(2) $[d(x), d(y)] + [x, y] \in Z(R)$ for all $x, y \in R$;
(3) R is commutative.

Moreover, if $d \neq 0$, then $[d(x), d(y)] \in Z(R)$ for all $x, y \in R$, implies that R is commutative.

3.8 Algebraic identities involving a pair of derivations

In [470, Theorem 3.7], it is proved that if a 2-torsion free prime ring with an involution of the second kind $(R, *)$ admits a nonzero derivation d such that $[d(x), x^*] \in Z(R)$ for all $x \in R$, then R is commutative. Now it is natural to ask what can we say if we consider the more general condition $d(x)x^* - x^*g(x) \in Z(R)$ for all $x, y \in R$. The purpose of the following theorem is to treat this problem.

Theorem 3.8.1 *[441, Theorem 1] Let $(R, *)$ be a 2-torsion free noncommutative prime ring with an involution of the second kind. If R admits derivations d and g satisfying $d(x)x^* - x^*g(x) \in Z(R)$ for all $x \in R$, then $d = g = 0$.*

Proof. Assume that

$$d(x)x^* - x^*g(x) \in Z(R) \quad \text{for all } x \in R. \tag{3.119}$$

Linearizing this equation, we can see that

$$d(x)y^* + d(y)x^* - x^*g(y) - y^*g(x) \in Z(R) \quad \text{for all } x, y \in R.$$

Accordingly,

$$d(x)y+d(y^*)x^*-x^*g(y^*)-yg(x)\in Z(R)\quad \text{for all } x,y\in R. \tag{3.120}$$

Substituting yh for y, where $h\in Z(R)\cap H(R)$, we obtain

$$d(x)yh+d(y^*)x^*h+y^*x^*d(h)-x^*g(y^*)h-x^*y^*g(h)-yg(x)h\in Z(R)\ \text{ for all } x,y\in R.$$

Using the last two equations, we find that

$$y^*x^*d(h)-x^*y^*g(h)\in Z(R)\quad \text{for all } x,y\in R. \tag{3.121}$$

Replacing y by h, where $h\in Z(R)\cap H(R)$, we get

$$x(d(h)-g(h))\in Z(R)\quad \text{for all } x\in R.$$

Therefore $x\in Z(R)$ or $g(h)=d(h)$. If $x\in Z(R)$ for all $x\in R$, the ring R must be commutative, is a contradiction. Suppose that $g(h)=d(h)$, equation (3.121) yields

$$[y,x]d(h)\in Z(R)\quad \text{for all } x,y\in R.$$

Then, either $[y,x]\in Z(R)$ so that R is commutative or $d(h)=0$. Since R is not commutative, then we get $d(h)=0=g(h)$ for all $h\in H(R)\cap Z(R)$. Replacing y by h in (3.120), it follows that $d(x)-g(x)\in Z(R)$ that is $D(x)\in Z(R)$ for all $x\in R$ where D is the derivation defined by $D(x)=d(x)-g(x)$. If $d\neq g$ then D is a nonzero centralizing derivation and R is commutative by Posner's theorem's, contradiction. Accordingly, we have $d=g$ and equation (3.119) reduces to

$$[d(x),x^*]\in Z(R)\quad \text{for all } x\in R.$$

Hence, d is $*$-centralizing in which case [470, Theorem 3.7] assures that $d=0$.

♣

Remark 3.8.2 *One can use Theorem 3.8.1 to characterize the kind of involution $*$. More precisely, if $(R,*)$ is a 2-torsion free noncommutative prime ring an involution with derivations d and g such that $d(x)x^*-x^*g(x)\in Z(R)$ for all $x\in R$ and either $d\neq 0$ or $g\neq 0$, then $*$ is of the first kind.*

Corollary 3.8.1 *Let $(R,*)$ be a 2-torsion free prime ring with an involution of the second kind and let d be a nonzero derivation of R. d is $*$-centralizing if and only if there exists a derivation g such that $d(x)x^*-x^*g(x)\in Z(R)$ for all $x\in R$.*

Corollary 3.8.2 *[470, Theorem 3.7] Let $(R,*)$ be a 2-torsion free prime ring with an involution of the second kind. If R admits a nonzero $*$-centralizing derivation, then R is commutative.*

Theorem 3.8.3 *[441, Theorem 2] Let $(R,*)$ be a 2-torsion free noncommutative prime ring with an involution of the second kind. There are no nonzero derivations d and g satisfying $d(x^*)x-x^*g(x)\in Z(R)$ for all $x\in R$.*

Proof. Suppose there exist nonzero derivations d and g such that

$$d(x^*)x - x^*g(x) \in Z(R) \quad \text{for all } x \in R. \tag{3.122}$$

Linearizing equation (3.122), we get,

$$d(x^*)y - x^*g(y) + d(y^*)x - y^*g(x) \in Z(R) \quad \text{for all } x, y \in R. \tag{3.123}$$

Writing yh instead of y in (3.123), where $h \in Z(R) \cap H(R) \backslash \{0\}$, we obtain

$$d(x^*)yh - x^*g(y)h - x^*yg(h) + d(y^*)xh + y^*xd(h) + y^*g(x)h \in Z(R) \quad \text{for all } x, y \in R.$$

Combining the last two equations, we obtain

$$y^*xd(h) - x^*yg(h) \in Z(R) \quad \text{for all } x, y \in R. \tag{3.124}$$

Taking $y = x$ in (3.124), yields,

$$x^*x(d(h) - g(h)) \in Z(R) \quad \text{for all } x \in R. \tag{3.125}$$

In view of the primeness hypothesis, it follows that $xx^* \in Z(R)$ or $d(h) = g(h)$.

If $xx^* \in Z(R)$ we conclude that R is commutative, a contradiction. Suppose that $d(h) = g(h)$, then equation (3.124) reduces to

$$(y^*x - x^*y)d(h) \in Z(R) \quad \text{for all } x, y \in R.$$

Since R is prime, we find that $y^*x - x^*y \in Z(R)$ or $d(h) = 0$. Suppose

$$y^*x - x^*y \in Z(R) \quad \text{for all } x, y \in R. \tag{3.126}$$

Replacing x by xs in (3.126), where $s \in Z(R) \cap S(R) \backslash \{0\}$, we obviously get

$$y^*xs + x^*ys \in Z(R) \quad \text{for all } x, y \in R.$$

According to the last two equations and using the 2-torsion freeness hypothesis, we arrive at $y^*x \in Z(R)$ for all $x, y \in R$, which implies that R is commutative. This is impossible. Assume that $d(h) = 0$ for all $h \in Z(R) \cap H(R) \backslash \{0\}$, then $g(h) = 0$, by [470, Fact 1] we get $d(z) = g(z) = 0$ for all $z \in Z(R)$. In equation (3.123), replacing y by h, we discover

$$d(x^*) - g(x) \in Z(R) \quad \text{for all } x \in R. \tag{3.127}$$

Putting s instead of x in (3.123), where $s \in Z(R) \cap S(R) \backslash \{0\}$, we get,

$$d(x^*) + g(x) \in Z(R) \quad \text{for all } x \in R.$$

Combining the last equation and using the 2-torsion freeness hypothesis, we can see that $d(x) \in Z(R)$ for all $x \in R$, then $d = 0$. Hence, the last equation becomes $g(x) \in Z(R)$ for all $x \in R$, therefore $g = 0$, a contradiction.

♣

Corollary 3.8.3 *Let $(R,*)$ be a 2-torsion free prime ring with an involution of the second kind. If R has nonzero derivations d and g, then the following assertions are equivalent:*
(1) $d(x^*)x - x^*g(x) \in Z(R)$ for all $x \in R$.
(2) R is a commutative integral domain.

In [470, Theorems 3.5, 3.8], it is proved that if $(R,*)$ is a 2-torsion free prime ring with involution provided with a derivation d which satisfies $d(x) \circ d(x^*) \pm x \circ x^* \in Z(R)$ for all $x \in R$ or $[d(x), d(x^*)] \pm x \circ x^* \in Z(R)$ for all $x \in R$, then R must be commutative. Motivated by [470], the following theorem treats the same identities for a pair of derivations.

Theorem 3.8.4 *[441, Theorem 3] Let $(R,*)$ be a 2-torsion free prime ring with an involution of the second kind, d and g derivations of R, then the following assertions are equivalent:*
(1) $d(x) \circ g(x^*) - x \circ x^* \in Z(R)$ for all $x \in R$
(2) $d(x) \circ g(x^*) + x \circ x^* \in Z(R)$ for all $x \in R$
(3) $[d(x), g(x^*)] - x \circ x^* \in Z(R)$ for all $x \in R$
(4) $[d(x), g(x^*)] + x \circ x^* \in Z(R)$ for all $x \in R$
(5) R is a commutative integral domain.

Proof. It is clear that (5) implies both (1), (2), (3) and (4).

In a nontrivial sense if $d = 0$ or $g = 0$ then the relations (1)–(4) reduce to $x \circ x^* \in Z(R)$ for all $x \in R$, and R is commutative by [470, Lemma 2.2]. Hence we may assume that $d \neq 0$ and $g \neq 0$.

Since $(1) \Longrightarrow (5)$ suppose that

$$d(x) \circ g(x^*) - x \circ x^* \in Z(R) \quad \text{for all } x \in R.$$

Linearizing this equation, we obtain

$$d(x) \circ g(y^*) + d(y) \circ g(x^*) - x \circ y^* - y \circ x^* \in Z(R) \quad \text{for all } x, y \in R. \tag{3.128}$$

Substituting yh for y, where $h \in Z(R) \cap H(R) \backslash \{0\}$, we get,

$$(d(x) \circ y^*)g(h) + (y \circ g(x^*))d(h) \in Z(R) \quad \text{for all } x, y \in R. \tag{3.129}$$

Substituting y with h, where $h \in Z(R) \cap H(R) \backslash \{0\}$, we obtain

$$d(x)g(h) + g(x^*)d(h) \in Z(R) \quad \text{for all } x \in R. \tag{3.130}$$

Putting $y = s$ in (3.129), where $s \in Z(R) \cap S(R) \setminus \{0\}$, we get,

$$-d(x)g(h) + g(x^*)d(h) \in Z(R) \quad \text{for all } x \in R.$$

Comparing the last two equations as expected we get

$$g(x)d(h) \in Z(R) \quad \text{for all } x \in R. \tag{3.131}$$

In view of the primeness of R, one can easily verify that

$$g(x) \in Z(R) \quad \text{or} \quad d(h) = 0.$$

If $g(x) \in Z(R)$ for all $x \in R$, then R is commutative. Suppose that $d(h) = 0$ for all $h \in Z(R) \cap S(R) \setminus \{0\}$, by [470, Fact 1], we are given that $d(z) = 0$ for all $z \in Z(R)$. Equation (3.130) reduces to

$$d(x)g(h) \in Z(R) \quad \text{for all } x \in R. \tag{3.132}$$

Since R is prime, we find that $d(x) \in Z(R)$ or $g(h) = 0$. If $d(x) \in Z(R)$ for all $x \in R$, it assures the commutativity of R. Now assume that $g(h) = 0$, by [[470], Fact 1] which means that $g(z) = 0$ for all $z \in Z(R)$. Replacing y by h in (3.128), it is easy to prove that

$$-x - x^* \in Z(R) \quad \text{for all } x \in R. \tag{3.133}$$

Replacing y by s in (3.128), where $s \in Z(R) \cap S(R) \setminus \{0\}$, we find that

$$x - x^* \in Z(R) \quad \text{for all } x \in R. \tag{3.134}$$

Adding relations (3.133) and (3.134) we can discover, $x \in Z(R)$. Hence, we conclude that R is commutative.

$(2) \Longrightarrow (5)$ Suppose that

$$d(x) \circ g(x^*) + x \circ x^* \in Z(R) \quad \text{for all } x \in R. \tag{3.135}$$

Substituting x with $x + y$ in (3.135), we get,

$$d(x) \circ g(y^*) + d(y) \circ g(x^*) + x \circ y^* + y \circ x^* \in Z(R) \quad \text{for all } x, y \in R. \tag{3.136}$$

Putting $y = yh$ in (3.136) where $h \in Z(R) \cap H(R) \setminus \{0\}$, and invoking (3.136), we get,

$$(d(x) \circ y^*)g(h) + (y \circ g(x^*))d(h) \in Z(R) \quad \text{for all } x, y \in R. \tag{3.137}$$

Using similar arguments as in the first case, we conclude that R is commutative.

Since $(3) \Longrightarrow (5)$, we assume

$$[d(x), g(x^*)] - x \circ x^* \in Z(R) \quad \text{for all } x \in R.$$

A linearization of this equation forces

$$[d(x) \circ g(y^*)] + [d(y) \circ g(x^*)] - x \circ y^* - y \circ x^* \in Z(R) \quad \text{for all } x, y \in R. \tag{3.138}$$

Replacing y with yh in (3.138) where $h \in Z(R) \cap H(R) \backslash \{0\}$, and using the last equation, we get,

$$[d(x), y^*]g(h) + [y, g(x^*)]d(h) \in Z(R) \quad \text{for all } x, y \in R. \tag{3.139}$$

Substituting ys for y in (3.139) where $s \in Z(R) \cap S(R) \backslash \{0\}$, we have

$$-[d(x), y^*]g(h) + [y, g(x^*)]d(h) \in Z(R) \quad \text{for all } x, y \in R. \tag{3.140}$$

Combining equations (3.140) and (3.139), we get,

$$[d(x), y^*]g(h) \in Z(R) \quad \text{for all } x, y \in R. \tag{3.141}$$

In light of primeness, it follows that either $[d(x), y^*] \in Z(R)$ or $g(h)$. If $[d(x), y^*] \in Z(R)$ for all $x, y \in R$, then R is commutative by Posner's Theorem. Hence we may assume that $g(h) = 0$ for all $h \in Z(R) \cap H(R) \backslash \{0\}$, by [470, Fact 1] we obtain $g(z) = 0$ for all $z \in Z(R)$. The relation (3.139) reduces to

$$[y, g(x^*)]d(h) \in Z(R) \quad \text{for all } x, y \in R. \tag{3.142}$$

Since R is prime, thereby, we obtain,

$$[y, g(x^*)] \in Z(R) \quad \text{or} \quad d(h) = 0$$

In the first case R is commutative by Posner's Theorem. Thus, we need only consider the case $d(h) = 0$ for all $h \in Z(R) \cap H(R) \backslash \{0\}$. By [470, Fact 1] we get $d(z) = 0$ for all $z \in Z(R)$. Replacing y by h in (3.138), we can see that

$$-x - x^* \in Z(R) \quad \text{for all } x \in R. \tag{3.143}$$

Replacing y by ys in (3.138), $s \in Z(R) \cap S(R) \backslash \{0\}$, we discover that

$$x - x^* \in Z(R) \quad \text{for all } x \in R. \tag{3.144}$$

Combining (3.143) and (3.144), we get $x \in Z(R)$. As a consequence of R must be commutative.

Since (4) $\Longrightarrow$ (5), suppose

$$[d(x), g(x^*)] + x \circ x^* \in Z(R) \quad \text{for all } x \in R.$$

Linearizing this equation, we discover,

$$[d(x), g(y^*)] + [d(y), g(x^*)] + x \circ y^* + y \circ x^* \in Z(R) \quad \text{for all } x, y \in R. \tag{3.145}$$

Substituting y with yh in (3.145), $h \in Z(R) \cap H(R) \backslash \{0\}$, and using (3.145) we get,

$$[d(x), y^*]g(h) + [y, g(x^*)]d(h) \in Z(R) \quad \text{for all } x, y \in R. \tag{3.146}$$

Since this equation is the same as (3.139), then reasoning as above we conclude that R is commutative. This completes our theorem.

♣

Corollary 3.8.4 *Let $(R, *)$ be a 2-torsion free prime ring with involution of the second kind, d and g be derivations of R, then the following assertions are equivalent:*
(1) $d(x) \circ g(y) - x \circ y \in Z(R)$ for all $x, y \in R$
(2) $d(x) \circ g(y) + x \circ y \in Z(R)$ for all $x, y \in R$
(3) $[d(x), g(y)] - x \circ y \in Z(R)$ for all $x, y \in R$
(4) $[d(x), g(y)] + x \circ y \in Z(R)$ for all $x, y \in R$
(5) R is a commutative integral domain.

The following example proves that the primeness hypothesis in Theorems 3.8.1 and 3.8.3 is necessary. In particular, our theorems cannot be extended to semiprime rings.

Example 3.8.1 *[441, Example 1] Let's consider $R = \mathbb{Q}[X] \times M_2(\mathbb{C})$, and define the involution on $M_2(\mathbb{C})$ by*

$$\tau : \quad M_2(\mathbb{C}) \longrightarrow M_2(\mathbb{C})$$
$$\begin{pmatrix} a & b \\ c & d \end{pmatrix} \longmapsto \begin{pmatrix} \bar{a} & \bar{c} \\ \bar{b} & \bar{d} \end{pmatrix}$$

Then it can be verified that $(M_2(\mathbb{C}), \tau)$ is a prime ring with an involution of the second kind. Let's define an involution $*$ of the second kind on R setting $(P, M)^* = (P, \tau(M))$ for all $(P, M) \in R$. Defining $d(P, M) = (P', 0)$ and $g(P, M) = (-P', 0)$ for all $(P, M) \in R$ it becomes clear that $(R, *)$ is a noncommutative semiprime ring and d, g are nonzero derivations of R satisfying the condition of Theorem 3.8.4. Additionally, we have $d(P, M)(P, M)^* - (P, M)^* g(P, M) \in Z(R)$ for all $(P, M) \in R$, however $d \neq g$.

The next example proves that "$*$ is of the second kind" which is necessary in Theorem 3.8.4.

Example 3.8.2 *[441, Example 2] Let us consider $R = M_2(\mathbb{Z})$ and $\begin{pmatrix} a & b \\ c & d \end{pmatrix}^* = \begin{pmatrix} d & -b \\ -c & a \end{pmatrix}$. It is can be verified that $(R, *)$ is a prime ring with an involution of the first kind. Moreover, we define derivations d and g by $d\begin{pmatrix} a & b \\ c & d \end{pmatrix} =$*

$\begin{pmatrix} 0 & -b \\ c & 0 \end{pmatrix}$ *and* $g\begin{pmatrix} a & b \\ c & d \end{pmatrix} = \begin{pmatrix} 0 & b \\ -c & 0 \end{pmatrix}$. *Then, d and g satisfy the condition* $d(x) \circ g(x^*) - x \circ x^* \in Z(R)$ *for all* $x \in R$. *However, R is a noncommutative ring.*

The following example proves that the primeness hypothesis in Theorem 3.8.4 is not superfluous. In particular, our theorem cannot be extended to semiprime rings.

Example 3.8.3 *[441, Example 3] Let us consider* $\mathcal{R} = R \times \mathbb{C}$, *where* R, $*$ *are in the preceding exaample. Obviously* $(\mathcal{R}, \sigma)$ *is a semiprime ring with an involution of the second kind where* $\sigma(x,z) = (x^*, \bar{z})$ *for all* $(x,z) \in \mathcal{R}$. *Furthermore, if d is the zero derivation and g any arbitrary derivation of* $\mathcal{R}$, *then* $d(X) \circ g(\sigma(X)) - X \circ \sigma(X) \in Z(\mathcal{R})$ *for all* $X \in \mathcal{R}$, *but* $\mathcal{R}$ *is not commutative.*

Chapter 4

Generalized Derivations in Rings

4.1 Introduction

The notion of generalized derivation was first introduced by Brešar [134] in 1991, and in 1998, Hvala [313] initiated an algebraic study of these so-called generalized derivations. Analysts have delved into studying these maps within the context of algebras on specific normed spaces. In [313], Hvala provided an algebraic exploration of generalized derivations of prime rings. Let R be a prime ring, and an additive mapping $F : R \to R$ is termed a generalized derivation of R if there exists a derivation d of R such that $F(xy) = F(x)y + xd(y)$ for all $x, y \in R$. Basic examples include derivations, generalized inner derivations (i.e., maps of the form $x \mapsto ax + xb$ for some $a, b \in R$), and right R-module mappings from R into itself. Hvala addressed three key problems in the study of generalized derivations: the product of two generalized derivations, generalized derivations satisfying $[F_1(x), F_2(x)] = 0$, and generalized derivations with nilpotent values.

In 1999, Lee [378] undertook a study on the extension problem of generalized derivations and provided a characterization for them. Based on this characterization, Lee extended Hvala's [313, Theorem 31] results to the context of semiprime rings without requiring torsion-free assumptions. He also achieved similar outcomes for generalized derivations with nilpotent values on Lie ideals or one-sided ideals. Lee [378] expanded Hvala's [313] definition of generalized derivations by defining them as additive mappings $F : I \to U$ such that $F(x) = F(x)y + xd(y)$ for all $x, y \in I$, where I is a dense right ideal of R, and d is a derivation from I into the right quotient ring U.

Furthermore, Lee demonstrated that every generalized derivation could be uniquely extended to a generalized derivation of U. Specifically, there exists $a \in U$ and a derivation d of U such that $F(x) = ax + d(x)$ for all $x \in U$. In the works of Kharachenko [345, 346], identities with derivations were described, providing potent tools for reducing differential identities to generalized polynomial identities. Therefore, to explore identities with generalized derivations, a corresponding theorem for such identities seems reasonable. In 2001, Lee and Shiue [391] established a version of Kharachenko's theorem for generalized derivations, obtaining results related to specific identities with generalized derivations. As applications to the structure theorem, Lee and Shiue [391] extended certain results obtained by Hvala [313, Theorems 1 and 2] to prime rings without the characteristic assumption.

Researchers have investigated the existence, uniqueness, and properties of derivations in different classes of rings, including prime rings, semiprime rings, and various types of algebras. The study of these derivations also involves exploring their impact on the structure of rings, such as their role in determining commutativity, centralizers, or other algebraic properties. A natural question arises: Do the results obtained for derivations also hold for generalized derivations on prime and semiprime rings, and various types of algebras? If so, these findings can be applied to the study of elementary operators on non-commutative Banach algebras. This chapter delves into the structure of rings with generalized derivations, generalized skew-derivations, b-generalized derivations, and their respective properties.

Overall, the study of generalized derivations represents a fruitful area of research within ring theory, providing insights into the intricate connections between algebraic structures and their associated derivations.

4.2 Product of two generalized derivations of rings

We begin our discussion with the generalized inner derivations definition:

Definition 37 (Generalized inner derivation) *An additive mapping $F : R \to R$ is called a* **generalized inner derivation** *if $F(x) = ax + xb$ for some $a, b \in R$.*

It is straight forward to note that if F is a generalized inner derivation, then for any $x, y \in R$,

$$F(xy) = F(x)y + x[y, b] = F(x)y + xI_b(y)$$

where I_b is an inner derivation.

Motivated by the above observations, Brešar [134] introduced the concept of a generalized derivation in rings as follows:

Definition 38 (Generalized derivation) *An additive mapping $F : R \to R$ is called a* **generalized derivation** *of a ring R if there exists a derivation $d : R \to R$ such that*

$$F(xy) = F(x)y + xd(y), \quad \text{where } x, y \in R.$$

Example 4.2.1 *Let S be any ring and $R = \left\{ \begin{pmatrix} a & b \\ 0 & 0 \end{pmatrix} \mid a, b \in S \right\}$. Define $F : R \to R$ such that $F(x) = 2e_{11}x - xe_{11}$. Then, F is a generalized derivation with the associated derivation $d(x) = e_{11}x - xe_{11}$.*

Thus, the notion of a generalized derivation encompasses both the idea of a derivation and a generalized inner derivation. Additionally, a generalized derivation with $d = 0$ includes the concept of a left multiplier, which is an additive map that satisfies $F(xy) = F(x)y$ for all $x, y \in R$.

Remark 4.2.1 *The following example is sufficient to demonstrate that a generalized derivation need not be a derivation.*

Example 4.2.2 *Let $R = \left\{ \begin{pmatrix} a & b \\ 0 & c \end{pmatrix} \mid a, b, c \in \mathbb{Z}_2 \right\}$. Define a map $F : R \to R$ by $F\begin{pmatrix} a & b \\ 0 & c \end{pmatrix} = \begin{pmatrix} a & 0 \\ 0 & 0 \end{pmatrix}$ and a derivation $d : R \to R$ by $d\begin{pmatrix} a & b \\ 0 & c \end{pmatrix} = \begin{pmatrix} 0 & b \\ 0 & 0 \end{pmatrix}$. Then it can be verified that F is a generalized derivation but not a derivation.*

Remark 4.2.2 *[134, Remark 1] Let R be a ring and $d : R \to R$ be a function and $F : R \to R$ be an additive mapping satisfying*

$$F(xy) = F(x)y + xd(y) \text{ for all } x, y \in R. \tag{4.1}$$

We intend to show that under rather mild assumptions d must be a derivation.

On one hand, we have

$$F(xyz) = F(x(yz)) = F(x)yz + xd(yz) \text{ for all } x, y, z \in R$$

and on the other hand,

$$F(xyz) = F((xy)z) = F(x)yz + xd(y)z + xyd(z) \text{ for all } x, y, z \in R.$$

Comparing the above two expressions, we obtain

$$x(d(yz) - d(y)z - yd(z)) = 0 \text{ for all } x, y, z \in R.$$

Similarly, comparing $F(x(y+z))$ in two ways, one can show that

$$x(d(y+z)-d(y)-d(z))=0 \text{ for all } x,y,z\in R.$$

If R has the property that $Ra=\{0\}$ implies that $a=0$. In particular, if R is a semiprime ring, then d is a derivation.

Remark 4.2.3 *Suppose a ring R has a unit element 1 and $x=1$ in (4.1). Then we get $F(y)=F(1)y+d(y)$ for all $y\in R$. Hence, every generalized derivation of R is an inner generalized derivation iff every derivation of R is an inner.*

We now express the essential remark. It can be confirmed through straightforward computations.

Remark 4.2.4 *Let R be a ring and $d_1,d_2:R\to R$ be a derivations and $F:R\to R$ be a generalized derivation. A mapping $\Delta=d_1d_2-F$ then satisfies the identity*

$$\Delta(xyz)-\Delta(xy)z-x\Delta(yz)+x\Delta(y)z=d_1(x)yd_2(z)+d_2(x)yd_1(z).$$

In 1998, Hvala [313] investigated the following result to extend a well-known theorem by Posner [487]. Posner's theorem states that if d_1 and d_2 are derivations of a prime ring with $char(R)\neq 2$ such that the iterate d_1d_2 is also a derivation of R, then at least one of d_1 or d_2 must be zero. Hvala derived the following theorem.

Theorem 4.2.5 *[313, Theorem 12] Let R be a prime ring of $char(R)\neq 2$ and $Q_r(R)$, the Martindale right ring of quotient with center C and central closure R_c. Suppose $F_1,F_2:R\to R$ are generalized derivations, the product $F_3=F_1F_2$ is a generalized derivation iff one of the following possibilities holds:*

- *(i) there exists $\gamma\in C$ such that either $F_1(x)=\gamma x$ or $F_2(x)=\gamma x$;*
- *(ii) there exists $a,b\in Q_r(R_c)$ such that either $F_1(x)=xa$ or $F_2(x)=xb$;*
- *(iii) there exists $a,b\in Q_r(R_c)$ such that either $F_1(x)=ax$ or $F_2(x)=bx$;*
- *(iv) there exists $a,b\in Q_r(R_c)$ and $\lambda,\mu\in C$ such that $F_1(x)=ax+xb$ and $F_2(x)=\lambda(x)+\mu(ax-xb)$.*

To establish the proof of the theorem mentioned above, we need the following results:

Lemma 4.2.6 *[138, Lemma 1] Let R be a prime ring with the right and symmetric Martindale ring of quotients $Q_r(R)$ and $Q_s(R)$ respectively with extended centroid C and central closure $R_c=RC$. If $a_i,b_i\in R$ satisfy $\sum a_ixb_i=0$ for all $x\in R$, then a_i's as well as b_i's are C-independent, unless all $a_i=0$ or all $b_i=0$.*

Another significant result that will emerge as a potent tool for the consideration of generalized derivations is the result of Brešar [138, Proposition 8].

Proposition 4.2.7 *[313, Proposition 1] Let R be a prime ring with the right and symmetric Martindale ring of quotients $Q_r(R)$ and $Q_s(R)$ respectively with extended centroid C and central closure $R_c = RC$. Suppose that*

$$\sum_{j=1}^{n} f_j(z)xa_j + \sum_{i=1}^{n} c_i z h_i(x) = 0 \text{ for all } x, z \in R,$$

where $a_j, c_i \in R$ and $f_j : R \to R$, $h_i : R \to R_c$ are random maps. If the sets $\{a_1, a_2, \ldots, a_n\}$ and $\{c_1, c_2, \ldots, c_k\}$ are C-independent then there exists $q_{ij} \in Q_r(R_c)$, $i = 1, 2, 3, \ldots, k$, $j = 1, 2, \ldots, n$, such that

$$f_j(z) = -\sum_{i=1}^{k} c_i z q_{ij}, \quad h_i(x) = \sum_{j=1}^{n} q_{ij} x a_j$$

for all $x, z \in R$, $i = 1, 2, \ldots, \quad k, j = 1, 2, \ldots, n$.

An examination of the proof reveals that the conclusion of Proposition 4.2.7 remains valid even if one of the a_i's or one of the c_i's is equal to 1, even in the absence of an identity element in R.

Lemma 4.2.8 *[313, Lemma 2] Let R be a prime ring with the right and symmetric Martindale ring of quotients $Q_r(R)$ and $Q_s(R)$ respectively with extended centroid C and central closure $R_c = RC$. Suppose $F : R \to R_c$ be an additive mapping satisfying $F(xy) = F(x)y$ for all $x, y \in R$. Then there exists $q \in Q_r(R_c)$ such that $F(x) = qx$ for all $x \in R$.*

Proof: Let us extend F from R to R_c according to $\overline{F}(\sum x_i\lambda_i) = \sum F(x_i)\lambda_i$, where $x_i \in R$ and $\lambda_i \in C$. In order to show that F is well defined it suffices to prove that $\sum x_i\lambda_i = 0$ yields $\sum F(x_i)\lambda_i = 0$. Let I be a nonzero ideal of R such that $\lambda_i I \subseteq R$ for every i. Pick $a \in I$ and note that the factors in the summation, $\sum x_i(\lambda_i a)$ lie in R. Therefore, we have $0 = \sum F(x_i)\lambda_i = (\sum F(x_i)\lambda_i)a$. Since this is true for all $a \in I$, we have $\sum F(x_i)\lambda_i = 0$. The fact that $\overline{F}(xy) = \overline{F}(x)y$ for all $x_i \in R_c$ can be verified by a direct computation. This proves that $\overline{F} : R_c \to R_c$ is a right R_c-module map, hence there exists $q \in Q_r(R_c)$ such that $\overline{F}(x) = qx$, $x \in R_c$. Since $\overline{F}$ is an extension of F, this proves the lemma.

♣

Lemma 4.2.9 *[313, Lemma 3] Let R be a prime ring with the right and symmetric Martindale ring of quotients $Q_r(R)$ and $Q_s(R)$ respectively with extended centroid C and central closure $R_c = RC$. Let R be a noncommutative ring. If $F : R \to C$ is a generalized derivation, then $F = 0$.*

Proof: Write $F(xy) = F(x)y + xd(y)$. As $F(xy)$ and $F(x)y$ commute with y, it follows that $xd(y)y = yxd(y)$, $x, y \in R$. Using Lemma 4.2.6, we obtain the following alternative, for every $y \in R$, we have either $y \in C$ or $d(y) = 0$. By the standard argument one of these must hold for all $y \in R$. Since the ring R is noncommutative, we have $d = 0$ and therefore $F(xy) = F(x)y, \ x, y \in R$. The fact that F maps into C and that R is a commutative ring makes this possible only when $F = 0$.

♣

Lemma 4.2.10 *[313, Lemma 4] Let R be a prime ring with the right and symmetric Martindale ring of quotients $Q_r(R)$ and $Q_s(R)$ respectively with extended centroid C and central closure $R_c = RC$. Suppose $a, b \in R$ and let $F : R \to R$ be defined by $F(x) = axb$. If F is a generalized derivation, then either $a \in C$ or $b \in C$.*

Proof: For $x, y \in R$, we have $axyb = axby + xd(y)$, where d is a derivation. Therefore

$$ax(yb - by) - xd(y) = 0 \text{ for all } x, y \in R.$$

Using Lemma 4.2.6 it follows at once that either $a \in C$ or $b \in C$.

♣

Proof of the Theorem 4.2.5: It is easy to verify that in each of these four cases, the product F_1F_2 is again a generalized derivation. Let us prove the converse.

Now, F_1, F_2 and $F_3 = F_1F_2$ are generalized derivations, so we have

$$F_i(xy) = F_i(x)y + xd_i(y), \ \text{ for all } x, y \in R, \ \textit{where } i = 1, 2, 3$$

for some derivation d_i, thus using $F_3(xy) = F_1(F_2(xy))$, we obtain

$$F_2(x)d_1(y) + F_1(x)d_2(y) + x(d_1d_2 - d_3)(y) = 0 \text{ for all } x, y \in R.$$

Replacing x by zx, we obtain,

$$F_1(z)xd_2(y) + F_2(z)xd_1(y) + z(d_2(x)d_1(y) + d_1(x)d_2(y) + x(d_1d_2 - d_3)(y)) = 0 \tag{4.2}$$

for all $x, y, z \in R$. Fix y and note that Proposition 4.2.7 gives us the following two possibilities:

(i) $d_1(y)$ and $d_2(y)$ are C-independent for all $y \in R$;

(ii) there exists $q_1, q_2 \in Q_r(R_c)$ such that $F_1(z) = -zq_1$ and $F_2(z) = -zq_2, \ z \in R$.

In case (ii) $a = -q_1$, $b = -q_2$ this is the second possibility (ii) of the theorem. Let us now consider case (i). Since $d_1(y)$ and $d_2(y)$ are C-independent, we have $[d_1(y), d_2(y)] = 0$ for all $y \in R$. As $char(R) \neq 2$, it follows from [361, Theorem 4], d_1 and d_2 are C-independent, i.e.,

$$\alpha_1 d_1 + \alpha_2 d_2 = 0 \text{ for some } \alpha_1, \ \alpha_2 \in C. \tag{4.3}$$

Let us first consider the case when $d_1 = 0$. This implies $F_1(xy) = F_1(x)y$ for all $x, y \in R$. It follows from Lemma 4.2.8 that there exists $a \in Q_r(R_c)$ such that $F_1(x) = ax$.

In this case, (i) can have an alternative form

$$azxd_2(y) - zxd_3(y) = 0 \text{ for all } x, y, z \in R. \tag{4.4}$$

In a similar manner we see that $d_2 = 0$ yields $F_2(x) = bx$ for $b \in Q_r(R_c)$ and all $x \in R$. This is third possibility (iii). If $d_2 \neq 0$ we can choose $x, y \in R$ such that $xd_2(y) \neq 0$. In this case, the relation (4.4) together with Lemma 4.2.6, gives us $a \in C$. Write $a = \gamma$ and we obtain $F_1(x) = \gamma x$, so that (i) holds.

Similarly we see that the possibility (i) appears in case $d_2 = 0$ and $d_1 \neq 0$; more precisely, we arrive at $F_2(x) = \gamma x$, $x \in R$.

Finally, let us consider the situation when $d_1 \neq 0$, $d_2 \neq 0$, so that $d_2 = rd_1$, for some nonzero $r \in R$. Define $F(x) = F_2(x) - rF_1(x)$ and note $F : R \to R_c$ and $F(xy) = F(x)y$, for all $x, y \in R$. Thus Lemma 4.2.8 implies the existence of $q \in Q_r(R_c)$ such that $F(x) = qx$. Therefore

$$F_2(x) = rF_1(x) + qx \quad \text{for all } x \in R. \tag{4.5}$$

Using this in (4.2), we obtain

$$(2rF_1(z) + qz)xd_1(y) + z(d_1(x)g(y) + xh(y)) = 0 \quad \text{for all } x, y, z \in R,$$

where $g(y) = -2rd_1(y)$ and $h = d_1d_2 - d_3$. Pick $y \in R$ such that $d_1(y) \neq 0$ and apply Proposition 4.2.7, then there exists $q' \in Q_r(R_c)$ such that

$$2rF_1(z) + qz = -zq'.$$

This gives us $F_2(x) = ax + xb$, where $a = -(2r)^{-1}q$, $b = -(2r)^{-1}q'$, $a, b \in Q_r(R_c)$. Using this result in (4.5), we obtain a similar result for the other function:

$$F_2(x) = (ra + q)x + xrb.$$

Now to compute the product F_1F_2, we obtain

$$F_3(x) = F_1(F_2(x)) = (ra^2 + aq)x + x(rb^2) + (2ra + q)xb.$$

Since both maps $x \mapsto ((ra^2 + aq)x + x(rb^2))$ and F_3 are generalized derivations so is their difference, i.e., the map $x \mapsto (2ra + q)xb$. In Lemma 4.2.10, we prove that this is possible iff one of the coefficients lies in C. Thus, we have either $2ra + q \in C$ or $b \in C$. It is easy to notice that the second possibility leads to the possibility (iii) of the theorem, so we consider the case $2ra + q = \lambda \in C$. Write $u = -r$, $q = \lambda + 2\mu a$, we obtain $F_2(x) = \lambda x + \mu(ax - xb)$. This is the fourth possibility (iv) of the theorem.

Remark 4.2.11 *In the case when the* $charR = 2$*, Theorem 4.2.5 is no longer true. If* F_2 *is an arbitrary generalized derivation and* $F_1(x) = \lambda x + \mu F_2(x)$*, for all* $\lambda, \mu \in C$*, then* F *and the product* F_1F_2 *are generalized derivation as well. Thus, we see that the choice of the generalized derivation whose product is a generalized derivation is limited to inner generalized derivations in case when* $char(R) \neq 2$ *and the choice is much wider when* $char(R) \neq 2$*. However, we will not investigate this isolated situation in this chapter. Therefore, in all the corollaries and the propositions during this section,* R *will be a prime ring and* $char(R) \neq 2$*.*

Corollary 4.2.1 *[313, Corollary 1] Let* R *be a prime ring with the right and symmetric Martindale ring of quotients* $Q_r(R)$ *and* $Q_s(R)$ *respectively with extended centroid* C *and central closure* $R_c = RC$*. If* $F_1, F_2 : R \to R$ *are generalized derivations, then* $F_1F_2 = 0$ *iff one of the following possibilities holds:*

(i) *either* $F_1 = 0$ *or* $F_2 = 0$*;*

(ii) *there exists* $a, b \in Q_r(R_c)$ *such that* $F_1(x) = xa$ *and* $F_2(x) = xb$ *and* $ba = 0$*;*

(iii) *there exists* $a, b \in Q_r(R_c)$ *such that* $F_1(x) = ax$ *and* $F_2(x) = bx$ *and* $ab = 0$*;*

(iv) *there exists* $a, b \in Q_r(R_c)$ *and* $\lambda, \mu \in C$ *such that* $F_1(x) = ax + xb$*,* $F_2(x) = \lambda x + \mu(ax - xb)$ *and* $\lambda a + \mu a^2 = \mu b^2 - \lambda b \in C$*.*

Proof: It is clear that in cases $(i) - (iv)$, we have $F_1F_2 = 0$. Let us prove the converse.

Since F_1F_2 is a (zero) generalized derivation, one of the four possibilities must holds from Theorem 4.2.5, i.e., (i) to (iv).

The possibility $F_1(x) = rx, r \in R$, implies $rF_2(x) = 0$ for all $x \in R$, therefore either $F_1 = 0$ or $F_2 = 0$. Then, $F_2(x) = rx$ for all $x \in C$ gives us $F_1(rx) = 0$ for all $x \in R$. Substituting xy for x in this relation, we get $rxd_1(y) = 0$ for all $x, y \in R$, which implies either $r = 0$. Therefore, $F_2 = 0$, or $d_1 = 0$. In case $r \neq 0$, the map $F_1 : R \to R$ is a right R-module homomorphism and there exists $q \in Q_r(R)$ such that $F_1(x) = qx$. But, then we have $qrx = rqx = 0$ for all $x \in R$. Therefore, $F_1 = 0$.

Since possibility (iii) of Theorem 4.2.5 holds, we have $F_1(F_2(x)) = xba = 0$, for all $x \in R$. Therefore, $ba = 0$. Similarly, we have $ab = 0$.

Let us finally consider case (iv) of Theorem 4.2.5. We have $F_1(F_2(x)) = (\lambda a + \mu a^2)x + x(\lambda b + \mu b^2) = 0$, for all $x \in R$, hence (using for instance, Lemma 4.2.6) $\lambda a + \mu a^2 = \mu b^2 - \lambda b \in C$.

♣

Corollary 4.2.2 *[313, Corollary 2] Let* R *be a prime ring with the right and symmetric Martindale ring of quotients* $Q_r(R)$ *and* $Q_s(R)$ *respectively with extended centroid* C *and central closure* $R_c = RC$*. Suppose* $F : R \to R$ *be general-*

ized derivations and $c, d \in R$. If we have $cF(x) + F(x)d = 0$ for all $x \in R$, then one of the following possibilities holds:

(*i*) *$c, d \in R$ and $c + d = 0$;*

(*ii*) *$c \in C$ and there exists $b \in Q_r(R_c)$ such that $F(x) = xb$ and $b(c+d) = 0$;*

(*iii*) *$d \in D$ and there exists $b \in Q_r(R_c)$ such that $F(x) = bx$ and $(c+d)b = 0$;*

(*iv*) *there exists $\lambda, \mu \in C$ such that $F(x) = \lambda x + \mu(cx - xd)$ and $\lambda c + \mu c^2 = \mu d^2 - \lambda d \in C$.*

Proof: Define $g(x) = cx + xd$ for all $x \in R$. Then we have $gF = 0$. According to Corollary 4.2.1, this is possible in one of the four cases. The possibility $g = 0$ is equivalent to $c, d \in C$, $c + d = 0$, and the possibility $F = 0$ is included in case (iv) of the Corollary 4.2.2 with $\lambda = \mu = 0$. Cases (ii) and (iii) of Corollary 4.2.1 can be easily transformed in cases (ii) and (iii) of Corollary 4.2.2. Similarly case (iv) of Corollary 4.2.1 gives us $F(x) = \lambda x + \mu(cx + xc)$. On knowing this, we write $cF(x) + F(x)d = 0$ and utilize Lemma 4.2.6 to derive $\lambda c + \mu c^2 = \mu d^2 - \lambda d \in C$.

♣

Remark 4.2.12 *One can observe that the Corollary 4.2.2 is of special interest in case $d = c$ or $d = -c$.*

Corollary 4.2.3 *[313, Corollary 3] Let R be a prime ring with the right and symmetric Martindale ring of quotients $Q_r(R)$ and $Q_s(R)$ respectively with extended centroid C and central closure $R_c = RC$. Suppose $F : R \to R$ is a generalized derivation and $c \in R - C$. If $F(x)c = cF(x)$ for all $x \in R$, then there exists $\lambda, \mu \in C$ such that $F(x) = \lambda x + \mu(cx + xc)$ and $\lambda c + \mu c^2 \in C$.*

Corollary 4.2.4 *[313, Corollary 4] Let R be a prime ring with the right and symmetric Martindale ring of quotients $Q_r(R)$ and $Q_s(R)$ respectively with extended centroid C and central closure $R_c = RC$. Suppose $F : R \to R$ is a generalized derivation and $c \in R - \{0\}$. If $F(x)c + cF(x) = 0$ for all $x \in R$, then there exists $\mu \in C$ such that $F(x) = \mu(cx - xc)$ and $\mu c^2 \in C$.*

It is well known that only the elements from the center $Z(R)$ of a prime ring R with $char(R) \neq 2$ can commute with the range of a nonzero derivation of R. This can be, for example, deduced from Posner's theorem mentioned above.

Remark 4.2.13 *Corollary 4.2.3 shows us that the result is not true for generalized derivations. For instance, if $c \in R - Z(R)$ and $c^2 = 0$, then $F(x) = cx + xc$ is a generalized derivation and c commutes with $F(R)$.*

However, the next result tells us that the choice of such elements is limited for nonzero generalized derivations, too.

Proposition 4.2.14 *[313, Proposition 2] Let R be a prime ring with the right and symmetric Martindale ring of quotients $Q_r(R)$ and $Q_s(R)$ respectively with extended centroid C and central closure $R_c = RC$. Suppose that $F : R \to R$ is a generalized derivation. If there exist elements $a,b \in R$ which commute with $F(R)$ such that $\{a,b,1\}$ is a C-independent set, then $F = 0$.*

Proof: It follows from the assumptions that we have $F(x)a = aF(x)$, $F(x)b = bF(x)$ for all $x \in R$. Since $\notin C$, Corollary 4.2.3 ensures the existence of $\lambda_a,\ \lambda_b,\ \mu_a,\ \mu_b \in C$ such that

$$F(x) = \lambda_a x + \ \mu_a(ax+xa) = \lambda_b x + \mu_b(bx+xb)\ x \in R, \tag{4.6}$$

and $\lambda_a a + \ \mu_a a^2 \in C,\ \lambda_b b + \mu_b b^2 \in C$. Relation (4.6) implies

$$((\lambda_a - \lambda_b) + \mu_a a + \mu_b b)x + x(\mu_a a + \mu_b b) = 0 \text{ for all } x \in R.$$

Lemma 4.2.6 tells us that $\mu_a a + \mu_b b \in C$. Since a,b and 1 are C-independent, we have $\mu_a = \mu_b = 0$. But then we have $\lambda_a a \in C$ and $\lambda_b b \in C$, which implies $\lambda_a = \lambda_b = 0$, therefore $F = 0$.

♣

In Theorem 4.2.5, Hvala considered two generalized derivations $F_1, F_2 : R \to R$ and described all the cases when the product $F_3 = F_1F_2$ is also a generalized derivation. In corollaries, until now, the author has analyzed the case when $F_3 = 0$. The following corollary discusses the case when $F_1 = F_2$.

Corollary 4.2.5 *[313, Corollary 4] Let R be a prime ring with the right and symmetric Martindale ring of quotients $Q_r(R)$ and $Q_s(R)$ respectively with extended centroid C and central closure $R_c = RC$. Suppose $F : R \to R$ is a generalized derivation. Then F^2 is a generalized derivation if and only if there exists $a \in Q_r(R_c)$ such that either $F(x) = ax$ for all $x \in R$, or $F(x) = xa$ for all $x \in R$. Moreover, we have $F^2 = 0$ iff F is of the form above and $a^2 = 0$.*

Proof: One approach is clear cut. Assume that F^2 is a generalized derivation and apply Theorem 4.2.5, $F = F_1 = F_2$. The possibilities $(i) - (iii)$ of the Theorem 4.2.5 can be unified in saying that F is either a left or right multiplication by an element from $Q_r(R_c)$. Let us now consider the possibility (iv). Theorem 4.2.5 gives us $F(x) = ax + xb = \lambda x + \mu(ax - xb)$, $a,b \in Q_r(R_c)$, $\lambda, \mu \in C$. From the equality above it follows that $((\mu - 1)a + \lambda)x - x(\mu + 1)b = 0, x \in R$, and Lemma 4.2.6 gives us $(\mu - 1)a + \lambda = (\mu + 1)b \in C$. This is possible only if either one of the elements a,b lie in C, or $\mu - 1 = \mu + 1 = 0$. Since $charR \neq 2$ the second result is not possible. So, F is again either a left or right multiplication by an element $a + b \in Q_r(R_c)$.

♣

A well-known result established by Herstein [297] asserts that if d is a derivation of a noncommutative prime ring R with $char(R) \neq 2$, then $[d(x), d(y)] = 0$ for all $x, y \in R$, implies $d = 0$. It is intuitive to speculate that if a prime ring R of characteristic different from 2 possesses a nonzero generalized derivation F such that $[F(x), F(y)] = 0$ for all $x, y \in R$, then R must be commutative. However, this conjecture does hold. However, this can be illustrated by the following example.

Example 4.2.3 *[114, Example] Let R be either the ring H of real quaternions or the subring K of H consisting of all elements $a+bi+cj+dk$, where a,b,c,d are integers. Define $F(x) = ix + xi$ for all $x \in R$. Then R is a noncommutative prime ring, and F is a generalized derivation such that $[F(x), F(y)] = 0$ for all $x, y \in R$.*

Example 4.2.4 *[114, Example] Let K be any field, and let R be the ring $M_2(K)$ of 2×2 matrices over K. Define $F(x) = cx + xc$, where c is either $e_{11} - e_{22}$ or e_{12}. It is easy to verify that, in either case, $[F(x), F(y)] = 0$ for all $x, y \in R$.*

Example 4.2.5 *[114, Example] Let R be the noncommutative prime ring $M_2(\mathbb{Z})$ and for arbitrary $x = \begin{bmatrix} a & b \\ c & d \end{bmatrix} \in R$, define $d(x) = \begin{bmatrix} 0 & -b \\ c & 0 \end{bmatrix}$. Define $F : R \to R$ by $F(x) = (e_{11} - e_{22})x + d(x)$. It is easily verified that d is a derivation on R, so that F is a generalized derivation and since $F\left(\begin{bmatrix} a & b \\ c & f \end{bmatrix}\right) = \begin{bmatrix} a & 0 \\ 0 & -d \end{bmatrix}$ $[F(x), F(y)] = 0$ for all $x, y \in R$. Note that F is the restriction to R of the map $\hat{F} : M_2(Q) \to M_2(Q)$ given by $\hat{F}(x) = cx + xc$, where $c = \frac{1}{2}e_{11} - \frac{1}{2}e_{22}$.*

However, Hvala [313] proved the following result:

Proposition 4.2.15 *[313, Proposotion 4] Let R be a noncommutative prime ring with the right and symmetric Martindale ring of quotients $Q_r(R)$ and $Q_s(R)$, extended centroid C and central closure $R_c = RC$. Suppose $F : R \to R$ be a nonzero generalized derivation satisfying $[F(x), F(y)] = 0$ for all $x, y \in R$. Then there exists $c \in R_c - C$ and additive map $\gamma, \phi : R \to C$ such that $F(x) = xc + cx$ and $F(x) = \gamma(x)c + \phi(x)$ for all $x \in R$, moreover, we have $c^2 \in C$.*

Proof: Since R is a commutative ring and F is nonzero, it cannot be map in the center $Z(R)$ according to Lemma 4.2.9. Therefore there exists $y_0 \in R$ such that $F(y_0) = \bar{c} \notin Z(R)$, we have $F(x)\bar{c} = \bar{c}F(x)$ for all $x \in R$. Corollary 4.2.3 tells us that $F(x) = \lambda x + \mu(\bar{c}x + x\bar{c})$, where $\lambda, \mu, \lambda\bar{c} + \mu\bar{c}^2 \in C$, $\mu \neq 0$. If we write $c = \mu\bar{c} + \frac{1}{2}\lambda$. We obtain $F(x) = cx + xc$, $c \notin C$ and $c - \mu\bar{c}$, $c^2 \in C$.

On the other hand, we already have two elements, namely c and 1, which commutes with $F(R)$, since $F = 0$, it follows from Proposition 4.2.14 that every element commuting with $F(R)$ is a C-combination of these two. Therefore, we have $F(x) = \gamma(x)c + \phi(x)$, where $\gamma, \phi : R \to R$ are additive maps. This concludes the proof.

♣

One might ponder whether an analogous condition like $F(x)F(y) - F(y)F(x) = 0$ could be explored using Corollary 4.2.1. This is indeed the case. However, at least when $charR \neq 2$, this relation is equivalent to $F(x^2) = 0$ for all $x \in R$. Hvala [313] investigated a more general scenario where $[F_1(x), F_2(y)] = 0$ and obtained the following theorem:

Theorem 4.2.16 *[313, Theorem 2] Let R be a noncommutative prime ring, $charR \neq 2$ and suppose $F_1, F_2 : R \to R$ be generalized derivations satisfying*

$$[F_1(x), F_2(x)] = 0 \textit{ for all } x \in R. \tag{4.7}$$

Then there exists $\lambda \in C$ such that $F_1(x) = \lambda F_2(x)$ for all $x \in R$.

Proof: Set

$$Y = \{y \in R \mid F_1(y), F_2(y) \text{ and } 1 \text{ are C-independent } \}$$

and $X = Y - R$ for all $x \in R$, therefore there exists $\alpha_i(x) \in C$, $i = 1,2,3$, not all zero, such that

$$\alpha_1(x)F_1(x) + \alpha_2(x) + \alpha_0(x) = 0. \tag{4.8}$$

The theorem will be proved in four steps.

Step 1. Suppose that there exists an element $x_0 \in X$ and $\lambda \in C$ such that $F_1(x_0) - \lambda F_2(x_0) \in C$. Then for every $x \in X$ we have either $F_1(x) - \lambda F_2(x) \in C$ or $[F_2(x), F_2(x_0)] = 0$.

Suppose that for $x \in X$, we have $[F_2(x), F_2(x_0)] \neq 0$. Of course, then $F_2(x) \notin C$, hence $\alpha_1(x) \neq 0$ and therefore $F_1(x) - \beta_x F_2(x) \in C$ for some $\beta_x \in C$.

The linearization of (4.7) gives

$$[F_1(x), F_2(y)] = [F_2(x), F_1(y)] \text{ for all } x, y \in R \tag{4.9}$$

and in particular $[F_1(x), F_2(x_0)] = [F_2(x), F_1(x_0)]$. According to the consideration above this relation could be rewritten as $\beta_x[F_2(x), F_2(x_0)] = \lambda[F_2(x), F_2(x_0)]$. Since the commutator is not zero as we obtain $\beta_x = \lambda$. Therefore, for all $x \in X$ one of the following alternatives must hold. Either $[F_2(x), F_2(x_0)] = 0$ or $F_1(x) - \lambda F_2(x) \in C$.

Step 2. Theorem 4.2.16 holds true under the additional assumption that

$$[F_1(x), F_1(y)] = 0, \qquad [F_2(x), F_2(y)] = 0 \text{ holds for all } x, y \in R. \tag{4.10}$$

We consider the relation (4.10) in Proposition 4.2.15. The proposition gives us

$$F_i(x) = xc_i + c_i x = r_i(x)c_i + \phi_i(x), \tag{4.11}$$

for all $i = 1,2$, $c_i \in R_c - C$ and $x \in R$. Moreover, we have $c_i^2 \in C$. Using (4.11) and (4.7), we see that $[c_1, c_2] = 0$. Next we have

$$0 = [F_1(x), F_2(x)] = [xc_1 + c_1x, \gamma_2(x)c_2 + \phi_2(x)] = \gamma_2(x)[xc_1 + c_1x, c_2].$$

It follows that for each $x \in R$, we have either $\gamma_2(x) = 0$ or $[xc_1 + c_1x, c_2] = 0$. Therefore, one of these must hold for all $x \in R$. Since F_2 does not map into C, $\gamma_2 \neq 0$, so, we have

$$xc_1c_2 + c_1xc_2 - c_2xc_1 - c_2c_1x = 0 \text{ for all } x \in R. \tag{4.12}$$

Applying Lemma 4.2.6 we see that $\{c_1c_2, c_1, c_2, 1\}$ is a C-independent set. Next, we will prove that $\{c_1, c_2, 1\}$ is a C-independent set. If that was not true, we would have $c_1c_2 = \alpha c_2 + \beta c_1 + \tau,\ \alpha, \beta, \tau \in C$. The relation (4.12) can be rewritten in the form

$$(c_1 + \alpha)xc_2 + (\beta - c_2)xc_1 - (\alpha c_2 + \beta c_1)x = 0.$$

Since $c_2 \notin C$, $\beta - c_2 \neq 0$, it follows from Lemma 4.2.6, $\{c_1, c_2, 1\}$ is a independent set. So, we have $c_2 = \lambda c_1 + \rho$ for all $\lambda, \rho \in C$. Using this in the expression (4.11) for F_2, we obtain

$$F_2(x) = \lambda F_1(x) + 2\rho x.$$

To conclude, we have proved that $2\rho = 0$. We have

$$0 = [F_1(x), F_2(x)] = 2\rho\gamma_1(x)[x, c_1].$$

Again we arrive at the alternative. For each $x \in R$, we have either $2\rho\gamma_1(x) = 0$ or $[x, c_1] = 0$. One of these relations must hold for all x. Since $\gamma \neq 0$, this is possible only if $2\rho = 0$. This proves step 2.

Step 3. Theorem 4.2.16 holds true in the case when $X = R$. It is clear from the assumption and from Lemma 4.2.9 that there exists an element $x_0 \in R$ such that $F_2(x_0) \notin C$. Therefore, $\alpha_1(x_0) \neq 0$. Write $\lambda = -\alpha_1(x_0)^{-1}\alpha_2(x_0)$ and note that $F_1(x_0) - \lambda F_2(x_0) \in C$. It follows from Step 1 that for every $x \in R$ one of the following alternatives either $[F_2(x), F_2(x_0)] = 0$ or $F_1(x) - \lambda F_2(x) \in C$, must hold. By the standard argument, one of these must hold for all $x \in R$. If $F(x) = F_1(x) - \lambda F_2(x) \in C$ for all $x \in R$, then F is a generalized derivation mapping into C. Therefore $F = 0$ according to Lemma 4.2.9. In this case we get $F_1 = \lambda F_2$ and Theorem 4.2.16 is proved. Suppose that the first alternative holds for all $x \in R$. Repeat this argument for all $x_0 \in R$ with $\alpha_1(x_0) \neq 0$ and note that in case $\alpha_1(x_0) = 0$, we have $F_2(x_0) \in C$. Therefore either the theorem is proved or we have $[F_1(x), F_2(x)] = 0$ for all $x \in R$. Reversing the role of F_1 and F_2 we get a similar result for F_1.

Step 4. $X = R$. Suppose on the contrary, Y is a nonempty set, say $y_0 \in Y$. This means that $a = F_1(y_0),\ b = F_2(y_0)$ and 1 are C-independent. Of course, (4.7) implies $ab = ba$. Note that F_i are generalized derivations, i.e., we have

$$F_i(xz) = F_i(x)z + xd_i(z) \tag{4.13}$$

for some derivations d_i, x, $z \in R$ and $i = 1,2$. The linearization of (4.7) gives (4.9). Taking y_0 for y in (4.9) and replacing x by xz in the relation, we get

$$(aF_2(x) - bF_1(x))z - F_2(x)za + F_1(x)zb + x(d_1(z)b - d_2(z)a) + axd_2(z) - bxd_1(z) = 0. \tag{4.14}$$

Denoting $G_1(x) = aF_2(x) - bF_1(x)$, $G_2(x) = -F_2(x)$, $G_3(x) = F_1(x)$, $H_1(z) = d(z)b - d_2(z)a$, $H_2(z) = d_2(z)$, $H_3(z) = -d_1(z)$ and $c_1 = 1$, $c_2 = a$, $c_3 = b$, we see that (4.14) is of the form $\sum_{i=1}^{3} c_i x H_i(z) = 0$ for all $x \in R$. Since $a,b,1$ are C-independent, the Proposition 4.2.7 ensures the existence of elements $q_{ij} \in Q_r(R_c)$, $i,j = 1,2,3$ such that $F_j(x) = -\sum_{i=1}^{3} c_i x q_{ij}$ and $H_i(z) = \sum_{i=1}^{3} q_{ij} z a_j$. Therefore, we have

$$F_1(x) = -xq_{13} - axq_{23} - bxq_{33} \tag{4.15}$$

$$F_2(x) = xq_{12} + axq_{22} + bxq_{32} \tag{4.16}$$

$$aF_2(x) - bF_1(x) = -xq_{11} - axq_{21} - bxq_{31} \tag{4.17}$$

$$d_1(z) = -q_{31}z - q_{32}za - q_{33}zb \tag{4.18}$$

$$d_2(z) = q_{21}z + q_{22}za + q_{23}zb \tag{4.19}$$

$$d_1(z)b - d_2(z)a = q_{11}z + q_{12}za + q_{13}zb. \tag{4.20}$$

Next we want to collect some more information about the elements q_{ij}.

Using the relations (4.15),(4.18) and (4.13) for $i = 1$ we obtain

$$x((q_{13} + q_{31})z + q_{32}za + q_{33}zb - zq_{13}) + ax(q_{23}z - zq_{23}) + bx(q_{33}z - zq_{33}) = 0$$

for all $x, z \in R$. Since $1, a$ and b are C-independent, it follows from Lemma 4.2.6 that coefficients appearing to the right of x in the relation above are zero for all $z \in R$.

This implies that q_{23}, $q_{33} \in C$ and

$$(q_{13} + q_{31})z + q_{32}za + q_{33}zb - zq_{13} = 0 \text{ for all } z \in R. \tag{4.21}$$

Similar arguments for F_2 give us q_{22}, $q_{32} \in C$ and

$$((q_{12} + q_{21})z + q_{22}za + q_{23}zb - zq_{12}) = 0 \text{ for all } z \in R. \tag{4.22}$$

Moreover, using (4.18) and the fact that d_1 is a derivation and q_{32}, $q_{33} \in C$, we obtain $(q_{32}xa + q_{33}xb + xq_{31})z = 0$ for all $x, y \in R$. Therefore, $q_{32}a + q_{33}b + q_{31} = 0$. Again, similar arguments for d_2 give us $q_{22}a + q_{23}b + q_{21} = 0$. Now, we can write (4.21) as $(q_{13} + q_{31})z - z(q_{13} + q_{31}) = 0$ for all $z \in R$, which gives $q_{13} + q_{31} \in C$. Similarly, we obtain $q_{12} + q_{21} \in C$ from (4.22).

There is some information hidden in (4.17). Using (4.15) and (4.16) and information about the coefficients q_{ij} we derived, (4.17) can be written in the form

$$xq_{11}+((q_{12}+q_{21})a+(q_{13}+q_{31})b+q_{22}a^2+(q_{23}+q_{32})ab+q_{33}b^2)x=0.$$

It follows from Lemma 4.2.6 that $q_{11}\in C$ and $-q_{11}=(q_{12}+q_{21})a+(q_{13}+q_{31})b+q_{22}a^2+(q_{23}+q_{32})ab+q_{33}b^2$. Note that relations $q_{32}a+q_{33}b+q_{31}=0$ and $q_{22}a+q_{23}b+q_{21}=0$ imply that the last three terms of the expression for $-q_{11}$ are equal to $-q_{21}a-q_{31}b$. Therefore, we have $q_{12}a+q_{13}b+q_{11}=0$.

Let us summarize the facts about the coefficients q_{ij}. The elements

$$q_{11},\ q_{22},\ q_{23},\ q_{32},\ q_{33},\ q_{13}+q_{31},\ q_{12}+q_{21}$$

lie in C, moreover, we have

$$q_{i1}+q_{i2}a+q_{i3}b=0 \quad i=1,2,3. \tag{4.23}$$

These facts, together with (4.15) and (4.16) enable us to express the maps in a more convenient form

$$F_1(x)=-x(q_{13}+q_{31})-x(q_{32}a+q_{33}b)-(q_{23}a+q_{33}b)x$$

$$F_2(x)=x(q_{12}+q_{21})+x(q_{22}a+q_{23}b)+(q_{22}a+q_{32}b)x.$$

Write $\zeta=-\frac{1}{2}(q_{13}+q_{31})$, $\xi=\frac{1}{2}(q_{12}+q_{21})$, $\alpha=-q_{23}$, $\beta=q_{23}$, $\gamma=q_{22}$, $r=-q_{23}$. Note that $\zeta,\xi,\alpha,\beta,\gamma$ and r lie in C. Defining $u=\zeta+\alpha a-\beta b$, $t=\zeta+ra+-\beta b$, $v=\xi+\gamma a-rb$ and $s=\xi+\gamma a-\alpha b$, we have

$$F_1(x)=ux+xt \qquad F_2(x)=vx+xs. \tag{4.24}$$

Note that $u,t,v,s\in R_c$. This is about all we could derive from the relation (4.14). However, all the coefficients in the expression of F_1 and F_2 depend on the choice of the element y_0 from the set Y. For other choices of $y\in Y$, we would get other elements $a(y),b(y)$, etc., and finally $u(y),t(y),v(y)$ and $s(y)$. When $y=y_0$ we will continue to omit the argument as in all the considerations up till now. For only $y\in Y$, we have

$$F_1(x)=ux+xt=u(y)+xt(y),\ \ F_2(x)=vx+xs=v(y)x+xs(y) \text{ for all } x\in R.$$

It follows from the Lemma 4.2.6 that the elements $u(y)-u$, $t(y)-t$, $v(y)-v$ and $s(y)-s$ lie in C. But then the differences $u(y)-u-t(y)+t$ and $v(y)-v-s(y)+s$ lie in C as well. According to the definition of the elements of above, we have

$$(\alpha(y)-r(y))a(y)-(\alpha-r)a\in C,\ \ (\alpha(y)-r(y))b(y)-(\alpha-r)b\in C.$$

Suppose that $\alpha=r$. Since $\alpha(y)\notin C$, we must have $\alpha(y)=r(y)$ for all $y\in Y$.

An argument of this kind proves that we have either $\alpha(y) = r(y)$ for all $y \in Y$ or $\alpha(y) \neq r(y)$ for all $y \in Y$. In the second case, we have

$$F_1(y) \in Ca + C, \qquad F_2(y) \in Cb + C \text{ for all } y \in Y. \tag{4.25}$$

Let us consider the case $\alpha = r$. We desire a result similar to (4.25). From the equality above it, follows $u = t$, $v = s$, and $F_1(x) = ux + xu$, $F_2(x) = vx + xv$. Moreover, the relation $u(y) - u$, $v(y) - v \in C$ gives us

$$\alpha(y)a(y) - \beta(y)b(y) \in C + Ca + Cb$$

$$\gamma(y)a(y) - \alpha(y)b(y) \in C + Ca + Cb.$$

Multiply the first relation by $\alpha(y)$ and second by $\beta(y)$ and sum up the relations so obtained, to get

$$(\alpha(y)^2 - \beta(y)\gamma(y))F_i(y) \in C + Ca + Cb$$

for $i = 1$ and all $y \in Y$. The relation for $i = 2$ could be derived in a similar manner.

Now we will prove that for $y \in Y$ the element $\alpha(y)^2 - \beta(y)\gamma(y)$ cannot be zero. This proves that for $y \in Y$, we have

$$F_1(y),\ F_2(y) \in C + Ca + Cb. \tag{4.26}$$

Assume on the contrary that $\alpha(y)^2 - \beta(y)\gamma(y) = 0$ for some $y \in Y$. In this case we have $\gamma(y)u(y) - \alpha(y)v(y) = \zeta(y)\gamma(y) - \alpha(y)\xi(y) = \rho \in C$. This implies that

$$\gamma(y)F_1(x) - \alpha(y)F_2(x) = 2\rho x. \tag{4.27}$$

If we have $\rho = 0$, then the fact that Y is not an empty set, gives us $\gamma(y) = \alpha(y) = 0$ and $F_2(x) = \xi(y)x$, $\xi(y) \neq 0$. In this case we see that $[F_i(x), x] = 0$, $x \in R$, $i = 1, 2$, that is, F_1 and F_2 are so called commuting maps. It is proved in [136] that F_1 and F_2 must be of the form $F_i(x) = \lambda_i(x) + \mu_i(x)$ where $\lambda_i \in C$ and $\mu_i : R \to C$ is an additive map. But then $\lambda_2 F_1 - \lambda_1 F_2$ maps into C, which contradicts the assumption that $Y \neq \phi$. This contradiction gives us $\rho \neq 0$. From this and (4.27) we obtain once again that F_1 and F_2 are commuting maps, which leads to the same contradiction.

Therefore, (4.26) holds for $y \in Y$ if $\alpha = r$. Since (4.25) implies (4.26), the later one holds for all $y \in Y$.

Next, we shall find all the C-combinations $F(x) = \rho_1 F_1(x) + \rho_2 F_2(x)$ which map the whole ring R into $C + Ca + Cb$. According to (4.24), we can write $F(x) = (\rho_1 u + \rho_2 v)x + x(\rho_1 t + \rho_2 s)$. We can consider F as a generalized inner derivation $F : R_c \to R_c$. Note that the rings R and R_c have the same extended centroid C. Since F maps R_c into $C + Ca + Cb$, a and b commute with $F(x)$ for all $x \in R_c$.

Apply Proposition 4.2.14 for the situation when the basic ring is R_c. Then we obtain $F=0$. Since Y is a nonempty set, this is possible only when $\rho_1=\rho_2=0$. Thus we have proved that no nonzero C-combination of F_1 and F_2 can map into $C+Ca+Cb$.

Now, we will move our attention from the set Y to the compliment $X=R-Y$. First see that F_2 can not map X into C. In this case F_2 would map the whole ring into $C+Ca+Cb$ according to (4.26), which is impossible as we saw above. Therefore $F_2(x_0)\notin C$ for some $x_0\in X$, which yields that $\alpha_1(x_0)\neq 0$ and, from (4.8), $F_1(x_0)-\lambda F_2(x_0)\in C$. Now step 1 tells us that every $x\in X$ holds for either $F_1(x)-\lambda F_2(x)\in C$ or $[F_2(x),F_2(x_0)]=0$. Note that (4.26) implies that we have $F_1(y)-F_2(y)\in C+Ca+Cb$ for all $y\in Y$. Therefore, for each $x\in R$, we have either $F_1(x)-\lambda F_2(x)\in C+Ca+Cb$ or $[F_2(x),F_2(x_0)]=0$. By a standard argument one of these must hold for all $x\in R$. Since only the zero combination of function F_1 and F_2 can map the ring into $C+C_a+C_b$, the second possibility must hold for all $x\in R$. Recalling (4.8), we see that a similar argument holds for every $x_0\in X$ with $\alpha_1(x_0)\neq 0$. In the case when $\alpha_1(x_0)=0$, we have $F_2(x_0)\in C$ and the conclusion is still true. Therefore, we have

$$[F_2(x),F_2(z)]=0,\quad x\in R,\ z\in X.$$

This means that the commutator is zero as soon as one of the elements lie in X. However, if both, x and z lie in Y, the commutator is zero as well because of (4.26). Therefore, we have $[F_2(x),F_2(y)]=0$ for all $x,y\in R$. Symmetrically we obtain $[F_1(x),F_1(y)]=0$ for all $x,y\in R$. This together with Step 2, contradicts the assumption that Y is a nonempty set. This proves the theorem.

♣

For 2-torsion-free prime rings with 1, Examples 4.2.3, 4.2.4, and 4.2.5 illustrate all possibilities, as demonstrated by the following result established by Bell and Rehman [114]:

Theorem 4.2.17 *[114, Theorem 3.4] Let R be a prime ring with 1 such that $char(R)\neq 2$. If R admits a nonzero generalized derivation F such that $[F(x),F(y)]=0$ for all $x,y\in R$, then one of the following holds:*

(i) R is commutative;

(ii) R is a noncommutative subring of a division ring Δ, and there exists $\delta\in\Delta$ such that $F(x)=\delta x+x\delta$ for all $x\in R$;

(iii) R is a noncommutative subring of a 2×2 total matrix ring M over a field, and there exists $m\in M$ such that $F(x)=mx+xm$ for all $x\in R$.

4.3 Generalized derivations of left faithful rings

Let R be a left faithful ring, that is, the left annihilator of R is zero. Let M_R be a right R-module. A submodule N of M_R is called **dense** if for $x, y \in M_R$ with $x \neq 0$, there exists $r \in R$ such that $yr \in N$ and $xr \neq 0$. A right ideal ρ is called a **dense right ideal** of R if ρ_R is a dense submodule of R_R. In terms of dense right ideals, the right Utumi quotient ring (i.e., the maximal right quotient ring) of R can be characterized as the ring $U(R)$ (up to isomorphisms fixing R) satisfying the following properties:

1. R is a subring of $U(R)$.
2. For each $a \in U(R)$, there exists a dense right ideal ρ of R such that $a\rho \subseteq R$.
3. If $a \in U(R)$ and $a\rho = 0$ for some dense right ideal ρ of R, then $a = 0$.
4. If ϕ is a right R-module map from a dense right ideal ρ of R into R, then there exists $a \in U(R)$ such that $\phi(x) = ax$ for all $x \in \rho$.

In 1999, Lee [378] expanded the concept of generalized derivations in rings in the following manner:

Definition 39 *[378] Let ρ be the dense right ideal of a ring R and $U(R)$ be the right Utumi quotient ring. $F : \rho \to U(R)$ is called a* **generalized derivation on** *ρ if there exists a derivation $d : \rho :\to U(R)$ such that $F(xy) = F(x)y + xd(y)$ for all $x, y \in \rho$.*

Lee [378] demonstrated that each generalized derivation on R can be uniquely extended to a generalized derivation of $U(R)$. Additionally, he provided a characterization of a generalized derivation as follows:

Theorem 4.3.1 *[378, Theorem] Every generalized derivation F on a dense right ideal of R can be extended to $U(R)$ and assumes the form $F(x) = ax + d(x)$ for some $a \in U(R)$ and a derivation d on $U(R)$.*

An extension ring S of R is called a right quotient ring of R if R_R is a dense submodule of S_R. Clearly, S is itself a left faithful ring and, moreover, $U(R)$ is a right quotient ring of R.

To prove Theorem 4.3.1, we need some basic properties concerning right quotient rings. We record these basic facts without proofs (for the proof refer to Proposition 2.1.7 of [104]).

Let S be a right quotient ring of a left faithful ring R. Then

Fact 1. For $x \in S$, there exists a dense right ideal ρ of R such that $x\rho \subseteq R$.

Fact 2. For $x \in S$, if $x\rho = \{0\}$ for some dense right ideal ρ of R, then $x = 0$.

Fact 3. There exists ring embedding $h : S \to U(R)$ such that $h(x) = x$ for all $x \in R$.

To establish the proof of the theorem mentioned above, we require the following lemma:

Lemma 4.3.2 *[378, Lemma] Let R be a left faithful ring and A be a subring of $U(R)$ such that A is a dense submodule of $U(R)_R$. Then there exists a ring isomorphism $\psi : U(R) \to U(R)$ satisfying $\psi(x) = x$ for all $x \in A$.*

Proof: We first note that A itself is a left faithful ring. Let $a \in A$ be such that $aA = \{0\}$. Then $a(A \cap R) = \{0\}$ implies $a = 0$ as $A \cap R$ is a dense right ideal of R. Thus $U(A)$ does exist. We break the proof into some steps.

Step 1. The ring $U(R)$ is a right quotient ring of A. Indeed, let $u, v \in U(R)$ with $u \neq 0$. First, there exists $r \in R$ such that $ur \neq 0$ and $vr \in A$ since A_R is a dense submodule of $U(R)_R$. Consider $ur, r \in U(R)$ with $ur \neq 0$. Then $urs \neq 0$ and $rs \in A$ for some $s \in R$. Set $t = rs \in A$. Then $ut \neq 0$ and $vt = vrs \in A$, proving Step 1.

Step 2. The ring $U(A)$ is a right quotient ring of $A \cap R$. Let $x, y \in U(A)$ with $x \neq 0$. Since $U(A)$ is a right quotient ring of A, we have $xa, ya \in A$ with $xa \neq 0$ for some $a \in A$. Consider $xa, a \in A \subseteq U(R)$ with $xa \neq 0$. Then there exists $r \in R$ such that $xar \neq 0$ and $ar \in R$. Since R_R is a dense submodule of $U(R)_R$. For $xar, yar \in A$ with $xar \neq 0$, there exists some $s \in R$ such that $xars \neq 0$ and $yars \in R$. Note that $yars \in A \cap R$ and $ars \in A \cap R$. Set $t = ars \in A \cap R$. Then $xt \neq 0$ and $yt \in A \cap R$. This proves that $U(A)$ is a right quotient ring of $A \cap R$, hence Step 2 is proved.

It follows from **Step 1** and Fact 3 that there exists a ring embedding $\phi : U(R) \to U(R)$ such that $\phi(x) = x$ for all $x \in A$. On the other hand, $A \cap R \subseteq \phi(R) \subseteq U(A)$, it follows from Step 2 that $U(A)$ is a right quotient ring of $\phi(R)$. Fact 3 says that there is a ring embedding $h : U(R) \to U(\phi(R))$ such that $h(\phi(r)) = \phi(r)$ for all $r \in R$. Since $\phi(R) \cong R$, there exists a ring isomorphism $g : U(\phi(R)) \to U(R)$ such that $g(\phi(r)) = r$ for all $r \in R$. Define $\psi = g \circ h$. Then $\psi : U(A) \to U(R)$ is a ring embedding satisfying $\psi(\phi(r)) = r$ for all $x \in R$.

Let $x \in U(A)$. By **Step 2**, $U(A)$ is a quotient ring of $A \cap R$. Thus, by the Fact 1, $x\rho \subseteq A \cap R$ for some dense right ideal of ρ of $A \cap R$. Let $x \in \rho$. Since $x\rho \subseteq A \cap R$, we have $\phi(xr) = xr \in R$ and hence $\psi(xr) = xr$. Thus, we have $(\phi \circ \psi)(xr) = xr$. On the other hand, $(\phi \circ \psi)(xr) = (\phi \circ \psi)(x)(\phi \circ \psi)(r) = (\phi \circ \psi)(x)r$. Hence $((\phi \circ \psi)(x) - x)r = 0$. So $((\phi \circ \psi)(x) - x)\rho = 0$ implying that $(\phi \circ \psi)(x) = x$ by Fact 2. This proves that $\phi \circ \psi$ is the identity map on $U(A)$ and so ψ is an epimorphism. Let $a \in A$, then $(\phi \circ \psi)(a) = a = \phi(a)$. This implies that $\psi(a) = a$. Since ψ is an injective map the theorem is proved.

♣

Lemma 4.3.3 *[378, Lemma] Every derivation from a dense right ideal of R into $U(R)$ can be uniquely extended to $U(R)$.*

Proof: We follow the argument of [375, Lemma 3]. Suppose, $d:\rho\to U(R)$ is a derivation and ρ is a dense right ideal of R. In view of Lemma 4.3.2, there exists a ring isomorphism ϕ from $U(\rho)$ onto $U(R)$ such that $\phi(x)=x$ for all $x\in\rho$. Set $\overline{d}=\phi^{-1}\circ d$. Let $a,b\in\rho$. Then $\overline{d}(ab)=\phi^{-1}(d(ab))=\phi^{-1}(d(a)b+ad(b))=\phi^{-1}(d(a))\phi^{-1}(b)+\phi^{-1}(a)\phi^{-1}(d(b))=\overline{d}(a)b+a\overline{d}(b)$. Therefore, $\overline{d}$ is a derivation from ρ into $U(\rho)$. By [372, Lemma 2], $\overline{d}$ can be uniquely extended to a derivation of $U(\rho)$, denoting the derivation by $\overline{d}$ also. Then it is clear that $\phi\circ\overline{d}\circ\phi^{-1}$ is a derivation of $U(R)$ extending the derivation $d:\rho\to U(R)$. Finally, we have to prove the uniqueness of the extension $\overline{d}$. It suffices to show that if d is a derivation on $U(R)$ vanishing on the dense right ideal of R, then $d=0$. Indeed, suppose $d(\rho_0)=0$ for some dense right ideal of R. Let $u\in U(R)$, choose the dense right ideal J of R such that $uJ\subseteq\rho_0$ and $J\subseteq\rho_0$. For $r\in J$, we have $0=d(ur)=ud(r)+d(u)r$, and so $d(u)J=\{0\}$. Thus $d(u)=0$. So $d=0$. This completes the proof.

♣

Proof of the Theorem 4.3.1 Suppose that $F:\rho\to U(R)$, where ρ is the dense right ideal of R. By definition, there exists a derivation d from ρ into $U(R)$ such that $F(xy)=F(x)y+xd(y)$ for all $x\in\rho$. By Lemma 4.3.3, the derivation d can be uniquely extended to $U(R)$. Denote this extension by d also.

Let $u\in U(R)$. Then there exists a right ideal ρ_0 of R contained in ρ such that $u\rho_0\subseteq\rho$. Define $\phi_u:\rho_0\to U(R)$ by $\phi_u(y)=F(uy)-ud(y)$ for all $y\in\rho_0$. We claim that ϕ_u is a right R-module map. Let $y\in\rho_0$ and $r\in\rho$. Then since $yr\in\rho_0$ and $uy\in\rho$, we have

$$\begin{aligned}\phi_u(yr) &= F(uyr)-ud(yr)\\ &= F(uy)r+uyd(r)-(ud(y)r+uyd(r))\\ &= (F(uy)-ud(y))r\\ &= \phi_0(y)r,\end{aligned}$$

as desired. Next, the map ϕ_u can be uniquely extended to a right $U(R)$- module map defined on $\rho_0U(R)$, say $\overline{\phi}_u$. Indeed, if $\sum_i x_iv_i\in\rho_0U(R)$, where $x_i\in\rho_0$, $v_i\in U(R)$, define $\overline{\phi}_u\sum_i x_iv_i=\sum\phi_u(x_i)v_i$. We have to prove the map is well defined. Suppose that $\sum_i x_iv_i=0$. Choose a dense right ideal J of R such that $v_iJ\subseteq\rho$ for each i. Let $z\in J$, then $0=\phi_u(\sum_i x_iv_iz)=\sum_i\phi_u(x_i(v_iz))=\sum_i\phi_u(x_i)v_iz$ and so $(\sum_i\phi_u(x_i)v_i)J=\{0\}$, implying $\sum_i\phi_u(x_i)v_i=0$. This proves $\overline{\phi}_u$ to be well defined. Clearly, $\rho_0U(R)$ is a dense right ideal of $U(R)$. Additionally, $\overline{\phi}_u$ is a right $U(R)$-module map. Since the right Utami quotient ring of $U(R)$ coincides with itself, the map $\overline{\phi}_u$ defines an element in $U(R)$, say $\overline{F}(u)$. Hence $\overline{F}$ is a map from $U(R)$ to $U(R)$ and possesses the following properties:

(i) $\overline{F}(x) = F(x)$ for all $x \in \rho$ and

(ii) If $u \in U(R)$ and $u\rho_0 \subseteq \rho$ for ρ_0 is a dense right ideal of R contained in ρ, then $F(uy) = \overline{F}(u)y + ud(y)$ for all $y \in \rho_0$.

Let $u, v \in U(R)$. We claim that $\overline{F}(uv) = \overline{F}(u)v + ud(v)$. First, choose ρ_0 as given in (ii). Choose the right ideal ρ_1 of R such that $v\rho_1 \subseteq \rho_0$ and $\rho_1 \subseteq \rho_0$. Let $r \in \rho_1$. Then using (ii), we have $F(u(vr)) = \overline{F}(u)vr + ud(vr) = \overline{F}(u)vr + ud(v)r + uvd(r)$. On the other hand, since $uv\rho_1 \subseteq \rho$, applying (ii) again we have $F((uv)r) = \overline{F}(uv)r + uvd(r)$.

Comparing the two formulae yields $(\overline{F}(uv) - \overline{F}(u)v - ud(v))r = 0$, $r \in \rho_1$. This implies that $\overline{F}(uv) = \overline{F}(u)v + ud(v)$ as desired.

Up to now, we have proved that $\overline{F}(xy) = \overline{F}(x)y + xd(y)$ for all $x, y \in U(R)$. In particular, let $x = 1$ and $a = \overline{F}(1)$. Then $\overline{F}(y) = ay + d(y)$ for all $y \in U(R)$. Since $\overline{F} = F$ on ρ, we see that $gF(y) = ay + d(y)$ for all $y \in \rho$. This completes the proof.

A natural question regarding Theorem 4.3.1 is: Whether a and d are uniquely determined by F. The answer is no. For instance,

Example 4.3.1 *Let $M_n(D)$ be $n \times n$ matrix over the division ring D and let $R = M_n(D)e_{11}$, where e_{ij} is the matrix with 1 in the (i, j)-entry only, for $1 \leq i, j \leq n$ and $n \geq 3$. Then R is a left faithful ring and $U(R) = M_n(D)$. Let $d = ad(e_{21})$ be the inner derivation on R induced by the element e_{21}. Define $F : R \to U(R)$ by $F(x) = e_{11}x + d(x)$ for all $x \in R$. Then g is a generalized derivation on R. Let $d = ad(e_{21} + e_{31})$. Since $Re_{31} = \{0\}$, we have $F(x) = (e_{11} - e_{31})x + d(x)$ for all $x \in R$, as desired.*

However, if R is a semiprime ring, a and d are uniquely determined by g. It suffices to prove that if $F = 0$ on ρ, a dense right ideal of R, then $a = 0$ and $d = 0$. Let $y \in \rho$ and $r \in R$. Then $0 = F(yr) = ayr + d(yr) = ayr + d(y)r + yd(y)$. Choose a dense right ideal ρ_0 of R such that $d(r)\rho_0 \subseteq R$. Then, for $t \in \rho_0$, we have $(d(r)ty)^2 = 0$ for all $y \in \rho$. Since R is semiprime, it follows from [294, Lemma 1.1] that $d(r)t_\rho$ for all $t \in \rho_0$. Thus $d(r)\rho_0\rho = 0$. Since $\rho \circ \rho$ is still a dense right ideal of R. Thus $d(r) = 0$ follows. Then $d(R) = 0$ and so $d = 0$. Now, the fact that $g(\rho) = \{0\}$ implies that $a\rho = \{0\}$ and so $a = 0$. This proves our claim. We record this fact in the following theorem.

Theorem 4.3.4 *[378, Theorem 4] Let R be a semiprime ring. Then every generalized derivation F on a dense right ideal of R is uniquely extended to $U(R)$ and assumes the form $F(x) = ax + d(x)$ for some $a \in U(R)$ and derivation d on $U(R)$. Moreover, a and d are uniquely determined by the generalized derivation g.*

In [345, 346], Kharchenko provided descriptions of identities involving derivations and his findings serve as potent instruments for transforming a differential identity into a generalized polynomial identity. Consequently, when delving into the investigation of identities featuring generalized derivations, it appears logical to seek an analogous theorem for such identities. Lee and Shiue [391] established a version of Kharchenko's theorem [346] specifically tailored for generalized derivations, and their result is outlined below:

Theorem 4.3.5 *[391, Theorem 1] Let R be a prime ring. If $\phi(X_i^{\Gamma_j})$ is a reduced generalized differential identity for a nonzero ideal of R, then $\phi(Z_{ij})$ is a generalized polynomial identity.*

The following result generalizes [346, Corollary 2] the case of algebraic generalized derivations.

Theorem 4.3.6 *[391, Corollary] Let R be a prime ring with an extended centroid C and a right Utumi quotient ring U, F a generalized derivation of U and $\phi(X)$ a monic polynomial over C of degree $n > 1$. Suppose that $\phi(F)$ is a generalized derivation of U and that either char $R = 0$ or $p > n$. Then both F and $\phi(F)$ are generalized as inner derivations.*

Proof: Write $\phi(X) = X^n + \beta_{n-1}X^{n-1} + \cdots + \beta_1 X + \beta_0 \in C[X]$. Suppose, on the contrary, that F is not a generalized inner derivation. By assumption, we write

$$\phi(F)(x) = F(x^n) + \beta_{n-1}F(x^{n-1}) + \cdots + \beta_1 F(x) + \beta_0 x \tag{4.28}$$

for all $x \in U$. If F and $\phi(F)$ are C-independent modulo G_{int}, where G_{int} is the C-subspace of the set of all generalized derivations of U, then $F, F^2, \ldots, F^n$ and $\phi(F)$ are distinct regular words as either char$R = 0$ or $p > n$. Applying Theorem 4.3.5 to equation (4.28), we see that

$$x_n + \beta_{n-1}x_{n-1} + \cdots + \beta_1 x_1 + \beta_0 x_0 = y$$

for $y, x_i \in U$, is a contradiction. This proves that $\phi(F) = \beta F + \mu$ for some $\beta \in C$ and μ is a generalized inner derivation. Thus, $\mu(x) = ax + bx$, where $a, b \in U$, hence

$$F^n(x) + \beta_{n-1}F^{n-1}(x) + \cdots + \beta_1 F(x) + \beta_0 x = \beta F(x) + ax + bx$$

for all $x \in U$. Since F is not generalized inner derivations, we can derive contradictions from the argument given as above. Thus, F is a generalized inner derivation. Now it is clear that the associated derivation of $\phi(F)$ is defined by a generalized linear map. By Kharchenko's theorem, the associated derivation of $\phi(F)$ is inner and so $\phi(F)$ is a generalized inner derivation.

♣

Hvala extended Posner's theorem by providing a characterization of generalized derivations F_1 and F_2 on a prime ring R with a characteristic not equal to 2, such that their product F_1F_2 is also a generalized derivation of R (see Theorem 4.2.5). Simultaneously, Hvala demonstrated that generalized derivations F_1 and F_2 are C-dependent if $[F_1(x), F_2(x)] = 0$ for all $x \in R$ (see Theorem 4.2.16). Lee and Shiue, leveraging Theorem 4.3.5, presented a comprehensive elucidation of Theorem 4.2.16, eliminating the prerequisite that char$R \neq 2$. Lee and Shiue [391], established the subsequent results as follows:

Theorem 4.3.7 *[391, Theorem 2] Let R be a prime ring with an extended centroid C and let F_1 and F_2 be generalized derivations of R. Then the product F_1F_2 is also a generalized derivation iff one of the following holds:*

(i) There exists $\lambda \in C$ such that either $F_1 = \lambda_l$ or $F_2 = \lambda_l$, where λ_l is left multiplied by λ;

(ii) Either $F_1, F_2 \in L(U)$ or $F_1, F_2 \in R(U)$, where $L(U)$ and $R(U)$ be the sets of all left and right multiplications of U, respectively;

(iii) There exist $\lambda, \mu \in C$ and $a, b \in U$ such that $F_1(x) = ax + bx$ and $F_2(x) = \lambda x + \mu(ax - xb)$ for all $x \in U$;

(iv) char$R = 2$ and there exist $\lambda, \mu \in C$ such that $F_2 = \lambda_l + F_1\mu$.

Theorem 4.3.8 *[391, Theorem 3] Let R be a noncommutative prime ring with an extended centroid C and let F_1 and F_2 be nonzero generalized derivations of R. Suppose that $[F_1(x), F_2(x)] = 0$ for all $x \in R$. Then there exists $\lambda \in C$ such that $F_2(x) = \lambda F_1(x)$ for all $x \in R$.*

4.4 An Engel condition with generalized derivations in rings

For elements $x, y \in R$ and every $k \geq 0$, define $[x,y]_0 = x$, $\quad [x,y]_1 = xy - yx$, and iteratively $[x,y]_k = [[x,y]_{k-1}, y]$ for $k > 1$. It is noteworthy that an **Engel condition** corresponds to a polynomial $[x,y]_k = \sum_{i=0}^{k}(-1)^i\binom{k}{i}y^i x y^{k-i}$ in non-commutative indeterminates x and y. A ring R satisfies an Engel condition if there exists a positive integer k such that $[x,y]_k = 0$. The Engel type identity with a derivation appeared in a well-known paper by Posner [487]. Since then, several authors have explored this type of identity with derivations acting on one-sided, two-sided, and Lie ideals of prime and semiprime rings (refer to [273, 307, 364, 377] for a partial bibliography).

In 1992, Daif and Bell [202, Theorem 3] established that if there exists a nonzero ideal I of a semiprime ring R and a derivation d of R such that $d([x,y]) = [x,y]$ for all $x, y \in I$, then $I \subseteq Z(R)$. If a semipirme ring is replaced with a prime ring, then the assumptions imply R is commutative. In 1998, Quadri et al. [497] extended this result to the class of generalized derivations of prime rings. Precisely, they proved the following result:

Theorem 4.4.1 *[497, Theorem 1] Let R be a prime ring, I a nonzero ideal of R. If R admits a generalized derivation F associated with a nonzero derivation d of R such that $F([x,y]) = [x,y]$, for all $x,y \in I$, then R is commutative.*

The following example illustrates that in the hypothesis of Theorem 4.4.1, if we replace the prime ring with a semiprime ring, then R may not be commutative, even for an ordinary derivation.

Example 4.4.1 *Let R_1 be an integral domain admitting a nonzero derivation d_1 and R_2 be any noncommutative prime ring with nonzero derivation d_2; $D = d_1 \oplus d_2$ is a derivation on R. The semiprime ring satisfies the hypothesis of Theorem 4.4.1. However, R is not commutative.*

Further, in the year 2002, Rehman [501] extended the Theorem 4.4.1 for a square closed Lie ideal and obtained the following:

Theorem 4.4.2 *[501, Theorem 3.3] Let R be a 2-torsion free prime ring and L a nonzero Lie ideal of R such that $u^2 \in L$ for all $u \in L$. If R admits a generalized derivation F with an associated derivation d such that $F([u,v]) = [u,v]$ for all $u,v \in L$, and if $F = 0$ or $d \neq 0$, then $L \subseteq Z(R)$.*

Brešar [136] initiated the investigation into cocentralizing derivations and demonstrated that a prime ring R possessing two derivations d and δ with $d(x)x - x\delta(x)$ for all $x \in R$ implies either $d = \delta = 0$ or R is commutative. In the work by Lanski [364], Posner's theorem was extended to a general Engel condition involving derivations, establishing that a prime ring R becomes commutative if there exists a nonzero derivation d satisfying $[d(x),x]_k = 0$ for all $x \in R$, where k is a fixed positive integer. Later, Lee and Shiue [389] provided a unified version of Brešar and Lanski's results, proving that if R is a prime ring with derivations d and δ, and $[d(x^m)x^n - x^s\delta(x^t), x^r]_k = 0$ for all $x \in R$ (with fixed positive integers m,n,s,t,r,k), then $d = \delta = 0$ or R is commutative. Numerous related results on cocentralizing derivations or Engel-type identities with derivations in prime rings can be found in [33, 160, 364, 366, 368, 370, 392, 398, 417, 420, 421, 422, 533, 536, 557]. In 2006, De Filippis [210] obtained Lanski's result for generalized derivations as follows:

Theorem 4.4.3 *[210, Theorem 1] Let R be a prime ring of characteristic differing from two with a right quotient ring U and an extended centroid C, $F \neq 0$ a generalized derivation of R, L a noncentral Lie ideal of R and $n \geq 1$. If $[F(u),u]^n = 0$ for all $u \in L$ then there exists an element $a \in C$ such that $F(x) = ax$ for all $x \in R$, unless when R satisfies s_4 and there exists an element $b \in U$ such that $F(x) = bx + xb$ for all $x \in R$.*

He presented an example that illustrates the limitation of expecting the conclusion that R is commutative, as follows:

Example 4.4.2 *[210, Example] Let $R = M_2(F)$, the ring of all 2×2 matrices over the field F and $a \in R$ such that $F(x) = xa + ax$ for all $x \in R$. If $L = [R,R]$, then we get $[F(u),u] = 0$ for all $u \in [R,R]$, because $u^2 \in F$, but obviously R is not commutative.*

Further, in 2008, Argaç et al. [33] investigated the Engel conditions $[F(x),x] = 0$ for all $x \in S$, a suitable subset of R, with F a generalized derivation of R as follows:

Theorem 4.4.4 *[33, Theorem 2] Let R be a prime ring, with an extended centroid C, F a nonzero generalized derivation of R, L a noncentral Lie ideal of R, $k \geq 1$ a fixed integer. If $[F(u),u]_k = 0$ for all $u \in L$, then either $F(x) = ax$, with $a \in C$ or R satisfies the standard identity S_4. Moreover in the latter case either $char(R) = 2$ or $char(R) \neq 2$ and $F(x) = ax + xb$, with $a, b \in Q$ and $a - b \in C$.*

In 2011, De Filippis and Huang [220] investigated the scenario where a generalized derivation F of a prime ring R satisfies $F([x,y])^n = [x,y]$ for all $x,y \in I$, where I is a nonzero ideal of R. They obtained the following theorem.

Theorem 4.4.5 *[220, Theorem 1] Let R be a prime ring, I a nonzero ideal of R and n a fixed positive integer. If R admits a generalized derivation F associated with a derivation d such that $(F([x,y]))^n = [x,y]$ for all $x,y \in I$, then either R is commutative or $n = 1$, $d = 0$ and F is the identity map on R.*

The importance of R being prime in the hypothesis is illustrated by the following example.

Example 4.4.3 *[220, Example] Let S be any ring and let $R = \left\{ \begin{pmatrix} a & b \\ 0 & 0 \end{pmatrix} | a,b \in S \right\}$ and let $I = \left\{ \begin{pmatrix} 0 & a \\ 0 & 0 \end{pmatrix} | a \in S \right\}$ be a nonzero ideal of R. We define a map $F : R \to R$ by $F(x) = 2e_{11}x - xe_{11}$. Then it is easy to see that F is a generalized derivation associated with a nonzero derivation $d(x) = [e_{11}, x]$. It is straightforward to check that F satisfies the property: $(F([x,y]))^n = [x,y]$ for all $x,y \in I$. However, R is not commutative.*

De Filippis and Huang extended Theorem 4.4.5 to semiprime rings in the following manner:

Theorem 4.4.6 *[220, Theorem 2] Let R be a semiprime ring and n a fixed positive integer. If R admits a generalized derivation F associated with a derivation d such that $(F([x,y]))^n = [x,y]$ for all $x,y \in I$, then either R is commutative or $n = 1$, $d(R) \subseteq Z(R)$, R contains a nonzero central ideal and $F(x) - x \in Z(R)$ for all $x \in R$.*

In 2013 Huang and Davvaz[309] generalized the Theorem 4.4.5 as follows:

Theorem 4.4.7 *[309, Theorem 1] Let R be a prime ring and m,n fixed positive integers. If R admits a generalized derivation F associated with a nonzero derivation d such that $F([x,y])^m = [x,y]^n$ for all $x,y \in R$, then R is commutative.*

On the other hand, it is worth noting the contribution of Giambruno et al. in 1994 [277], where they investigated the commutativity of rings that satisfy $([x,y]_k)^n = [x,y]_k$. This naturally leads to the question of the commutativity of R when the relation $F([x,y]_k)^n = ([x,y]_k)^l$ holds, where F represents a generalized derivation associated with a nonzero derivation d of R, and l,n,k are fixed positive integers. In 2016, Raza et al. provided an answer to this question for the case of a prime underlying ring, yielding the following result.

Theorem 4.4.8 *[500, Theorem 2.3] Let R be a prime ring with characteristics not equal to two and I be a nonzero ideal of R. If R admits a nonzero generalized derivation F associated with a nonzero derivation d such that $F([x,y]_k)^n = ([x,y]_k)^l$ for all $x,y \in I$, where l,n,k are fixed positive integers, then R is commutative.*

This result generalizes all the above results for a general derivation. If we take $k = 1$, then we obtain the Theorem 4.4.7.

Recently, Lee and Zhou [394] successfully characterized cocommuting generalized derivations and proved the following.

Theorem 4.4.9 *[394, Theorem 1.1] Let R be a noncommutative prime ring such that $R \ncong M_2(GF(2))$. Let C be the extended centroid of R and let Q be the symmetric Martindale right ring of quotients of R. Suppose that F and F_1 are generalized derivations of R such that $F(x^m)x^n = x^n F_1(x^m)$ for all $x \in R$, where m,n are fixed positive integers. Then the following two conditions hold: (1) There exists $a \in Q$ such that $F(x) = xa$ and $F_1(x) = ax$ for all $x \in R$; (2) Either $a \in C$ or x^m and x^n are linearly dependent over C for every $x \in R$.*

In [6], Albaş et al. extended the result of Lanksi [364] to generalized derivations as follows.

Theorem 4.4.10 *[6, Theorem 1] Let R be a noncommutative prime ring with extended centroid C. Suppose that F is a generalized derivation of R such that $[F(x^r),x^r]_k = 0$ for all $x \in R$, where r,k are fixed positive integers. Then there exists $a \in C$ such that $F(x) = ax$ for all $x \in R$.*

Recently, Demir and Argaç [232] established that if R is a noncommutative prime ring with the right Utumi quotient ring U and F, a generalized derivation of R satisfying $[F(x^m)x^n,x^r]_k = 0$ for all $x \in R$, where n,m,r,k are fixed positive integers. Then, the following results were obtained.

Theorem 4.4.11 *[232, Therem 1] Let R be a noncommutative prime ring and k,m,n,r fixed positive integers. If there exists a generalized derivation F of R such that $[F(x^m)x^n,x^r]_k = 0$ for all $x \in R$, then there exists an element $a \in U$ such that $F(x) = xa$ for all $x \in R$.*

Theorem 4.4.12 *[232, Therem 2] Let R be a noncommutative prime ring, I a non-zero left ideal of R and k,m,n,r fixed positive integers. If there exists a generalized derivation F of R such that $[F(x^m)x^n,x^r]_k = 0$ for all $x \in I$, then there exists $a \in U$ such that $F(x) = xa$ for all $x \in R$ except when $R \cong M_2(GF(2))$ and $I[I,I] = \{0\}$.*

They demonstrated that Theorems 4.4.11 and 4.4.12 do not hold in semiprime rings with the following example.

Example 4.4.4 *[232, Example 2] Let F be a field and $\lambda = GF(2)$. Consider the semiprime ring $R = \begin{pmatrix} \lambda & \lambda & 0 \\ \lambda & \lambda & 0 \\ 0 & 0 & F \end{pmatrix}$. Let, $I = \begin{pmatrix} \lambda & 0 & 0 \\ \lambda & 0 & 0 \\ 0 & 0 & F \end{pmatrix}$ be the left ideal of R. If $a = \begin{pmatrix} 1 & 1 & 0 \\ 0 & 0 & 0 \\ 0 & 0 & 1 \end{pmatrix}$, $b = \begin{pmatrix} 0 & 0 & 0 \\ 0 & 0 & 0 \\ 0 & 0 & \alpha \end{pmatrix}$ for $\alpha \in F$ fixed, one can easily see that $[ax^2 + [b,x]x,x] = 0$ for all $x \in I$, since $uv(u+v) = 0$ for all $u,v \in \lambda$. Then $F(x) = ax + [b,x] = (a+b)x - xb$ is a generalized derivation such that $[F(x)x,x] = 0$ for all $x \in I$. But $a+b \notin C = \left\{ \begin{pmatrix} \beta & 0 & 0 \\ 0 & \beta & 0 \\ 0 & 0 & \mu \end{pmatrix} | \beta \in \lambda, \quad \mu \in F \right\}$.*

Later, Alahmadi et al. [4] proved the following.

Theorem 4.4.13 *[4, Theorem 1.2] Let R be a noncommutative prime ring with an extended centroid C. Suppose that F is a generalized derivation of R and δ is a derivation of R such that $[F(x^m x^n + x^n\delta(x^m)),x^r]_k = 0$ for all $x \in R$, where m,n,r,k are fixed positive integers. Then $\delta = 0$ and there exists $a \in U$, the left Utumi quotient ring of R, such that $F(x) = xa$ for all $x \in R$.*

It is natural to wonder how Theorem 4.4.13 can be extended to the case when both the generalized derivation F and the derivation d act on left ideals and Lie ideals of R. Dhara and De Filippis [245] answered this question and obtained the following results.

Theorem 4.4.14 *[245, Theorem 1.1] Let R be a noncommutative prime ring and I be a nonzero left ideal of R. Let U be the left Utumi quotient ring of R and $C = Z(U)$ be the extended centroid of R. Let F be a generalized derivation of R and d be a derivation of R such that $[F(x^m x^n + x^n d(x^m)),x^r]_k = 0$ for all $x \in R$, where m,n,r,k are fixed positive integers. Then one of the following holds:*

(1) *There exist* $b, c \in U$ *and* $\beta \in C$ *such that* $F(x) = xb$ *for all* $x \in R$ *and* $d(x) = [c, x]$ *for all* $x \in R$, *with* $I(c - \beta) = 0$;

(2) *There exists* $a \in U$ *such that* $F(x) = xa$ *for all* $x \in R$ *and* d *is an outer derivation of* R, *with* $d(I) \subseteq IC$ *and* $I[I, I] = 0$; *(3)* $R \cong M_2(GF(2))$ *and* $I[I, I] = 0$.

Theorem 4.4.15 *[245, Theorem 1.2] Let* R *be a noncommutative prime ring of characteristic differing from 2,* L *be a noncentral Lie ideal of* R. *Let* U *be the left Utumi quotient ring of* R *and* $C = Z(U)$ *be the extended centroid of* R. *Let* F *be a generalized derivation of* R *and* d *be a derivation of* R *(which is not necessarily associated to* F*) such that* $[F(u^m)u^n + u^n d(u^m), u^r]_k = 0$ *for all* $u \in L$, *where* m, n, r, k *are fixed positive integers. Then one of the following holds:*

(1) $d = 0$ *and there exists* $\lambda \in C$ *such that* $F(x) = \lambda x$, *for all* $x \in R$;

(2) R *satisfies* s_4, *the standard identity of degree* 4;

(3) $char(R) = p > 0; R \subset M_t(C)$, *the ring of* $t \times t$ *matrices over* $C, d = 0$ *and there is an element* $a \in U$ *such that* $F(x) = ax$, *for any* $x \in R$. *Moreover, there exists an integer* $s \geq 1$ *such that, for any* $u \in [R, R]$ *there exists* $\lambda \in C$ *for which* $u^{rp^s+m} = \lambda u^m$ *and* $u^{rp^s+n} = \lambda u^n$.

Very recently, in 2020, Liu [423] gave a common generalization of Theorem 4.4.9, Theorem 4.4.10, and Theorem 4.4.13 as follows:

Theorem 4.4.16 *[423, Theorem 1.1] Let* R *be a noncommutative prime ring with an extended centroid* C *and maximal right ring of quotients* Q. *Let* I *be a nonzero ideal of* R *and let* F *and* F_1 *be two generalized derivations of* R. *Suppose that* $[F(x^m)x^n - x^s F_1(x^t), x^r]_k = 0$ *for all* $x \in I$, *where* m, n, s, t, r, k *are fixed positive integers. Then one of the following conditions holds:*

1. *There exist* $\alpha, \beta \in C$ *such that* $F(x) = \alpha x$ *and* $F_1(x) = \beta x$ *for all* $x \in R$;
2. $m = s$, $n = t$ *and there exist* $a \in Q$ *and* $\alpha \in C$ *such that* $F(x) = xa$ *and* $F_1(x) = (a + \alpha)x$ *for all* $x \in R$;
3. C *is a finite field,* $R = Q = M_l(C)$, *the* $l \times l$ *matrix ring over* C *for some integer* $l \geq 2$ *and there exist* $a, u \in R$ *such that* $F(x) = xa$ *and* $F_1(x) = ux$ *for all* $x \in R$.

For the case $R \cong M_l(C)$, the $l \times l$ matrix ring over a finite field C, there exist generalized derivations F and G of R which satisfy the assumption in Theorem 4.4.16 but they are not of the forms described in Theorem 4.4.16(1) or Theorem 4.4.16(2). Liu [423] provided the following example illustrating the same.

Example 4.4.5 *[423, Example] Let* $R = M_l(F)$ *be the* $l \times l$ *matrix ring over a finite field* F, *where* $l \geq 2$ *is an integer; let* $\{e_{ij} \mid 1 \leq i, j \leq l\}$ *be the set of the usual matrix units in* R *and let* I *be the identity matrix of* R. *Then,* R *is a* *noncommutative prime ring with an extended centroid $C = Z(R) = F \cdot I_l$ and maximal right ring of quotients $Q = R = M_l(F)$. Let $a = e_{11}$ and $u = e_{12}$. Then, $a \notin C = F \cdot I_l$ and $u \notin C \cdot a + C$. Since R is a finite ring, in view of the proof of [394, Remark 1], there exist positive integers p, q with $q < p$ such that $x^q = x^p$ for all $x \in R$. Let $r = p - q$. Then, $x^q = x^{q+r} = x^{q+2r} = x^{q+3r}$ for all $x \in R$. Let $m = q$, $n = q + r$, $s = q + 2r$ and $t = q + 3r$. Then, $x^q = x^m = x^n = x^s = x^t$ for all $x \in R$. Hence, $[x^m a x^n, x^r] = [x^q a x^q, x^r] = x^q a x^{q+r} - x^{q+r} a x^q = x^q a x^q - x^q a x^q = 0$ for all $x \in R$. Similarly, we have $[x^m u x^n, x^r] = [x^s u x^t, x^r] = 0$ for all $x \in R$. In particular, $[x^m a x^n - x^s u x^t, x^r] = 0$ and $[x^m a x^n - x^m u x^n, x^r] = 0$ for all $x \in R$. Let $F : R \to R$ and $F_1 : R \to R$ be the generalized derivations of R defined by $F(x) = xa$ and $F_1(x) = ux$ for all $x \in R$. Then, we have $[F(x^m)x^n - x^s F_1(x^t), x^r] = 0$ and $[F(x^m)x^n - x^m G(x^n), x^r] = 0$ for all $x \in R$ but F and G are not of the forms described in Theorem 4.4.16 1 or Theorem 4.4.16 2.*

The following results are immediate consequences of Theorem 4.4.16.

Corollary 4.4.1 *[423, Corollary 1.3] Let R be a noncommutative prime ring with extended centroid C. Suppose that F is a generalized derivation of R and δ is a derivation of R such that $[F(x^m)x^n + x^n \delta(x^m), x^r]_k = 0$ for all $x \in R$, where m, n, r, k are fixed positive integers. Then $\delta = 0$ and one of the following conditions holds: (1) There exists $a \in C$ such that $F(x) = ax$ for all $x \in R$; (2) C is a finite field, $R \cong M_l(F)$, the $l \times l$ matrix ring over C for some integer $l \geq 2$ and there exists $a \in R$ such that $F(x) = xa$ for all $x = R$.*

Corollary 4.4.2 *[423, Corollary 3.4] Let R be a noncommutative prime ring and I a nonzero ideal of R. Let F be a generalized derivation of R with d as its associated derivation and let $0 \neq \gamma \in C$. Suppose that $[F(x^m)x^n + \gamma x^s F(x^t), x^r]_k = 0$ for all $x \in R$, where m, n, s, t, r, k are fixed positive integers. Then $d = 0$ and there exists $\alpha \in C$ such that $F(x) = \alpha x$ for all $x \in R$.*

Proof: Let $h = -\gamma F$. Then, h is a generalized derivation of R with $\delta = -\gamma d$ as its associated derivation. By assumption, we have $[F(x^m)x^n + \gamma x^s F(x^t), x^r]_k = 0$ for all $x \in R$. By Theorem 4.4.16, $\delta = -\gamma d = 0$ and there exist $a, u \in Q$ such that $F(x) = xa$ and $h(x) = ux$ for all $x \in R$. From $\delta = -\gamma d = 0$ and $0 \neq \gamma \in C$, it follows that $d = 0$. Next using $ux = h(x) = -\gamma F(x)$ for all $x \in R$ and $0 \neq \gamma \in C$, we get $F(x) = (-\gamma^{-1}u)x$ for all $x \in R$. Hence,$(-\gamma^{-1}u)x = F(x) = xa$ for all $x \in R$, implying that $(-\gamma^{-1}u)x = xa$ for all $x \in R$ and thus for all $x \in Q$ by [104, Theorem 6.4.4]. Setting $x = 1$, we get $(-\gamma^{-1}u) = a$ and then $ax = xa$ for all $x \in Q$. This implies that $a \in C$. Consequently, $F(x) = xa = ax$ for all $x \in R$, where $a \in C$, proving the corollary.

♣

In [241, Theorem 3.6], Dhara and Ali demonstrated that if R is a $n!$-torsion free semiprime ring and F is a generalized derivation of R with the associated derivation d such that $F(x)x^n = x^n F(x)$ for all $x \in I$, where I is a nonzero ideal of R, and n is a fixed positive integer, then either $d(I)=0$ or $d(I)$ is contained in a nonzero central ideal of R. On the other hand, Alahmadi et al. [4, Theorem 1.1] showed that if R is a noncommutative prime ring with an extended centroid C and F is a generalized derivation of R such that $[F(x^m)x^n+x^nF(x^m),x^r]_k=0$ for all $x\in R$, where m,n,r,k are fixed positive integers, then there exists $a\in C$ such that $F(x)=ax$ for all $x\in R$. As an application of Theorem 4.4.16, Liu generalized the above two results as follows:

Corollary 4.4.3 *[423, Corollary 1.4] Let R be a semiprime ring with extended centroid C and maximal right ring of quotients Q. Let I be a nonzero ideal of R and let F be a generalized derivation of R. Suppose that $[F(x^m)x^n+\gamma x^sF(x^t),x^r]_k=0$ for all $x\in I$, where m,n,s,t,r,k are fixed positive integers and $\gamma\in\{1,-1\}$. Then there exist $a\in Q$ and a derivation d of R such that $F(x)=ax+d(x)$ for all $x\in R$, $[a,QIQ]=0$ and $\langle d(I)\rangle\subseteq Z(R)$, where $\langle d(I)\rangle$ denotes the ideal of R generated by $d(I)$. Particularly, in case $I=R$, we have $a\in C$ and $d(R)\subseteq Z(R)$.*

Proof: By Theorem 4 [378], there exist $a\in Q$ and a derivation d of R such that $F(x)=ax+d(x)$ for all $x\in R$. It is known that d can be uniquely extended to a derivation of Q [372, Lemma 2]. B denotes the complete Boolean algebra of idempotents in C [97]. Fix a maximal ideal P of B. Then, PQ is a prime ideal of Q, invariant under all derivations of Q [97]. Let $\overline{d}$ be the canonical derivation of $\overline{Q}=Q/PQ$ induced by d. Then, $\overline{d}$ is a derivation of the prime ring $\overline{Q}$ defined by $\overline{d}(\overline{x})=\overline{d(x)}$ for $x\in Q$, where $\overline{x}=x+PQ\in\overline{Q}$. Moreover, $\overline{F}$ is a generalized derivation of $\overline{Q}$ defined by $\overline{F}(\overline{x})=\overline{a}\cdot\overline{x}+\overline{d}(\overline{x})$ for $x\in Q$. By assumption, $[F(x^m)x^n+\gamma x^sF(x^t),x^r]_k=0$ for all $x\in RIR$ and hence for all $x\in QIQ$ by [372, Theorem 3]. So, we have

$$[\overline{F}(\overline{x}^m)\overline{x}^n+\gamma\overline{x}^s\overline{F}(\overline{x}^t),\overline{x}^r]_k=0 \tag{4.29}$$

for all $x\in QIQ$. Applying Corollary 4.4.2 to (4.29), we see that $\overline{QIQ}=\overline{0}$ or $\overline{Q}$ is commutative or $\overline{d}=0$ and there exists an element α in the extended centroid of $\overline{Q}$ such that $\overline{F}(\overline{x})=\alpha\overline{x}$ for all $x\in Q$, implying that $\overline{a}=\alpha$ as $\alpha\overline{x}=\overline{F}(\overline{x})=\overline{a}\cdot\overline{x}+\overline{d}(\overline{x})=\overline{a}\cdot\overline{x}$ for all $x\in Q$. Hence

$$\overline{QIQ}=\overline{0}\quad\text{or}\quad[\overline{Q},\overline{Q}]=\overline{0}\quad\text{or}\quad\overline{d}(\overline{Q})=\overline{0}\quad\text{and}\quad[\overline{a},\overline{Q}]=\overline{0}. \tag{4.30}$$

Suppose $\overline{QIQ}=\overline{0}$, then $\overline{I}=\overline{0}$ and then $\overline{d}(\overline{I})=\overline{0}$. Consequently, by (4.30) we have $[\overline{a},\overline{QIQ}]=\overline{0}$ and $[\overline{Qd(I)Q},\overline{Q}]=\overline{0}$. This implies that $[a,QIQ]\subseteq PQ$ and $[Qd(I)Q,Q]\subseteq PQ$. Since $\bigcap\{PQ|P$ is any maximal ideal of $B\}=0$ [97], we conclude that $[a,QIQ]=0$ and $[Qd(I)Q,Q]=0$, implying that $Qd(I)Q\subseteq C$. Hence, $\langle d(I)\rangle\subseteq Qd(I)Q\cap R\subseteq C\cap R=Z(R)$. Suppose $I=R$. Then, $\langle d(I)\rangle\subseteq Z(R)$ and $[a,QRQ]=0$. So $d(R)\subseteq Z(R)$ and $[a,Q]=0$ by [372, Theorem 3], implying that $a\in C$. This proves the corollary.

♣

4.5 Generalized skew derivations and their related problems in rings

Definition 40 *An additive map $\delta : R \longrightarrow R$ is said to be an* **α-derivation** *of R if*

$$\delta(xy) = \delta(x)y + \alpha(x)\delta(y)$$

for all $x, y \in R$, α is called an associated automorphism of R. For brevity, an α-derivation is called a **skew derivation.**

When $\alpha = 1$, the identity maps of R (skew derivation) are merely ordinary derivations.

Example 4.5.1 *It is easy to see that*

$$\begin{aligned}(\alpha - I_R)xy &= \alpha(x)\alpha(y) - xy \\ &= \alpha(x)\alpha(y) - \alpha(x)y + \alpha(x)y - xy \\ &= (\alpha - I_R)(x)y + \alpha(x)(\alpha - I_R)(y).\end{aligned}$$

This shows that $\alpha - I_R$ is an example of the skew derivation over a ring R.

In the realm of pure algebra and operator algebras, extensive exploration has been conducted on skew derivations and generalized derivations of associative algebras. A natural question emerges: is there a unified definition encompassing skew and generalized derivations? Based on this idea, Cheng and Wei [177] propose a novel definition that collectively generalizes the preceding definitions. This concept is termed as the **generalized skew derivation** and is articulated as follows:

Definition 41 *[177, Definition 2.1] Let R be a ring and α be an automorphism of R. An additive map $G : R \to R$ is called a* **generalized skew derivation** *of R if there exits a skew derivation δ of R with the associated automorphism α such that*

$$G(xy) = \alpha(x)G(y) + \delta(x)y$$

for all $x, y \in R$.

Example 4.5.2 *Let $\mathcal{A}$ be an associative algebra and α be an automorphism of $\mathcal{A}$. Then an additive map G such that $G(xy) = \alpha(x)G(y)$ or $G(xy) = G(x)\alpha(y)$ for all $x, y \in \mathcal{A}$ is a generalized skew derivation. In particular, if $\alpha = I_{\mathcal{A}}$, then $G(xy) = xG(y)$ or $G(xy) = G(x)y$ for all $x, y \in \mathcal{A}$. This shows that all right or left multipliers mapping on $\mathcal{A}$ are generalized skew derivations of $\mathcal{A}$.*

In [165], the modified definition of a generalized skew derivation was proposed by Chang and is defined as:

Definition 42 *An additive map $G : R \longrightarrow R$ is said to be a* **right generalized α-derivation** *of R if there exists a skew derivation δ of R with an associated automorphism α such that*

$$G(xy) = G(x)y + \alpha(x)\delta(y)$$

for all $x, y \in R$. The map δ is called an associated skew derivation of G and automorphism α is called the associated automorphism of G. For brevity, the right generalized α-derivation is called a generalized skew derivation.

Based on the provided definition, it becomes evident that general outcomes regarding generalized skew derivations can yield valuable and potent corollaries concerning derivations, generalized derivations, and skew derivations. To illustrate this, we present the following examples of generalized skew derivations.

Example 4.5.3 *Let S be any ring and let $R = \left\{ \begin{pmatrix} a & b \\ 0 & 0 \end{pmatrix} : a, b \in S \right\}$. Define a map $G : R \longrightarrow R$ by $G(x) = 2e_{11}x - xe_{11}$ for all $x \in R$, where e_{ij} is a unit matrix having 1 at ij_{th} entry and zero elsewhere. Clearly, G is a generalized skew derivation on R induced by an associated skew derivation δ defined by $\delta(x) = e_{11}x - xe_{11}$ and an associated automorphism $\alpha : R \longrightarrow R$ given by $\alpha \begin{pmatrix} a & b \\ 0 & 0 \end{pmatrix} = \begin{pmatrix} a & -b \\ 0 & 0 \end{pmatrix}$.*

Example 4.5.4 *Let $\mathcal{A}$ be an associative algebra and $a, b \in \mathcal{A}$ such that one of them is not equal to zero and let α be an automorphism of $\mathcal{A}$. Define $G : \mathcal{A} \longrightarrow \mathcal{A}$ by $G(x) = ax + \alpha(x)b$ for all $x \in \mathcal{A}$. Then for $x, y \in \mathcal{A}$, we have $G(x+y) = G(x) + G(y)$ and*

$$\begin{aligned} G(xy) &= axy + \alpha(xy)b \\ &= axy + \alpha(x)\alpha(y)b = axy - \alpha(x)by + \alpha(x)by + \alpha(x)\alpha(y)b \\ &= (ax + \alpha(x)b)y + \alpha(x)(\alpha(y)b - by) \\ &= G(x)y + \alpha(x)\delta(y). \end{aligned}$$

Here, $\delta(y) = \alpha(y)b - by$. It is easy to check that d is an inner skew derivation induced by the element $b \in \mathcal{A}$; such generalized skew derivations are called **generalized skew inner derivations**.

Remark 4.5.1 *Let $\mathcal{A}$ be an associative algebra. All derivations, generalized derivations and skew derivations are also generalized skew derivations.*

Lemma 4.5.2 [177, Lemma 2.6] *Let $\mathcal{A}$ be an associative algebra with unity 1 and $G : \mathcal{A} \to \mathcal{A}$ be a linear map. Then the following statements are equivalent:*

1. *G is a generalized skew derivation of* $\mathcal{A}$.
2. *There exists a skew derivation d of* $\mathcal{A}$ *with an associated automorphism* α *such that* $G(x) = \alpha(x)F(1) + \delta(x)$ *for all* $x \in \mathcal{A}$.
3. *The equality* $G(xyz) = G(xy)z - \alpha(x)G(y)z + \alpha(x)G(yz)$ *holds for all* $x, y, z \in \mathcal{A}$.

Lemma 4.5.3 [177, Lemma 2.7] *Let R be a prime ring with unity 1 and G be a generalized skew derivation of R with the associated automorphism* α *and the associated skew derivation d. Then G is a generalized inner skew derivation of R iff d is an inner skew derivation of R.*

Proof: If G represents an inner generalized skew derivation of R, then there exist two fixed elements a and b in R such that $G(x) = ax + \alpha(x)b$ for all x in R. This deduction stems from Lemma 4.5.2, implying $G(x) = \alpha(x)G(1) + \delta(x)$ for all x in R. Consequently, $ax + \alpha(x)b = \alpha(x)G(1) + \delta(x)$ holds for all x in R, leading to $\delta(x) = ax - \alpha(x)(G(1) - b)$ for all x in R. Thus, $\delta(xy) = axy - \alpha(x)\alpha(y)(G(1) - b)$ for all x, y in R.

Observe that $\delta(xy) = \delta(x)y + \alpha(x)\delta(y) = axy - \alpha(x)(G(1) - b)y + \alpha(x)ay - \alpha(x)\alpha(y)(G(1) - b)$. This implies $\alpha(x)(G(1) - b)y = \alpha(x)ay$, for all $x, y \in R$, suggesting $(G(1) - b - a)R = 0$. Since R is a prime ring, $G(1) - b = a$. Consequently, $\delta(x) = ax - \alpha(x)(G(1) - b) = ax - \alpha(x)a$ for all x that lie in R. This demonstration establishes that δ is an inner skew derivation of R.

Conversely, if δ represents an inner skew derivation, then there exists s that lies in R such that $\delta(x) = sx - \alpha(x)s$ for all x that lie in R. By Lemma 4.5.2, it follows that $G(x) = \alpha(x)G(1) + sx - \alpha(x)s = sx - \alpha(x)(s - G(1))$ for all x lying in R. This illustrates that G is an inner generalized skew derivation of R.

♣

Cheng and Wei [177] addressed the extension problem of generalized skew derivations on semiprime rings. This problem can be formulated as follows:

Theorem 4.5.4 [177, Theorem 3.1] *Let R be a semiprime ring, K be an essential ideal of R and G be a generalized skew derivation on K with the associated automorphism* α *and the associated skew derivation* δ. *Then G can be uniquely extended to a generalized skew derivation of* R_F *and that of Q. Where* R_F *is a left Martindale quotient ring and Q is a symmetric Martindale quotient ring.*

Also in [177], they also discussed generalized skew derivations with nilpotent values on prime rings, which is a natural generalization of Lemma 1 in [378], which can be stated as

Theorem 4.5.5 [177, Theorem 4.1] *Let R be a prime ring, K be an essential ideal of R and G be a generalized skew derivation on K with the associated*

automorphism α and the associated skew derivation d. If $(G(x))^n = 0$ for all $x \in K$, where n is a fixed positive integer, then $\mu = 0$.

The authors make certain assumptions in the work by Lee and Liu (referenced as [382]). Specifically, they consider a prime algebra R over a field $\mathbb{F}$, an automorphism σ of R, and a nonzero generalized σ-derivation (generalized skew derivation) G of R with an associated additive map δ. The assumption is that for every element x in R, $G(x)$ is algebraic over $\mathbb{F}$ and has a bounded degree m. Further, the authors present a unified framework encompassing skew and generalized derivations. This unification is achieved by defining the latter in terms of an additive map, replacing the concept of the skew derivation as given by Cheng and Feng in [177].

Definition 43 *[382] Let R be a prime ring with an automorphism σ. An additive map $G : R \to R$ is called a* **generalized σ-derivation (or a generalized skew derivation)** *if there exists an additive map $\delta : R \to R$ such that $G(xy) = \sigma(x)G(y) + \delta(x)y$ for all $x, y \in R$. It is clear that δ is uniquely determined by G, which is called the associated additive map of G.*

In [382], the authors initiate the proof establishing that the associated additive map δ serves as a σ-derivation (skew derivation). They also provide the characterization of a generalized σ-derivation (or generalized skew derivation). In fact, they proved that

Theorem 4.5.6 [382, Theorem 2.1] *Let R be a prime ring with an automorphism σ. Suppose that $G : R \to R$ is a generalized skew derivation. Then its associated additive map δ is a σ-derivation (skew derivation) of R. Moreover, there exists $b \in R_F$ such that $G(x) = \delta(x) + \sigma(x)b$ for all $x \in R$.*

Proof: Let $x, y, z \in R$. By definition, we have

$$G(x(yz)) = \sigma(x)G(yz) + \delta(x)yz = \sigma(x)(\sigma(y)G(z) + \delta(y)z) + \delta(x)yz.$$

On the other hand, we have

$$G((xy)z) = \sigma(xy)G(z) + \delta(xy)z.$$

Comparing the above two equations, we see that $\delta(xy)z = \sigma(x)\delta(y)z + \delta(x)yz$. By the primeness of R, $\delta(xy) = \sigma(x)\delta(y) + \delta(x)y$. Thus, δ is a σ-derivation (skew derivation) of R. Define $h(x) = F(x) - \delta(x)$ for all $x \in R$. Then by a direct computation, $h(xy) = \sigma(x)h(y)$ and so $\sigma^{-1}(h(xy)) = x\sigma^{-1}(h(y))$ for all $x \in R$. That is, the map $x \in R \to \sigma^{-1}(h(x))$ is a left R-module map. Thus, there exists $c \in R_F$ such that $\sigma^{-1}(h(y)) = yc$ for all $y \in R$. So, $h(y) = \sigma(y)\sigma(c)$ for all $y \in R$. We are done by setting $b = \sigma(c) \in R_F$. This proves the theorem.

♣

Lemma 4.5.7 *[382, Lemma 2.2] The ring R is a GPI-ring.*

Lemma 4.5.8 *[382, Lemma 2.2] Let R be a prime algebra over a field $\mathbb{F}$ and let G be a nonzero generalized σ-derivation (generalized skew derivation) of R. Suppose that $G(x)$ is algebraic over $\mathbb{F}$ with a bounded degree for all $x \in R$, then R is a strongly primitive ring if R contains a nontrivial idempotent.*

Proof: Suppose on the contrary that R is not a strongly primitive ring. Let e be a nontrivial idempotent in R. For $r \in R$ we note that $(r - r\sigma(e))\sigma(e) = 0$. In view of Lemma 4.5.9, we see that $(r - r\sigma(e))\delta(e) = 0$. Thus $\delta(e) = \sigma(e)\delta(e)$. On the other hand, $\sigma(e)\sigma((\sigma(r) - e\sigma(r))) = 0$, so, by Lemma 4.5.9 again, $\sigma(e)\delta((\sigma(r) - e\sigma(r))) = 0$. That is, $\sigma(e)\delta(e)\sigma(r) = 0$ for all $r \in R$ and so $\sigma(e)\delta(e) = 0$. Thus, $\delta(e) = 0$ follows.

♣

Definition 44 *A ring R is called* **strongly primitive** *if R is a primitive ring with a minimal right ideal eR, where $e^2 = e \in R$, such that eRe is a finite-dimensional central division algebra.*

Theorem 4.5.9 *[382, Lemma 2.3] Suppose that R is not a strongly primitive ring. If $y\sigma(a) = 0$ for some $a, y \in r$, then $y\delta(a) = 0$.*

Proof: We set

$$\lambda = l_R(\sigma(a)) = \{x \in R | x\sigma(a) = 0\},$$

the left annihilator of $\sigma(a)$ in R. Let $y \in \lambda$. By assumption $g(ay) = \delta(ay) + \sigma(ay)b$ is F-algebraic of degree $\leq m$. Thus, there exists $\beta_1, \ldots, \beta_{m-1} \in F$ such that

$$0 = (\delta(ay) + \sigma(ay)b)^m + \beta_1(\delta(ay) + \sigma(ay)b)^{m-1} + \cdots + \beta_{m-1}(\delta(ay) + \sigma(ay)b). \quad (4.31)$$

Note that

$$y(\delta(ay) + \sigma(ay)b)^k = y(\sigma(a)\delta(y) + \delta(a)y + \sigma(a)\sigma(y)b)^k = y(\delta(a)y)^k$$

for each $k \geq 1$. Thus multiplying (4.31) by $\sigma(a)$ from the right side and by y from the left side, we see that

$$(y\delta(a))^{m+1} + \beta_1(y\delta(a))^m + \cdots + \beta_{m-1}(y\delta(a))^2 = 0. \quad (4.32)$$

It follows from (4.32) that $\lambda\delta(a)$ is a left ideal of R which is F-algebraic of bounded degree $m+1$. But R is not a strongly primitive ring, by [380, Lemma 1]; thus, we have $\lambda\delta(a) = 0$. In particular, $y\delta(a) = 0$ follows.

♣

Theorem 4.5.10 *[382, Lemma 2.5] Let V be a vector space over a division ring D and $a, b \in End_D(V)/\{0\}$. Suppose that va and vb are D-dependent for each $v \in V$. If either $dim_D Va \geq 2$ and $dim_D Vb \geq 2$, then there exists $\beta \in K$, the centre of D, such that $b = \beta a$, where K is a field.*

Proof: We may assume, without loss of generality, that $dim_D Va \geq 2$. Choose $u, w \in V$ such that va and wa are D-independent. Then, by assumption, there exist $\beta, \mu, \zeta \in D$ such that $vb = \beta va, wb = \mu wa$ and $(v+w)b = \zeta(v+w)a$. Thus $(\zeta - \beta)(va) + (\zeta - \mu)(wa) = 0$. It follows from the D-independence of va and wa that $\zeta = \mu = \beta$. Let $z \in V$. If $za \neq 0$, then either va and za are D-independent or wa and za are D-independent. In either case, $zb = \beta(za)$. This proves the fact that there exists $\beta \in D$ such that $vb = \beta(va)$ for all $v \in V$. Let $\eta \in D$ and $v \in V$. Then $(\eta v)b = \beta((\eta v)a) = \beta\eta(va)$, and on the other hand, $(\eta v)b = \eta(v)b = \eta\beta(va)$. Hence, $\beta\eta = \eta\beta$ for all $\eta \in D$. That is, $\beta \in K$, the centre of D.

♣

Lemma 4.5.11 *[382, Lemma 2.6] Suppose,* $(\sigma(x)a - cx)^m = 0$ *for all* $x \in R$*, where* σ *is an automorphism of* R*,* $a, c \in R$ *and* m *is a fixed positive integer. Then* $\sigma(x)a - cx = 0$ *for all* $x \in R$.

Theorem 4.5.12 *[382, Theorem 2.7] Let* R *be a prime ring with a (generalized skew derivation) generalized* σ*-derivation* G*. Suppose that* $(G(x))^m = 0$ *for all* $x \in J$*, a nonzero ideal* J *of* R*, where* m *is a fixed positive integer. Then* $G = 0$.

Proof: We write $G(x) = \delta(x) + \sigma(x)b$ for $x \in R$, where $\delta : R \to R$ is the associated σ-derivation of G. By [170, Theorem 2],

$$(\delta(x) + \sigma(x)b)^m = 0 \tag{4.33}$$

for all $x \in R_F$. If δ is X-outer, then by [170, Theorem 1] we have $(y + \sigma(x)b)^m = 0$ for all $x, y \in R$. In particular $y^m = 0$ for all $y \in R$. By [294, Levitzki's Lemma], this is a contradiction. Now assume that δ is x-inner. we write $\delta(x) = \delta(x)c - cx$ for all $x \in R$, where $c \in Q$. We write (4.33) as

$$(\sigma(x)(b+c) - cx)^m = 0$$

for all $x \in R$ and so for all $x \in R_F$ [187, Theorem 1]. By Lemma 4.5.11, $\sigma(x)(b+c) - cx = 0$ for all $x \in R_F$. Thus $g = 0$ follows. This proves the theorem.

♣

In reference [252], De Filippis established that if R is a prime algebra over a field $\mathbb{F}$, and d is a nonzero derivation of R such that for every x in R, the product xd is algebraic over $\mathbb{F}$ with a bounded degree, then R is a strongly primitive ring. Lee and Lee later extended De Filippis's theorem to both the right ideal and the Lie ideal cases, as outlined in [380, Theorems 1 and 2]. Skew derivations and generalized derivations emerge as natural extensions of derivations. In [382], Lee and Liu further extend De Filippis's theorem by unifying the two cases through a generalized skew derivation. This extension can be expressed as:

Theorem 4.5.13 [382, Main Theorem] *Let* R *be a prime algebra over a field* $\mathbb{F}$ *and let* G *be a nonzero generalized* σ*-derivation (generalized skew derivation)* *of R. Suppose, $G(x)$ is algebraic over $\mathbb{F}$ of bounded degree for all $x \in R$. Then R is a strongly primitive ring.*

Proof: By Lemma 4.5.7, R is a prime GPI-ring. By Lemma 4.5.8, we may assume that R has no nontrivial idempotents. Let I be a nonzero proper ideal of R, then $G(\sigma^{-1}(I)I) \subseteq IG(I) + \sigma^{-1}G(I)I \subseteq I$. For x $\in \sigma^{-1}(I)I$, by [293, Lemma 1.3.2], we have either $G(x)$ is invertible or $(G(x))^m = 0$. Since $G(x) \in I$ and I is a proper ring, then $G(x)$ cannot be invertible. Thus $(G(x))^m = 0$ for all $x \in \sigma^{-1}(I)I$. In view of Theorem 4.5.12, we have $G = 0$, which is a contradiction. This means that R is a simple ring. Thus $R = RC$. Sine R is a prime GPI-ring, by Martindale's Theorem [447], R itself is a strongly primitive ring. This completes the proof.

♣

In [379], Lee started with a definition of generalized skew derivations, generalizing both skew derivations and generalized derivations and gave a characterization of generalized skew derivations by acting on zero products. Precisely, if $g, \delta : R \to R$ are additive maps such that $\sigma(x)g(y) + \delta(x)y = 0$ for all $x, y \in R$ with $xy = 0$, where σ is an automorphism of R, then both g and δ are characterized as specific generalized σ-derivations (generalized skew derivation) on a nonzero ideal of R.

Definition 45 *An additive map $G : R \to Q_{ml}$ is called a* **generalized σ-derivation (generalized skew derivation)**, *where σ is an automorphism of R, if there exists an additive map $\delta : R \to Q_{ml}$ such that $G(xy) = \sigma(x)G(y) + \delta(x)y$ for all $x, y \in R$. It is clear that σ is uniquely determined by G, which is called the associated additive map of G.*

The subsequent theorem provides a characterization of generalized skew derivations based on their action on specific subsets (such as zero products or local actions) within the ring R. This can be formulated as:

Theorem 4.5.14 [379, Theorem 1.1] *Let R be a prime ring with symmetric Martindale quotient ring Q. Suppose that Q contains a nontrivial idempotent and that $g, \delta : R \to R$ are additive maps. If $\sigma(x)g(y) + \delta(x)y = 0$ for all $x, y \in R$ with $xy = 0$, where σ is an automorphism of R, there exists a nonzero ideal N of R and a σ-derivation $d : R \to Q$ such that*

$$g(x) = d(x) + \sigma(x)b$$

and

$$\delta(x) = d(x) + ax$$

for all $x \in N$, where $b \in Q_{ml}$ and $a \in Q_{mr}$. In addition, we can take $N = R$ if $eR \cup Re \subseteq R$ for some nontrivial idempotent $e \in Q$.

As an immediate consequence of Theorem 4.5.14, we have the following generalization of Chebotar, Ke and Lee's theorem *[176, Theorem 2]*:

Corollary 4.5.1 [176, Corollary 1.3] *Let R be a prime ring with extended centroid C and symmetric Martindale quotient ring Q. Suppose that Q contains a nontrivial idempotent. If $\delta : R \to R$ is an additive map such that $x\delta(y)+\delta(x)y=0$ for $x,y \in R$ with $xy=0$, then, there exists a nonzero ideal N of R such that $\delta(x)=d(x)+\lambda x$ for all $x \in N$, where $\lambda \in C$ and $d : R \to Q$ is a derivation. In addition, we can take $R=A$ if $eR \cup Re \subseteq R$ for some nontrivial idempotent $e \in Q$.*

The subsequent outcome presents Lu and Li's theorem *[329, Theorem 6]* within the framework of prime rings:

Corollary 4.5.2 [379, Corollary 1.3] *Let R be a prime ring with extended centroid C and $c \in R$. Suppose that R possesses a nontrivial idempotent. If $\delta : R \to R$ is an additive map such that $x\delta(y)+\delta(x)y+xcy=0$ for $x,y \in R$ with $xy=0$, then there exist a derivation $d : R \to RC$ and $\mu \in C$ such that $\delta(x)=d(x)+(\mu-C)x$ for all $x \in R$.*

Proof: By assumption, we have $x\delta(y)+\delta(x)y+xcy=0$ for $x,y \in R$ with $xy=0$. In view of Theorem 4.5.14, there exist a derivation $d : R \to Q$, and $\mu \in Q_{ml}$ such that $\delta(x)=d(x)+ax$ and $\delta(c)+cx=d(x)+x\mu$ for all $x \in R$. Choose a dense right ideal ρ and a dense left ideal λ of R such that $a\rho \subseteq R$ and $\lambda\mu \subseteq R$. Let $x \in \rho, z \in R$ and $y \in \lambda$. Then $xyz \in \rho R\lambda \subseteq \rho \cup \lambda$. Thus $(a+c)x, y\mu \in R$ and $((a+c)x)zy = xz(y\mu)$. By Martindale's Lemma [447], y and $y\mu$ are C-dependent for $y \in \lambda$. It is now easy to prove that $\mu \in C$. Thus $a=\mu-c$ follows, so $\delta(x)=d(x)+(\mu-c)x$ for all $x \in R$. In particular, we have $d(R) \subseteq RC$, proving the corollary.

♣

The following application is relevant to generalized polynomial identities. An additive map $f : R \to R$ is called an elementary operator if there exist finitely many $a_i, b_i \in RC$ such that $f(x)=\sum_i a_i x b_i$ for all $x \in R$.

Corollary 4.5.3 *[379, Corollary 1.4] Let R be a prime ring with an extended centroid C. Suppose that its symmetric Martindale quotient ring Q contains a nontrivial idempotent. If f and g are two elementary operators of R satisfying $xf(y)+g(x)y=0$ for $x,y \in R$ with $xy=0$, then there exist $a,b,q \in RC$, such that $f(y)=[q,y]+yb$ and $g(x)=[q,x]+ax$ for all $x,y \in R$.*

In [455], Mayne proved the analogous result for nonidentity centralizing automorphisms. Several authors in the literature have obtained many related generalizations. See, for instance, [364, 366, 384]. In [207], De Filippis provided the proof that follows. Let R be a prime ring of characteristic not equal to 2, d a nonzero derivation and L a noncentral Lie ideal of R. For $a \in R$, if

$a[d(u),u] = 0$ for all $u \in L$, then $a = 0$. That is the left annihilator of the set $\{[d(u),u]/u \in L\}$. Also, Shiue in [518] generalized this result by imposing the condition: $a[d(u),u]_k = 0$ for all $u \in L$, where k is a fixed positive integer. Chuang et al. extend Shiue's result to skew derivations. Then Chang [165] extends further to the so-called generalized skew derivations. Which can be stated as:

Theorem 4.5.15 [165, Main Theorem] *Let R be a noncommutative prime ring and $a \in R$. Suppose that G is a right generalized skew derivation of R such that $a[G(x),x]_k = 0$ for all $x \in R$, where k is a fixed positive integer. Then $a = 0$ or there exists $s \in C$ such that $G(x) = sx$ for all $x \in R$ except when $R = M_2(GF(2))$.*

In 2014, Rehman and Raza [502] proved the following result: Let R be a prime ring, I a nonzero ideal of R and $n \geq 1$ a fixed integer. If R admits a skew derivation δ associated with an automorphism σ of R such that $\delta([x,y]) = [x,y]_n$ for all $x,y \in I$, then R is commutative. In view of the above discussion, it would be natural to ask what is the structure of a ring that admits a generalized skew derivation G and satisfies the identity $G([x,y]) = [x,y]_n$. Recently [510] extends the above mentioned theorem of Rehman and Raza [502] systematically by using the theory of generalized polynomial identities. The main result in [510] is stated as follows:

Theorem 4.5.16 *Let R be a prime ring with $char(R) \neq 2$ and L be a Lie ideal of R. If $G : R \to R$ is a generalized skew derivation with an associated automorphism α and skew derivation δ such that $G([x,y]) = [x,y]_n$ for all $x,y \in L$ and fixed integer $n \geq 0$, then either $L \subseteq Z(R)$ or $n = 1$ and $G = I_R$, is the identity mapping of R.*

In [298], Herstein proved that if D is an inner derivation of a prime ring R such that $D(x)^n = 0$ (resp. $D(x)^n$ is central) for all $x \in R$; then $D = 0$, (resp. $dim_C RC \leq 4$ where C is the extended centroid of R, which is the field in prime ring). Later, Giambruno and Herstein [278] extended this result to an arbitrary derivation D for the nilpotent case and then Herstein [299] took care of the case where powers are central. In [378], Lee extended this result to generalized derivations when R is a prime ring. Motivated by these results, Chang in [167] and [169] extend further to the right generalized skew derivation (generalized skew derivation). Consequently, this result also covers the case of a skew derivation. This can be stated as:

Theorem 4.5.17 *[167, Main Theorem] Let R be a prime ring and L a noncommutative Lie ideal of R. Suppose that G is a right generalized skew derivation (generalized σ-derivation) of R associated with a σ-derivation such that $G(x)^n = 0$ for all $x \in L$, where n is a fixed positive integer. Then $G = 0$.*

Theorem 4.5.18 *[169, Main Theorem] Let R be a prime ring with center Z and L a noncommutative Lie ideal of R. Suppose that G is a right generalized skew derivation (generalized σ-derivation) of R associated with a σ-derivation such* *that $G(x)^n \in Z$ for all $x \in L$, where n is a fixed positive integer. Then $G = 0$ unless $dim_C RC = 4$.*

In 2014, Liu et al. [424] continued this investigation and studied generalized skew derivations with nilpotent values on left ideals. As usual, they denote it as $J(R)$ the Jacobson radical of R. We say that R is semisimple if $J(R) = 0$. The first main result reads as follows:

Theorem 4.5.19 [424, Theorem 1.1] *Let R be a unital ring, L a nonzero left ideal of R and G a generalized skew derivation (generalized σ-derivation) of R, where σ is an automorphism of R. Suppose that $G(x)^{n(x)} \in J(R)$ for all $x \in L$, where $n(x)$ is a positive integer depending on x. Then $LG(L) \subseteq J(R)$ and $G(L)G(L) \subseteq J(R)$. In particular, $G(R) \subseteq J(R)$ if $L = R$.*

Let V be a right vector space over a division ring D and let $End(V_D)$ denote the ring of all D-linear transformations on V. A subring R of $End(V_D)$ is said to be dense if given any finitely many D-independent $v_i \in V$ and arbitrary $w_i \in V$, where $1 \leq i \leq n$, there exists $r \in R$ such that $rv_i = w_i$ for $1 \leq i \leq n$. As an application of Theorem 4.5.19, they proved the following theorem.

Theorem 4.5.20 [424, Theorem 1.2] *Let R be a unital dense subring of $End(V_D)$, where V is a right vector space over a division ring D, L a nonzero left ideal of R and G a generalized σ-derivation of R, where σ is an automorphism of R. Suppose that $G(x)^{n(x)} = 0$ for all $x \in L$, where $n(x)$ is a positive integer depending on x. Then there exist $a, b \in End(V_D)$ with $\sigma(L)b = 0$ and $L(a) = 0$ such that $G(x) = ax + \sigma(x)b$ for all $x \in R$. Consequently, $LG(L) = 0$ and $G(L)G(L) = 0$.*

Let X be a Banach space over the field F of real or complex numbers; let $B(X)$ be the algebra of all bounded linear operators on X and let $F(X)$ be the subalgebra of all bounded linear operators on X of finite rank. A subalgebra A of $B(X)$ is standard provided A contains $F(X)$. Using Theorem 4.5.19, they obtained:

Theorem 4.5.21 [424, Theorem 1.3] *Let $\mathfrak{A}$ be a unital standard operator algebra on a Banach space X over the field $\mathbb{F}$ of real or complex numbers, L a nonzero left ideal of A and G a generalized σ-derivation of $\mathfrak{A}$, where σ is an automorphism of A. Suppose that $G(x)^{n(x)} = 0$ for all $x \in L$, where $n(x)$ is a positive integer depending on x. Then there exists $a, b \in B(X)$ with $\sigma(L)b = 0$ and $La = 0$ such that $G(x) = ax + \sigma(x)b$ for all $x \in A$. Consequently, $LG(L) = 0$ and $G(L)G(L) = 0$.*

Many results in the literature indicate how the global structure of a ring R is often tightly connected to the behaviour of additive mappings defined on R. A well-known result of Posner [487] states that if d is a derivation of R such that $[d(x), x] \in Z(R)$, for any $x \in R$, then either $d = 0$ or R is commutative. Later in [371], Carini et al. obtained the following:

Theorem 4.5.22 *Let R be a prime ring of characteristic not equal to 2, L a non-central Lie ideal of R, d a non-zero derivation of R such that $[d(u),u]^n \in Z(R)$, for any $u \in L$. Then R satisfies s_4. In particular, if d satisfies $[d(u),u]^n = 0$, for any $u \in L$, then R is commutative.*

In 2006, De Filippis [208] considered a similar situation in which a generalized derivation replaces the derivation d. Similarly, Wang in [543] also finds a similar situation in the case the derivation d is replaced by a non-trivial automorphism σ of R. Motivated by the above cited results, Carini et al. [160] continued this line of investigation, and they examined what happens in case $G \neq 0$ is a generalized skew derivation of R such that $[G(x),x]^n = 0$ for all $x \in R$ and $n \geq 1$. More precisely, they obtained the following result:

Theorem 4.5.23 [160, Theorem 4] *Let R be a non-commutative prime ring of characteristic $\neq 2$ with extended centroid C, $G \neq 0$ a generalized skew derivation of R, and $n \geq 1$ such that $[G(x),x]^n = 0$, for all $x \in R$. Then there exists an element $\lambda \in C$ such that $G(x) = \lambda x$ for all $x \in R$.*

A great deal of work has been done regarding derivations with nilpotent values in prime and semiprime rings. The first such result which inspired these works is owed to Herstein in [298], who proved that if R is a prime ring and d an inner derivation of R such that $d(x)^n = 0$ for all $x \in R$, where n is a fixed positive integer, then $d = 0$. Later in [378], Lee showed that if F is a generalized derivation of R and L a noncommutative Lie ideal of R such that $F(x)^n = 0$ for all $x \in L$, where n is a fixed positive integer, then $F = 0$. Following this line of investigation, Dhara and DeFilippis in [243] established that if a prime ring R admits a generalized derivation F and L is a noncommutative Lie ideal of R satisfying $u^sF(u)u^t = 0$ for all $u \in L$, where $s,t \geq 0$ are fixed integers then $F(x) = 0$ for all $x \in R$ unless $char(R) = 2$ and R satisfies s_4, the standard identity in four variables. Motivated by the above cited results, Ashraf et al. [53] considered the situation when R satisfies the identity $u^mG(u)^tu^n = 0$ for all $u \in L$, where m,n,t are positive fixed integers, L is a non-central Lie ideal of R, and G is a generalized skew derivation defined on R. The definition of a generalized skew derivation is a unified notion of skew and generalized derivations, which are considered classical additive mappings of non-associative algebras that have been investigated by many mathematicians from various perspectives. The main result in [53] stated as follows:

Theorem 4.5.24 [53, Theorem 1] *Let R be a prime ring with characteristic not equal to 2, C its extended centroid, L a noncentral Lie ideal of R and $m,n,t \geq 1$ fixed integers. Suppose that G is a nonzero generalized skew derivation of R such that $u^mG(u)^tu^n = 0$ for all $u \in L$, Then $dim_CRC = 4$.*

As a consequence of the above theorem, they also proved the following:

Theorem 4.5.25 [53, Theorem 2] *Let R be a prime ring of characteristic not equal to 2, C its extended centroid, I a nonzero ideal of R and $m,n,t \geq 1$ fixed integers. Suppose that G is a generalized skew derivation of R such that $x^m G(x)^t x^n = 0$ for all $x \in I$. Then $F = 0$.*

Theorem 4.5.26 [53, Theorem 3] *Let R be a semiprime ring of characteristic different from 2, and $m,n \geq 1$ fixed integers. Suppose that G is a generalized skew derivation of R such that $x^m G(x)x^n = 0$ for all $x \in R$. Then $F = 0$.*

Furthermore, Ashraf [35] introduced a decomposition theorem for rings that satisfies either $yx = x^m f(xy)x^n$ or $xy = x^m f(xy)x^n$, where m and n are non-negative integers, and $f(x) \in x^2 Z[X]$. This decomposition theorem facilitates the determination of the commutativity of the ring R. In light of this, several related ring properties are considered

(1) Let $m \geq 0, n \geq 0, p \geq 1$ be fixed integers, and L be a Lie ideal of a prime ring R which admits a generalized skew derivation G satisfying $vu = u^m G(uv)^p u^n$ for all $u, v \in L$.

(2) Let $m \geq 0, n \geq 0, p \geq 1$ be fixed integers, and L be a Lie ideal of a prime ring R which admits a generalized skew derivation G satisfying $uv = u^m G(uv)^p u^n$ for all $u, v \in L$.

Under a mild assumption, Ashraf et al. [52] demonstrate that if R admits a generalized skew derivation G satisfying either of the above conditions, then either L is central or $char(R) = 2$, $R \subseteq M_2(C)$, the ring of 2×2 matrices over C, L is commutative and $u^2 \in Z(R)$ for all $u \in L$. Their investigation is confined explicitly to condition (1), and it is noted that a similar result can be established for rings satisfying property 2. They present the following theorems:

Theorem 4.5.27 [52, Theorem 1] *Let R be a prime ring, G a generalized skew derivation associated with a skew derivation δ and an automorphism α. Assume that there exist $t \geq 1$ and fixed integersm, $n \geq 0$ fixed such that*

$$[y_1,y_2][x_1,x_2] = [x_1,x_2]^m G([x_1,x_2][y_1,y_2])^t [x_1,x_2]^n$$

holds for all $x_1,x_2,y_1,y_2 \in R$. Then R is commutative.

Theorem 4.5.28 [52, Theorem 2] *Let R be a prime ring, $Z(R)$ its center, C its extended centroid, L a Lie ideal of R, G a generalized skew derivation associated with a skew derivation δ and an automorphism α. Assume that there exist $t \geq 1$ and fixed integers $m,n \geq 0$ fixed such that*

$$vu = u^m G(uv)^t u^n$$

for all $u, v \in L$. Then either L is central or $char(R) = 2$, $R \subseteq M_2(C)$, the ring of 2 $\times$ 2 matrices over C, L is commutative and $u^2 \in Z(R)$ for all $u \in L$.

Proof: Let L be a Lie ideal of R and assume L is not central. In case L is not commutative, then $0 \neq I = R[L,L]R$ is a two-sided ideal of R. Then $0 \neq [I,R] \subseteq L$. According to our hypothesis,

$$vu = u^m G(uv)^t u^n$$

holds for all $u \in [I,R]$. Since I,R and Q_r satisfy the same generalized polynomial identities with automorphisms, Q_r satisfies

$$[y_1,y_2][x_1,x_2] = [x_1,x_2]^m G([x_1,x_2][y_1,y_2])^t [x_1,x_2]^n$$

and by Theorem [52], it follows the contradiction that R is commutative. Thus L is a commutative Lie ideal of R and, by [371, Lemma 2, Theorem 4], it follows that $char(R) = 2$, $R \subseteq M_2(C)$ and $u^2 \in Z(R)$ for any $u \in L$, as required.

♣

In 2021, De Filippis [215] proves the following theorem:

Theorem 4.5.29 [208, Theorem 1.1] *Let R be a prime ring of characteristic not equal to 2, Q_r be its right Martindale quotient ring and C be its extended centroid and L be a non-central Lie ideal of R. Let G and G_1 be two non-zero generalized skew derivations of R such that $G(G_1(x)x) = 0$ for all $x \in L$. Then one of the following holds:*

1. *There exist $a,c \in Q_r$ such that $G(x) = ax$ and $G_1(x) = cx$, for all $x \in R$, with $ac = 0$.*

2. *$R \subseteq M_2(C)$ is the ring of 2×2 matrices over C, and there exist $a,b,c,q \in Q_r$ such that q is an invertible element of Q_r, $G(x) = ax + qxq^{-1}b, G_1(x) = cx$ for all $x \in R$, with $ac + qcq^{-1}b = 0$.*

3. *$R \subseteq M_2(C)$ and there exists $a,b,c \in Q_r$ and α automorphisms of R, such that $G(x) = ax + \alpha(x)b, G_1(x) = cx$ for all $x \in R$, with $ac = \alpha(c)b = 0$. Moreover, in this sense α is not an automorphism of R.*

Recently, De Fillips et al. [230] starting from Bergen's initial idea and following the research line associated with the study of generalized skew derivations with nilpotent values and proved the following:

Theorem 4.5.30 [230, Theorem 1.1] *Let R be a prime ring of characteristic not equal to 2, Q_r its right Martindale quotient ring, L a non-central Lie ideal of R (i.e., $[L,R] \neq 0$), $n \geq 1$ a fixed integer, G and G_1 two generalized skew derivations of R, associated with the same automorphism α of R. If $(G(xy) - G_1(x)y)^n = 0$ for any $x,y \in L$, then there exists $a \in Q_r$ such that $G(x) = G_1(x) = ax$ for any $x \in R$, unless when R satisfies the standard polynomial identity $s_4(x_1,\ldots,x_4)$.*

4.6 Annihilators of the generalized skew derivation

In abstract algebra, an annihilator of a ring is a set of its elements that when multiplied by any element of the ring, produce the additive identity of the ring. Let R be a ring and S be a subset of R. The left annihilator of S is the set of all elements x in R such that $xs = 0$ for every $s \in S$. The right annihilator of S is the set of all elements x in R such that $sx = 0$ for every $s \in S$. The two-sided annihilator of S is the intersection of the left and right annihilators of S. The concept of an annihilator plays a crucial role in studying ideals and modules over a ring. For instance, an element r of a ring R is said to be a left zero divisor if a nonzero element s exists in R such that $rs = 0$. The set of all left zero divisors of R is precisely the left annihilator of R, denoted by $Ann(R)$.

Similarly, an element r of a ring R is said to be a right zero divisor if a nonzero element s exists in R such that $sr = 0$. The collection of all right zero divisors of R is precisely termed the right annihilator of R, denoted by $Ann'(R)$. The concept of an annihilator proves significant across various branches of algebra, encompassing ring, module and representation theories.

In the module theory, for instance, the annihilator of a module is defined analogously as the set of all elements within the ring that yield zero when acting on the module. The notion of annihilator is a valuable tool for exploring the structure of rings and modules and establishing theorems pertaining to them. In the forthcoming section, we explore the study of generalized skew derivations with specific annihilating conditions.

In 2015, Dhara and Mondal [249] investigated an annihilating condition on commutators and established the following result:

Theorem 4.6.1 [249, Theorem 1.2] *Let R be a prime ring with right Martindale quotient ring Q_r and extended centroid C, $F \neq 0$ a generalized derivation of R and $n \geq 1$ such that $F(x)[F(x),x]^n = 0$ for all $x \in R$. Then there exists $\lambda \in C$ such that $F(x) = \lambda x$ for all $x \in R$, unless when $R \subseteq M_2(C)$ and $char(R) = 2$.*

Theorem 4.6.2 [249, Theorem 1.1] *Let R be a prime ring with a right Martindale quotient ring Q_r and an extended centroid C, $F \neq 0$ a generalized derivation of R, L a noncentral Lie ideal of R and $n \geq 1$ such that $F(x)[F(x),x]^n = 0$ for all $x \in L$. Then either there exists $\lambda \in C$ such that $F(x) = \lambda x$ for all $x \in R$, or $R \subseteq M_2(C)$ and there exists $a \in Q_r$ and $\lambda \in C$ such that $F(x) = ax + xa + \lambda x$ for any $x \in R$, unless when $R \subseteq M_2(C)$ and $char(R) = 2$.*

De Filippis continue this line of investigation and examines what happens in the case $G \neq 0$ is a generalized skew derivation of R such that $G(x)[G(x),x]^n = 0$ for all $x \in S$, where S is an appropriate subset of R and $n \geq 1$ is a fixed integer. The result he obtained is stated as following:

Theorem 4.6.3 [213, Theorem 1.7] *Let R be a prime ring of characteristic not equal to 2 and 3, Q_r its right Martindale quotient ring, C its extended centroid, G a nonzero generalized skew derivation of R, L a non-central Lie ideal of R and $n \geq 1$ a fixed positive integer. If $G(x)[G(x),x]^n = 0$ for all $x \in L$, then either there exists $\lambda \in C$ such that $G(x) = \lambda x$ for all $x \in R$, or $R \subseteq M_2(C)$ and there exist $a \in Q_r$ and $\lambda \in C$ such that $G(x) = ax + xa + \lambda x$ for any $x \in R$.*

Any investigation of derivations in prime rings from the algebraic point of view is interesting. In [171], Chang and Lin consider the situation when $d(u)u^n = 0$ for all $u \in \rho$, where ρ is a non-zero right ideal of R and d is a non-zero derivation of R. They show that if R is a prime ring and n is a fixed positive integer, then $d(\rho)\rho = 0$, and if $u^n d(u) = 0$ for all $u \in \rho$, then $R \cong M_2(F)$, the 2×2 matrices over a field F of two elements.

Dhara and Sharma in their work [246], extended the results mentioned above. Specifically, for a prime ring R of characteristic not equal to 2, a non-zero derivation d of R, an element $a \in R$, a non-central Lie ideal L of R, and fixed integers s,t,n, such that $ax^s d(x)^n x^t \in Z(R)$ for all $x \in L$, either $a = 0$ or $R \subseteq M_2(K)$ for a field K.

Continuing along this investigative path, Dhara and De Filippis obtained a result [243] with a similar essence to the ones mentioned earlier by replacing the derivation d with a generalized derivation F. By incorporating the definition of a generalized derivation, they established that if $u^s F(u) u^t = 0$ for all u in a non-commutative Lie ideal L of R, where $s \geq 0$ and $t \geq 0$ are fixed integers $F(x) = 0$ for all $x \in R$ unless $char(R) = 2$ and R satisfies S_4.

Furthermore, De Filippis [214] expanded upon the aforementioned results by introducing an annihilating condition and subsequently proving the following theorem:

Theorem 4.6.4 [214, Theorem 1.2] *Let R be a prime ring of characteristic not equal to 2, Q_r be its right Martindale quotient ring, and C be its extended centroid. Suppose that G is a generalized skew derivation of R, L a non-central Lie ideal of R, $0 \neq a \in R, m \geq 0$ and fixed integers, $n,s \geq 1$. If*

$$a(u^m G(u) u^n)^s = 0$$

for all $u \in L$, then either $R \subseteq M_2(C)$, is the ring of 2×2 matrices over C, or $m = 0$, and there exists $b \in Q_r$ such that $G(x) = bx$, for any $x \in R$, with $ab = 0$.

In [202], Daif and Bell proved that if R is a semi-prime ring, I a nonzero ideal of R, and d a derivation of R such that $d([x,y]) = [x,y]$ for all $x,y \in I$, then $I \subseteq Z(R)$. This result has been extended in [301] when $d([x,y]) \pm [x,y] \in Z(R)$ for all $x,y \in I$. Consequently, it was proved that $I \subseteq Z(R)$. Later, in [497], the case when the derivation d was replaced by a generalized derivation F was studied. It was proved that if R is a prime ring and I a nonzero ideal of R such that

$F([x,y]) \pm [x,y] = 0$ (respectively, $F(xy+yx) \pm (xy+yx) = 0$) for all $x,y \in I$, then R is commutative. The set $\mathfrak{T} = \{f(xy) - yx : x,y \in S\}$ has been studied in several articles presented in existing literature. Here, f is a suitable additive map on R and S is a subset of R.

In [75], Ashraf and Rehman have shown that if R is a prime ring and d is a derivation of R such that $d(xy) \pm yx \in Z(R)$ (respectively $d(xy) \pm xy \in Z(R)$) for all $x,y \in I$, a nonzero ideal of R, then ring is commutative. Later, in [45], it was proved that the same results hold if the additive map F is a generalized derivation of R. More precisely, in [17], it has been proved that if R is a prime ring and I is a nonzero right ideal of R, $n \geq 1$, and F a generalized derivation of R, such that $(F(xy) - yx)^n = 0$ for all $x,y \in I$, then $[I,I]I = 0$, moreover $F(x) = ax + xb$ and there exist $\alpha, \beta \in C$, such that $(a-\alpha)I = 0, (b-\beta)I = 0$, and $((a+b)x - x)I = 0$ for all $x \in I$. In particular, if $I = R$, then F is an identity map and R is commutative.

In [16], Ali et al. have continued the study of analogue problems on Lie ideals of a prime ring involving a generalized skew derivation F of R instead of a generalized derivation and proved the following theorem:

Theorem 4.6.5 *Let R be a prime ring, C the extended centroid of R, L a Lie ideal of R, G a nonzero generalized skew derivation of R with an associated automorphism α and $n \geq 1$ a fixed integer. If $(G(xy) - yx)^n = 0$ for all $x,y \in L$, then L is commutative, and one of the following statements holds:*

1. *Either L is central.*
2. *$R \subseteq M_2(C)$ is the 2×2 matrix ring over C, with $char(R) = 2$.*

Following this line of investigation, Sharma et al. [517] have examined what happens in case $(G(x)G(y) - yx)^n = 0$ for all $x,y \in L$, where $n \geq 1$ is a fixed integer, G is a generalized skew derivation of R and L is a non-central Lie ideal of R. More precisely, they proved the following:

Theorem 4.6.6 [517, Theorem 1.1] *Let R be a prime ring with its right Martindale quotient ring Q, L a non-central Lie ideal of R and $n \geq 1$ a fixed integer. If G is a generalized skew derivation of R such that $(G(x)G(y) - yx)^n = 0$ for all $x,y \in L$, then $char(R) = 2$ and $R \subseteq M_2(C)$, the ring of 2×2 matrices over C.*

Recently, De Filippis et al. generalized the result in [229] to the case when two different generalized skew derivations act on the non-central Lie ideal L. They also introduced an annihilating condition. To be more precise, they proved the following:

Theorem 4.6.7 *[229, Theorem 1.1] Let R be a prime ring, Q_r its right Martindale quotient ring, L a non-central Lie ideal of R, $n \geq 1$ a fixed integer, G and G_1 two generalized skew derivations of R with the same associated automor-*

phism, $p \in R$ *a fixed element. If* $p(G(x)G(y) - G_1(y)x)^n = 0$*, for any* $x, y \in L$*, then there exist* $a, c \in Q_r$ *such that* $G(x) = ax$ *and* $G_1(x) = cx$*, for any* $x \in R$*, with* $pa = pc = 0$*, unless when* R *satisfies the standard polynomial identity* $s_4(x_1, \ldots, x_4)$.

In [519], Shutaj and Khan proved that if R is a prime ring with a generalized derivation F with associated nonzero derivation d, a noncentral Lie ideal L and $a \in R$ such that $a(d(u)^{l_1} F(u)^{l_2} d(u)^{l_3} \cdots F(u)^{l_k})^n = 0$ for all $u \in L$, then either $a = 0$ or R satisfies s_4, where $l_1, l_2, \ldots, l_k$ are fixed nonnegative integers all not being zero and n a positive fixed integer. Motivated by [519], as an extension of the previous results, Eroglu and Argaç generalized the result in [519] to generalized skew derivations and obtained the following:

Theorem 4.6.8 [519, Main Theorem] *Let* R *be a prime ring with a maximal right ring of quotients* Q *and a noncommutative Lie ideal* L*. Assume that* G *and* G_1 *are generalized skew derivations (generalized* σ*-derivations) with the same associated* σ*-derivation* δ *of* R *and* $0 \neq p \in R$ *such that*

$$p(G(u)^{l_1} G_1(u)^{l_2} G(u)^{l_3} \mu(u)^{l_4} \cdots G_1(u)^{l_k})^n = 0$$

for all $u \in L$*, where* $l_1, l_2, \ldots, l_k$ *are fixed integers with* $l_1 \neq 0$ *and* $n>0$ *a fixed integer. Then* $d = 0$*, so, there exist* $a, c \in Q$ *such that* $G_1(x) = cx$ *and* $G(x) = ax$ *for all* $x \in R$*. Moreover, either* $pc = o$ *or* $a = 0$ *unless* R *satisfies* s_4.

4.7 Generalized skew derivations on multilinear polynomials

The well-known Posner theorem [487] states that if $[d(x), x] \in Z(R)$ for all $x \in R$, then R must be commutative. In [360], Lanski extended Posner's result to Lie ideals. More precisely, he showed that if d is a nonzero derivation of R such that $[d(x), x] \in Z(R)$ for all x in a Lie ideal L of R, then either L is central in R or $char(R) = 2$ and R satisfies the standard polynomial identity $S_4(x_1, \ldots, x_4)$ of degree 4. Several authors studied the relationship between the structure of R and the behaviour of an additive mapping f that satisfies the Engel-type condition $[f(x), x]_k = 0$.

On the other hand, for a prime ring R of characteristic not equal to 2, any noncentral Lie ideal contains the set $\{[x_1, x_2] : x_1, x_2 \in I\}$ of all evaluations of the polynomial $[x_1, x_2]$ in a two-sided ideal I of R. For this reason, many researchers in this area analyzed in detail the case when the Lie ideal is replaced by the set of all evaluations of a polynomial $f(x_1, \ldots, x_n)$ and $[d(f(x_1, \ldots, x_n)), f(x_1, \ldots, x_n)]_k$ is a differential identity for some ideal of R.

Lee and Lee proved in [383] that if $f(x_1, \ldots, x_n)$ is a multilinear polynomial and d is a nonzero derivation of R such that $[d(x), x]_k = 0$ for all

$x \in \{f(x_1,\dots,x_n);x_1,\dots,x_n \in R\}$, then $f(x_1,\dots,x_n)$ must be central-valued in R unless $char(R)=2$ and R satisfies $s_4(x_1,\dots,x_4)$. Moreover, Lee extended this result in [377] to any polynomial by removing the multilinearity hypothesis. De Filippis and Di Vincenzo consider another related generalization in [217]. They describe what happens if a generalized derivation of R replaces the derivation d with the associated derivation d. De Filippis [211] obtained a related result with a specific annihilator condition on a generalized derivation acting on a multilinear polynomial based on these definitions. Let R be a prime ring of characteristic not equal to 2, U a symmetric Utumi quotient ring, and C its extended centroid. Let $f(x_1,\dots,x_n)$ be a non-central multilinear polynomial over C with n non-commuting variables and $0 \neq a \in R$. Suppose that $G: R \to R$ is a non-zero generalized derivation satisfying the condition

$$a[G(f(r_1,\dots,r_n)),f(r_1,\dots,r_n)]=0$$

for all $r_1,\dots,r_n \in R$. Then either there exists $\lambda \in C$ such that $G(x)=\lambda x$ for all $x \in R$, or there exist $q \in U$ and $\lambda \in C$ such that $G(x)=(q+\lambda)x+xq$ for all $x \in R$ and $f(x_1,\dots,x_n)^2$ is central-valued on R. In 2012, De Filippis and Wei continue this line of investigation of the set

$$S=\{[G(f(x_1,\dots,x_n)),f(x_1,\dots,x_n)]|x_1,\dots,x_n \in R\}$$

for a generalized skew derivation G of R instead of a generalized derivation. The standard approach in studying the aforementioned set S is to examine its size. For this, it is reasonable to study its left annihilator in R. They proved the following theorem.

Theorem 4.7.1 *[223, Main Theorem 1.1.] Let R be a prime ring of characteristic not equal to 2, Q_r be its right Martindale quotient ring and C be its extended centroid. Suppose that G is a non-zero generalized skew derivation of R and $f(x_1,\dots,x_n)$ is a non-central multilinear polynomial over C with n non-commuting variables. If there exists a non-zero element a of R such that*

$$a[G(f(r_1,\dots,r_n)),f(r_1,\dots,r_n)]=0$$

for all $r_1,\dots,r_n \in R$, then one of the following holds:

1. *There exists $\lambda \in C$ such that $G(x)=\lambda x$ for all $x \in R$;*
2. *There exist $q \in Q_r$ and $\lambda \in C$ such that $G(x)=(q+x)x+xq$ for all $x \in R$ and $f(x_1,\dots,x_n)^2$ is central-valued on R.*

A reasonable criterion for examining the size of S is to study the left annihilator and centralizer of S in R. Following this line of investigation, it is proved in [223] that if $S \neq 0$, then the left annihilator of S in R is zero. In 2017, Albaş et al. [7] proved that the centralizer of S in R is trivial, too; that is, it coincides with the center of R. They established the following result.

Theorem 4.7.2 *Let R be a prime ring of characteristic not equal to 2, Q be the right Martindale quotient ring of R, and C be the extended centroid of R. Suppose that G is a nonzero generalized skew derivation of R and $f(x_1,\ldots,x_n)$ is a noncentral multilinear polynomial over C with n noncommuting variables. Let $f(R)=\{f(r_1,\ldots,r_n):r_i\in R\}$ be the set of all evaluations of $f(x_1,\ldots,x_n)$ in R, while $S=\{[G(f(x_1,\ldots,x_n)),f(x_1,\ldots,x_n)]|x_1,\ldots,x_n\in R\}$, and let $C_R(S)$ be the centralizer of S in R, i.e., $C_R(S)=\{a\in R:[a,x]=0\ \forall x\in S\}$. If $S\neq 0$, then $C_R(S)=Z(R)$.*

In 2019, Dhara [239] extends the above result by considering central values, i.e., $[a,[G(f(r_1,\ldots,r_n)),f(r_1,\ldots,r_n)]]\in Z(R)$ for all $r_1,\ldots,r_n\in R$, and then determines all possible forms of the mapping G. More precisely, the main objective is to prove the following theorem:

Theorem 4.7.3 *[239, Theorem 1.1] Let R be a noncommutative prime ring of characteristic not equal to 2 with extended centroid C and $f(X_1,\ldots,X_n)$ a multilinear polynomial over C, non central-valued on R. Suppose that G is a generalized skew derivation of R and $a\in R-C$ such that*

$$[a,[G(f(r)),f(r)]]\in Z(R)$$

for all $r=(r_1,\ldots,r_n)\in R^n$. Then one of the following holds:

1. *There exists $\lambda\in C$ such that $G(x)=\lambda x$ for all $x\in R$;*
2. *There exist $b\in Q_r$ and $\lambda\in C$ such that $G(x)=bx+xb+\lambda x$ for all $x\in R$ and $f(X_1,\ldots,X_n)^2$ is central valued on R.*

Corollary 4.7.1 *[239, Corollary 1.2] Let R be a prime ring of characteristic not equal to 2, Q_r be the right Martindale quotient ring of R, and C be the extended centroid of R. Suppose that G is a nonzero generalized skew derivation of R and $f(X_1,\ldots,X_n)$ is a noncentral multilinear polynomial over C with n noncommuting variables. Let $f(R)=\{f(r_1,\ldots,r_n):r_i\in R\}$ be the set of all evaluations of $f(x_1,\ldots,x_n)$ in R, while $A=\{[G(f(r_1,\ldots,r_n)),f(r_1,\ldots,r_n)]:r_i\in R\}$, and let $S=\{a\in R:[a,x]\in Z(R),\forall x\in A\}$. If $A\neq 0$, then $S=Z(R)$.*

Corollary 4.7.2 *[239, Corollary 1.3] Let R be a noncommutative prime ring of characteristic not equal to 2 with extended centroid C and $f(X_1,\ldots,X_n)$ a multilinear polynomial over C, non central-valued on R. Suppose that G is a generalized skew derivation of R such that $[G(f(r)),f(r)]\in Z(R)$ for all $r=(r_1,\ldots,r_n)\in R^n$. Then one of the following holds:*

1. *There exists $\lambda\in C$ such that $G(x)=\lambda x$ for all $x\in R$;*
2. *There exist $b\in Q_r$ and $\lambda\in C$ such that $G(x)=bx+xb+\lambda x$ for all $x\in R$ and $f(X_1,\ldots,X_n)^2$ is central-valued on R.*

The following theorem is an application of Theorem 4.7.3, which is a generalization of the main result in [221] to a generalized skew-derivation case.

Theorem 4.7.4 *[239, Theorem 1.4] Let R be a noncommutative prime ring of characteristic not equal to 2 with right Martindale quotient ring Q_r and extended centroid C. Suppose that $f(X_1,\dots,X_n)$ is a multilinear polynomial over C, noncentral-valued on R and denoted by $f(R)$, the set of all the evaluations of $f(X_1,\dots,X_n)$ on R. If G and G_1 are two generalized skew derivations of R such that $[[G(u),u],[G_1(v),v]] \in Z(R)$ for all $u,v \in f(R)$, then one of the following holds:*

1. *There exists $\lambda \in C$ such that $G(x) = \lambda x$ for all $x \in R$;*
2. *There exists $\mu \in C$ such that $G_1(x) = \mu x$ for all $x \in R$;*
3. *$f(X_1,\dots,X_n)^2$ is central-valued on R and there exist $b \in Q_r$ and $\lambda \in C$ such that $G(x) = bx+xb+\lambda x$ for all $x \in R$ or there exist $c \in Q_r$ and $\mu \in C$ such that $G_1(x) = cx+xc+\mu x$ for all $x \in R$.*

Wong [543] considered two different derivations and obtained the following result. Let k be a commutative ring with unity, R a prime K-algebra with center $Z(R)$ and $f(x_1,\dots,x_n)$ a multilinear polynomial over K, not central-valued on R. Suppose that d and δ are derivations of R such that $d(f(x_1,\dots,x_n))f(x_1,\dots,x_n) - f(x_1,\dots,x_n)\delta(f(x_1,\dots,x_n)) \in Z(R)$ for all $x_1,\dots,x_n$ in some non-zero ideal I of R. Then, either $d = \delta = 0$, or $d = -\delta$ and $f(x_1,\dots,x_n)^2$ is central-valued on R, except when $char(R) = 2$ and R satisfies $s_4(x_1,x_2,x_3,x_4)$. Liu [417] developed Wong's result to one-sided ideals and studied the situation when

$$d(f(x_1,\dots,x_n))f(x_1,\dots,x_n) - f(x_1,\dots,x_n)\delta(f(x_1,\dots,x_n)) \in Z(R)$$

for all $x_1,\dots,x_n$ in a certain left ideal of R. The analogous result for $\delta = d$ (or $\delta = -d$) were also obtained.

The other related generalizations are considered in [30] and [216]. More precisely, the results describe what happens if generalized derivations replace d and δ. The above observations motivated Carini et al. [160] to study the set

$$S = \{G(f(x_1,\dots,x_n))f(x_1,\dots,x_n) - f(x_1,\dots,x_n)G_1(f(x_1,\dots,x_n)) | x_1,\dots,x_n \in R\}$$

where G and G_1 were generalized skew derivations. In fact, they proved the following result:

Theorem 4.7.5 *[160, Theorem 1.1] Let R be a prime ring of characteristic not equal to from 2, Q_r be its right Martindale quotient ring and C be its extended centroid. Suppose that G, G_1 are generalized skew derivations of R and $f(x_1,\dots,x_n)$ is a non-central multilinear polynomial over C with n non-commuting variables. If $G(f(r_1,\dots,r_n))f(r_1,\dots,r_n) - f(r_1,\dots,r_n)$ $G_1(f(r_1,\dots,r_n)) \in C$ for all $r_1,\dots,r_n \in R$, then one of the following holds:*

(a) There exists $v \in Q_r$ such that $g(x) = xv$ and $G_1(x) = vx$ for all $x \in R$;

(b) $f(x_1,\ldots,x_n)^2$ *is central-valued on R, and there exists a non-zero derivation d of R and* $v, v' \in Q_r, \lambda, \mu \in C$ *such that* $G(x) = (v+\lambda)x + x(v' + \mu) + d(x), G_1(x) = v'x + xv - d(x)$ *for all* $x \in R$;

(c) $f(x_1,\ldots,x_n)^2$ *is central-valued on R, and there exists* $v, v' \in Q_r, \lambda, \mu \in C$ *such that* $G(x) = (v+\lambda)x + x(v' + \mu), G_1(x) = v'x + xv$ *for all* $x \in R$;

(d) *R satisfies the standard polynomial identity* $s_4(x_1,x_2,x_3,x_4)$ *of degree* 4, *and there exist* $v, v' \in Q_r, \lambda, \mu \in C$ *such that* $G(x) = vx + xv' + \lambda x$ *and* $G_1(x) = v'x - xv + \mu x$ *for all* $x \in R$;

(e) *R satisfies* $s_4(x_1,x_2,x_3,x_4)$, *and there exists a non-zero derivation d of R and* $v, v' \in Q_r$ *with* $v + v' \in C$ *such that* $G(x) = vx + d(x)$ *and* $G_1(x) = xv' - d(x)$ *for all* $x \in R$;

(f) *R satisfies* $s_4(x_1,x_2,x_3,x_4)$, *and there exists a non-zero derivation d of R and* $v, v' \in Q_r$ *with* $v - v' \in C$ *such that* $G(x) = xv + d(x)$ *and* $G_1(x) = vx' - d(x)$ *for all* $x \in R$.

4.8 b-generalized derivations

Let σ be an automorphism of R. An Additive map $\delta : R \to R$ is called a *(right) σ-derivation* if $\delta(xy) = x\delta(y) + \delta(x)\sigma(y)$ for $x, y \in R$. A basic example of σ-derivations are derivations and $\sigma - 1$ derivations. Given $b \in R$, the map $x \mapsto xb - b\sigma(x)$ for $x \in R$ obviously defines a σ-derivation, which is called the *inner σ-derivation* induced by b. It is clear that any σ-derivation of R can be uniquely extended to a σ-derivation of $Q_{mr}(R)$. In [382], Lee and Liu gave a common generalization of both generalized derivations and (skew derivation) σ-derivations. An additive map $G : R \to R$ is called a **right generalized σ-derivation** if there exists an additive map $\delta : R \to R$ such that $G(xy) = xG(y) + \delta(x)\sigma(y)$ for all $x, y \in R$. It is clear that δ is uniquely determined by the map of G. The additive map δ is called the associated map G. In a recent paper [354], Kosan and Lee proposed the following new definition:

Definition 46 *Let* $\delta : R \to Q_{mr}(R)$ *be an additive mapping and* $b \in Q_{mr}(R)$. *An additive mapping* $H : R \to Q_{mr}(R)$ *is called a* **(left) b-generalized derivation** *with associated map* δ *if* $H(xy) = H(x)y + bx\delta(y)$ *for all* $x, y \in R$.

Definition 47 *Let* $\delta : R \to Q_{mr}(R)$ *be an additive mapping and* $b \in Q_{mr}(R)$ *an additive map. An additive map* $H : R \to Q_{mr}(R)$ *is called a* **(right) b-generalized derivation** *with associated map* δ *if* $H(xy) = xH(y) + \delta(x)yb$ *for all* $x, y \in R$.

Clearly, a generalized derivation is a 1-generalized derivation and a right generalized σ-derivation is a right u^{-1} generalized derivation if $\sigma(x) = uxu^{-1}$ for $x \in R$, where u is a unit element in $Q_{mr}(R)$. In the same paper it is proved that if R is a prime ring, then δ is a derivation of R. For simplicity of notation, this mapping

H is called a b-**generalized derivation** with associated pair (b,δ). Clearly, any generalized derivation with associated derivation δ is a b-generalized derivation with associated pair $(1,\delta)$. Similarly the mapping $x \mapsto ax + b[x,c]$ for $a,b,c \in Q_{mr}(R)$, is a b-generalized derivation with associated pair $(b, ad(c))$, where $ad(c)(x) = [x,c]$ denotes the inner derivation of R induced by the element c. More generally, the mapping $x \mapsto ax + qxc$ for $a,q,c \in Q_{mr}(R)$, is a *b-generalized derivation* with associated pair $(q, ad(c))$. This mapping is called an **inner** b-**generalized derivation**. Moreover, if $\alpha \in Aut(R)$, with $\alpha(x) = qxq^{-1}$ for q an invertible element in $Q_{mr}(R)$, and H is the inner generalized skew derivation with associated automorphism α, then δ is a b-**generalized derivation** with associated pair $(q, ad(q^{-1}b))$, for a suitable element $b \in Q_{mr}(R)$.

Before stating the main result of Kosan and Lee [354], we recall some basic properties of idempotent C. We write $\mathbf{B}$ for the set of all idempotents in C. The set $\mathbf{B}$ forms a Boolean algebra with respect to the operation $e + h := e + h - 2eh$ and $e \cdot h := eh$ for all $e,h \in \mathbf{B}$. It is complete with respect to the partial order $e \leq h$ (defined by $eh = e$) in the sense that any subset S of $\mathbf{B}$ has a supremum $\bigvee S$ and an infimum $\bigwedge S$. Given a subset S of $Q_{mr}(R)$, we define $E[S]$ to be the infimum of idempotent $e \in B$ such that $ex = x$ for all $x \in S$. If $S = \{s\}$, we substitute $E[S]$ by $E[b]$. Note that, for $a,b \in Q_{mr}(R), aRb = 0$ iff $E[a]E[b] = 0$. By the characterization, it is easy to see that if a b-generalized derivation H has associated maps δ and δ', then $E[b]\delta(x) = E[b]\delta'(x)$ for all $x \in R$. Kosan and Lee [354] obtained the following result.

Theorem 4.8.1 [354, Theorem 1.2] *Let R be a semiprime ring, $b \in Q_{mr}$, and let $H : R \to Q_{mr}(R)$ be a b-**generalized derivation** with associated map δ. Suppose that $H(x)^n = 0$ for all $x \in R$, where n is a positive integer. Then there exists $q \in Q_{mr}(R)$ such that $E[b]\delta(x) = [q,x]$ for $x \in R$, where $H(x) = -bxq$ for $x \in R$, and $qb = 0$.*

Theorem 4.8.2 [354, Theorem 1.3] *Let R be a semiprime ring, $b \in Q_{ml}$, and let $H : R \to Q_{ml}(R)$ be a right b-generalized derivation with associated map δ. Suppose that $H(x)^n = 0$ for all $x \in R$, where n is a positive integer. Then there exists $q \in Q_{ml}(R)$ such that $E[b]\delta(x) = [q,x]$ for $x \in R$, where $H(x) = qxb$ for $x \in R$, and $bq = 0$.*

In [533], Vukman studied the connection of Engel-type conditions and derivations and showed that a prime ring R is commutative if either $[d(x),x]_2 = 0$ for all $x \in R$ and char$(R) \neq 2$ or if $[d(x),x]_2$ is central for all $x \in R$ and $char(R) \neq 2,3$, where d is a nonzero derivation of R. Instead of using an elementary but lengthy calculation in [533], Lanski [364] applied the theory of differential identities to obtain a more general Engel condition as follows:

Theorem 4.8.3 *[364, Theorem 1] Let R be a prime ring, I a nonzero ideal of R and d a nonzero derivation of R. If $[d(x),x]_k = 0$ for all $x \in I$ and $k \geq 0$ fixed, then R is commutative.*

Argaç et al. [33] obtained the analogous result of Theorem 4.8.3 for generalized derivations. In [421], Liu extends Theorem 4.8.3 to b-generalized derivations. Precisely, it can be stated as:

Theorem 4.8.4 [421, Theorem 1.1] *Let R be a noncommutative prime ring, $b \in Q$, I a nonzero ideal of R and H a b-generalized derivation of R. If $[H(x^m),x^n]_k = 0$ for all $x \in I$, where m,n,k are fixed positive integers, then there exists $\lambda \in C$ such that $H(x) = \lambda cx$ for all $x \in R$ unless $R \cong M_2(GF(2))$, the 2×2 matrix ring over the Galois field $GF(2)$ of two elements.*

For the case $R \cong M_2(GF(2))$, there exist b-generalized derivations of R which satisfy the conditions in Theorem 4.8.4, but they are not of the form described in Theorem 4.8.4.

Example 4.8.1 *Let $R = M_2(GF(2))$ be the 2×2 matrix ring over the Galois field $GF(2)$ of two elements. As usual, $\{e_{ij} | i,j = 1,2\}$ denotes a set of matrix units in R. Let H be the b-generalized derivation of R defined by $H(x) = ax + bxc$ for $x \in R$, where $a = e_{11}+e_{21}, b = e_{21}$ and $c = e_{11}+e_{12}$. Note that $x^6 \in GF(2).(e_{11}+e_{22})$ for all $x \in R$ $\{e_{11},e_{22},e_{11}+e_{12},e_{11}+e_{21},e_{22}+e_{12},e_{11}+e_{21}\}$. It is easy to see that $[g(x),x^6] = 0$ for all $x \in R$, but g is not of the form described in Theorem 4.8.4.*

If G is a generalized σ-derivation (generalized skew derivation) of R with the associated σ-derivation (skew derivation) δ of R and σ is X-inner with $\sigma(x) = uxu^{-1}$ for all $x \in R$, where $u \in Q$, then it is easy to see that H is a U-generalized derivation of R with the associated map $u^{-1}\delta$. As an application of Theorem 4.8.4, we have the following result:

Theorem 4.8.5 [421, Corollary 1.2] *Let R be a noncommutative prime ring, let I be a nonzero ideal of R and let H be a generalized σ-derivation of R with an X-inner automorphism σ of R. Suppose that $[H(x^m),x^n]_k = 0$ for all $x \in I$, where m,n,k are fixed positive integers. Then there exists $\lambda \in C$ such that $H(x) = \lambda x$ for all $x \in R$.*

In [241], Dhara and Ali proved that if R is a $n!$-torsion free semiprime ring, I a nonzero ideal of R and F a generalized derivation of R with the associated derivation d of R such that $[F(x),x^n] = 0$ for all $x \in I$, where $n \geq 1$ is a fixed integer, then either $d(I) = 0$ or $d(I)$is contained in a nonzero central ideal of R. Finally, using Theorem 4.8.3 Liu [421] generalized this result to generalized b-derivations without the torsion freeness restriction. This is stated as:

Theorem 4.8.6 [421, Theorem 1.3] *Let R be a semiprime ring, let I be a nonzero ideal of R and let H be a b-generalized derivation of R with the associated map*

d, where $b \in Q$ is a unit. Suppose that $[H(x^m), x^n]_k = 0$ for all $x \in I$, where m, n, k are fixed positive integers. Then there exists $a \in Q$ such that $H(x) = ax + bd(x)$ for all $x \in R$, $[a, QIQ] = 0$ and $Qd(I)Q \subseteq C$.

When treating an additive map of a ring R, the main goal is to characterize the map or to state some structural results related to the ring itself. The works that have been done related to power values of some additive maps so far were initiated in [278]. Giambruno and Herstein proved that if R is a semiprime ring and d is a derivation of R such that $d(x)^n = 0$ for all $x \in R$, where $n \geq 1$ is a fixed positive integer, then $d = 0$. Several researchers have worked on the generalization of this problem. For instance, in [133] Brešar proved that if R is a prime ring with a nonzero derivation d and a is an element of R such that $ad(x)^n = 0$ for all $x \in R$, where $n \geq 1$ is a fixed positive integer, then $a = 0$ in case char$R \neq (n-1)!$. In [386], Lee and Lin obtained the same result on some noncentral Lie ideals of R without assuming the characteristic restriction. In [166] Chang proved that if R is a prime ring and δ is a right generalized (α, β)-derivation of R such that $a\delta(x)^n = 0$ for all $x \in R$ and some $a \in R$ where $n \geq 1$ is a positive integer, then $\delta = 0$ or $a = 0$. Motivated by these results, Yarbil [562] extends these results to b-generalized derivations. This is stated as:

Theorem 4.8.7 *Let R be a prime ring, H be a nonzero right b-generalized derivation of R associated with the map d of R and s, b be nonzero elements in R. If $s(H(x))^n = 0$, where n is a fixed positive integer, then there exist $c \in Q_{ml}$ and $\beta \in C$ such that $d(x) = ad_c(x), H(x) = (c+\beta)xb$ and either $s(c+\beta) = 0$ or $b(c+\beta) = 0$.*

In the year 2002, Ashraf and Rehman [79] proved that if R is a prime ring, I a nonzero ideal of R and d a derivation of R such that $d(x \circ y) \pm x \circ y = 0$, for all $x, y \in I$, then R is commutative. Later, in [501], this result was also extended to the case of generalized derivations. More precisely, Rehman proved that if a prime ring R admits a generalized derivation F associated with derivation d such that $F(x \circ y) \pm x \circ y = 0$ for all $x, y \in I$, a nonzero ideal I of R, and if $F = 0$ or $d \neq 0$, then R is commutative. The authors applied an elementary calculation to obtain commutativity theorems instead of differential identities in the above results. Therefore, extending some results regarding derivations and generalized derivations to b-generalized derivations should be interesting. Motivated by these observations, Huang [308], extended some commutativity results for derivations and generalized derivations to b-generalized derivations. This is stated as:

Theorem 4.8.8 [308, Theorem 3.1] *Let R be a prime ring and I be a nonzero ideal of R. Suppose that H is a nonzero b-generalized derivation of R such that $H(x \circ y) = 0$ for all $x, y \in I$. Then R is commutative.*

Corollary 4.8.1 [308, Corollary 3.2] *Let R be a prime ring and I be a nonzero ideal of R. Suppose that H is a nonzero b-generalized derivation of R such that $H(x^2) = 0$ for all $x \in I$. Then R is commutative.*

Proof: By assumption, we get $H(x^2) = 0$ for all $x \in I$. Linearization of the above equation gives, $H(x^2) + H(y^2) + H(x \circ y) = 0$ for all $x, y \in I$. This equation reduces to $H(x \circ y) = 0$ for all $x, y \in I$. It follows from Theorem 4.8.8 that R is commutative.

♣

Theorem 4.8.9 [308, Theorem 3.3] *Let R be a prime ring and I be a nonzero ideal of R. Suppose that H is a nonzero b-generalized derivation of R such that $[H(x), x] = 0$ for all $x \in I$. Then R is commutative.*

Theorem 4.8.10 [308, Theorem 3.4] *Let R be a prime ring and I be a nonzero ideal of R. Suppose that $H \neq I_{id}$ is a nonzero b-generalized derivation of R such that $H(x \circ y) = x \circ y$ for all $x, y \in I$. Then R is commutative.*

Corollary 4.8.2 [308, Corollary 3.5] *Let R be a prime ring and I be a nonzero ideal of R. Suppose that $H \neq I_{id}$ is a b-generalized derivation of R such that $H(x^2) = x^2$ for all $x \in I$. Then R is commutative.*

Theorem 4.8.11 [308, Theorem 3.6] *Let R be a prime ring and I be a nonzero ideal of R. Suppose that $H \neq I_{id}$ is a b-generalized derivation of R such that $H(x \circ y) + x \circ y = 0$ for all $x, y \in I$. Then R is commutative.*

Proof: Since H is a b-generalized derivation of R such that $H(x \circ y) + x \circ y = 0$, then $-H(x \circ y) - x \circ y = 0$ for all $x, y \in I$. This implies that $-H$ is also a b-generalized derivation of R satisfying $(-H)(x \circ y) = x \circ y$ for all $x, y \in I$. In view of Theorem 4.8.10 , we get the required result.

♣

Corollary 4.8.3 [308, Corollary 3.7] *Let R be a prime ring and I be a nonzero ideal of R. Suppose that $H \neq I_{id}$ is a b-generalized derivation of R such that $H(x^2) + x^2 = 0$ for all $x \in I$. Then R is commutative.*

The following example shows that the above results cannot be proved for semiprime rings.

Example 4.8.2 [308, Example 3.8] *Let R_1 and R_2 be prime rings, where R_1 is noncommutative and put $R = R_1 \oplus R_2$. It is clear that R is a semiprime ring. Let $H_1 : R_1 \to R_2$ be a nonzero b-generalized derivation satisfying the hypothesis of Theorem 4.8.8, that is, $H_1(x \circ y) = 0$ for all $x, y \in R_1$. Then the mapping $H : R \to R$ defined by $H(r_1, r_2) = (H_1(r_1), 0)$ is a nonzero $(b, 0)$-generalized derivation of R such that $H(X \circ Y) = 0$ for all $X, Y \in R$. However, R is not commutative.*

In 2004, Albaş and Argaç [5] have extended the result obtained by Bell and Kappe [112] to the case of generalized derivations as follows:

Let R be a prime ring. Suppose that $d : R \to R$ is a nonzero generalized derivation of R associated with a derivation α.

1. If d acts as a homomorphism on R and $d \neq 0$, then $d = I_d$.

2. If d acts as an antihomomorphism on R and $d \neq 0$, then $d = I_d$.

More recently, Dhara [238] has extended the result obtained by Albas and Argaç [5] to the case of semiprime rings and showed that if F is a nonzero generalized derivation of a semiprime ring R associated with a derivation d satisfying the relation $F(xy) = F(x)F(y)$ for all $x, y \in I$, a nonzero ideal of R, then $d(I) = 0$ and F is a commuting left multiplier mapping on I. In particular, it has been proved that if R is a prime ring, then $d = 0$ and F is the identity mapping of R. Moreover, it was shown that if $F(xy) = F(y)F(x)$ for all $x, y \in I$, then either $d(I) = 0$ or R contains a nonzero central ideal. In particular, it was proved that if R is a prime ring, then R is commutative, and F is a left multiplier mapping for R. Motivated by the above results, Pehlivan and Albas [483] investigated a b-generalized derivation H on a prime ring R satisfying suitable algebraic conditions. Now, we present the following necessary remark, which is frequently used in the results obtained in [483]:

Remark 4.8.12 [483, Remark 2.1] *Let R be a prime ring with the right Martindale quotient ring Q and $b \in Q$. If H is a b-generalized derivation of R with the associated derivation d, then the map $f = -H : R \to Q$ defined by $(-H)(x) = -H(x)$ for all $x \in R$ is also a b-generalized derivation of R with the associated map $-d$.*

Proof: By the definition of a generalized b-derivation, we know that

$$H(xy) = H(x)y + bxd(y)$$

for all $x, y \in R$. Also, it is clear that, since d is a derivation, we have

$$(-d)(xy) = -d(xy) = -d(x)y - xd(y) = (-d)(x)y + x(-d)(y)$$

for all $x, y \in R$. This implies that $-d$ is also a derivation of R. Hence, we get

$$(-H)(xy) = -H(xy) = -H(x)y - bxd(y) = (-H)(x)y + bx(-d)(y)$$

for all $x, y \in R$, which means that $f = -H$ is also a b-generalized derivation of R with the associated derivation $-d$.

♣

The following theorem is a generalization of Theorem 3.1 of [5]:

Theorem 4.8.13 [483, Theorem 2.1] *Let R be a noncommutative prime ring, let Q be the right Martindale quotient ring of R, let C be the extended centroid of R, and let $b \in Q$. Suppose that H is a b-generalized derivation of R such that $H([x,y]) = 0$ for all $x,y \in R$. Then $H = 0$.*

The following results are motivated by Albaş and Argaç [5] (Theorem 3.2 and Corollary 3.3). These theorems can be formulated as follows:

Theorem 4.8.14 [483, Theorem 2.2] *Let R be a noncommutative prime ring, let Q be the right Martindale quotient ring of R, and let $b \in Q$. Suppose that H is a nonzero b-generalized derivation of R. If either $H([x,y]) = [x,y]$ for all $x,y \in R$ or $H([x,y]) = [x,y]$ for all $x,y \in R$, then either $H = I_d$ or $H = -I_d$, respectively.*

Corollary 4.8.4 [483, Corollary 2.1] *Let R be a noncommutative prime ring, let Q be the right Martindale quotient ring of R, and let $b \in Q$. Suppose, H is a nonzero b-generalized derivation of R. If either $H([x,y]) = [x,y]$ or $H([x,y]) = -[x,y]$ holds for all $x,y \in R$, then either $H = I_d$ or $H = -I_d$, respectively.*

Proof: By the hypothesis, we can assume that R admits a b-generalized derivation H such that $H([x,y]) = [x,y]$ or $H([x,y]) = -[x,y]$ for all $x,y \in R$. Therefore, for each $y \in R$ we consider two subsets $N_y = \{x \in R : H[x,y] = [x,y]\}$ and $K_y = \{x \in R : H[x,y] = -[x,y]\}$ of R. Then N_y and K_y are two additive subgroups of $(R,+)$ such that $(R,+) = (N_y,+) \cup (K_y,+)$. Further, since a group cannot be the union of two proper subgroups, for each $y \in R$, we get either $R = N_y$ or $R = K_y$. By using the same procedure as above, we obtain either $R = \{y \in R : R = N_y\}$ or $R = \{y \in R : R = K_y\}$. Finally, by using Theorem 4.8.14, we arrive at the required conclusion.

♣

Theorem 4.8.15 [483, Theorem 2.3] *Let R be a noncommutative prime ring, let Q be the right Martindale quotient ring of R, and let $b \in Q$. Suppose that H is a nonzero b-generalized derivation of R. If one of the following conditions holds:*

(*i*) *$H(xy) = xy$ for all $x,y \in R$*

(*ii*) *$H(xy) = -xy$ for all $x,y \in R$*

(*iii*) *either $H(xy) = xy$ or $H(xy) = -xy$ for all $x,y \in R$,*

then either $H = I_d$ or $H = -I_d$, respectively.

Proof: (*i*) Assume that H is a nonzero b-generalized derivation of R satisfying the equality $H(xy) = xy$ for all $x,y \in R$. Moreover, by the hypothesis we also have $H(yx) = yx$ for all $x,y \in R$. Comparing the last relations, we obtain $H([x,y]) = [x,y]$ for all $x,y \in R$. Further, by using Theorem 4.8.14, we get $H = I_d$.

(ii) Assume that H is a nonzero b-generalized derivation of R such that $H(xy) = -xy$ for all $x, y \in R$. By using the same arguments as in (i) and Theorem 4.8.14, we obtain $H = -I_d$. This yields the desired result.

(iii) For each $x \in R$, we set $N_x = \{y \in R : H(xy) = xy\}$ and $K_x = \{y \in R : H(xy) = -xy\}$. In the same way as in the proof of Corollary 4.8.4, by using (i) and (ii), we have either $H - I_d$ or $H = -I_d$.

♣

Theorem 4.8.16 [483, Theorem 2.4] *Let R be a prime ring, let Q be the right Martindale quotient ring of R, and let $b \in Q$. Suppose that H is a nonzero b-generalized derivation of R. If $H(xy) = \mp yx$ for all $x, y \in R$, then R is commutative or $H = I_d$ or $H = -I_d$, respectively.*

Now, the following theorem is a generalization of Theorem 2.1 in [45]:

Theorem 4.8.17 [483, Theorem 2.5] *Let R be a prime ring, let Q be the right Martindale quotient ring of R, $Z(R)$ be the center of R, C be the extended centroid of R, and $b \in Q$. Suppose that H is a nonzero b-generalized derivation of R. If*

$$H(xy) \mp xy \in Z(R)$$

for all $x, y \in R$, then R is commutative; otherwise, either $H = I_d$ or $H = -I_d$, respectively.

Theorem 4.8.18 [483, Theorem 2.6] *Let R be a prime ring, let Q be the right Martindale quotient ring of R, let $Z(R)$ be the center of R, let C be the extended centroid of R and let $b \in Q$. Suppose that H is a nonzero b-generalized derivation of R. If $H(xy) \mp yx \in Z(R)$ for all $x, y \in R$. Then R is commutative.*

The following theorems can be regarded as generalizations of Theorem 3.4 of [483]:

Theorem 4.8.19 [483, Theorem 2.7] *Let R be a prime ring, Q be the right Martindale quotient ring of R, C be the extended centroid of R, and $b \in Q$. Suppose, H is a nonzero b-generalized derivation of R such that $H(xy) = H(x)H(y)$ for all $x, y \in R$. Then either there exists $q \in Q(RC)$ such that $H(x) = -bxq$ and $d(x) = [q, x]$ for all $x \in R$ with $qb = -1$ or F is the identity map on R.*

Theorem 4.8.20 [483, Theorem 2.8] *Let R be a prime ring with characteristic not equal to 2, Q be the right Martindale quotient ring of R, C be the extended centroid of R, and $b \in Q$. Suppose that H is a nonzero b-generalized derivation of R such that $H(xy) = H(y)F(x)$ for all $x, y \in R$. Then R is commutative and H is the identity map of R.*

4.9 b-generalized skew derivations

In [354], Kosan and Lee introduced a novel definition which was examined in Section 4.8. This prompts an inquiry into the existence of a unified definition encompassing both b-generalized and generalized skew derivations. Building on this concept, De Filippis and Wei [227] proposed a new definition that collectively generalizes the two aforementioned definitions:

Definition 48 *[227, Definition 1.1] Let R be an associative algebra, $b \in Q$, $d : R \to R$ a linear mapping and α be an automorphism of R. A linear mapping $H : R \to R$ is called a b-***generalized skew derivation** *of R, with associated term (b,α,d) if $H(xy) = H(x)y + b\alpha(x)d(y)$ for all $x,y \in R$.*

Firstly, we would like to give some examples of additive maps which are b-generalized skew derivations:

Example 4.9.1 *[227, Example 2.1.] Let R be an associative algebra and α be an automorphism of R. The following mapping $H : R \to R$ such that $H(x) = ax + b\alpha(x)c$ is a b-generalized skew derivation of R with associated term (b,α,d), where a,b and c are fixed elements in R and $d(x) = \alpha(x)c - cx$ for all $x \in R$. Indeed, for all $x,y \in R$,*

$$\begin{aligned} H(xy) &= axy + b\alpha(xy)c \\ &= axy - b\alpha(x)cy + b\alpha(x)cy + b\alpha(x)\alpha(y)c \\ &= (ax + b\alpha(x)c)y + b\alpha(x)(\alpha(y)c - cy) \\ &= H(x)y + b\alpha(x)d(y). \end{aligned}$$

where $d(y) = (\alpha(y)c - cy)$ is an inner skew derivation of R induced by the element $c \in R$, with an associated automorphism α. Such b-generalized skew derivations are called **inner b-generalized skew derivations**.

Example 4.9.2 *[227, Example 2.2.] Let R be an associative algebra, I_R be the identical mapping of R, α an automorphism of R, and $a,b \in R$. Then $H(x) = b(\alpha - aI_R)(x)$ is a b-generalized skew derivation of R with associated term $(b,\alpha,\alpha - I_R)$. Indeed, for all $x,y \in R$ we get,*

$$\begin{aligned} H(xy) &= b(\alpha - I_R)(xy) = b\alpha(x)\alpha(y) - baxy \\ &= b\alpha(x)\alpha(y) - b\alpha(x)y + b\alpha(x)y - baxy \\ &= b(\alpha - I_R)(x)y + b\alpha(x)(\alpha - I_R)(y) \\ &= H(x)y + b\alpha(x)d(y), \end{aligned}$$

where $d(x) = \alpha(x) - x$ for all $x \in R$. Note that d is a skew derivation of R in this case.

Example 4.9.3 *[227, Example 2.3.] Let R be an associative algebra, α an automorphism of R, d a skew derivation associated with α, a and $b \in R$. Then*

$H(x) = b(\alpha - d)(x)$ is a b-generalized skew derivation of R with associated term $(b, \alpha, \alpha - I_R - d)$. Indeed, for all $x, y \in R$ we get,

$$\begin{aligned} H(xy) &= b(\alpha - d)(xy) = b\alpha(x)\alpha(y) - bd(x)y - b\alpha(x)d(y) \\ &= b\alpha(x)\alpha(y) - b\alpha(x)y + b\alpha(x)y - bd(x)y - b\alpha(x)d(y) \\ &= H(x)y + b\alpha(x)(\alpha(y) - y - d(y)). \end{aligned}$$

Here, the skew derivation $\delta = \alpha(x) - x - d(x)$ is the additive mapping associated with F.

Example 4.9.4 *[227, Example 2.4.] Let R ba an associative algebra, H be a b-generalized derivation of R with associated pair (b,d), where d is a derivation and b is an invertible element of R. Then the mapping*

$$G : R \to R, \ \ x \to aH(x)$$

is a b-generalized skew derivation of R. Indeed, for all $x, y \in R$,

$$\begin{aligned} G(xy) &= aH(xy) \\ &= aH(x)y + abxd(y) \\ &= aH(x)y + abxb^{-1}bd(y) \\ &= G(x)y + a\alpha(x)bd(y). \end{aligned}$$

Example 4.9.5 *[227, Example 2.5.] Let R be an associative algebra, α an automorphism of R, δ a skew derivation associated with α, G a generalized skew derivation associated with α and d, and $b \in R$. Then $G(x) = b(\alpha - H)(x)$ is a b-generalized skew derivation of R with associated term $(b, \alpha, \alpha - I_R - \delta)$. Indeed, for all $x, y \in R$, we obtain*

$$\begin{aligned} G(xy) = b(\alpha - H)(xy) &= b\alpha(x)\alpha(y) - bH(x)y - b\alpha(x)\delta(y) \\ &= b\alpha(x)\alpha(y) - b\alpha(x)y + b\alpha(x)y - bH(x)y - b\alpha(x)\delta(y) \\ &= G(x)y + b\alpha(x)(\alpha(y) - y - \delta(y)). \end{aligned}$$

According to the above examples, we can conclude that general results about b-generalized skew derivations may give useful and powerful corollaries about derivations, generalized derivations, skew derivations and generalized skew derivations. Many results in the literature indicate how the global structure of a ring R is often tightly connected to the behaviour of additive mappings defined on R. A well-known result of Posner [487] states: If d is a derivation of R, such that $[d(x), x] \in Z(R)$, for any $x \in R$, then either $d = 0$ or R is commutative. In [365], Lanski generalizes the Posner's result to a Lie ideal. Later, in [155], the authors prove: If R is a prime ring of characteristic not equal to 2, L a non-central Lie ideal of R, d a non-zero derivation of R, such that $[d(u), u]^n \in Z(R)$ for any $u \in L$, then R satisfies s_4, the standard identity of degree 4. In particular, if d satisfies $[d(u), u] = 0$ for any $u \in L$, then R is commutative.

Wang [543] considered the case when the derivation d is replaced by a nontrivial automorphism σ of R and proved that if R is a prime ring having center Z, with $char(R) \geq n$ or $char(R) = 0$, L a non-central Lie ideal of R and σ a non-trivial automorphism of R, such that $[u^{\sigma}, u]^n \in Z$ for all $u \in L$, then R satisfies S_4.

In [159], the authors continued this line of investigation and examined what happens in the case, when the involved additive map is a generalized skew derivation. They proved: If R is a non-commutative prime ring of characteristic not equal to 2, $G \neq 0$ a generalized skew derivation of R, and $n \geq 1$, such that $[G(x), x]^n = 0$, for all $x \in R$, then there exists an element $\lambda \in C$, such that $G(x) = \lambda x$, for all $x \in R$.

By taking motivation from the above results, De Filippis and Wei [227] extended all previously cited results to b generalized skew derivations. The results they obtained are as follows:

Theorem 4.9.1 *[227, Theorem 1.5] Let R be a non-commutative prime ring of characteristic not equal to 2 and 3, Q its right Martindale quotient ring, C its extended centroid, $H \neq 0$ a b-generalized skew derivation of R, L a non-central Lie ideal of R, $0 \neq a \in R$ and $n \geq 1$ such that $a[H(x), x]^n = 0$ for all $x \in L$. Then, either there exists an element $\lambda \in C$, such that $H(x) = \lambda x$, for all $x \in R$ or R satisfies $s_4(x_1, x_2, \ldots, x_4)$, the standard identity of degree 4, and there exist $\lambda \in C$ and $b \in Q$, such that $H(x) = bx + xb + \lambda x$, for all $x \in R$.*

Theorem 4.9.2 *[227, Theorem 1.6] Let R be a non-commutative prime ring, $Z(R)$ its center, C its extended centroid, $H \neq 0$ a b-generalized skew derivation of R, $0 \neq a \in R$ and $n \geq 1$ such that $a[H(x), x]^n \in C$, for all $x \in R$. Assume that $char(R) = 0$ or $char(R) \geq n$, then either there exists an element $\lambda \in C$ such that $H(x) = \lambda x$ for all $x \in R$ or R satisfies s_4.*

Now we prove the following vital remarks on b-generalized skew derivations:

Remark 4.9.3 *Let R be a prime ring, $\alpha \in Aut(R)$, $0 \neq b \in Q$, $d : R \to R$ be an additive mapping of R and H be the b-generalized skew derivation of R with associated term (b, α, d). Then d is a skew derivation of R with associated automorphism α.*

Proof: For any $x, y \in R$, we have

$$H(xyz) = H(xy)z + b\alpha(xy)d(z) = H(x)yz + b\alpha(x)d(y)z + b\alpha(x)\alpha(y)d(z)$$

and

$$H(xyz) = H(x)yz + b\alpha(x)d(yz).$$

By comparing the preceding relations, it follows that

$$0 = b\alpha(x)d(y)z + b\alpha(x)\alpha(y)d(z) - b\alpha(x)d(yz).$$

That is

$$b\alpha(R)(d(yz)-d(y)z-\alpha(y)d(z))=0. \tag{4.34}$$

Therefore, by the primeness of R and since $b\neq 0$, we get $d(yz)=d(y)z+\alpha(y)d(z)$, for all $y,z\in R$, as required.

♣

Remark 4.9.4 *Let R be a prime ring, $\alpha\in Aut(R)$, $b\in Q$, $d:R\to R$ be an additive mapping of R and H be the b-generalized skew derivation of R with associated term (b,α,d). Then H can be uniquely extended to Q and assumes the form $H(x)=ax+bd(x)$, where $\alpha\in Q$.*

Proof: In case $b=0$, then $H(xy)=H(x)y$, for all $x,y\in R$. Thus, F is a centralizer (that is, H is a generalized derivation) and the result follows from Theorem 4.3.1.

Let us consider the case of $b\neq 0$. Define $T:R\to R$ such that $T(x)=H(x)-bd(x)$. Since d is a skew derivation of R, we have

$$T(xy)=H(x)y+b\alpha(x)d(y)-bd(x)y-b\alpha(x)d(y)=(H(x)-bd(x))y=T(x)y$$

for all $x,y\in R$. For any $x\in Q$, there exists an ideal I_x of R, such that $xI_x\subseteq R$. Thus, T can be extended to Q by $T(xy)=T(x)y$, for all $y\in I_x$. Since $H(x)=T)(x)+bd(x)$ and both T and d can be uniquely extended to Q, we infer that H can be uniquely extended to Q.

Moreover, for any $x\in R$, $H(x)=H(1.x)=H(1)x+bd(x)=ax+d(x)$, where $a=H(1)\in Q$.

♣

Lanski [365] generalizes Posner's result [487] by replacing the element $x\in R$ with an element of a noncentral Lie ideal L of R. He proved that if $[d(x),x]_k=0$ for all $x\in L$ and $k\geq 1$ is a fixed integer, then $char(R)=2$ and R satisfies s_4, the standard polynomial identity of degree 4. In [33], Argaç et al. continue the line of investigation concerning the Engel conditions $[F(x),x]_k=0$ for all $x\in S$, where S is a suitable subset of R and F is an additive mapping of R. They study the case when $S=L$ denotes a noncentral Lie ideal of R, and F is a generalized derivation on R and prove that if $[F(u),u]_k=0$ holds for all $u\in L$, then either $F(x)=ax$, with $a\in C$ or R satisfies the standard identity s_4. Moreover, in the latter case either $char(R)=2$ or $char(R)\neq 2$ and $F(x)=ax+xb$, with $a,b\in Q_r$ and $a-b\in C$. Also, the last result is extended to skew derivations and generalized skew derivations (in [181] and [168], respectively). Motivated by the above results, De Filippis and Wei [225] characterized b-generalized skew derivations which satisfy Engel conditions on Lie ideals of prime rings. They proved the following:

Theorem 4.9.5 *[225, Theorem 1.2] Let R be a prime ring of characteristic not equal to 2, Q_r be its right Martindale quotient ring and C be its extended centroid, H be a b-generalized skew derivation of R, L be a non-central Lie ideal of R. If $[H(u),u]_k = 0$ for all $u \in L$, where $k \geq 1$ is a fixed positive integer, then one of the following statements holds:*

1. *There exists $\lambda \in C$ such that $H(x) = \lambda x$, for all $x \in R$.*
2. *$R \subseteq M_2(C)$, is the ring of 2×2 matrices over C, and there exists $a \in Q_r$ and $\lambda \in C$ such that $H(x) = ax + xa + \lambda x$, for all $x \in R$.*

In 2009, De Filippis and Tammam El-Sayiad [224] extended Lanski's result [365] to generalized derivations as follows.

Theorem 4.9.6 *Let R be a non-commutative prime ring of characteristic not equal to 2, U its Utumi quotient ring and C its extended centroid. Suppose, L is a non-central Lie ideal of R, F a nonzero generalized derivation of R. If*

$$[[F(x),x],F(x)] \in Z(R),$$

for all $x \in L$, then one of the following holds:

1. *There exists $\lambda \in C$ such that $F(x) = \lambda x$, for all $x \in R$.*
2. *R satisfies $s_4(x_1,\ldots,x_4)$, the standard polynomial identity on 4 non-commuting variables, and there exist $a \in U$ and $\lambda \in C$ such that $F(x) = ax + xa + \lambda x$, for all $x \in R$.*

In [240], Dhara extended the latter result to the case of generalized skew derivations. In fact, Dhara proves the following theorem for generalized skew derivations:

Theorem 4.9.7 *Let R be a non-commutative prime ring of characteristic not equal to 2 and 3, Q_r its right Martindale quotient ring and C its extended centroid. Suppose L is a non-central Lie ideal of R, G a nonzero generalized skew derivation of R. If*

$$[[G(x),x],G(x)] \in Z(R),$$

for all $x \in L$, then one of the following holds:

1. *There exists $\lambda \in C$ such that $G(x) = \lambda x$, for all $x \in R$.*
2. *$R \subseteq M_2(C)$, the ring of 2×2 matrices over C, and there exists $a \in Q_r$ and $\lambda \in C$ such that $G(x) = ax + xa + \lambda x$, for all $x \in R$.*

Recently, Carini and Scudo [158] continued this line of investigation and examined what happens when a nonzero generalized skew derivation of a prime ring R is replaced with a nonzero b-generalized skew derivation.

The results they obtained have a similar flavor to the previously mentioned results due to De Filippis and Tammam El-Sayiad [224] and Dhara [241]. They have proved the following theorem.

Theorem 4.9.8 *[158, Theorem 1.3] Let R be a non-commutative prime ring of characteristic not equal to 2 and 3, Q_r its right Martindale quotient ring and C its extended centroid. Suppose that L is a non-central Lie ideal of R, H a nonzero b-generalized skew derivation of R. If*

$$[[H(x),x],H(x)] \in Z(R),$$

for all $x \in L$, then one of the following holds:

1. *There exists $\lambda \in C$ such that $H(x) = \lambda x$, for all $x \in R$.*
2. *$R \subseteq M_2(C)$, the ring of 2×2 matrices over C, and there exists $a \in Q_r$ and $\lambda \in C$ such that $H(x) = ax + xa + \lambda x$, for all $x \in R$.*

4.10 b-generalized skew derivations on multilinear polynomials

We have already discussed the definition of a b-generalized skew derivation (X-generalized skew derivation) in Section 4.9. In this section, we discuss b-generalized skew derivations on multilinear polynomials. According to the above definition, we conclude that general results on X-generalized skew derivations may give useful and powerful corollaries on derivations, generalized derivations, skew derivations, and generalized skew derivations.

In the year 2018, De Filippis and Feng Wei [225] investigated, the algebraic point of view, the set $P(G,S) = \{G(x)x : x \in S\}$, where $G: R \to R$ is an additive map of R and S is an appropriate subset of R. In [390, Theorem 2] Lee and Shiue proved that if $f(x_1,\ldots,x_n)$is a multilinear polynomial over C in n non-commuting indeterminates, $f(R) = \{f(r_1,\ldots,r_n) : r_1,\ldots,r_n \in R\}$ and G is a nonzero derivation of R, such that $P(G,f(R)) \subseteq C$, then $f(x_1,\ldots,x_n)$ is central-valued on R, unless when $char(R) = 2$ and R satisfies s_4. Demir and Argaç [233] extend the above results to the case of generalized derivations acting on multilinear polynomials, and they showed that if G is a nonzero generalized derivation of R such that $P(G,f(R)) \subseteq C$, then either $f(x_1,\ldots,x_n)$ is central-valued on R or there exists $b \in C$ such that $G(x) = bx$ for all $x \in R$ and $f(x_1,\ldots,x_n)^2$ is central-valued on R, unless when $char(R) = 2$ and R satisfies s_4. In [31], the same authors study the properties of the set $P(G,f(R))$ related to its left annihilator. More precisely, if $char(R) = 2$, G is a nonzero generalized derivation of R and $a \in R$ such that $ax = 0$, for any $x \in P(G,f(R))$, then either $a = 0$ or there exists $b \in C$ such that $G(x) = bx$ for all $x \in R$ and $ab = 0$, unless when $f(x_1,\ldots,x_n)$ is central-valued on R.

All the previous results imply that the subset $P(G,S)$ is rather large in R, where G is a nonzero generalized derivation of R and S is the set of all the evaluations of a non-central multilinear polynomial over C. In [156], Carina et al. proved that if δ is a nonzero derivation of R, G a nonzero generalized derivation of R, $f(x_1,\ldots,x_n)$ a non-central multilinear polynomial over C such that $\delta(x) = 0$, for all $x \in P(G, f(R))$, then $f(x_1,\ldots,x_n)^2$ is central-valued on R and there exist $a, b \in U$ such that $G(x) = ax$ and $\delta(x) = [b,x]$, for any $x \in R$, with $[a,b] = 0$. In [160], Carina et. al proved that if δ is a nonzero derivation of R, G a nonzero generalized derivation of R, $f(x_1,\ldots,x_n)$ a non-central multilinear polynomial over C such that $\delta(x) = 0$, for all $x \in P(G, f(R))$, then $f(x_1,\ldots,x_n)^2$ is central-valued on R and there exist $a, b \in U$ such that $G(x) = ax$ and $\delta(x) = [b,x]$, for any $x \in R$, with $[a,b] = 0$. In 2016, Argaç and Dhara in [239] extend the result contained in [156] to the case F and G are generalized derivations of R, $f(x_1,\ldots,x_n)$ is a non-central multilinear polynomial over C such that $F(x) = 0$, for all $x \in P(G, f(R))$. In this case, they give a complete description of the form of F and G. In light of the results mentioned above, it would seem natural to question the relationship between the form of the X-generalized skew derivation (b-generalizedskew derivation) G and the algebraic properties satisfied by the set $P(G,S)$, where S is the set of all the evaluations of a non-central multilinear polynomial over C. In this sense, De Filippis and Wei [226] proved the following theorem.

Theorem 4.10.1 *[226, Theorem 1.2] Let R be a prime ring of characteristic not equal to 2, Q_r be its right Martindale quotient ring and C be its extended centroid, H a nonzero X -generalized skew derivation of R, and S be the set of the evaluations of a multilinear polynomial $f(x_1,\ldots,x_n)$ over C with n non-commuting variables. Let $u, v \in R$ such that $uH(x)x + H(x)xv = 0$ for all $x \in S$. Then one of the following statements holds:*

(a) *$v \in C$ and there exist $a, b, c \in Q_r$ such that $H(x) = ax + bxc$ for all $x \in R$ with $(u+v)a = (u+v)b = 0$;*

(b) *$f(x_1,\ldots,x_n)^2$ is central-valued on R and there exists $a \in Q_r$ such that $H(x) = ax$ for all $x \in R$ with $ua + av = 0$.*

In [250], Dhara et al. consider a pair of generalized derivations and studied the identity of multilinear polynomials. In fact, they proved the following result: Let R be a prime ring of characteristic not equal to 2 with Utumi quotient ring U, C the extended centroid of R, F and F_1 two non-zero generalized derivations of R and $f(x_1,\ldots,x_n)$ a multilinear polynomial over C, which is not central valued on R. If $[F(u)u, F_1(v)v] = 0$ for all $u, v \in f(R)$, then there exist $a, b \in U$ such that $F(x) = ax$ and $F_1(x) = bx$ for all $x \in R$ with $[a,b] = 0$ and $f(x_1,\ldots,x_n)^2$ is central valued on R. Recently, Das et al. [205] generalized the preceding result and proved the following: Let R be a non-commutative prime ring of characteristic not equal to 2, U the Utumi quotient ring of R, C the extended centroid of R and $f(x_1,\ldots,x_n)$ a non-central multilinear polynomial over C. If d is a nonzero derivation of R and F is a non-zero generalized derivation of R such that $d(F(u)u) \in Z(R)$ for all $u \in f(R)$ then $f(x_1,\ldots,x_n)^2$ is central valued on R, and there exists $b \in U$ such that $F(x) = bx$ for all $x \in R$ with $d(b) \in C$. Motivated by the above results, in 2022, Prajapati and Gupta [488] considered a b-generalized skew derivation and proved the following theorem:

Theorem 4.10.2 *[488, Main Theorem] Let R be a prime ring of characteristic not equal to 2 with Utumi quotient ring U, extended centroid C and $f(x_1,\ldots,x_n)$ a multilinear polynomial over C, which is not central valued on R. Suppose $P \in R$ and H is a nonzero b-generalized skew derivation of R such that $[p,H(u)u] \subset C$ for all $u \in f(R)$: Then one of the following holds:*

(a) $p \in C$;

(b) *$f(x_1,\ldots,x_n)^2$ is central valued and there exists $Q \in U$ such that $H(x) = Qx$ for all $x \in R$ and $[P,Q] \in C$.*

In particular, we have the following corollary to Theorem 4.10.2.

Corollary 4.10.1 *[488, Corollary 1.5] Let R be a prime ring of characteristic not equal to 2 with Utumi quotient ring U, extended centroid C and $f(x_1,\ldots,x_n)$ a multilinear polynomial over C, which is not central valued on R. Suppose $P \in R$ and H is a non-zero b-generalized skew derivation of R such that $[p,H(u)u] = 0$ for all $u \in f(R)$: Then one of the following holds:*

(a) $p \in C$;

(b) *$f(x_1,\ldots,x_n)^2$ is central valued and there exists $Q \in U$ such that $H(x) = Qx$ for all $x \in R$ and $[P,Q] = 0$.*

Corollary 4.10.2 *[488, Corollary 1.6] Let R be a prime ring of characteristic not equal to 2 with Utumi quotient ring U, extended centroid C and $f(x_1,\ldots,x_n)$ a multilinear polynomial over C, which is not central valued on R. Suppose $P \in R$ and H is a non-zero b-generalized skew derivation of R such that $H(u)u \in C$ for all $u \in f(R)$. Then $f(x_1,\ldots,x_n)^2$ is central valued and there exists $Q \in C$ such that $H(x) = Qx$ for all $x \in R$.*

In [156], Carini and De Filippis proved that if R is a prime ring of $char(R) \neq 2$, U the Utumi ring of quotients, δ a non-zero derivation of R, F a non-zero generalized derivation of R, and $f(x_1,\ldots,x_n)$ a non-central multilinear polynomial over C such that $\delta(F(f(X))f(X)) = 0$ for all $X = (x_1,\ldots,x_n) \in R^n$, then there exist $a,b \in U$ such that $G(x) = ax$ and $\delta(x) = [b,x]$ for all $x \in R$, with $[b,a] = 0$ and $f(x_1,\ldots,x_n)^2$ central-valued on R. Further, Dhara and Argaç [242] extended the above result replacing derivation δ with another generalized derivation F_1, that is, $F_1(F(f(X))f(X)) = 0$ for all $X = (x_1,\ldots,x_n) \in I^n$, and then gave the complete description of the additive maps F and F_1, where I is a non-zero two-sided ideal of R.

In [244], Dhara and De Filippis gave the complete description of generalized derivations F and F_1 when $F^2(f(X))f(X) - f(X)F_1^2(f(X)) = 0$ for all $X = (x_1, \ldots, x_n) \in I^n$, where I is a non-zero two-sided ideal of R. Then in [231], De Filippis et al. studied the situation replacing F^2 and F_1^2 with two b-generalized skew derivations in a prime ring.

In [248], Dhara et al. studied the case $d(F^2(f(X))f(X)) = f(X)F_1^2(f(X))$ for all $X = (x_1, \ldots, x_n) \in I^n$ and obtained the map forms, F and F_1, where F and F_1 are two generalized derivations. By continuing this line of investigation, in 2023, Bera et al. [123] studied the above situation by replacing F^2 and F_1^2 with two different b-generalized skew derivations and d with an inner derivation in a prime ring. More precisely, they proved the following theorem.

Theorem 4.10.3 *[123, Theorem 1.1] Let R be a prime ring of characteristic not equal to 2, $Z(R)$ its center, Q_r its right Martindale quotient ring, C its extended centroid, $f(x_1, \ldots, x_n)$ a non-central multilinear polynomial over C, $b \in Q_r$ and H, H_1 two b-generalized skew derivations of R. If there exists $a \in R$ such that*

$$[a, H(f(r_1, \ldots, r_n))f(r_1, \ldots, r_n)] = f(r_1, \ldots, r_n)H_1(f(r_1, \ldots, r_n))$$

for any $r_1, \ldots, r_n \in R$, then $f(x_1, \ldots, x_n)^2$ is central-valued on R and there exist $p, q \in Q_r$ such that $H(x) = px, H_1(x) = xq$ for all $x \in R$ with $[a, p] = q$.

In particular, if $H_1 = 0$, we have the following corollary.

Corollary 4.10.3 *[123, Corollary 1.2] Let R be a prime ring of $char(R) \neq 2$, Q_r its right Martindale quotient ring, C its extended centroid and $b \in Q_r$. Also let $f(x_1, \ldots, x_n)$ be a multilinear polynomial over C which is not central-valued on R. Suppose that H is a b-generalized skew derivation of R and $a \in R$ such that*

$$[a, H(f(r_1, \ldots, r_n))f(r_1, \ldots, r_n)] = 0$$

for all $x_1, \ldots, x_n \in R$. Then there exists $p \in Q_r$ such that $H(x) = px$ for all $x \in R$ with $f(x_1, \ldots, x_n)^2$ central-valued on R and $[a, p] = 0$.

Corollary 4.10.4 *[123, Corollary 1.3] Let R be a prime ring of $char(R) \neq 2$, Q_r its right Martindale quotient ring, C its extended centroid and $b \in Q_r$. Also let $f(x_1, \ldots, x_n)$ be a multilinear polynomial over C which is not central-valued on R. Suppose that H and H_1 are two b-generalized skew derivations of R such that $[H(u)u, H_1(v)v] = 0$ for all $u, v \in f(R)$, where $f(R) = \{f(x_1, \ldots, x_n) | x_1, \ldots, x_n \in R\}$. Then there exist $p, q \in Q_r$ such that $H(x) = px, H_1(x) = qx$ for all $x \in R$ with $f(x_1, \ldots, x_n)^2$ central-valued on R and $[p, q] = 0$.*

4.11 Open problems

This section addresses unresolved issues, emphasizing significant questions and challenges that remain of interest to the readers. By examining these open problems, our goal is to elucidate the intricate landscape of generalized derivations in rings and algebras and to encourage further exploration in this captivating field of study.

Open Problem 4.11.1 *Let n,m,k be nonzero fixed positive integers. Next, let R be a (semi-)prime ring, Q_r its right Martindale quotient ring, C its extended centroid, and let $H : R \to R$ and $H_1 : R \to R$ be b-generalized derivations of R satisfying any one of the differential identities*

(i) $[H(x^m),x^n]^k = 0$ *or* $(H(x^m) \circ x^n)^k = 0$,

(ii) $[H(x^m),x^n] = [x^m,x^n]$ *or* $H(x^m) \circ x^n = x^m \circ x^n$,

(iii) $[H(x^m),G(x^n)]_k = 0$ *or* $(H(x^m) \circ H_1(x^n))_k = 0$,

(iv) $H([x^m,x^n]) = 0$ *or* $H(x^m \circ x^n) = 0$,

(v) $H(x^m)x^n \pm x^n H_1(x^m) = 0$.

Then, determine the structures of R, H and H_1.

Open Problem 4.11.2 *Let n,m,k be nonzero fixed positive integers. Next, let R be a (semi-)prime ring, Q_r its right Martindale quotient ring, C its extended centroid, and let $H : R \to R$ and $H_1 : R \to R$ be b-generalized derivations of R satisfying any one of the differential identities*

(i) $[H(x),x]^n \in Z(R)$ *or* $(H(x) \circ x)^n \in Z(R)$,

(ii) $[[H(x^m),x^n],H(x^l)] \in Z(R)$,

(iii) $[H(x^m),x^n]_k = 0$ *or* $(H(x^m) \circ x^n)_k = 0$,

(iv) $[H(x^m),H_1(x^n)]_k = 0$ *or* $(H(x^m) \circ H_1(x^n))_k$,

(v) $H([x^m,x^n]) = [x,y]$ *or* $H(x^m \circ x^n) = 0$.

Then, determine the structures of R, H and H_1.

Open Problem 4.11.3 *Let $n \geq 1$ be a fixed integer. Next, let R be a (semi-) prime ring, Q_r its right Martindale quotient ring, C its extended centroid, and let $H : R \to R$ and $H_1 : R \to R$ be b-generalized derivations of R satisfying*

$$p(H(x)H(y) - H_1(y)x)^n = 0.$$

What can we say about the structures of R, H and H_1?

Open Problem 4.11.4 *Let R be a (semi-)prime ring, Q_r its right Martindale quotient ring, and C its extended centroid. Suppose $H : R \to R$ be a nonzero b-generalized skew derivation of R and $f(x_1,\ldots,x_n)$ be a non-central multilinear polynomial with n non-commuting variables. If there exists a nonzero element a of R satisfying*

$$a[H(f(r_1,\ldots,r_n)),f(r_1,\ldots,r_n)] = 0$$

for all $r_1,\ldots,r_n \in R$. What can we say about the structures of R and H?

Open Problem 4.11.5 *Let R be a (semi-)prime ring, Q_r its right Martindale quotient ring, and C its extended centroid. Suppose $f(X_1,\ldots,X_n)$ be a multilinear polynomial over C that is noncentral-valued on R. Denote the set of all the evaluations of $f(X_1,\ldots,X_n)$ on R by $f(R)$. If $H : R \to R$ and $H_1 : R \to R$ are two b-generalized skew derivations of R satisfying $[[H(u),u],[H_1(v),v]] \in Z(R)$ for all $u,v \in f(R)$. What can we say about the structures of R, H and H_1?*

Open Problem 4.11.6 *Let R be a (semi-)prime ring, Q_r its right Martindale quotient ring, and C its extended centroid. Suppose $f(X_1,\ldots,X_n)$ be a multilinear polynomial over C that is noncentral-valued on R. Denote the set of all the evaluations of $f(X_1,\ldots,X_n)$ on R by $f(R)$. Let $H : R \to R$ and $H_1 : R \to R$ be two b-generalized skew derivations of R satisfying*

(i) $[a,[H(f(r)),f(r)]] \in Z(R)$ *for all* $r = (r_1,\ldots,r_n) \in R^n$ *and* $a \in R - C$.

(ii) $[[H(u),u],[H_1(v),v]] \in Z(R)$ *for all* $u,v \in f(R)$.

What can we say about the structures of R, H and H_1?

Chapter 5

Higher Derivations in Rings and Algebras

5.1 Introduction

This chapter builds a framework that gives new insights into various seemingly unrelated problems and also demystifies some problems that were previously known to have efficient solutions. Rings are fundamental algebraic objects with associated natural operations of addition and multiplication. We tackle two well-known problems using the framework on rings and algebras namely additivity of certain classes of higher derivable mappings in rings and the characterization of various kinds of higher derivations in the settings of triangular algebras, generalised matrix algebras, von Neumann algebras, operator algebras, unital algebras, and more.

Let $\mathbb{N}$ be the set of non-negative integers, $\mathcal{A}$ an algebra over $\mathbb{Q}$, and $d : \mathcal{A} \to \mathcal{A}$ a derivation. If we define a sequence $\{d_n\}_{n\in\mathbb{N}}$ of $\mathbb{Q}$-linear mappings on $\mathcal{A}$ by $d_0 = id_{\mathcal{A}}$ (the identity mapping on $\mathcal{A}$) and $d_n = \frac{1}{n!}d^n$, then for each $n \in \mathbb{N}$, d_n satisfies the condition $d_n(ab) = \sum\limits_{i+j=n} d_i(a)d_j(b)$ for all $a, b \in \mathcal{A}$.

Motivated by the above observation, Hasse and Schmidt [287] introduced the notion of a higher derivation as follows:

Definition 49 *Let $\mathbb{N}$ be the set of non-negative integers and $D = \{d_n\}_{n\in\mathbb{N}}$ be a family of additive mappings on a ring R such that $d_0 = I_R$ (the identity mapping on R). Then, D is said to be a higher derivation on R if for each $n \in \mathbb{N}$, $d_n(ab) = \sum\limits_{i+j=n} d_i(a)d_j(b)$ for all $a, b \in R$.*

We now provide some examples of higher derivations in rings.

Example 5.1.1 *Let R be any ring. Define $d_0 = I_R$, the identity mapping on R and for any $a \in R$, $d_n(x) = (-1)^n(xa^n - axa^{n-1}) = (-1)^n(xa - ax)a^{n-1}$ for any $n \geq 1$, $x \in R$. Then, it is easy to verify that $D = \{d_n\}_{n\in\mathbb{N}}$ is a higher derivation on R.*

Example 5.1.2 *Let $\mathbb{F}$ be any field and in $\mathbb{F}[x]$, $d_i(x^j) = \binom{j}{i}x^{j-i}$, where $\binom{j}{i} = 0$ if $i \geq j$. Then, $D = \{d_i\}_{i\geq 0}$ is a higher derivation on $\mathbb{F}[x]$.*

Example 5.1.3 *For any field $\mathbb{F}$, char$(R) \neq 2$, let $R = \mathbb{F}[x]\langle y,z\rangle$ denote the free ring with coefficients in $\mathbb{F}[x]$ in the indeterminates y,z (where y and z commute with x but not with each other). This is a 2-torsion free prime ring. Thus, $Z(R) = \mathbb{F}[x]$. Let $D = \{d_n\}_{n\in\mathbb{N}}$ be a sequence of additive mappings of R such that $d_0 = I_R$ and, for every $n \geq 1$, $d_n(x) = p_n(x)$, where $p_n(x) \in \mathbb{F}[x]$, $d_n(y) = q_n(x,y,z)$, where $q_n(x,y,z) \in R$; $d_n(z) = r_n(x,y,z)$, where $q_n(x,y,z) \in R$; $d_n(a^j) = \sum_{i=0}^{n} d_i(a^{j-1})d_{n-i}(a)$, for all $j \geq 2$, where $a = x,y$ or z. Then, D is a higher derivation of R.*

5.2 Jordan higher derivations in rings

Inspired by Herstein's concepts of Jordan derivation, triple derivation, and Jordan triple derivation, Ferrero and Haetinger [267] introduced the higher forms of these ideas.

Definition 50 *Let $\mathbb{N}$ be the set of non-negative integers and $D = \{d_n\}_{n\in\mathbb{N}}$ be a family of additive mappings on R such that $d_0 = I_R$ (the identity mapping on R). Then, D is said to be*

(*i*) *a triple higher derivation on R if for each $n \in \mathbb{N}$,*

$$d_n(abc) = \sum_{i+j+k=n} d_i(a)d_j(b)d_k(c)$$

for all $a,b,c \in R$;

(*ii*) *a Jordan higher derivation on R if for each $n \in \mathbb{N}$,*

$$d_n(a^2) = \sum_{i+j=n} d_i(a)d_j(a)$$

for all $a \in R$.

(*iii*) *a Jordan triple higher derivation on R if for each $n \in \mathbb{N}$,*

$$d_n(aba) = \sum_{i+j+k=n} d_i(a)d_j(b)d_k(a)$$

for all $a,b \in R$.

Analogously, if U is a Lie ideal of a ring R, then D is said to a triple higher derivation or a Jordan higher derivation or a Jordan triple higher derivation of U respectively into R in case the above corresponding conditions are satisfied for all $a,b,c \in U$.

Ferrero and Haetinger [267] and Haetinger [281] extended Herstein's tremendous result [289] to higher derivations for prime and semiprime rings. More precisely, Haetinger obtained the following theorems.

Theorem 5.2.1 *[281, Theorem 2.1.10] Let R be 2-torsion free prime ring and $\Delta = \{\delta_n\}_{n\in\mathbb{N}}$ be a Jordan higher derivation on R. Then, Δ is higher derivation.*

Theorem 5.2.2 *[281, Theorem 2.2.1] Let R be 2-torsion free semiprime ring and $\Delta = \{\delta_n\}_{n\in\mathbb{N}}$ be a Jordan higher derivation on R. Then, Δ is a higher derivation.*

In the year 1984, Awtar [89] proved that if U is a Lie ideal of a 2-torsion free prime ring satisfying $u^2 \in U$ for all $u \in U$ and $\delta : R \to R$ is an additive map such that $\delta|_U$ is a Jordan derivation of U into R then $\delta|_U$ is a higher derivation of U into R. This result has been extended by Haetinger [283] as the following theorem.

Theorem 5.2.3 *[283, Theorem 2.1] Let R be prime ring of characteristic not equal to 2 and U be an admissible Lie ideal of R. Then, every Jordan higher derivation of U into R is a higher derivation of U into R.*

It is natural to ask whether the above theorem holds for Lie ideals in a semiprime ring. The following example illustrates this fact.

Example 5.2.1 *Let $\mathbb{F}$ be a field and $\mathcal{S}$ be the ring of all polynomials over $\mathbb{F}$ in non-commuting inderterminates x and y with nonzero constant terms. Further, let $\mathcal{T}$ be any semiprime algebra over $\mathbb{F}$ and put $R = \mathcal{S} \times \mathcal{T}$. Thus $U = \mathcal{S} \times \{0\}$ is a Lie ideal of R such that $u^2 \in U$ for all $u \in U$. Define $\delta : R \to R$ by $\delta(p,t) = (a-b)t_0 \in \mathcal{T}$, where $p \in \mathcal{S}, t \in \mathcal{T}$, a (resp. b) is the coefficient at xy (resp. yx) in the polynomial p and $t_0 \neq 0$ is a fixed element in $\mathcal{T}$. Then, it is easy to show that $\delta((p,t)^2) = 0$ for every $(p,t) \in R$. However, $\delta(xy,0) = t_0 \neq 0$. Hence, $\delta|_U$ is a Jordan derivation and not a derivation of U into R. Now, we further assume that $t^2 = 0$, so that $\delta(u)^2 = 0$ for every $u \in U$ and define $\delta'_0 = \mathcal{I}_R$, the identity mapping on R, $\delta'_1 = \delta$ and $\delta'_n = 0$ for every $n \geq 2$. Then, $\Delta = \{\delta'_n\}_{n\in\mathbb{N}}$ is a Jordan higher derivation which is not a higher derivation of U into R.*

In the year 2002, Ferrero and Haetinger [267] extended the results of Herstein [289] and Brešar [131, 132], to the case of higher derivations, and proved the following results.

Theorem 5.2.4 *[267, Theorem 1.2] Let R be a 2-torsion-free semiprime ring (resp. prime ring of characteristic not equal to 2 and U an admissible Lie ideal* *of R). Then, every Jordan triple higher derivation of R (resp. of U into R) is a higher derivation of R (resp. of U into R).*

Theorem 5.2.5 *[267, Theorem 1.3] Let R be a 2-torsion-free ring and U a Lie ideal of R. Then, every Jordan higher derivation of R (resp. of U into R) is a Jordan triple higher derivation of R (resp. of U into R).*

Salih and Haetinger [508] extended the the latter's result [281] to the case of triple higher derivations to obtain the following Theorems.

Theorem 5.2.6 *[508, Theorem 6] Every Jordan triple higher derivation of a 2-torsion free prime ring R is triple higher derivation of R.*

Theorem 5.2.7 *[508, Theorem 7] Every Jordan triple higher derivation of a 2-torsion free prime ring R is a higher derivation of R.*

Let σ, τ be endomorphisms of ring R. In the following, we give the (σ,τ)-version of higher derivations, Jordan higher derivations and Jordan triple higher derivations. Note that taking $\sigma,\tau = I_R$, the respective mappings are particular cases of the mappings defined below.

Definition 51 *Let R be a ring and σ, τ endomorphisms of R. For a family $D = \{\delta_i\}_{i\in\mathbb{N}}$ of additive mappings $\delta_i : R \to R$, $\delta_0 = I_R$, the identity mapping on R is said to be*

(i) a (σ,τ)-higher derivation if for each $n \in \mathbb{N}$

$$\delta_n(ab) = \sum_{i+j=n} \delta_i(\sigma^{n-i}(a))\delta_j(\tau^{n-j}(b))$$

holds for all $a,b \in R$

(ii) a Jordan (σ,τ)-higher derivation if for each $n \in \mathbb{N}$

$$\delta_n(a^2) = \sum_{i+j=n} \delta_i(\sigma^{n-i}(a))\delta_j(\tau^{n-j}(a))$$

holds for all $a \in R$

(iii) a Jordan triple (σ,τ)-higher derivation if for each $n \in \mathbb{N}$

$$\delta_n(aba) = \sum_{i+j+k=n} \delta_i(\sigma^{n-i}(a))\delta_j(\sigma^k\tau^i(b))\delta(\tau^{n-k}(a))$$

holds for all $a,b \in R$.

We now pause to look at an example of a (σ,τ)-higher derivation on R.

Example 5.2.2 *Let R be an algebra over the field of rationals $\mathbb{Q}$ and σ,τ be endomorphisms of R. Define $\delta_n = \frac{\eta^n}{n!}$ for all $n \in \mathbb{N}$, where η is a (σ,τ)-derivation on R such that $\eta\sigma = \sigma\eta$ and $\eta\tau = \tau\eta$. Consider the sequence $\Delta = \{\delta_n\}_{n\in\mathbb{N}}$.*

Using mathematical induction, it is easy to verify that Δ is (σ,τ) a higher derivation.

The above definitions suggest that every (σ,τ)-higher derivation on R is a Jordan (σ,τ)-higher derivation on R but the converse need not be true in general. It is also worth mentioning that in the above example if η is assumed to be a Jordan (σ,τ)-derivation on R which is not a (σ,τ)-derivation on R, then it is equally easy to find a Jordan (σ,τ)-higher derivation on R which is not a (σ,τ)-higher derivation on R.

A Jordan (σ,τ)-higher derivation might not always be a (σ,τ)-higher derivation, as we previously indicated. Below is an answer from Ashraf and his coauthors[67] to this.

Theorem 5.2.8 *[67, Theorem 2.2] Let R be a 2-torsion free ring and σ,τ be commuting endomorphisms of R such that τ is one-one and onto. If R has a commutator which is not a right zero divisor, then every Jordan (σ,τ)-higher derivation on R is a (σ,τ)-higher derivation on R.*

Theorem 5.2.9 *[67, Theorem 2.3] Let R be a non-commutative prime ring with char $(R)\neq 2$ and σ,τ, commuting endomorphisms of R such that τ is one to one and onto. Then, every Jordan (σ,τ)-higher derivation on R is a (σ,τ)-higher derivation on R.*

Later, Ashraf and Khan [65] obtained a similar kind of result for a semiprime ring.

Theorem 5.2.10 *[65, Theorem 2.1] Let R be a 2-torsion free semiprime ring. Then, every Jordan triple (σ,τ)-higher derivation of R is a (σ,τ)-higher derivation of R.*

Definition 52 *Let U be a Lie ideal of a ring R. Then, a family $D=\{\delta_i\}_{i\in\mathbb{N}}$ of additive mappings $\delta_i : R\to R$, $\delta_0 = I_R$ the identity mapping on R, is said to be a higher (U,R)-derivation if for every $n\in\mathbb{N}$,*

$$\delta_n(ur+su) = \sum_{i+j=n} \{\delta_i(u)\delta_j(r)+\delta_i(s)\delta_j(u)\}$$

holds for all $u\in U, r,s\in R$.

Example 5.2.3 *Let $\mathcal{S}$ be a commutative ring of char$(\mathcal{S})=2$ and consider $R=\mathcal{M}_2(\mathcal{S})$. Let $U=\left\{\begin{pmatrix} a & b \\ b & a\end{pmatrix} \,\middle|\, a,b\in\mathcal{S}\right\}$. Then, U is a Lie ideal of R.*

For any $n\in\mathbb{N}$, define $\delta_n : R\to R$ by $\delta_n\begin{pmatrix} a & b \\ c & d\end{pmatrix}=\begin{pmatrix} 0 & -b \\ nd & 0\end{pmatrix}$ for $n=1,2$ and $\delta\begin{pmatrix} a & b \\ c & d\end{pmatrix}=\begin{pmatrix} 0 & 0 \\ 0 & 0\end{pmatrix}$ for all $n\geq 3$ and $\begin{pmatrix} a & b \\ c & d\end{pmatrix}\in R$. Then, $D=\{\delta_i\}_{i\in\mathbb{N}}$ is a higher (U,R)-derivation.

The following result presents the form of a higher (U,R)-derivation on a ring R, obtained by Faraj and his co-authors in [259].

Theorem 5.2.11 *[259, Theorem 4.5] Let R be a prime ring, $char(R) \neq 2$, U be an admissible Lie ideal of R and $D = \{\delta_i\}_{i\in\mathbb{N}}$ be a higher (U,R)-derivation of R. Then, $\delta_n(ur) = \sum\limits_{i+j=n} \delta_i(u)\delta_j(r)$ for all $u \in U, r \in R$.*

5.3 Jordan (triple) higher derivable maps

Suppose that R is a ring with a nontrivial idempotent e_1. Then, $e_2 = 1 - e_1$. Note that R need not have an identity element. Put $e_iRe_j = R_{ij}$ for any $i,j \in \{1,2\}$. Then, by Peirce decomposition[1] [484] of R, we have $R = R_{11} \oplus R_{12} \oplus R_{21} \oplus R_{22}$. Throughout this section, the notation a_{ij} will denote an arbitrary element of R_{ij} and any element $a \in R$ can be expressed as $a = a_{11} + a_{12} + a_{21} + a_{22}$, where $a_{ij} \in R_{ij}$.

Definition 53 *A map $d : R \to R$ on a ring R (without the assumption of additivity) is said to be derivable if $d(ab) = d(a)b + ad(b)$ holds for all $a,b \in R$.*

Definition 54 *A map $d : R \to R$ on a ring R (without the assumption of additivity) is said to be Jordan derivable (resp. Jordan triple derivable) if $d(a^2) = d(a)a + ad(a)$ (resp. $d(aba) = d(a)ba + ad(b)a + abd(a)$) holds for all $a,b \in R$.*

Note that every derivable map on a ring R need not be additive.

Example 5.3.1 *Let $R = \left\{ \begin{pmatrix} 0 & a \\ 0 & 0 \end{pmatrix} \,\middle|\, a \in \mathbb{Z} \right\}$ be the ring of all 2×2 matrices over $\mathbb{Z}$, the ring of integers. Define $d : R \to R$ such that $d\begin{pmatrix} 0 & a \\ 0 & 0 \end{pmatrix} = \begin{pmatrix} 0 & a^2 \\ 0 & 0 \end{pmatrix}$. Then, it can be seen that $d(xy) = d(x)y + xd(y)$, for all $x,y \in R$ but $d(x+y) \neq d(x) + d(y)$.*

Characterizing the interrelation between the multiplicative and the additive structures of a ring is an interesting topic and has received attention from many mathematicians. If a ring R contains a nontrivial idempotent, it is kind of surpris-

[1] If e is an idempotent in a ring R (not necessarily with an identity), then we can decompose R as the direct sum of four components (subrings), each of them related to e. Concretely,

$$R = eRe \oplus eR(1-e) \oplus (1-e)Re \oplus (1-e)R(1-e);$$

note that here $(1-e)R$ is just notation that means $\{x - ex \mid x \in R\}$. This holds similarly for $R(1-e)$. This decomposition is known as the Peirce decomposition of R with respect to e.

ing that every multiplicative bijective[2] map from a ring R onto an arbitrary ring R' is automatically additive. This result was given by Martindale III in a well presented paper [444]. More precisely, he proved the Theorem.

Theorem 5.3.1 *[444, Theorem] Let R be a ring containing a family* $\{e_\alpha : \alpha \in \Lambda\}$ *of idempotents which satisfy,*

(i) $aR = \{0\}$ *implies* $a = 0$,

(ii) *if* $e_\alpha Ra = \{0\}$ *for each* $\alpha \in \Lambda$, *then* $a = 0$ *(and hence* $Ra = \{0\}$ *implies* $a = 0$*),*

(iii) *for each* $\alpha \in \Lambda$, $e_\alpha a e_\alpha R(1 - e_\alpha) = \{0\}$ *implies* $e_\alpha a e_\alpha = 0$.

Then, any multiplicative bijective map from R onto an arbitrary ring R' *is additive.*

Over the last few decades a lot of work has been done on the additivity of mappings on rings. These mappings include multiplicative mappings, (Jordan) derivable mappings, Jordan (triple) derivable mappings, and so on. Many authors have made important contributions to the related topics (see [327, 400, 429, 430, 447], where further references can be found). Inspired by the above result given by Martindale III [444], Daif [200] obtained the additivity of any derivable mapping of R on certain conditions imposed on a ring R. Further F. Lu [431] proved that every Jordan derivable map d on a 2-torsion free unital prime ring containing a nontrivial idempotent is a derivation. He also proved that each Jordan triple derivable map d on same outlining structure is a Jordan triple derivation. Moreover, it is not only a Jordan triple derivation but also a derivation. It is obvious that the condition of derivable maps at some point is much weaker than the condition of being a derivation. Hence it attracted the attention of many mathematicians who obtained some interesting results in this area.

Motivated by the above result due to Lu [431], in the year 2011, Jing and Lu [327] characterized Jordan derivable maps and Jordan triple derivable maps to a larger class of rings and proved the following theorems.

Theorem 5.3.2 *[327, Theorem 1.2] Let R be a ring containing a nontrivial idempotent satisfying the following conditions for* $i, j, k \in \{1, 2\}$*:*

(i) *If* $a_{ij}x_{jk} = 0$ *for all* $x_{jk} \in R_{jk}$, *then* $a_{ij} = 0$.

(ii) *If* $x_{ij}a_{jk} = 0$ *for all* $x_{ij} \in R_{ij}$, *then* $a_{jk} = 0$.

(iii) *If* $a_{ii}x_{ii} + x_{ii}a_{ii} = 0$ *for all* $x_{ii} \in R_{ii}$, *then* $a_{ii} = 0$.

[2] A bijective map $f : R \to R$ is said to be a multiplicative bijective if f satisfies $f(ab) = f(a)f(b)$, and for $a, b \in R$.

If a mapping $d : R \to R$ satisfies

$$d(ab+ba) = d(a)b + ad(b) + d(b)a + bd(a)$$

for all $a, b \in R$, then d is additive. Additionally, if R is 2-torsion free, then d is a Jordan derivation.

Theorem 5.3.3 *[327, Theorem 1.3] Suppose R is a ring containing a nontrivial idempotent and satisfies the following conditions:*

(i) *If $a_{11}x_{12} = 0$ for all $x_{12} \in R_{12}$, then $a_{11} = 0$.*

(ii) *If $x_{12}a_{22} = 0$ for all $x_{12} \in R_{12}$, then $a_{22} = 0$.*

(iii) *If $x_{ij}ax_{ij} = 0$ for all $x_{ij} \in R_{ij}$, $(i, j \in \{1,2\})$, then $a_{ji} = 0$.*

If a mapping $d : R \to R$ satisfies

$$d(aba) = d(a)ba + ad(b)a + abd(a)$$

for all $a, b \in R$, then d is additive.

Motivated by the concept of derivable mappings and higher derivation, the notion of higher derivable and Jordan higher derivable mappings have been introduced by Ashraf and Parveen in [69].

Definition 55 *A family $D = \{d_n\}_{n \in \mathbb{N}}$ of mappings $d_n : R \to R$ (without the assumption of additivity) is said to be*

(i) *a higher derivable on R if $d_n(ab) = \sum_{i+j=n} d_i(a)d_j(b)$ holds for each $n \in \mathbb{N}$ and for all $a, b \in R$.*

(ii) *a Jordan higher derivable on R if $d_n(a^2) = \sum_{i+j=n} d_i(a)d_j(a)$ holds for each $n \in \mathbb{N}$ and for all $a \in R$.*

The additivity of Jordan higher derivable mappings in rings has been obtained by Ashraf and Parveen [69]. Of course, the results obtained require few assumptions on a ring R, the underlying ring R having a nonzero idempotent. In fact, they proved the following theorem:

Theorem 5.3.4 *[69, Theorem 2] Let R be a ring containing a nontrivial idempotent satisfying the following conditions for $i, j, k \in \{1,2\}$:*

(i) *If $a_{ij}x_{jk} = 0$ for all $x_{jk} \in R_{jk}$, then $a_{ij} = 0$.*

(ii) *If $x_{ij}a_{jk} = 0$ for all $x_{ij} \in R_{ij}$, then $a_{jk} = 0$.*

(iii) *If $a_{ii}x_{ii} + x_{ii}a_{ii} = 0$ for all $x_{ii} \in R_{ii}$, then $a_{ii} = 0$.*

If the family $D = \{d_n\}_{n\in\mathbb{N}}$ of mappings $d_n : R \to R$ such that $d_0 = I_R$ satisfy

$$d_n(ab+ba) = \sum_{p+q=n} \Big(d_p(a)d_q(b) + d_p(b)d_q(a) \Big) \text{ for all } a,b \in R,\ n \in \mathbb{N}, \quad (5.1)$$

then $D = \{d_n\}_{n\in\mathbb{N}}$ is the family of additive mappings. In addition, if R is 2-torsion free, then $D = \{d_n\}_{n\in\mathbb{N}}$ is a Jordan higher derivation.

Remark 5.3.5 *[327, Lemma 1.5] Let R be a 2-torsion free semiprime ring with a nontrivial idempotent, then: (iii) if $a_{ii}x_{ii} + x_{ii}a_{ii} = 0$ for all $x_{ii} \in R_{ii}$, then $a_{ii} = 0$.*

If the underlying ring is semiprime, applying the above remark and the fact that every Jordan higher derivation on a 2-torsion free semiprime ring is a higher derivation (see [267]), we have;

Corollary 5.3.1 *Let R be a 2-torsion free semiprime ring with a nontrivial idempotent satisfying following conditions for $i, j \in \{1,2\}$;*

(i) If $a_{ii}x_{ij} = 0$ for all $x_{ij} \in R_{ij}$, $(i \neq j)$, then $a_{ii} = 0$.

(ii) If $x_{ji}a_{ii} = 0$ for all $x_{ji} \in R_{ji}$, $(i \neq j)$ then $a_{ii} = 0$.

If the family $D = \{d_n\}_{n\in\mathbb{N}}$ of mappings $d_n : R \to R$ such that $d_0 = I_R$ satisfy

$$d_n(ab+ba) = \sum_{p+q=n} (d_p(a)d_q(b) + d_p(b)d_q(a))$$

for all $a,b \in R$, then $D = \{d_n\}_{n\in\mathbb{N}}$ is the family of additive mappings. Moreover, $D = \{d_n\}_{n\in\mathbb{N}}$ is a higher derivation.

Application of the fact that every Jordan higher derivation on a prime ring of characteristic not equal to 2 is a higher derivation (see the main Theorem of [268]), we establish the following corollary.

Corollary 5.3.2 *Let R be a prime ring of characteristic not equal to 2 which has a nontrivial idempotent element. If the family $D = \{d_n\}_{n\in\mathbb{N}}$ of mappings $d_n : R \to R$ such that $d_0 = I_R$ satisfy*

$$d_n(ab+ba) = \sum_{p+q=n} (d_p(a)d_q(b) + d_p(b)d_q(a))$$

for all $a,b \in R$, then $D = \{d_n\}_{n\in\mathbb{N}}$ is the family of additive mappings. Moreover, $D = \{d_n\}_{n\in\mathbb{N}}$ is a higher derivation.

We shall conclude by giving an application of Theorem 5.3.4 in operator algebras[3]. Since every standard operator algebra[4] is prime, we can easily establish Corollary 5.3.3.

[3] An operator algebra is an algebra of continuous linear operators on a topological vector space with the multiplication given by the composition of mappings.

[4] Standard operator algebra is a subalgebra of $B(X)$ which contains $F(X)$, where $B(X)$ is the algebra of all bounded linear operators and $F(X)$ is the ideal of all finite rank operators in $B(X)$.

Corollary 5.3.3 *Let $\mathcal{A}$ be a standard operator algebra in a Banach space[5] X with dimension is > 1. If the family $D = \{d_n\}_{n\in\mathbb{N}}$ of mappings $d_n : \mathcal{A} \to \mathcal{A}$ such that $d_0 = I_{\mathcal{A}}$ satisfy*

$$d_n(AB+BA) = \sum_{p+q=n} (d_p(A)d_q(B) + d_p(B)d_q(A))$$

for all $A, B \in \mathcal{A}$, then $D = \{d_n\}_{n\in\mathbb{N}}$ is the family of additive mappings. Moreover, $D = \{d_n\}_{n\in\mathbb{N}}$ is a higher derivation.

Definition 56 *A family $D = \{d_n\}_{n\in\mathbb{N}}$ of mappings $d_n : R \to R$ (without the assumption of additivity) such that $d_0 = I_R$, is said to be a Jordan triple higher derivable map if*

$$d_n(aba) = \sum_{p+q+r=n} d_p(a)d_q(b)d_r(a),$$

holds for all $a, b \in R$ and for each $n \in \mathbb{N}$.

Note that if for every $n \in \mathbb{N}$, d_n is an additive map, then every Jordan triple higher derivable map is called a Jordan triple higher derivation.

Also, a Jordan triple higher derivable map need not be additive. Therefore, it is important to investigate the additivity of such mappings. In view of the definition of Jordan triple higher derivable mappings it is natural to question that under what conditions does Jordan triple higher derivable map become additive?

To accomplish this goal certain conditions on an arbitrary ring have been imposed by Ashraf and Parveen [70] and it is shown that if the family $D = \{d_n\}_{n\in\mathbb{N}}$ of mappings $d_n : R \to R$ is a Jordan triple higher derivable for each $n \in \mathbb{N}$, then $D = \{d_n\}_{n\in\mathbb{N}}$ is the family of additive mappings. In fact, they proved the following result:

Theorem 5.3.6 *[70, Theorem 2.1] Suppose R is a ring containing a nontrivial idempotent and satisfies the following conditions:*

(i) *If $a_{11}x_{12} = 0$ for all $x_{12} \in R_{12}$, then $a_{11} = 0$.*

(ii) *If $x_{12}a_{22} = 0$ for all $x_{12} \in R_{12}$, then $a_{22} = 0$.*

(iii) *If $x_{ij}ax_{ij} = 0$ for all $x_{ij} \in R_{ij}$, $(i, j \in \{1,2\})$, then $a_{ji} = 0$.*

If the family $D = \{d_n\}_{n\in\mathbb{N}}$ of mappings $d_n : R \to R$ such that $d_0 = I_R$ satisfies

$$d_n(aba) = \sum_{p+q+r=n} d_p(a)d_q(b)d_r(a) \quad \text{for all } a,b \in R \text{ and for each } n \in \mathbb{N}, \tag{5.2}$$

then $D = \{d_n\}_{n\in\mathbb{N}}$ is the family of additive mappings.

[5] *A Banach space is a vector space X over the field $\mathbb{R}$ of real numbers, or over the field $\mathbb{C}$ of complex numbers, which is equipped with a norm and which is complete with respect to that norm, that is to say, for every Cauchy sequence $\{x_n\} \subseteq X$, there exists an element $x \in X$ such that $\lim_{n\to\infty} x_n = x$, or equivalently: $\lim_{n\to\infty} \| x_n - x \|_X = 0$.*

Note that every Jordan triple higher derivation on a 2-torsion free semiprime ring is a higher derivation (see [268]).

Corollary 5.3.4 *Let R be a 2-torsion free semiprime ring with a nontrivial idempotent satisfying the following conditions for $i, j \in \{1,2\}$;*

(i) *If $a_{11}x_{12} = 0$ for all $x_{12} \in R_{12}$, then $a_{11} = 0$,*

(ii) *If $x_{12}a_{22} = 0$ for all $x_{12} \in R_{12}$, then $a_{22} = 0$,*

(iii) *If $x_{ij}ax_{ij} = 0$ for all $x_{ij} \in R_{ij}$, $(i \neq j)$, then $a_{ji} = 0$.*

If the family $D = \{d_n\}_{n\in\mathbb{N}}$ of mappings $d_n : R \to R$ such that $d_0 = I_R$ satisfies

$$d_n(aba) = \sum_{p+q+r=n} d_p(a)d_q(b)d_r(a)$$

for all $a, b \in R$ and for each $n \in \mathbb{N}$. Then, $D = \{d_n\}_{n\in\mathbb{N}}$ is the family of additive mappings. Moreover, $D = \{d_n\}_{n\in\mathbb{N}}$ is a higher derivation.

Applying the well-known result that every Jordan triple higher derivation on a prime ring of characteristic not equal to 2 is a higher derivation (see [268]), one establishes the following:

Corollary 5.3.5 *Let R be a prime ring of characteristic not equal to 2 which admits a nontrivial idempotent element. If the family $D = \{d_n\}_{n\in\mathbb{N}}$ of mappings $d_n : R \to R$ such that $d_0 = I_R$ satisfies*

$$d_n(aba) = \sum_{p+q+r=n} d_p(a)d_q(b)d_r(a)$$

for all $a, b \in R$ and for each $n \in \mathbb{N}$, then $D = \{d_n\}_{n\in\mathbb{N}}$ is the family of additive mappings. Moreover, $D = \{d_n\}_{n\in\mathbb{N}}$ is a higher derivation.

Since every standard operator algebra is prime, we can easily state the following:

Corollary 5.3.6 *Let $\mathcal{A}$ be a standard operator algebra in a Banach space $\mathfrak{X}$ whose dimension is ¿ 1. If the family $D = \{d_n\}_{n\in\mathbb{N}}$ of mappings $d_n : \mathcal{A} \to \mathcal{A}$ such that $d_0 = I_{\mathcal{A}}$ satisfies*

$$d_n(aba) = \sum_{p+q+r=n} d_p(a)d_q(b)d_r(a)$$

for all $a, b \in \mathcal{A}$ and for each $n \in \mathbb{N}$. Then, $D = \{d_n\}_{n\in\mathbb{N}}$ is the family of additive mappings. Moreover, $D = \{d_n\}_{n\in\mathbb{N}}$ is a higher derivation.

5.4 Lie (triple) higher derivable maps

The work discussed in the previous section encourages mathematicians to investigate resembling problem concerning Lie products.

Definition 57 *A map $d : R \to R$ (need not be additive) is said to be a Lie derivable if $d([a,b]) = [d(a),b] + [a,d(b)]$ holds for all $a,b \in R$.*

Note that every Lie derivable map need not be additive. Let $d : R \to R$ be a Lie derivation and $g : R \to Z(R)$ be any map (not necessarily additive) with $g([R,R]) = \{0\}$, then $d+g$ is a Lie derivable, which is not a Lie derivation.

It is a well-known result due to Martindale [444] that every multiplicative bijective map from a prime ring containing a nontrivial idempotent onto an arbitrary ring is necessarily additive. Further many results were obtained in this area.

There have been several results on the characterizations of Lie derivations on rings. The first characterization of Lie derivations is due to Martindale [443] who proved the following result in the year 1964.

Theorem 5.4.1 *[443, Theorem 2] Let R be a primitive ring*[6] *of characteristic not equal to 2 containing a nontrivial idempotent. Then, every Lie derivation L of R into itself is of the form $L = D+T$, where D is an ordinary derivation of R into a primitive ring $\bar{R}$ containing R and T is an additive mapping of R into the centre of $\bar{R}$ that maps commutators into zero.*

Further, Brešar [137] obtained the above characterization of Lie derivations on prime rings. By the motivation of derivable maps and Jordan derivable maps defined on a ring containing a nontrivial idempotent, in the year 2010, Lu and Liu [434] introduced the concept of Lie derivable maps and extended the result given by Brešar [137] on operator algebras for Lie derivable mappings. Later on in the year 2012, Jing and Lu [328] generalized the above result on Lie derivable mappings to prime rings containing a nontrivial idempotent. In fact, they proved the following:

Theorem 5.4.2 *[328, Theorem 2.21] Let R be a prime ring containing a nontrivial idempotent. Suppose that a mapping $\delta : R \to R$ satisfies*

$$\delta([a,b]) = [\delta(a),b] + [a,\delta(b)]$$

for all $a,b \in R$. Then, there exists an element $z_{a,b}$ (depending on a and b) in the center of R such that $d(a+b) = d(a) + d(b) + z_{a,b}$.

Definition 58 *A family $D = \{L_n\}_{n\in\mathbb{N}}$ of mappings $L_n : R \to R$ (without the assumption of additivity) is said to be a Lie higher derivable mapping on R if*

$$L_n([a,b]) = \sum_{i+j=n} [L_i(a), L_j(b)]$$

for all $a,b \in R$ and for all $n \in \mathbb{N}$. In addition, if $D = \{L_n\}_{n\in\mathbb{N}}$ is a family of additive mappings on R, then D is said to be a Lie higher derivation on R.

---

[6] *A ring R is said to be a primitive ring iff it has a faithful simple R-module. A R-module M is called a faithful module if $Ann_R(M) = \{0\}$, where $Ann_R(M) = \{r \in R | rm = 0, \text{ for all } m \in M\}$. The simple modules over a ring R are the modules over R that have no non-zero proper submodules.*

This notion was introduced by Ashraf and Parveen [71]. Many algebraists have made significant contributions to the related topics (see [180, 324, 416, 561, 563] for references).

In the previous section, we discussed the additivity of Jordan higher derivable mappings on rings. This work stimulates the investigation of an analogous problem concerning Lie higher derivable maps on a prime ring containing a nontrivial idempotent. In the year 2016, Ashraf and Parveen [71] were able to prove that under certain given conditions every Lie higher derivable map becomes nearly additive.

Theorem 5.4.3 *[71, Theorem 2.1] Let R be a prime ring containing a nontrivial idempotent. Suppose that $D = \{d_n\}_{n\in\mathbb{N}}$ is a family of mappings $d_n : R \to R$ (not necessarily additive) satisfying*

$$d_n([a,b]) = \sum_{p+q=n} [d_p(a), d_q(b)] \tag{5.3}$$

for all $a, b \in R$ and for each $n \in \mathbb{N}$. Then, there exists an element $z_{a,b}$ (depending on a and b) in $Z(R)$, such that

$$d_n(a+b) = d_n(a) + d_n(b) + z_{a,b}.$$

Definition 59 *A map $d : R \to R$ (without the assumption of additivity) is called a Lie triple derivable if $d([[a,b],c]) = [[d(a),b],c] + [[a,d(b)],c] + [[a,b],d(c)]$, holds for all $a, b, c \in R$.*

Particularly, if d is additive, then the above-defined mapping is called a *Lie triple derivation*. It is not difficult to observe that any derivation is a Lie derivation and any Lie derivation is a Lie triple derivation. However, the converse statements are in general not true, which can be seen by the following example.

Example 5.4.1 *Let $d : R \to R$ be a derivation on a ring R and $g : R \to Z(R)$ be an additive mapping with $g([R,R]) = \{0\}$. Then, $d+g$ is a Lie derivation, which is not necessarily a derivation. If $h : R \to Z(R)$ is an additive mapping such that $h([[R,R],R]) = \{0\}$, then $d+h$ is a Lie triple derivation on R, which is not necessarily a Lie derivation on R.*

Recently, Li et al. [399] studied Lie triple derivable mappings on rings and obtained the following result:

Theorem 5.4.4 *[399, Theorem 2.1] Let R be a ring containing a nontrivial idempotent element satisfying the following condition:*

$$\textit{If } a_{11}b_{12} = b_{12}a_{22} \textit{ for all } b_{12} \in R_{12}, \textit{ then } a_{11} + a_{22} \in Z(R). \tag{$*$}$$

Suppose that $d : R \to R$ is Lie triple derivable. Then, there exists a $z_{a,b}$ (depending on a and b) in $Z(R)$ such that $d(a+b) = d(a) + d(b) + z_{a,b}$, for all $a, b \in R$.

In [72], the authors have generalized the above theorem and obtained that the condition assumed in the above-mentioned theorem given by Li et al. [399] is superfluous. In order to illustrate the results in the literature focused on Lie triple derivable maps and Lie higher derivations on rings the author [72] introduced the concept of Lie triple higher derivable mappings on rings.

Definition 60 *A family $D = \{d_n\}_{n\in\mathbb{N}}$ of mappings $d_n : R \to R$ (need not be additive) such that $d_0 = I_R$, is said to be a Lie triple higher derivable map if*

$$d_n([[a,b],c]) = \sum_{p+q+r=n} [[d_p(a), d_q(b)], d_r(c)],$$

holds for all $a,b,c \in R$ and for each $n \in \mathbb{N}$.

In addition, if d_n is additive for each $n \in \mathbb{N}$, then $D = \{d_n\}_{n\in\mathbb{N}}$ is said to be a Lie triple higher derivation.

In view of the definition of Lie triple higher derivable mappings, it is natural to question under what conditions does a Lie triple higher derivable map becomes additive? The following result addresses this problem;

Theorem 5.4.5 *[72, Theorem 2.1] Let R be a ring having a nontrivial idempotent element. Assume that $D = \{d_n\}_{n\in\mathbb{N}}$ is a family of Lie triple higher derivable maps $d_n : R \to R$. Then, $d_n(a+b) = d_n(a) + d_n(b) + z_{a,b}$, for all $a,b \in R$, where $z_{a,b} \in Z(R)$ is a central element depending on a and b.*

5.5 Generalized higher derivations on rings

In this section, we address the generalized version of the mappings defined in the previous section.

Definition 61 *An additive mapping $\Delta : R \to R$ is said to be a generalized derivation if there exists a derivation $\delta : R \to R$ such that $\Delta(ab) = \Delta(a)b + a\delta(b)$ holds for all $a,b \in R$.*

Definition 62 *An additive mapping $\Delta : R \to R$ is said to be a generalized Jordan derivation if there exists a Jordan derivation $\delta : R \to R$ such that $\Delta(a^2) = \Delta(a)a + a\delta(a)$ holds for all $a \in R$.*

Remark 5.5.1 *It is easy to see that if we take $\Delta = \delta$ then a generalized derivation (resp., generalized Jordan derivation) is a derivation (resp. Jordan derivation).*

Example 5.5.1 *Let R be a ring and $a,b \in R$ be such that one of them in not zero. Define $\Delta(x) = ax + xb$ for all $x \in R$. Then, for all $x,y \in R$, we have*

$\Delta(x+y) = \Delta(x) + \Delta(y)$ *and*

$$\begin{aligned}\Delta(x^2) &= ax^2 + x^2 b \\ &= (ax+xb)x + x(-bx - x(-b)) \\ &= \Delta(x)x + x\delta(x),\end{aligned}$$

where δ defined by $\delta(x) = -bx - x(-b)$ for all $x \in R$ is a Jordan derivation on R. That is, Δ is a generalized Jordan derivation on R.

Definition 63 *An additive mapping $\Delta : R \to R$ is said to be a Jordan generalized derivation if there exists a derivation $\delta : R \to R$ such that $\Delta(a^2) = \Delta(a)a + a\delta(a)$ holds for all $a \in R$.*

Remark 5.5.2 *Since every derivation is a Jordan derivation, therefore, a Jordan generalized derivation is a generalized Jordan derivation but not conversely as a Jordan derivation need not necessarily be a derivation in general.*

The notion of generalized derivations on a ring R which was first introduced by Brešar in [134] is related to a derivation of R. In [468], Nakajima defined another category of generalized derivations without using derivations, and gave some categorical properties of those generalized derivations. However, these two notions of generalized derivations coincide when R has unity. Later, Hvala studied generalized derivations in [313]. Ashraf and Nadeem [76] proved that in a 2-torsion free ring R having a commutator which is not a zero divisor, every generalized Jordan derivation is a generalized derivation on R.

Motivated by the above results, Jing and Lu [326] introduced the concept of generalized Jordan (triple) derivations on rings and obtained the following results.

Definition 64 *An additive mapping $\Delta : R \to R$ is called a generalized Jordan triple derivation if there exits a Jordan triple derivation $\delta : R \to R$ such that $\Delta(aba) = \Delta(a)ba + a\delta(b)a + ab\delta(a)$ holds for all $a, b \in R$.*

Remark 5.5.3 *Every generalized Jordan triple derivation is a Jordan triple derivation whenever $\Delta = \delta$.*

Theorem 5.5.4 *[326, Theorem 2.5] Let R be a 2-torsion free prime ring, then every generalized Jordan derivation on R is a generalized derivation.*

Theorem 5.5.5 *[326, Theorem 3.5] Let R be a 2-torsion free prime ring, then every generalized Jordan triple derivation on R is a generalized derivation.*

They also conjectured that in case $\Delta : R \to R$, where R is a 2-torsion free semiprime ring, it is either a generalized Jordan derivation or generalized Jordan derivation, is a generalized derivation. Vukman [538] answered these conjectures in case of semiprime rings in the affirmative and obtained the following results.

Theorem 5.5.6 *[538, Theorem 1] Let R be a 2-torsion free semiprime ring and let* $\Delta : R \to R$ *be a generalized Jordan derivation. In this case,* Δ *is a generalized derivation.*

Theorem 5.5.7 *[538, Theorem 2] Let R be a 2-torsion free semiprime ring and let* $\Delta : R \to R$ *be a generalized Jordan triple derivation. In this case,* Δ *is a generalized derivation.*

Ashraf and Nadeem [76] proved the following.

Theorem 5.5.8 *Let R be a 2-torsion free ring such that R has a commutator which is not a zero divisor. Then, every Jordan generalized derivation on R is a generalized derivation on R.*

Later in the year 2003, Ashraf et al. [82] established another set of conditions in terms of nonzero square-closed Lie ideals and proved the following.

Theorem 5.5.9 *[82, Theorem] Let R be a 2-torsion free prime ring and U a nonzero ideal of R such that* $u^2 \in U$ *for all* $u \in U$*. If* Δ *is an additive mapping of R into itself satisfying* $\Delta(u^2) = \Delta(u)u + u\delta(u)$ *for all* $u \in U$*, then* $\Delta(uv) = \Delta(u)v + u\delta(v)$ *for all* $u, v \in U$.

Corollary 5.5.1 *[82, Corollary] Let R be a 2-torsion free prime ring and* $\Delta : R \to R$ *be a Jordan generalized derivation. Then,* Δ *is a generalized derivation on R.*

Also, they provided the following example which shows that the primeness is necessary in the hypotheses of the above theorem.

Example 5.5.2 *[82, Example] Let* $\mathcal{S}$ *be a ring such that the square of the elements in R is zero, but the product of some elements in* $\mathcal{S}$ *is nonzero. Consider* $R = \left\{ \begin{pmatrix} x & y \\ y & x \end{pmatrix} \,\middle|\, x, y \in \mathcal{S} \right\}$*. Now, define a map* $\Delta : R \to R$ *by* $\Delta \begin{pmatrix} x & y \\ 0 & 0 \end{pmatrix} = \begin{pmatrix} 0 & x \\ 0 & 0 \end{pmatrix}$*. Then with derivation* $\delta = 0$ *on R and* $U = R$*, it can easily be seen that* $\Delta(a^2) = \Delta(a)a = \Delta(a)b = 0$ *for all* $a, b \in R$ *but* $\Delta(ab) \neq 0$ *for some* $a, b \in R$.

Remark 5.5.10 *Since it is easy to see that every prime ring with characteristic not equal to two is 2-torsion free, the above theorem or a 2-torsion free prime ring can be restated for a prime ring of characteristic not equal to two.*

Let σ, τ be two endomorphisms of a ring R. The concept of generalized derivations, generalized Jordan derivations and generalized Jordan triple derivations have been extended to generalized (σ, τ)-derivations, generalized Jordan (σ, τ)-derivations and generalized Jordan triple (σ, τ)-derivations, respectively, as follows.

Definition 65 *Let R be a ring and σ, τ be two endomorphisms of R. An additive mapping $\Delta : R \to R$ is said to be a generalized (σ, τ)-derivation on R if there exists a (σ, τ)-derivation $\delta : R \to R$ such that $\Delta(ab) = \Delta(a)\tau(b) + \sigma(a)\delta(b)$ for all $a, b \in R$.*

Definition 66 *An additive mapping $\Delta : R \to R$ is said to be a generalized (σ, τ)-inner derivation if $\Delta(x) = \sigma(x)b + a\tau(x)$ holds for all $x \in R$ and for some fixed $a, b \in R$.*

Note that if Δ is a generalized (σ, τ)-inner derivation then a simple calculation shows that $\Delta(xy) = \Delta(x)\delta_{-b} + \Delta(x)\tau(y)$, where δ_{-b} is a (σ, τ)-derivation for some fixed $b \in R$. That is, every generalized (σ, τ)-inner derivation is a generalized (σ, τ)-derivation. However the converse of this statement is not true in general.

Example 5.5.3 *Let $\mathcal{S}$ be any ring and consider $R = \left\{ \begin{pmatrix} x & y \\ 0 & z \end{pmatrix} \middle| \, x, y, z \in \mathcal{S} \right\}$. Define an additive mapping $\Delta : R \to R$ by $\Delta \begin{pmatrix} x & y \\ 0 & z \end{pmatrix} = \begin{pmatrix} x & 0 \\ 0 & z \end{pmatrix}$ and endomorphisms $\sigma, \tau : R \to R$ by $\sigma \begin{pmatrix} x & y \\ 0 & z \end{pmatrix} = \begin{pmatrix} 0 & 0 \\ 0 & z \end{pmatrix}$ and $\tau \begin{pmatrix} x & y \\ 0 & z \end{pmatrix} = \begin{pmatrix} x & 0 \\ 0 & 0 \end{pmatrix}$, respectively. Routine exercise shows that Δ is a generalized (σ, τ)-derivation on R with associated (σ, τ)-derivation $\delta : R \to R$ such that $\delta \begin{pmatrix} x & y \\ 0 & z \end{pmatrix} = \begin{pmatrix} 0 & 0 \\ 0 & z \end{pmatrix}$.*

Definition 67 *An additive mapping $\Delta : R \to R$ is said to be a generalized Jordan (σ, τ)-derivation if there exists a Jordan (σ, τ)-derivation $\delta : R \to R$ satisfying $\Delta(a^2) = \Delta(a)\tau(a) + \sigma(a)\Delta(a)$ holds for all $a \in R$.*

Definition 68 *An additive mapping $\Delta : R \to R$ is said to be a generalized Jordan triple (σ, τ)-derivation if there exists a Jordan triple (σ, τ)-derivation $\delta : R \to R$ satisfying $\Delta(aba) = \Delta(a)\sigma(ba) + \tau(a)\delta(b)\sigma(a) + \tau(ab)\delta(a)$ holds for all $a, b \in R$.*

It is obvious that every generalized (σ, τ)-derivation on a ring R is a generalized Jordan (σ, τ)-derivation on R, however, the converse is not true in general.

Ashraf and Wafa introduced the concept of generalized (σ, τ)-derivation and generalized Jordan (σ, τ)-derivation on a ring R in [49] and proved, if $\mathcal{S}$ be an arbitrary ring and R be a noncommutative ring, σ, τ are homomorphisms of R to $\mathcal{S}$. Let $\mathcal{M}$ is a 2-torsion free R-bimodule. Then supposing τ is onto and $mRx = \{0\}$ with $m \in \mathcal{M}, x \in R$, implies either $m = 0$ or $x = 0$. Then, generalized Jordan (σ, τ)-derivation $\Delta : \mathcal{S} \to R$ is a generalized (σ, τ)-derivation on $\mathcal{S}$. Motivated by this result, in the year 2004, Ashraf et al. [44] obtained the following result for a noncommutative Lie ideal of R.

Theorem 5.5.11 *[44, Theorem 2.1] Let R be a 2-torsion free prime ring and U a noncommutative Lie ideal of R such that $u^2 \in U$ for all $u \in U$. Suppose that σ, τ be two endomorphisms of R such that σ is one-one, onto and δ is a (σ, τ)-derivation of R. If $\Delta : R \to R$ is a generalized Jordan (σ, τ)-derivation on U, then Δ is a generalized (σ, τ)-derivation on U.*

Corollary 5.5.2 *[44, Corollary 2.1] Let R be a 2-torsion free noncommutative prime ring and let $\Delta : R \to R$ be a generalized Jordan derivation on R. Then, Δ is a generalized derivation on R.*

If the underlying ring R is arbitrary, then we have the following.

Theorem 5.5.12 *[44, Theorem 2.3] Let R be a 2-torsion free ring and U a Lie ideal of R such that $u^2 \in U$ for all $u \in U$. Suppose that σ, τ be two endomorphisms of R such that σ is one-one, onto and δ is a (σ, τ)-derivation of R. If $\Delta : R \to R$ is a generalized Jordan (σ, τ)-derivation on U, then Δ is a generalized (σ, τ)-derivation on U.*

Corollary 5.5.3 *[44, Corollary 2.3] Let R be a 2-torsion free ring and let $\Delta : R \to R$ be a generalized Jordan derivation. If R has a commutator which is not a zero divisor then Δ is a generalized derivation on R.*

Later, a more general result was obtained in [47] which unifies and extends several well-known results present in the literature.

Theorem 5.5.13 *[47, Theorem 2.1] Let R be a ring and U a Lie ideal of R such that $u^2 \in U$. Let σ, τ be endomorphisms of R such that σ is one-one and onto and $\mathcal{M}$ be a 2-torsion free R-bimodule. Suppose that $mRx = 0$ with $m \in \mathcal{M}, x \in R$ implies that either $m = 0$ or $x = 0$. If $\Delta : R \to \mathcal{M}$ is a generalized Jordan (σ, τ)-derivation on U, then Δ is a generalized (σ, τ)-derivation on U.*

Let R be a ring and U a Lie ideal of the ring R. The notion of a (U,R)-derivation on R can be extended to a generalized (U,R)-derivation on R (see [259]).

Definition 69 *Let U be a Lie ideal of a ring R. Then, an additive mapping $\Delta : R \to R$ is said to be a generalized (U,R)-derivation of R if there exists a (U,R)-derivation δ such that $\Delta(ur + su) = \Delta(u)r + u\delta(r) + \Delta(s)u + s\delta(u)$ for all $u \in U, r, s \in R$.*

Example 5.5.4 *Let $\mathcal{S}$ be a commutative ring and consider the ring $R = \left\{ \begin{pmatrix} x & y \\ z & w \end{pmatrix} \;\middle|\; x,y,z,w \in \mathcal{S} \right\}$. Let $U = \left\{ \begin{pmatrix} x & y \\ y & x \end{pmatrix} \;\middle|\; x,y \in \mathcal{S} \right\}$. Then, U is a Lie ideal of R. Define $\Delta, \delta : R \to R$ by $\Delta \begin{pmatrix} x & y \\ z & w \end{pmatrix} = \begin{pmatrix} x & 0 \\ 0 & -w \end{pmatrix}$ and*

$\Delta\begin{pmatrix} x & y \\ z & w \end{pmatrix} = \begin{pmatrix} 0 & -y \\ z & 0 \end{pmatrix}$ *for all* $\begin{pmatrix} x & y \\ z & w \end{pmatrix} \in R$. *Then, can be verified that* δ *is a* (U,R)*-derivation and* Δ *is a generalized* (U,R)*-derivation.*

In the following theorem, Faraj et al. [259] discussed a generalized (U,R)-derivation on prime rings.

Theorem 5.5.14 *[259, Theorem 3.5] Let* R *be a prime ring,* $char(R) \neq 2$, U *be a square closed Lie ideal of* R *and* Δ *be a generalized* (U,R)*-derivation of* R*. Then,* $\Delta(ur) = \Delta(u)r + u\delta(r)$ *for all* $u \in U, r \in R$.

In the previous section, we have gone through various generalizations of Herstein's result [289] and discussed higher derivations on rings. Inspired by the concept of generalized derivations, in the present section, we shall discuss generalized higher derivations on rings.

Definition 70 *Let* $G = \{g_i\}_{i\in\mathbb{N}}$ *be a family of additive mappings* $g_i : R \to R$ *such that* $g_0 = I_R$*, the identity mapping on* R*. Then,* G *is said to be a*

(i) *Generalized higher derivation if there exists a higher derivation* $D = \{d_i\}_{i\in\mathbb{N}}$ *of* R *such that for each* $n \in \mathbb{N}$,
$$g_n(ab) = \sum_{i+j=n} g_i(a)d_j(b)$$
for all $a,b \in R$,

(ii) *Jordan generalized higher derivation if there exists a higher derivation* $D = \{d_i\}_{i\in\mathbb{N}}$ *of* R *such that for each* $n \in \mathbb{N}$,
$$g_n(a^2) = \sum_{i+j=n} g_i(a)d_j(a)$$
for all $a \in R$,

(iii) *Jordan generalized triple higher derivation if there exists a higher derivation* $D = \{d_i\}_{i\in\mathbb{N}}$ *of* R *such that for each* $n \in \mathbb{N}$,
$$g_n(aba) = \sum_{i+j=n} g_i(a)d_j(b)d_k(a)$$
for all $a,b \in R$.

Further, let U be a Lie ideal of R. If $D = \{d_i\}_{i\in\mathbb{N}}$ is a higher derivation of U into R and the above conditions hold for all $a,b \in U$, then the respective mappings defined above are called *generalized higher derivations* of U into R, *Jordan generalized higher derivation* of U into R and *Jordan generalized triple higher derivations* of U into R.

In this direction, Cortes and Haetinger [196] were the first to introduce the notion of generalized higher derivations on rings and obtained a couple of results. Actually, they extended the result of Ashraf and Haetinger [76] as follows.

Theorem 5.5.15 *[196, Theorem 1.3] Let R be a 2-torsion free ring. If R has a commutator which is not a right nonzero divisor and U is a square closed Lie ideal of R, then every Jordan generalized higher derivation of U into R is a generalized higher derivation of U into R.*

Remark 5.5.16 *Note that $U = R$ is a square closed Lie ideal of R. Therefore, the above theorem is true for a generalized Jordan higher derivation on R.*

In particular, taking $g_i = d_i, i \in \mathbb{N}$, they obtained the following corollary.

Corollary 5.5.4 *[196, Corollary 1.4] Let R be a 2-torsion free ring. If R has a commutator which is not a right nonzero divisor and U is a square closed Lie ideal of R, then every Jordan higher derivation of U into R is a higher derivation of U into R.*

We provide a couple of well-known examples of rings that have a commutative nonzero divisor.

Example 5.5.5 *One can consider any noncommutative ring without zero divisors, the matrix algebra over a division ring, and also every noncommutative prime Goldie ring.*

Example 5.5.6 *Consider the 2×2 matrix algebra $\mathcal{M}_2(D)$ over a domain D with 1. Let E_{ij} be the usual matrix units. Then, the commutator $[E_{12}, E_{21}] = E_{11} - E_{22}$ is invertible. When char$(D) \neq 2$, then in $\mathcal{M}_3(D)$ we have $[E_{12} + E_{23}, E_{21} - E_{32}] = E_{11} - 2E_{22} + E_{33}$, is a nonzero divisor.*

Clearly variations of this will work for $\mathcal{M}_n(D)$, where D is a noncommutative domain (just consider 2×2 block diagonal matrices and a 3×3 block at the bottom if n is odd). This block matrix idea will also work for the ring of all (countably infinite) row and column finite matrices over a domain. In rings like $R = \frac{\mathbb{F}\langle x,y \rangle}{\langle x^2 \rangle}$, the zero divisors must lie in xR or Rx, so $[x,y]$ is regular.

Note that the above corollary states the result of [267, Corollary 1.4] without the semiprimality condition. The following example is due to Brešar which shows that the semiprimality and right nonzero divisor commutator assumptions are independent of each other.

Example 5.5.7 *A semiprime ring may not contain a commutator nonzero divisor (after all, take commutative semiprime ring, or more generally, a semiprime ring R containing a nonzero central idempotent element $e \in R$ such that eR is commutative). Conversely, a ring may contain a commutator nonzero divisor, but is not a semiprime ring. For example, let $R = \mathcal{T}(\mathcal{A}_1)$ be the ring of all the 2×2 upper triangular matrices whose entries are the elements from the Weyl algebra $\mathcal{A}_1$ (polynomials in x, y such that $xy - yx = 1$). Then, R is not a semiprime ring, but the commutator of scalar matrices generated by x and y is the identity matrix.*

Furthermore, Cortes and Haetinger [196] generalize [267, Theorem 1.3] as follows.

Theorem 5.5.17 *[196, Theorem 1.5] Let R be a 2-torsion free ring and U a Lie ideal of R. Then, every generalized Jordan higher derivation of U into R is a Jordan generalized triple higher derivation of U into R.*

Definition 71 *Let $G = \{g_i\}_{i\in\mathbb{N}}$ be a family of additive mappings $g_i : R \to R$ such that $g_0 = I_R$, the identity mapping on R. Then, G is said to be*

(i) A generalized higher Jordan derivation if there exists a higher Jordan derivation $D = \{d_i\}_{i\in\mathbb{N}}$ of R such that for each $n \in \mathbb{N}$,

$$g_n(a^2) = \sum_{i+j=n} g_i(a)d_j(a)$$

for all $a \in R$,

(ii) A generalized Jordan triple higher derivation if there exists a Jordan triple higher derivation $D = \{d_i\}_{i\in\mathbb{N}}$ of R such that for each $n \in \mathbb{N}$,

$$g_n(aba) = \sum_{i+j=n} g_i(a)d_j(b)d_k(a)$$

for all $a, b \in R$.

Moreover, let U be a Lie ideal of R. If $D = \{d_i\}_{i\in\mathbb{N}}$ is a Jordan (triple) higher derivation of U into R and the above conditions hold for all $a, b \in U$, then the respective mappings defined above are called generalized Jordan higher derivations of U into R and generalized Jordan triple higher derivations of U into R.

Jung in his research paper [337] investigated that the notions of generalized higher derivations, generalized Jordan higher derivations and generalized Jordan triple higher derivations on a 2-torsion free prime ring are equivalent to each other by showing the following result which generalized Jing and Lu's result [326, Theorem 3.5].

Theorem 5.5.18 *[337, Theorem] Every generalized Jordan triple higher derivations on a 2-torsion free prime ring is a generalized higher derivation.*

Wei and Xiao [553] extend the above result to the case of semiprime rings.

Theorem 5.5.19 *[553, Theorem 2.7] Let R be a 2-torsion free semiprime ring. Then, every generalized Jordan triple higher derivation on R is a generalized higher derivation on R.*

It is interesting to question, under what condition, a generalized Jordan higher derivation on a ring R is a generalized Jordan triple higher derivation on R? The following theorem answers this question which is actually the common generalization of [267, Theorem 1.3] and [267, Corollary 1.4], respectively.

Theorem 5.5.20 *[553, Theorem 2.10] Let R be a 2-torsion free ring and $\mathcal{L}$ a Lie ideal of R. Then, every generalized Jordan higher derivation of R (resp. $\mathcal{L}$ into R) is a generalized Jordan triple higher derivation of R (resp. $\mathcal{L}$ into R).*

Corollary 5.5.5 *[553, Corollary 2.11] Let R be a semiprime ring of characteristics not equal to 2 (resp. prime ring of characteristics not equal to 2 and $\mathcal{L}$ be a non-central Lie ideal of R). Then, every generalized Jordan higher derivation on R (resp. $\mathcal{L}$ into R) is a generalized higher derivation of R (resp. $\mathcal{L}$ into R).*

Definition 72 *Let σ, τ be endomorphisms of a ring R. A family $G = \{g_i\}_{i\in\mathbb{N}}$ of additive mappings $g_i : R \to R$, $g_0 = I_R$ the identity mapping on R, is said to be*

(i) A generalized (σ,τ)-higher derivation of R if there exists a (σ,τ)-higher derivation $D = \{\delta_i\}_{i\in\mathbb{N}}$ of R such that

$$g_n(ab) = \sum_{i+j=n} g_i(\sigma^{n-i}(a))\delta_j(\tau^{n-j}(b))$$

holds for all $a, b \in R$.

(ii) A generalized Jordan (σ,τ)-higher derivation of R if there exists a (σ,τ)-higher derivation $D = \{\delta_i\}_{i\in\mathbb{N}}$ of R such that

$$g_n(a^2) = \sum_{i+j=n} g_i(\sigma^{n-i}(a))\delta_j(\tau^{n-j}(a))$$

holds for all $a \in R$.

(iii) A generalized Jordan triple (σ,τ)-higher derivation of R if there exists a (σ,τ)-higher derivation $D = \{\delta_i\}_{i\in\mathbb{N}}$ of R such that

$$g_n(aba) = \sum_{i+j+k=n} g_i(\sigma^{n-i}(a))\delta_j(\sigma^k\tau^i(b))\delta_k(\tau^{n-k}(a))$$

holds for all $a, b \in R$.

Further, let U be a Lie ideal of R. If $D = \{\delta_i\}_{i\in\mathbb{N}}$ is a (σ,τ)-higher derivation of U into R and the above conditions hold for all $a, b \in U$, then the respective mappings defined above are called *generalized (σ,τ)-higher derivation* of U into R, *generalized Jordan (σ,τ)-higher derivation* of U into R and *generalized Jordan triple (σ,τ)-higher derivation* of U into R, respectively.

Example 5.5.8 *Let R be an algebra over the field of rational numbers and σ, τ be the endomorphisms of R. Define $g_n = \frac{\eta^n}{n!}$ for all $n \in \mathbb{N}$, where η is a generalized (σ,τ)-derivation on R with associated (σ,τ)-derivation δ such that $\eta\sigma = \sigma\eta$ and $\delta\tau = \tau\delta$. Consider the sequence $G = \{g_n\}_{n\in\mathbb{N}}$, then this defines a generalized (σ,τ)-higher derivation with associated (σ,τ)-higher derivation $D = \{\frac{\delta^n}{n!}\}_{n\in\mathbb{N}}$ on R.*

If we choose a generalized Jordan (σ, τ)-derivation on R which is not a generalized (σ, τ)-derivation on R then one can easily construct an example of a generalized Jordan (σ, τ)-higher derivation on R wich is not a generalized (σ, τ)-higher derivation on R.

In the year 2002, Haetinger [283] proved that if R is a prime ring of characteristic not equal to 2 and U a square closed Lie ideal of R such that $U \nsubseteq Z(R)$. Then, every Jordan higher derivation of U into R is a higher derivation of U into R. The following result obtained by Ashraf and Khan [64] shows that the above result still holds for an arbitrary square closed Lie ideal of R.

Theorem 5.5.21 *[64, Theorem 2.1] Let R be a prime ring such that $char(R) \neq 2$ and U a square closed Lie ideal of R. Suppose that σ, τ are endomorphisms of R such that $\sigma\tau = \tau\sigma$ and τ is one-one and onto. Then, every generalized Jordan (σ, τ)-higher derivation of U into R is a generalized (σ, τ)-higher derivation of U into R.*

Also, in the following corollaries, the authors [64] settled the conjecture given in [44] for a square closed Lie ideal while another one improves the main theorems of [67] and [283].

Corollary 5.5.6 *[44, Corollary 2.1] Let R be a 2-torsion free prime ring and U a square closed Lie ideal of R. Suppose that θ, ϕ are endomorphisms of R such that ϕ is one-one and onto. If $\Delta : R \to R$ is a generalized Jordan (θ, ϕ)-derivation on U, then Δ is a generalized (θ, ϕ)-derivation on U.*

Corollary 5.5.7 *[44, Corollary 2.2] Let R be a prime ring of characteristic not equal to 2 and U a square closed Lie ideal of R. Then, every Jordan higher derivation of U into R is a higher derivation of U into R.*

In the above theorem if the underlying ring is arbitrary, then they proved the following.

Theorem 5.5.22 *[64, Theorem 2.2] Let R be a 2-torsion free ring and U be a square closed Lie ideal of R. Suppose that σ, τ are endomorphisms of R such that $\sigma\tau = \tau\sigma$ and τ is one-one and onto. If U has a commutator which is not a right zero divisor, then every generalized Jordan (σ, τ)-higher derivation of U into R is a generalized (σ, τ)-higher derivation of U into R.*

It is an easy exercise to show that on a 2-torsion free ring R, every generalized (σ, τ)-higher derivation of a square closed Lie ideal U into R is a generalized Jordan triple (σ, τ)-higher derivation of U into R. However the converse statement is not true in general.

Ashraf and Khan [66] obtained the conditions on a ring R so that the converse of the above statement holds. Actually, they obtained the following result which generalizes many findings already present in the literature (for example see [44, 64, 67, 148, 150, 267, 289]).

Theorem 5.5.23 *[66, Theorem 5] Let R be a prime ring of characteristics not equal to 2 and let U be a square closed Lie ideal of R such that $U \not\subseteq Z(R)$. Then, every generalized Jordan triple (σ,τ)-higher derivation of U into R is a generalized (σ,τ)-higher derivation of U into R.*

Later, Ashraf and Khan [65] extended the above result in the case of semiprime rings and proved the following:

Theorem 5.5.24 *[65, Theorem 2.3] Let R be a 2-torsion free semiprime ring. Then, every generalized Jordan triple (σ,τ)-higher derivation of R is a generalized (σ,τ)-higher derivation of R.*

Definition 73 *Let U be a Lie ideal of a ring R and $G=\{\xi_i\}_{i\in\mathbb{N}}$ be a family additive mappings $\xi_i : R \to R$ such that $\xi_0 = I_R$, the identity mapping on R. Then, G is said to be a generalized higher (U,R)-derivation if there exists a higher (U,R)-derivation $D=\{\delta_i\}_{i\in\mathbb{N}}$ such that for each $n\in\mathbb{N}$,*

$$\xi_n(ur+su) = \sum_{i+j=n} \{\xi_i(u)\delta_j(r) + \xi_i(s)\delta_j(u)\}$$

holds for all $u \in U, r,s \in R$.

Example 5.5.9 *Let $R=\mathcal{M}_2(\mathcal{S})$ be over a commutative ring $\mathcal{S}$ with $char(\mathcal{S})=2$. Let $G=\{\xi_i\}$ be a family of additive mappings $\xi_i : R \to R$ defined by $\xi_i\begin{pmatrix} x & y \\ z & w \end{pmatrix} = \begin{pmatrix} nx & 0 \\ 0 & -w \end{pmatrix}$ for $n=1,2$ and $\Delta\begin{pmatrix} x & y \\ z & w \end{pmatrix} = \begin{pmatrix} 0 & 0 \\ 0 & 0 \end{pmatrix}$ for all $n \geq 2$, for all $\begin{pmatrix} x & y \\ z & w \end{pmatrix} \in R$. Next, for any $n\in\mathbb{N}$, define $\delta_n : R \to R$ by $\Delta\begin{pmatrix} x & y \\ z & w \end{pmatrix} = \begin{pmatrix} 0 & -y \\ nz & 0 \end{pmatrix}$ for $n=1,2$ and $\Delta\begin{pmatrix} x & y \\ z & w \end{pmatrix} = \begin{pmatrix} 0 & 0 \\ 0 & 0 \end{pmatrix}$ for all $n \geq 2$, for all $\begin{pmatrix} x & y \\ z & w \end{pmatrix} \in R$. Then, one can verify that $G=\{\xi_i\}_{i\in\mathbb{N}}$ is a generalized higher (U,R)-derivation on R with associated higher (U,R)-derivation $D=\{\delta_i\}_{i\in\mathbb{N}}$.*

Faraj et al. [259] studied generalized higher (U,R)-derivations on ring R for an admissible Lie ideal U of R and obtained the following result which generalizes [89, Theorem].

Theorem 5.5.25 *[259, Theorem 5.1] Let R be a prime ring, $char(R)=2$, U an admissible Lie ideal of R and $G=\{\xi_i\}_{i\in\mathbb{N}}$ a generalized higher (U,R)-derivation of R with an associated higher (U,R)-derivation $D=\{\delta_i\}_{i\in\mathbb{N}}$. Then, for each $n\in\mathbb{N}$, $\xi_n(ur)=\sum_{i+j=n}\xi_i(u)\delta_j(r)$ holds for all $u\in U, r\in R$.*

Definition 74 *An additive mapping $G : R \to R$ is called a generalized Lie derivation (resp. generalized Lie triple derivation) if there exists a Lie derivation (resp. Lie triple derivation) $L : R \to R$ such that $G([a,b]) = [G(a),b] + [a,L(b)]$ (resp. $G([[a,b],c]) = [[G(a),b],c] + [[a,L(b)],c] + [[a,b],L(c)]$) holds for all $a,b,c \in R$.*

Note that if $G,L : R \to R$ are not necessarily additive in the above definition then G is said to be a *generalized Lie derivable* (resp. *generalized Lie triple derivable*) mapping on R.

It was furthermore, inspired by the concept of a generalized higher derivation (see Cortes and Haetinger [196]). The concept of a generalized higher derivable mapping in rings was first presented by Ashraf and Akhtar [2].

Definition 75 *A family $\Delta = \{G_n\}_{n\in\mathbb{N}}$ of mappings $G_n : R \to R$ (not necessarily additive) such that $G_0 = I_R$, the identity mapping on R, is said to be*

(*i*) *A generalized higher derivable mapping on R if there exists a higher derivable map $\mathfrak{L} = \{L_n\}_{n\in\mathbb{N}}$ on R such that $G_n(ab) = \sum_{i+j=n} G_i(a)L_j(b)$, holds for all $a,b \in R$ and for each $n \in \mathbb{N}$.*

(*ii*) *A generalized Lie higher derivable mapping on R if there exists a Lie higher derivable map $\mathfrak{L} = \{L_n\}_{n\in\mathbb{N}}$ on R such that $G_n([a,b]) = \sum_{i+j=n} [G_i(a),L_j(b)]$, holds for all $a,b \in R$ and for each $n \in \mathbb{N}$.*

(*iii*) *A generalized Lie triple higher derivable mapping on R if there exists a Lie triple higher derivable map $\mathfrak{L} = \{L_n\}_{n\in\mathbb{N}}$ on R such that $G_n([[a,b],c]) = \sum_{i+j+k=n} [[G_i(a),L_j(b)],L_k(c)]$, holds for all $a,b,c \in R$ and for each $n \in \mathbb{N}$.*

If $\Delta = \{G_n\}_{n\in\mathbb{N}}$ is the family of additive mappings $G_n : R \to R$ and $\mathfrak{L} = \{L_n\}_{n\in\mathbb{N}}$ is also a sequence of additive mappings $L_n : R \to R$ in the above definitions then Δ is said to be a *generalized higher derivation*, *generalized Lie higher derivation* and *generalized Lie triple higher derivation* on R respectively.

Very recently, Ashraf and Akhtar [2] examined the additivity of generalized Lie triple derivable mappings on rings. In fact they proved the following:

Theorem 5.5.26 *[2, Theorem 2.1] Let R be a ring containing a nontrivial idempotent element. Suppose that $G : R \to R$ is a generalized Lie triple derivable mapping with an associated Lie triple derivable mapping $L : R \to R$ such that $G([[a,b],c]) = [[G(a),b],c] + [[a,L(b)],c] + [[a,b],L(c)]$ for all $a,b,c \in R$. Then, there exists $z_{a,b}$ (depending on a and b) in $Z(R)$ such that $G(a+b) = G(a) + G(b) + z_{a,b}$, for all $a,b \in R$.*

Further, this result was extended to generalized Lie triple higher derivable maps.

Theorem 5.5.27 *[2, Theorem 3.1] Let R be a ring containing a nontrivial idempotent element. Suppose that* $\Delta = \{G_n\}_{n\in\mathbb{N}}$ *is the family of mappings* $G_n : R \to R$ *such that* $G_0 = I_R$*, the identity mapping of R, satisfying*

$$G_n([[a,b],c]) = \sum_{i+j+k=n} [[G_i(a),L_j(b)],L_k(c)]$$

for all $a,b,c \in R$ *and for each* $n \in \mathbb{N}$*, where* $\mathfrak{L} = \{L_n\}_{n\in\mathbb{N}}$ *is the family of associated Lie triple higher derivable mappings* $L_n : R \to R$*. Then, there exists* $z_{a,b}$ *(depending on a and b) in* $Z(R)$ *such that* $G_n(a+b) = G_n(a) + G_n(b) + z_{a,b}$.

5.6 Lie-type higher derivations on algebras

In this section, we shall discuss the concept of Lie-type higher derivations and its generalizations on different kinds of algebras. Let R be a commutative ring with the identity and $\mathcal{A}$ be an algebra over R. For any $a_1, a_2, \ldots, a_n \in \mathcal{A}$, let us consider the following sequence of polynomials:

$$\begin{aligned} p_1(a_1) &= a_1, \\ p_2(a_1,a_2) &= [p_1(a_1),a_2] = [a_1,a_2], \\ &\vdots \qquad\qquad \vdots \\ p_n(a_1,a_2,\ldots,a_n) &= [p_{n-1}(a_1,a_2,\ldots,a_{n-1}),a_n] \quad (n \geq 2). \end{aligned}$$

The polynomial $p_n(a_1,a_2,\ldots,a_n)$ is said to be the $(n-1)$*th commutator* $(n \geq 2)$.

Definition 76 *An R-linear mapping* $L : \mathcal{A} \longrightarrow \mathcal{A}$ *is called a Lie n-derivation* $(n \geq 2)$ *if*

$$L(p_n(a_1,a_2,\ldots,a_n)) = \sum_{i=1}^{n} p_n(a_1,\ldots,a_{i-1},L(a_i),a_{i+1},\ldots,a_n)$$

holds for all $a_1,a_2,\ldots,a_n \in \mathcal{A}$.

Remark 5.6.1 *In particular, a Lie 2-derivation (that is, an R-linear mapping* $L : \mathcal{A} \longrightarrow \mathcal{A}$ *that satisfies* $L([a_1,a_2]) = [L(a_1),a_2] + [a_1,L(a_2)]$ *for all* $a_1,a_2 \in \mathcal{A}$*) is called a Lie derivation and a Lie 3-derivation (that is, an R-linear mapping* $L : \mathcal{A} \longrightarrow \mathcal{A}$ *that satisfies* $L([[a_1,a_2],a_3]) = [[L(a_1),a_2],a_3] + [[a_1,L(a_2)],a_3] + [[a_1,a_2],L(a_3)]$ *for all* $a_1,a_2,a_3 \in \mathcal{A}$*) is said to be a Lie triple derivation. Lie 2-derivations, Lie 3-derivations and Lie n-derivations are collectively referred to as Lie-type derivations. If the mapping L is not necessarily linear in the above definition, then it is said to be a nonlinear Lie n-derivation. Obviously, each derivation is a Lie n-derivation for each* $n \geq 2$ *and each Lie n-derivation is a* $(n+k(n-1))$*-derivation for every* $k \in \mathbb{N}$ *(see [276]). However, the converse statements are not true in general. For example, if* $\delta : \mathcal{A} \to \mathcal{A}$ *is a derivation* *and $\gamma : \mathcal{A} \to Z(\mathcal{A})$ is an R-linear mapping such that $\gamma([\mathcal{A},\mathcal{A}]) = 0$, then $\delta + \gamma$ is a Lie derivation, which is not necessarily a derivation. Similarly, if $\gamma : \mathcal{A} \to Z(\mathcal{A})$ is an R-linear mapping such that $\gamma([[\mathcal{A},\mathcal{A}],\mathcal{A}]) = 0$, then $\delta + \gamma$ is a Lie triple derivation, which is not necessarily a Lie derivation. These types of maps are standard examples of Lie (triple) derivations. In general, a Lie n-derivation $L : \mathcal{A} \longrightarrow \mathcal{A}$ is said to be of the standard or proper form if $L = \delta + \gamma$ for some derivation $\delta : \mathcal{A} \longrightarrow \mathcal{A}$ and for some R-linear map $\gamma : \mathcal{A} \longrightarrow \mathcal{Z}(\mathcal{A})$, such that $\gamma(p_n(\mathcal{A},\mathcal{A},\ldots,\mathcal{A})) = 0$.*

Lie-type derivations in different backgrounds have been extensively studied by many authors over the past few decades (see, for examples, [2, 3, 39, 117, 120, 121, 180, 255, 276, 284, 324, 407, 408, 409, 416, 419, 443, 460, 461, 491, 544, 546, 549, 560, 563, 570]). In most of the cases, the authors considered the question when do all Lie (type) derivations on an algebra have the standard form. In the year 1964, Martindale [443] obtained the first characterization of Lie derivations on primitive rings and proved that under certain restrictions every Lie derivation on a primitive ring is of the standard form. Motivated by this result, Miers [460, 461] studied Lie (triple) derivations on von Neumann algebras with no central abelian summands. Recall that a von Neumann algebra is a weakly closed $*$-subalgebra of the algebra of all bounded linear operators acting on a complex Hilbert space which contains the identity operator.

Theorem 5.6.2 *[460, Theorem 3] Let $\mathcal{M}$ be a von Neumann algebra with no central abelian summands. If $L : \mathcal{M} \to \mathcal{M}$ is a Lie derivation, then $L(x) = [a,x] + \gamma(x)$, where $a \in \mathcal{M}$ and γ is a linear map from $\mathcal{M}$ into $Z(\mathcal{M})$ which annihilates brackets of operators.*

Theorem 5.6.3 *[461, Theorem 1] Let $L : \mathcal{M} \to \mathcal{M}$ be a Lie triple derivation, where $\mathcal{M}$ is a von Neumann algebra with no central abelian summands. There exists an operator $a \in \mathcal{M}$ such that $L(x) = [a,x] + \gamma(x)$, where $\gamma : \mathcal{M} \to Z(\mathcal{M})$ is a linear map which annihilates brackets.*

The concept of Lie n-derivations was introduced by Abdullaev [1], where forms of Lie n-derivations of certain von Neumann algebras were described. The following result is a generalization of Theorems 5.6.2 and 5.6.3.

Theorem 5.6.4 *[1, Theorem] Let $\mathcal{M}$ be a von Neumann algebra of operators on a Hilbert space. Then, every Lie n-derivation $L : \mathcal{M} \to \mathcal{M}$ can be decomposed to $L = \delta + \gamma$, where δ is an ordinary derivation on $\mathcal{M}$ and γ is a linear map from $\mathcal{M}$ into its center which annihilates all $(n-1)$th commutators of $\mathcal{M}$.*

The structure of nonlinear Lie derivations on von Neumann algebras was studied by Bai and Du [92].

Theorem 5.6.5 *[92, Main Theorem] Let $\mathcal{M}$ be a von Neumann algebra with no central summands of type I_1 (=central abelian projection). If $L : \mathcal{M} \to \mathcal{M}$ is a* *nonlinear Lie derivation, then L is of the form $\delta + \gamma$, where δ is an additive derivation of $\mathcal{M}$ and γ is a mapping of $\mathcal{M}$ into its center $\mathcal{Z}(\mathcal{M})$ which maps commutators into zero.*

Motivated by the powerful and elegant results of Miers [460, 461], Fošner et al. [276] studied nonlinear Lie-type derivations of von Neumann algebras and proved the following result, generalizing the result of Bai and Du [92].

Theorem 5.6.6 *[276, Theorem 2.3] Let $\mathcal{M}$ be a von Neumann algebra without abelian central summands of type I_1 and $L : \mathcal{M} \to \mathcal{M}$ be a nonlinear Lie n-derivation of $\mathcal{M}$. Then, L is of the standard form, that is, $L = \delta + \gamma$, where δ is an additive derivation of $\mathcal{M}$ and γ is a central-valued mapping which annihilates each $(n-1)$th commutator of $\mathcal{M}$.*

Besides operator algebras, Lie-type derivations have been studied on various algebras such as triangular algebras and generalized matrix algebras. Let us now discuss Lie-type derivations on these algebras.

Triangular algebras

Many properties of some linear mappings such as automorphisms, derivations, Lie derivations, and commuting mappings on triangular algebras are explored in Cheung's dissertation [179]. Let us recall some basic facts about triangular algebras, which were first introduced by Chase [172] and later explored by Cheung [179].

Let $\mathcal{A}$ and $\mathcal{B}$ be unital algebras over a commutative ring R with identity, and $\mathcal{M}$ be a nonzero $(\mathcal{A},\mathcal{B})$-bimodule. Then, the set

$$\mathrm{Tri}(\mathcal{A},\mathcal{M},\mathcal{B}) = \left\{ \begin{pmatrix} a & m \\ 0 & b \end{pmatrix} \middle| \, a \in \mathcal{A}, m \in \mathcal{M}, b \in \mathcal{B} \right\}$$

forms an R-algebra with respect to the following matrix-like operations:

$$\begin{pmatrix} a & m \\ 0 & b \end{pmatrix} + \begin{pmatrix} a' & m' \\ 0 & b' \end{pmatrix} = \begin{pmatrix} a+a' & m+m' \\ 0 & b+b' \end{pmatrix}$$

and

$$\begin{pmatrix} a & m \\ 0 & b \end{pmatrix} \begin{pmatrix} a' & m' \\ 0 & b' \end{pmatrix} = \begin{pmatrix} aa' & am'+mb' \\ 0 & bb' \end{pmatrix}.$$

Definition 77 *An R-algebra $\mathfrak{A}$ is called a triangular algebra if there exist two R-algebras $\mathcal{A},\mathcal{B}$ and a nonzero $(\mathcal{A},\mathcal{B})$-bimodule $\mathcal{M}$ such that $\mathfrak{A}$ is isomorphic to* $\mathrm{Tri}(\mathcal{A},\mathcal{M},\mathcal{B})$.

Examples of triangular algebras

Let us describe the three classical examples of triangular algebras: upper triangular matrix algebras, block upper triangular matrix algebras and nest algebras. For other examples of triangular algebras see [285, 554] and references therein.

(i) **Upper triangular matrix algebras**: Let R be a commutative ring with identity. We denote the set of all $p \times q$ matrices over R by $M_{p\times q}(R)$ and denote the set of all $n \times n$ upper triangular matrices over R by $T_n(R)$. For $n \geq 2$ and each $1 \leq k \leq n-1$, the upper triangular matrix algebra $T_n(R)$ can be written as a triangular algebra of the form

$$T_n(R) = \begin{pmatrix} T_n(R) & M_{k\times n-k}(R) \\ 0 & T_{n-k}(R) \end{pmatrix}.$$

(ii) **Block upper triangular matrix algebras**: Let R be a commutative ring with identity. For each positive integer n and each positive integer m with $m \leq n$, we denote an ordered m-vector of positive by $\overline{d} = (d_1, \ldots, d_i, \ldots, d_m) \in \mathbb{N}^m$ such that $n = d_1 + \cdots + d_i + \cdots + d_m$. The block upper triangular matrix algebra $B_n^{\overline{d}}(R)$ is a subalgebra of $M_n(R)$ of the form

$$B_n^{\overline{d}}(R) = \begin{pmatrix} M_{d_1}(R) & \ldots & M_{d_1\times d_i}(R) & \ldots & M_{d_1\times d_m}(R) \\ & \ddots & \vdots & & \vdots \\ & & M_{d_i}(R) & \ldots & M_{d_i\times d_m}(R) \\ & & & \ddots & \vdots \\ & & & & M_{d_m}(R) \end{pmatrix}.$$

If $n \geq 2$ and $B_n^{\overline{d}}(R) \neq M_n(R)$, then $B_n^{\overline{d}}(R)$ is a triangular algebra and can be represented as

$$B_n^{\overline{d}}(R) = \begin{pmatrix} B_n^{\overline{d}_1}(R) & M_{n\times(n-j)}(R) \\ O_{(n-j)\times j} & B_{n-j}^{\overline{d}_2}(R) \end{pmatrix},$$

where $1 \leq j < m$ and $\overline{d}_1 \in \mathbb{N}^j, \overline{d}_2 \in \mathbb{N}^{m-j}$. Note that the full matrix algebra $M_n(R)$ and the upper triangular matrix algebra $T_n(R)$ are two special examples of block upper triangular matrix algebras.

(iii) **Nest algebras**: Let $\mathbf{H}$ be a complex Hilbert space and $B(\mathbf{H})$ be the algebra of all bounded linear operators on $\mathbf{H}$. Let I be an index set. A nest is a set $\mathcal{N}$ of closed subspaces of $\mathbf{H}$ satisfying the following conditions:

(*i*) $\{0\}, \mathbf{H} \in \mathcal{N}$;

(*ii*) If $N_1, N_2 \in \mathcal{N}$, then either $N_1 \subseteq N_2$ or $N_2 \subseteq N_1$;

(*iii*) If $\{N\}_{i\in I} \subseteq \mathcal{N}$, then $\bigcap_{i\in I} N_i \in \mathcal{N}$;

(*iv*) If $\{N\}_{i\in I} \subseteq \mathcal{N}$, then the norm closure of the linear span of $\bigcup_{i\in I} N_i$ also lies in $\mathcal{N}$.

If $N = \{0, \mathbf{H}\}$, then $\mathcal{N}$ is called a trivial nest, otherwise it is called a nontrivial nest. The nest algebra associated with the nest $\mathcal{N}$, denoted by $T(\mathcal{N})$, is the algebra

$$T(\mathcal{N}) = \{T \in B(\mathbf{H}) \mid TN \subseteq N \text{ for all } N \in \mathcal{N}\}.$$

A nontrivial nest algebra is a triangular algebra. Indeed, if $N \in \mathcal{N} \setminus \{0, \mathbf{H}\}$ and E is the orthogonal projection onto N, then $\mathcal{N}_1 = E(\mathcal{N})$ and $\mathcal{N}_2 = (1-E)(\mathcal{N})$ are nests of N and $N^{\perp}$, respectively. Moreover, $T(\mathcal{N}_1) = ET(\mathcal{N})E, T(\mathcal{N}_2) = (1-E)T(\mathcal{N})(1-E)$ are nest algebras and

$$T(\mathcal{N}) = \begin{pmatrix} T(\mathcal{N}_1) & ET(\mathcal{N})(1-E) \\ 0 & T(\mathcal{N}_2) \end{pmatrix}.$$

Note that any finite-dimensional nest algebra is isomorphic to a complex block upper triangular matrix algebra. We refer the reader to [206] for the general theory of nest algebras.

Definition 78 *Let $\mathfrak{A} = \mathrm{Tri}(\mathcal{A}, \mathcal{M}, \mathcal{B})$ be a triangular algebra. We define two natural projections $\pi_{\mathcal{A}} : \mathfrak{A} \to \mathcal{A}$ and $\pi_{\mathcal{B}} : \mathfrak{A} \to \mathcal{B}$ by*

$$\pi_{\mathcal{A}} \begin{pmatrix} a & m \\ 0 & b \end{pmatrix} = a \text{ and } \pi_{\mathcal{B}} \begin{pmatrix} a & m \\ 0 & b \end{pmatrix} = b.$$

Next, we recall the definition of faithful bimodules.

Definition 79 *A left (resp. right) $\mathcal{A}$-module $\mathcal{M}$ is said to be faithful if $a = 0$ is the only element in $\mathcal{A}$ satisfying $a\mathcal{M} = 0$ (resp. $\mathcal{M}a = 0$). An $(\mathcal{A}, \mathcal{B})$-bimodule $\mathcal{M}$ is called a faithful bimodule if it is both a faithful left $\mathcal{A}$-module and a faithful right $\mathcal{B}$-module.*

The center of a triangular algebra $\mathfrak{A} = \mathrm{Tri}(\mathcal{A}, \mathcal{M}, \mathcal{B})$ is given in the next proposition.

Proposition 5.6.7 *[179, Theorem 1.4.4] Let $\mathcal{A}$ and $\mathcal{B}$ be unital algebras and $\mathcal{M}$ be a faithful $(\mathcal{A}, \mathcal{B})$-bimodule. Then, the center of $\mathfrak{A} = \mathrm{Tri}(\mathcal{A}, \mathcal{M}, \mathcal{B})$ is given by*

$$Z(\mathfrak{A}) = \left\{ \begin{pmatrix} a & 0 \\ 0 & b \end{pmatrix} \mid a \in \mathcal{A}, b \in \mathcal{B}, am = mb \text{ for all } m \in \mathcal{M} \right\}.$$

Furthermore, $\pi_{\mathcal{A}}(Z(\mathfrak{A})) \subseteq Z(\mathcal{A})$ and $\pi_{\mathcal{B}}(Z(\mathfrak{A})) \subseteq Z(\mathcal{B})$, and there exists a unique algebra isomorphism $\tau : \pi_{\mathcal{A}}(Z(\mathfrak{A})) \to \pi_{\mathcal{B}}(Z(\mathfrak{A}))$ such that $am = m\tau(a)$ for all $m \in \mathcal{M}$ and $a \in \pi_{\mathcal{A}}(Z(\mathfrak{A}))$.

Proof. First, we show that the center of a triangular algebra $\mathfrak{A} = \text{Tri}(\mathcal{A}, \mathcal{M}, \mathcal{B})$ is given by

$$Z(\mathfrak{A}) = \left\{ \begin{pmatrix} a & 0 \\ 0 & b \end{pmatrix} \mid a \in Z(\mathcal{A}), b \in Z(\mathcal{B}), am = mb \text{ for all } m \in \mathcal{M} \right\}. \quad (5.4)$$

Suppose that $a \in Z(\mathcal{A})$, $b \in Z(\mathcal{B})$ and $am = mb$ for all $m \in \mathcal{M}$. Then, for any $\begin{pmatrix} a' & m' \\ 0 & b' \end{pmatrix} \in \mathfrak{A}$, we have

$$\begin{aligned} \begin{pmatrix} a & 0 \\ 0 & b \end{pmatrix} \begin{pmatrix} a' & m' \\ 0 & b' \end{pmatrix} &= \begin{pmatrix} aa' & am' \\ 0 & bb' \end{pmatrix} = \begin{pmatrix} a'a & m'b \\ 0 & b'b \end{pmatrix} \\ &= \begin{pmatrix} a' & m' \\ 0 & b' \end{pmatrix} \begin{pmatrix} a & 0 \\ 0 & b \end{pmatrix}. \end{aligned}$$

Hence $\begin{pmatrix} a & 0 \\ 0 & b \end{pmatrix} \in Z(\mathfrak{A})$. Conversely, assume that $\begin{pmatrix} a & m \\ 0 & b \end{pmatrix} \in Z(\mathfrak{A})$. Then, we have

$$\begin{pmatrix} a & m \\ 0 & b \end{pmatrix} = \begin{pmatrix} 1 & 0 \\ 0 & 0 \end{pmatrix} \begin{pmatrix} a & m \\ 0 & b \end{pmatrix} = \begin{pmatrix} a & m \\ 0 & b \end{pmatrix} \begin{pmatrix} 1 & 0 \\ 0 & 0 \end{pmatrix} = \begin{pmatrix} a & 0 \\ 0 & 0 \end{pmatrix}.$$

This implies that $m = 0$. Now, for any m', we have

$$\begin{pmatrix} 0 & am' \\ 0 & 0 \end{pmatrix} = \begin{pmatrix} a & 0 \\ 0 & b \end{pmatrix} \begin{pmatrix} 0 & m' \\ 0 & 0 \end{pmatrix} = \begin{pmatrix} 0 & m' \\ 0 & 0 \end{pmatrix} \begin{pmatrix} a & 0 \\ 0 & b \end{pmatrix} = \begin{pmatrix} 0 & m'b \\ 0 & 0 \end{pmatrix}.$$

This implies that $am' = m'b$ for all $m' \in \mathcal{M}$. Therefore, the assertion (5.4) follows.

Now, we show that if $\mathcal{M}$ is a faithful $(\mathcal{A}, \mathcal{B})$-bimodule, then the conditions $a \in Z(\mathcal{A})$ and $b \in Z(\mathcal{B})$ in the expression (5.4) become redundant.

Suppose that $am = mb$ for all $m \in \mathcal{M}$. Then, for any $a' \in \mathcal{A}$, we see that

$$(aa' - a'a)m = a(a'm) - a'(am) = (a'm)b - a'(mb) = 0$$

for all $m \in \mathcal{M}$. Since $\mathcal{M}$ is a faithful left $\mathcal{A}$-module, it follows that $aa' = a'a$ for $a' \in \mathcal{A}$. Thus, $a \in Z(\mathcal{A})$. Similarly, one can show that $b \in Z(\mathcal{A})$.

It follows from the expression (5.4) that $\pi_{\mathcal{A}}(Z(\mathfrak{A})) \subseteq Z(\mathcal{A})$ and $\pi_{\mathcal{B}}(Z(\mathfrak{A})) \subseteq Z(\mathcal{B})$. Next, we show that there exists a unique algebra isomorphism $\tau : \pi_{\mathcal{A}}(Z(\mathfrak{A})) \to \pi_{\mathcal{B}}(Z(\mathfrak{A}))$ such that $am = m\tau(a)$ for all $m \in \mathcal{M}$ and $a \in \pi_{\mathcal{A}}(Z(\mathfrak{A}))$.

Observe that for any $a \in \pi_{\mathcal{A}}(Z(\mathfrak{A}))$, there exists a unique $b \in \pi_{\mathcal{B}}(Z(\mathfrak{A}))$ such that $\begin{pmatrix} a & 0 \\ 0 & b \end{pmatrix} \in Z(\mathfrak{A})$. Suppose there exists another b' satisfying $\begin{pmatrix} a & 0 \\ 0 & b' \end{pmatrix} \in$

$Z(\mathfrak{A})$, then we have $mb' = am = mb$ for all $m \in \mathcal{M}$ and hence $b' = b$ as $\mathcal{M}$ is a faithful right $\mathcal{B}$-module. Therefore, such a mapping τ exists and is unique. Furthermore, it is straightforward to check that τ is an algebra isomorphism. This completes the proof of the proposition.

From now on, it is assumed that $\mathfrak{A} = \mathrm{Tri}(\mathcal{A}, \mathcal{M}, \mathcal{B})$ is a triangular algebra consisting of two unital algebras $\mathcal{A}, \mathcal{B}$ and a faithful $(\mathcal{A}, \mathcal{B})$-bimodule $\mathcal{M}$. Let us state a condition on an algebra which is sometimes very useful while studying mappings on triangular algebras. Suppose that $\mathcal{A}$ is an algebra such that

$$Z(\mathcal{A}) = \{a \in \mathcal{A} \mid [[a,x],y] = 0 \text{ for all } x, y \in \mathcal{A}\}. \tag{5.5}$$

Condition (5.5) is equivalent to saying that there exist no nonzero central inner derivations on $\mathcal{A}$. Some standard examples of algebras satisfying (5.5) are triangular algebras, full matrix algebras and (semi-)prime algebras (see [560] for details).

In the year 2003, Cheung [180] initiated the study of Lie derivations on triangular algebras and obtained sufficient conditions under which every Lie derivation on a triangular algebra is of the standard form. More precisely, Cheung proved the following result:

Theorem 5.6.8 *[180, Theorem 11] Let $\mathfrak{A} = \mathrm{Tri}(\mathcal{A}, \mathcal{M}, \mathcal{B})$ be a triangular algebra and $\mathcal{L} : \mathfrak{A} \to \mathfrak{A}$ be a Lie derivation. If $\pi_{\mathcal{A}}(Z(\mathfrak{A})) = Z(\mathcal{A})$ and $\pi_{\mathcal{B}}(Z(\mathfrak{A})) = Z(\mathcal{B})$, then $\mathcal{L}$ is of the standard form, i.e., $\mathcal{L}$ is of the form $\mathcal{L} = \delta + \gamma$, where $\delta : \mathfrak{A} \to \mathfrak{A}$ is a derivation and $\gamma : \mathfrak{A} \to \mathfrak{A}$ is a R-linear mapping vanishing on all commutators of $\mathfrak{A}$.*

The above result has been generalized in different ways (see [116, 121, 324, 330, 416, 560, 561, 563]). For example, Xiao and Wei in [560] extended Cheung's result to the case of Lie triple derivations of triangular algebras.

Theorem 5.6.9 *[560, Theorem 2.1] Let $\mathfrak{A} = \mathrm{Tri}(\mathcal{A}, \mathcal{M}, \mathcal{B})$ be a 2-torsion free triangular algebra and $\mathcal{L} : \mathfrak{A} \to \mathfrak{A}$ be a Lie triple derivation. Suppose that $\mathfrak{A}$ satisfies the following conditions:*

(i) $Z(\mathcal{A}) = \pi_{\mathcal{A}}(Z(\mathfrak{A}))$ and $Z(\mathcal{B}) = \pi_{\mathcal{B}}(Z(\mathfrak{A}))$, and

(ii) $\mathcal{A}$ or $\mathcal{B}$ satisfies (5.4).

Then, $\mathcal{L}$ is of the standard form. More precisely, there exist a derivation $\delta : \mathfrak{A} \to \mathfrak{A}$ and an R-linear mapping $\gamma : \mathfrak{A} \to Z(\mathfrak{A})$ vanishing on all second commutators of $\mathfrak{A}$ such that $\mathcal{L} = \delta + \gamma$.

The structure of nonlinear mappings on rings or algebras has been studied by many authors (see [55, 178, 561, 563] and references therein). Chen and Zhang [178] described nonlinear Lie derivations of upper triangular matrix algebras. Motivated by the above work, Yu and Zhang [563] extended Cheung's result to nonlinear Lie derivations of triangular algebras.

Theorem 5.6.10 *[563, Theorem 2.1] Let* $\mathfrak{A} = \mathrm{Tri}(\mathcal{A}, \mathcal{M}, \mathcal{B})$ *be a triangular algebra and* $L : \mathfrak{A} \to \mathfrak{A}$ *a nonlinear Lie derivation. If* $\pi_{\mathcal{A}}(Z(\mathfrak{A})) = Z(\mathcal{A})$ *and* $\pi_{\mathcal{B}}(Z(\mathfrak{A})) = Z(\mathcal{B})$*, then* L *is the sum of an additive derivation* δ *on* $\mathfrak{A}$ *and a map* γ *from* $\mathfrak{A}$ *into its center* $Z(\mathfrak{A})$ *sending each commutator to zero.*

Ji et al. [324] proved a similar result for a nonlinear Lie triple derivation of triangular algebras.

Theorem 5.6.11 *[324, Theorem 2.1] Let* $\mathfrak{A} = \mathrm{Tri}(\mathcal{A}, \mathcal{M}, \mathcal{B})$ *be a 2-torsion free triangular algebra. Suppose that* $\mathfrak{A}$ *satisfies the following:*

(i) $\pi_{\mathcal{A}}(Z(\mathfrak{A})) = Z(\mathcal{A})$ *and* $\pi_{\mathcal{B}}(Z(\mathfrak{A})) = Z(\mathcal{B})$,

(ii) $\mathcal{A}$ *or* $\mathcal{B}$ *satisfies* (5.4).

Then, every nonlinear Lie triple derivation $L : \mathfrak{A} \to \mathfrak{A}$ *is the sum of an additive derivation* δ *on* $\mathfrak{A}$ *and a map* γ *from* $\mathfrak{A}$ *into its center* $Z(\mathfrak{A})$ *which annihilates all Lie triple products.*

Generalizing the notion of a nonlinear Lie (triple) derivation, Benkovič and Eremita [121] introduced the notion of a nonlinear (multiplicative) Lie n-derivation of an algebra (ring) and discussed nonlinear (multiplicative) Lie n-derivations of triangular algebras (rings), which generalize the results of [324, 563].

Theorem 5.6.12 *[121, Theorem 5.9] Let* $\mathfrak{A} = \mathrm{Tri}(\mathcal{A}, \mathcal{M}, \mathcal{B})$ *be a* $(n-1)$*-torsion free triangular algebra such that:*

(i) $Z(\mathcal{A}) = \pi_{\mathcal{A}}(Z(\mathfrak{A}))$ *and* $Z(\mathcal{B}) = \pi_{\mathcal{B}}(Z(\mathfrak{A}))$*, and*

(ii) $\mathcal{A}$ *or* $\mathcal{B}$ *satisfies* (5.4).

Then, any nonlinear Lie n-derivation $L : \mathfrak{A} \to \mathfrak{A}$ *has the standard form, that is,* $L = \delta + \gamma$*, where* $\delta : \mathfrak{A} \to \mathfrak{A}$ *is an additive derivation and* $\gamma : \mathfrak{A} \to Z(\mathfrak{A})$ *is a mapping that vanishes on all* $(n-1)$*th commutators of* $\mathfrak{A}$*.*

Generalized matrix algebras

Let R be a commutative ring with identity. A Morita context consists of two R-algebras $\mathcal{A}$ and $\mathcal{B}$, two bimodules ${}_{\mathcal{A}}\mathcal{M}_{\mathcal{B}}$ and ${}_{\mathcal{B}}\mathcal{N}_{\mathcal{A}}$, and two bimodule homomorphisms $\zeta_{\mathcal{MN}} : \mathcal{M} \underset{\mathcal{B}}{\otimes} \mathcal{N} \to \mathcal{A}$ and $\chi_{\mathcal{NM}} : \mathcal{N} \underset{\mathcal{A}}{\otimes} \mathcal{M} \to \mathcal{B}$ satisfying the following commutative diagrams:

$$
\begin{array}{ccc}
\mathcal{M}\underset{\mathcal{B}}{\otimes}\mathcal{N}\underset{\mathcal{A}}{\otimes}\mathcal{M} & \xrightarrow{\zeta_{mn}\otimes I_m} & \mathcal{A}\underset{\mathcal{A}}{\otimes}\mathcal{M} \\
\Big\downarrow {\scriptstyle I_m\otimes\chi_{nm}} & & \Big\downarrow {\scriptstyle\cong} \\
\mathcal{M}\underset{\mathcal{B}}{\otimes}\mathcal{B} & \xrightarrow{\cong} & \mathcal{M}
\end{array}
\qquad \text{and} \qquad
\begin{array}{ccc}
\mathcal{N}\underset{\mathcal{A}}{\otimes}\mathcal{M}\underset{\mathcal{B}}{\otimes}\mathcal{N} & \xrightarrow{\chi_{nm}\otimes I_n} & \mathcal{B}\underset{\mathcal{B}}{\otimes}\mathcal{N} \\
\Big\downarrow {\scriptstyle I_n\otimes\zeta_{mn}} & & \Big\downarrow {\scriptstyle\cong} \\
\mathcal{N}\underset{\mathcal{A}}{\otimes}\mathcal{A} & \xrightarrow{\cong} & \mathcal{N}.
\end{array}
$$

Let us write this Morita context as $(\mathcal{A},\mathcal{B},\mathcal{M},\mathcal{N},\zeta_{mn},\chi_{nm})$.

If $(\mathcal{A},\mathcal{B},\mathcal{M},\mathcal{N},\zeta_{mn},\chi_{nm})$ is a Morita context, then the set

$$\begin{pmatrix} \mathcal{A} & \mathcal{M} \\ \mathcal{N} & \mathcal{B} \end{pmatrix} = \left\{ \begin{pmatrix} a & m \\ n & b \end{pmatrix} \;\middle|\; a\in\mathcal{A}, m\in\mathcal{M}, n\in\mathcal{N}, b\in\mathcal{B} \right\}$$

forms an R-algebra with respect to the following matrix-like operations:

$$\begin{pmatrix} a & m \\ n & b \end{pmatrix} + \begin{pmatrix} a' & m' \\ n' & b' \end{pmatrix} = \begin{pmatrix} a+a' & m+m' \\ n+n' & b+b' \end{pmatrix}$$

and

$$\begin{pmatrix} a & m \\ n & b \end{pmatrix} \begin{pmatrix} a' & m' \\ n' & b' \end{pmatrix} = \begin{pmatrix} aa'+mn' & am'+mb' \\ na'+bn' & nm'+bb' \end{pmatrix},$$

where $mn' = \zeta_{mn}(m\otimes n')$ and $nm' = \chi_{nm}(n\otimes m')$. Such an R-algebra is called a *generalized matrix algebra* and it is denoted by $G = G(\mathcal{A},\mathcal{M},\mathcal{N},\mathcal{B})$. This kind of algebra was first introduced by Sands in [509], where the author investigated various radicals of rings occurring in Morita contexts.

Examples of generalized matrix algebras

Observe that if $\mathcal{N}=\{0\}$, then the generalized matrix algebra $G = G(\mathcal{A},\mathcal{M},\mathcal{N},\mathcal{B})$ exactly degenerates to the triangular algebra $\mathrm{Tri}(\mathcal{A},\mathcal{M},\mathcal{B})$. Besides triangular algebras, let us see some more examples of generalized matrix algebras.

(i) **Full Matrix Algebras**: Let $\mathcal{A}$ be a unital R-algebra and $M_n(\mathcal{A})$ be the algebra of all $n\times n$ matrices over $\mathcal{A}$ with $n\geq 2$. Then, the full matrix algebra $M_n(\mathcal{A})$ can be written as a generalized matrix algebra of the form

$$M_n(\mathcal{A}) = \begin{pmatrix} \mathcal{A} & M_{1\times(n-1)}(\mathcal{A}) \\ M_{(n-1)\times 1}(\mathcal{A}) & M_{(n-1)}(\mathcal{A}) \end{pmatrix}.$$

(ii) **Unital algebras with nontrivial idempotents**: Let R be a commutative ring with identity and $\mathcal{A}$ be a unital algebra over R. Suppose there exists a nontrivial idempotent $e \in \mathcal{A}$, and $f = 1 - e$. One can easily construct the following generalized matrix algebra:

$$G = \begin{pmatrix} e\mathcal{A}e & e\mathcal{A}f \\ f\mathcal{A}e & f\mathcal{A}f \end{pmatrix} = \left\{ \begin{pmatrix} eae & ecf \\ fde & fbf \end{pmatrix} \mid a,b,c,d \in \mathcal{A} \right\}.$$

Moreover, it is easy to see that the R-linear mapping $\xi : \mathcal{A} \to G$ defined by

$$a \mapsto \begin{pmatrix} eae & eaf \\ fae & faf \end{pmatrix}$$

is an isomorphism from $\mathcal{A}$ to G. This shows that any unital algebra with nontrivial idempotents is isomorphic to a generalized matrix algebra.

From now on, we shall assume that $G = G(\mathcal{A}, \mathcal{M}, \mathcal{N}, \mathcal{B})$ is a generalized matrix algebra with faithful $(\mathcal{A}, \mathcal{B})$-bimodule $\mathcal{M}$. We define two natural projections $\pi_{\mathcal{A}} : G \to \mathcal{A}$ and $\pi_{\mathcal{B}} : G \to \mathcal{B}$ by

$$\pi_{\mathcal{A}} : \begin{pmatrix} a & m \\ n & b \end{pmatrix} \mapsto a \text{ and } \pi_{\mathcal{B}} : \begin{pmatrix} a & m \\ n & b \end{pmatrix} \mapsto b.$$

Then, in view of [559, Lemmas 3.1 & 3.2], the center of G is given by

$$Z(G) = \left\{ \begin{pmatrix} a & 0 \\ 0 & b \end{pmatrix} \middle| \; a \in \mathcal{A}, b \in \mathcal{B}, am = mb, na = bn \text{ for all } m \in \mathcal{M}, n \in \mathcal{N} \right\}.$$

Moreover, $\pi_{\mathcal{A}}(Z(G)) \subseteq Z(\mathcal{A})$, $\pi_{\mathcal{B}}(Z(G)) \subseteq Z(\mathcal{B})$ and there exists a unique algebra isomorphism $\eta : \pi_{\mathcal{A}}(Z(G)) \to \pi_{\mathcal{B}}(Z(G))$ such that $am = m\eta(a)$ and $na = \eta(a)n$ for all $a \in \pi_{\mathcal{A}}(Z(G))$, $m \in \mathcal{M}$, $n \in \mathcal{N}$.

In the year 2012, Du and Wang [255] investigated the standard form Lie derivations of generalized matrix algebras and proved the following theorem.

Theorem 5.6.13 *[255, Theorem 1] Let $G = G(\mathcal{A}, \mathcal{M}, \mathcal{N}, \mathcal{B})$ be a generalized matrix algebra. Suppose that*

(i) *$Z(\mathcal{A}) = \pi_{\mathcal{A}}(Z(G))$ and $Z(\mathcal{B}) = \pi_{\mathcal{B}}(Z(G))$;*

(ii) *Either $\mathcal{A}$ or $\mathcal{B}$ does not contain nonzero central ideals.*

If L is an R-linear map of G such that $L([X,Y]) = [L(X),Y] + [X,L(Y)]$ for all $X,Y \in G$ with $XY = 0$, then there exists a derivation δ of G and an R-linear map $\gamma : G \to Z(G)$ vanishing at commutators $[X,Y]$ with $XY = 0$ such that $L(X) = \delta(X) + \gamma(X)$ for all $X \in G$.

Ashraf and Akhtar [37] gave the structure of Lie triple derivations of generalized matrix algebras and used them to obtain some necessary and sufficient conditions for a Lie triple derivation on a generalized matrix algebra to be proper.

Theorem 5.6.14 *[37, Theorem 2.1] Let $G = G(\mathcal{A},\mathcal{M},\mathcal{N},\mathcal{B})$ be a generalized matrix algebra. An R-linear mapping $L : G \to G$ is a Lie triple derivation on G if and only if L has the form,*

$$L\begin{pmatrix} a & m \\ n & b \end{pmatrix} = \begin{pmatrix} A_{11}(a)+B_{11}(b)-mn_0-m_0n & am_0-m_0b+C_{12}(m)+D_{12}(n) \\ n_0a-bn_0+C_{21}(m)+D_{21}(n) & A_{22}(a)+B_{22}(b)+n_0m+nm_0 \end{pmatrix}$$

for all $\begin{pmatrix} a & m \\ n & b \end{pmatrix} \in G$*, where* $m_0 \in \mathcal{M}$*,* $n_0 \in \mathcal{N}$*;* $A_{11} : \mathcal{A} \to \mathcal{A}$*,* $B_{11} : \mathcal{B} \to [\mathcal{A},\mathcal{A}]'$*,* $C_{12} : \mathcal{M} \to \mathcal{M}$*,* $D_{12} : \mathcal{N} \to \mathcal{M}$*,* $C_{21} : \mathcal{M} \to \mathcal{N}$*,* $D_{21} : \mathcal{N} \to \mathcal{N}$*,* $A_{22} : \mathcal{A} \to [\mathcal{B},\mathcal{B}]'$*,* $B_{22} : \mathcal{B} \to \mathcal{B}$ *are R-linear mappings satisfying the following conditions:*

(i) A_{11} *is a Lie triple derivation on* $\mathcal{A}$ *and* $A_{11}(mn) - B_{11}(nm) = C_{12}(m)n + mD_{21}(n)$;

(ii) B_{22} *is a Lie triple derivation on* $\mathcal{B}$ *and* $B_{22}(nm) - A_{22}(mn) = nC_{12}(m) + D_{21}(n)m$;

(iii) $C_{12}(am) = A_{11}(a)m - mA_{22}(a) + aC_{12}(m)$, $C_{12}(mb) = C_{12}(m)b + mB_{22}(b) - B_{11}(b)m$, $C_{21}(am) = C_{21}(m)a$ *and* $C_{21}(mb) = bC_{21}(m)$;

(iv) $D_{12}(na) = aD_{12}(n)$, $D_{12}(bn) = D_{12}(n)b$, $D_{21}(na) = D_{21}(n)a + nA_{11}(a) - A_{22}(a)n$ *and* $D_{21}(bn) = B_{22}(b)n - nB_{11}(b) + bD_{21}(n)$;

(v) $[[A_{22}(a),b_1],b_2] = 0$ *and* $[[B_{11}(b),a_1],a_2] = 0$.

Theorem 5.6.15 *[37, Theorem 2.2] Let $G = G(\mathcal{A},\mathcal{M},\mathcal{N},\mathcal{B})$ be a generalized matrix algebra. A Lie triple derivation $L : G \to G$ (of the form given in Theorem 5.6.14) is proper if and only if the following conditions hold:*

(i) $B_{11}(\mathcal{B}) \subseteq \pi_{\mathcal{A}}(Z(G))$, $A_{22}(\mathcal{A}) \subseteq \pi_{\mathcal{B}}(Z(G))$,

(ii) $B_{11}(nm) \oplus A_{22}(mn) \in Z(G)$ *for all* $m \in \mathcal{M}$, $n \in \mathcal{N}$,

(iii) $D_{12} = 0$ *and* $C_{21} = 0$.

Wang and Wang [546] extended Benkovič's result [121, Theorem 5.9] to the case of generalized matrix algebras.

Theorem 5.6.16 *[546, Theorem 1] Let $G = G(\mathcal{A},\mathcal{M},\mathcal{N},\mathcal{B})$ be a $(n-1)$-torsion free generalized matrix algebra such that:*

(i) $Z(\mathcal{A}) = \pi_{\mathcal{A}}(Z(\mathfrak{A}))$ *and* $Z(\mathcal{B}) = \pi_{\mathcal{B}}(Z(\mathfrak{A}))$, *and*

(ii) *Either* $\mathcal{A}$ *or* $\mathcal{B}$ *does not contain nonzero central ideals,*

(iii) $\mathcal{A}$ *or* $\mathcal{B}$ *satisfies* (5.4),

(iv) *If n is odd and* $\mathcal{N} \neq \{0\}$, *then for each* $m \in \mathcal{M}$ *the condition* $m\mathcal{N} = \{0\} = \mathcal{N}m$ *implies* $m = 0$,

(v) *If n is odd, then for each $t \in \mathfrak{N}$ the condition $\mathfrak{M}t = \{0\} = t\mathfrak{M}$ implies $t = 0$.*

Then, any nonlinear Lie n-derivation $L : G \to G$ has the standard form.

Let $\mathcal{A}$ be a unital algebra (over a commutative ring with identity). Let us assume that $\mathcal{A}$ has an idempotent $e \neq 0, 1$ and let f denote the idempotent $1 - e$. In this case, $\mathcal{A}$ can be represented in the so-called Peirce decomposition form

$$\mathcal{A} = e\mathcal{A}e + e\mathcal{A}f + f\mathcal{A}e + f\mathcal{A}f$$

where $e\mathcal{A}e$ and $f\mathcal{A}f$ are subalgebras with identity elements e and f, respectively, $e\mathcal{A}f$ is an $(e\mathcal{A}e, f\mathcal{A}f)$-bimodule and $f\mathcal{A}e$ is an $(f\mathcal{A}f, e\mathcal{A}e)$-bimodule. We further assume that $\mathcal{A}$ satisfies

$$\begin{aligned} eae \cdot e\mathcal{A}f = \{0\} = f\mathcal{A}e \cdot eae \text{ implies } eae = 0, \\ e\mathcal{A}f \cdot faf = \{0\} = faf \cdot f\mathcal{A}e \text{ implies } faf = 0, \end{aligned} \tag{5.6}$$

for all $a \in \mathcal{A}$. A Jordan derivation $\delta : \mathcal{A} \to \mathcal{A}$ is said to be a *singular Jordan derivation* if

$$\delta(e\mathcal{A}e) = \{0\}, \ \delta(e\mathcal{A}f) \subseteq f\mathcal{A}e, \ \delta(f\mathcal{A}e) \subseteq e\mathcal{A}f, \ \delta(f\mathcal{A}f) = \{0\}.$$

It is obvious that singular Jordan derivations on $\mathcal{A}$ are zero when $\mathcal{A}$ is a triangular algebra, since $f\mathcal{A}e = \{0\}$. Some specific examples of unital algebras with nontrivial idempotents having the property (5.6) are triangular algebras, matrix algebras and prime (and hence in particular simple) algebras with nontrivial idempotents.

Recall that generalized matrix algebras are in fact isomorphic to unital algebras with nontrivial idempotents. The authors have used the assumption that the bimodule $e\mathcal{A}f$ from the Peirce decomposition (5.6) is faithful as a left $e\mathcal{A}e$-module and as a right $f\mathcal{A}f$-module. It is worth mentioning that the assumption (5.6) is milder than the faithfulness of the bimodule $e\mathcal{A}f$.

In Theorem 5.6.16, Wang and Wang studied nonlinear Lie n-derivations on generalized matrix algebras. Benkovič [120] dropped the assumptions (iv) and (v) from Theorem 5.6.12 and studied Lie triple derivations of a unital algebra $\mathcal{A}$ with the property (5.6).

Theorem 5.6.17 *[120, Theorem 5.3] Let $\mathcal{A}$ be a unital 2-torsion free algebra with a nontrivial idempotent e satisfying* (5.6). *Lets assume that*

(i) *$Z(e\mathcal{A}e) = Z(A)e$ and $Z(f\mathcal{A}f) = Z(A)f$;*

(ii) *Either $e\mathcal{A}e$ or $f\mathcal{A}f$ contains no central ideal;*

(iii) *Either $e\mathcal{A}e$ or $f\mathcal{A}f$ satisfies* (5.4).

Then, every Lie triple derivation $L : \mathcal{A} \to \mathcal{A}$ is of the form $L = d + \delta + \gamma$, where $d : \mathcal{A} \to \mathcal{A}$ is a derivation, $\delta : \mathcal{A} \to \mathcal{A}$ is a singular Jordan derivation, and $\gamma : \mathcal{A} \to Z(A)$ is an R-linear map that vanishes on $[[\mathcal{A}, \mathcal{A}], \mathcal{A}]$.

Wang [544] extended the above result to the case of Lie n-derivation of unital algebras with the property (5.6) and proved the following result:

Theorem 5.6.18 *[544, Theorem 2.1] Let $\mathcal{A}$ be a unital $(n-1)$-torsion free algebra with a nontrivial idempotent e satisfying* (5.6). *Let us assume that*

(i) $Z(e\mathcal{A}e) = Z(A)e$ *and* $Z(f\mathcal{A}f) = Z(A)f$;

(ii) Either $e\mathcal{A}e$ or $f\mathcal{A}f$ contains no central ideal;

(iii) Either $e\mathcal{A}e$ or $f\mathcal{A}f$ satisfies (5.4).

Then, every Lie n-derivation $L : \mathcal{A} \to \mathcal{A}$ is of the form $L = d + \delta + \gamma$, where $d : \mathcal{A} \to \mathcal{A}$ is a derivation, $\delta : \mathcal{A} \to \mathcal{A}$ is a singular Jordan derivation, and $\gamma : \mathcal{A} \to Z(A)$ is an R-linear map that vanishes on $p_n(\mathcal{A}, \mathcal{A}, \ldots, \mathcal{A})$.

Motivated by the notion of generalized derivations, the notion of generalized Lie-type derivations can also be introduced as follows:

Definition 80 *An R-linear mapping $G_L : \mathcal{A} \longrightarrow \mathcal{A}$ is said to be a generalized Lie n-derivation if there exists a Lie n-derivation $L : \mathcal{A} \longrightarrow \mathcal{A}$ such that*

$$G_L(p_n(a_1, a_2, \ldots, a_n)) = p_n(G_L(a_1), a_2, \ldots, a_n) + \sum_{i=2}^{n} p_n(a_1, \ldots, a_{i-1}, L(a_i), a_{i+1}, \ldots, a_n) \text{ holds for all } a_1, a_2, \ldots, a_n \in \mathcal{A}.$$

Remark 5.6.19 *If we drop the linearity condition in the above definition, then G_L is called a nonlinear generalized Lie n-derivation with associated nonlinear Lie n-derivation L. In particular, a generalized Lie 2-derivation is called a generalized Lie derivation and a generalized Lie 3-derivation is said to be a generalized Lie triple derivation. Generalized Lie 2-derivations, generalized Lie 3-derivations and generalized Lie n-derivations are collectively referred to as generalized Lie-type derivations. Note that every Lie n-derivation is a generalized Lie n-derivation (set $G_L = L$). However, the mapping $x \mapsto \lambda x$, where $\lambda \in Z(A)$ is an example of a generalized Lie n-derivation (set $G_L(x) = \lambda x$ for all $x \in \mathcal{A}$ and $L = 0$) which is not a Lie n-derivation.*

Motivated by the results obtained by Benkovič [116], Li and Shen [416] and Ji et al. [324], Ashraf and Jabeen [55] obtained the structure of linear as well as nonlinear generalized Lie triple derivations on triangular algebras and proved that under certain assumptions every nonlinear generalized Lie triple derivation is the sum of an additive generalized derivation and a central map vanishing on all Lie triple products.

Theorem 5.6.20 *[55, Theorem 2.1] Let $\mathfrak{A} = \mathrm{Tri}(\mathcal{A}, \mathcal{M}, \mathcal{B})$ be a 2-torsion free triangular algebra. Suppose that $\mathfrak{A}$ satisfies the following conditions:*

(i) $\pi_{\mathcal{A}}(Z(\mathfrak{A})) = Z(\mathcal{A})$ *and* $\pi_{\mathcal{B}}(Z(\mathfrak{A})) = Z(\mathcal{B})$,

(ii) $\mathcal{A}$ *or* $\mathcal{B}$ *satisfies* (5.4).

Then, every nonlinear generalized Lie triple derivation $G_L : \mathfrak{A} \to \mathfrak{A}$ is the sum of an additive generalized derivation and a map from $\mathfrak{A}$ into its center $Z(\mathfrak{A})$ which annihilates all Lie triple products.

Lin [407] extended the above result to the case of nonlinear generalized Lie n-derivations of triangular algebras and proved the following theorem.

Theorem 5.6.21 *[407, Theorem 3.3] Let $\mathfrak{A} = \mathrm{Tri}(\mathcal{A}, \mathcal{M}, \mathcal{B})$ be a $(n-1)$-torsion free triangular algebra. Suppose that $\mathfrak{A}$ satisfies the following conditions:*

(i) $\pi_{\mathcal{A}}(Z(\mathfrak{A})) = Z(\mathcal{A})$ *and* $\pi_{\mathcal{B}}(Z(\mathfrak{A})) = Z(\mathcal{B})$,

(ii) $\mathcal{A}$ *or* $\mathcal{B}$ *satisfies* (5.4).

Then, every nonlinear generalized Lie n-derivation $G_L : \mathfrak{A} \to \mathfrak{A}$ is the sum of an additive generalized derivation and a map from $\mathfrak{A}$ into its center $Z(\mathfrak{A})$ which annihilates all $(n-1)$th commutators.

Akhtar et al. [3] described the form of generalized Lie triple derivations of generalized matrix algebras and obtained some necessary and sufficient conditions for a generalized Lie triple derivation on a generalized matrix algebra to be the sum of a generalized derivation and a linear central mapping annihilating all second commutators.

Theorem 5.6.22 *[3, Corollary 3.1] Let $G = G(\mathcal{A}, \mathcal{M}, \mathcal{N}, \mathcal{B})$ be a generalized matrix algebra. An R-linear mapping $G_L : G \to G$ is a generalized Lie triple derivation with associated Lie triple derivation L on G (of the form given in Theorem 5.6.14) iff G_L has the form,*

$$G_L\begin{pmatrix} a & m \\ n & b \end{pmatrix} = \begin{pmatrix} R_{11}(a) + S_{11}(b) - mn_0 - m_0n & am_0 - m_0b + T_{12}(m) + U_{12}(n) \\ n_0a - bn_0 + T_{21}(m) + U_{21}(n) & R_{22}(a) + S_{22}(b) + n_0m + nm_0 \end{pmatrix}$$

for all $\begin{pmatrix} a & m \\ n & b \end{pmatrix} \in G$, *where* $m_0 \in \mathcal{M}$, $n_0 \in \mathcal{N}$; $R_{11} : \mathcal{A} \to \mathcal{A}$, $S_{11} : \mathcal{B} \to \mathcal{A}$, $T_{12} : \mathcal{M} \to \mathcal{M}$, $U_{12} : \mathcal{N} \to \mathcal{M}$, $T_{21} : \mathcal{M} \to \mathcal{N}$, $U_{21} : \mathcal{N} \to \mathcal{N}$, $R_{22} : \mathcal{A} \to \mathcal{B}$ *and* $S_{22} : \mathcal{B} \to \mathcal{B}$ *are R-linear mappings satisfying the following conditions:*

(i) R_{11} *is a generalized Lie triple derivation on* $\mathcal{A}$ *and* $R_{11}(mn) - S_{11}(nm) = T_{12}(m)n + mD_{21}(n)$;

(ii) S_{22} *is a generalized Lie triple derivation on* $\mathcal{B}$ *and* $S_{22}(nm) - R_{22}(mn) = U_{21}(n)m + nC_{12}(m)$;

(*iii*) $T_{12}(am) = R_{11}(a)m - mR_{22}(a) + aC_{12}(m)$, $T_{12}(mb) = T_{12}(m)b + mB_{22}(b) - B_{11}(b)m$, $T_{21}(am) = C_{21}(m)a$ *and* $T_{21}(mb) = bT_{21}(m)$;

(*iv*) $U_{12}(na) = aU_{12}(n)$, $U_{12}(bn) = D_{12}(n)b$, $U_{21}(na) = U_{21}(n)a + nA_{11}(a) - A_{22}(a)n$ *and* $U_{21}(bn) = S_{22}(b)n - nS_{11}(b) + bD_{21}(n)$;

(*v*) $[[R_{22}(a), b_1], b_2] = 0$ *and* $[[S_{11}(b), a_1], a_2] = 0$.

Theorem 5.6.23 *[3, Theorem 4.2] Let $G = G(\mathcal{A}, \mathcal{M}, \mathcal{N}, \mathcal{B})$ be a generalized matrix algebra. A generalized Lie triple derivation G_L on G (of the form given in Theorem 5.6.22) is proper iff there exist linear mappings $\ell_{\mathcal{A}} : \mathcal{A} \to Z(\mathcal{A})$ and $\ell_{\mathcal{B}} : \mathcal{B} \to Z(\mathcal{B})$ satisfying the following conditions:*

(*i*) $\ell_{\mathcal{A}}(a) \oplus R_{22}(a) \in Z(G)$ *and* $S_{11}(b) \oplus \ell_{\mathcal{B}}(b) \in Z(G)$ *for all* $a \in \mathcal{A}$, $b \in \mathcal{B}$;

(*ii*) $\ell_{\mathcal{A}}(mn) = S_{11}(nm)$ *and* $R_{22}(mn) = \ell_{\mathcal{B}}(nm)$ *for all* $m \in \mathcal{M}$, $n \in \mathcal{N}$,

(*iii*) $U_{12}(n) = 0$ *and* $T_{21}(m) = 0$ *for all* $m \in \mathcal{M}$ *and* $n \in \mathcal{N}$.

Motivated by the intensive work of Benkovič [115], Ashraf and Ansari [51] investigated nonlinear generalized Lie n-derivations on unital algebras (precisely, on unital rings) with nontrivial idempotents with property (5.6), which not only extended Theorems 5.6.20 and 5.6.21 but also obtained a description of nonlinear generalized Lie n-derivations on some operator algebras.

Let $G_L : \mathcal{A} \to \mathcal{A}$ be a nonlinear generalized Lie n-derivation with associated nonlinear Lie n-derivation $L : \mathcal{A} \to \mathcal{A}$. Set $\Phi = G_L - L$. Then, it follows that Φ satisfies

$$\Phi(p_n(a_1, a_2, \ldots, a_n)) = p_n(\Phi(a_1), a_2, \ldots, a_n) \tag{5.7}$$

for all $a_1, a_2, \ldots, a_n \in \mathcal{A}$. Therefore, to characterize a nonlinear generalized Lie n-derivation G_L, it suffices to consider a mapping Φ satisfying (5.7).

Remark 5.6.24 *Any mapping $\Phi : \mathcal{A} \to \mathcal{A}$ satisfying* (5.7) *is called a nonlinear Lie n-centralizer. In particular, a nonlinear Lie 2-centralizer is said to be a nonlinear Lie centralizer. In the year 2019, Fošner and Jing [270] introduced the notion of a nonlinear Lie centralizer and investigated the additivity of nonlinear Lie centralizers of triangular rings. Recently, Liu studied nonlinear Lie centralizers of generalized matrix algebras in [419]. In addition, Lie centralizers have been studied in [316].*

Following [115, Proposition 2.2], Ashraf and Ansari [51] proved the following proposition which characterizes a mapping Φ satisfying (5.7) (that is, nonlinear Lie n-centralizer) on unital algebras with nontrivial idempotents.

Proposition 5.6.25 *Let $\mathcal{A}$ be a unital algebra with a nontrivial idempotent e satisfying* (5.6). *Suppose that*

(*i*) $Z(e\mathcal{A}e) = Z(\mathcal{A})e$ *and* $Z(f\mathcal{A}f) = Z(\mathcal{A})f$,

(ii) *Either $e\mathcal{A}e$ or $f\mathcal{A}f$ does not contain nonzero central ideals.*

If a mapping $\Phi : \mathcal{A} \to \mathcal{A}$ satisfies

$$\Phi(p_n(a_1, a_2, \ldots, a_n)) = p_n(\Phi(a_1), a_2, \ldots, a_n)$$

for all $a_1, a_2, \ldots, a_n \in \mathcal{A}$, then $\Phi(a) = \lambda a + \gamma(a)$ for all $a \in \mathcal{A}$, where $\lambda \in Z(\mathcal{A})$ and $\gamma : \mathcal{A} \to Z(\mathcal{A})$ is a mapping such that $\gamma(p_n(a_1, a_2, \ldots, a_n)) = 0$ for all $a_1, a_2, \ldots, a_n \in \mathcal{A}$.

Now, applying Proposition 5.6.25, we are in a position to state and prove the following theorem which gives a description of nonlinear generalized Lie n-derivations on unital algebras.

Theorem 5.6.26 *[51, Theorem 3.1] Let $\mathcal{A}$ be a unital algebra with a nontrivial idempotent e satisfying* (5.6). *Let us assume that*

(i) $Z(e\mathcal{A}e) = Z(A)e$ *and* $Z(f\mathcal{A}f) = Z(A)f$;

(ii) *Either $e\mathcal{A}e$ or $f\mathcal{A}f$ contains no central ideals.*

Then, every nonlinear generalized Lie n-derivation $G_L : \mathcal{A} \to \mathcal{A}$ is of the form $G_L(a) = \lambda a + \delta(a)$ for all $a \in \mathcal{A}$, where $\lambda \in Z(A)$ and $\delta : \mathcal{A} \to \mathcal{A}$ is a nonlinear Lie n-derivation.

Proof. Lets assume that $G_L : \mathcal{A} \to \mathcal{A}$ is a nonlinear generalized Lie n-derivation with an associated nonlinear Lie n-derivation $L : \mathcal{A} \to \mathcal{A}$. Then, $\Phi = G_L - L$ is a mapping satisfying

$$\Phi(p_n(a_1, a_2, \ldots, a_n)) = p_n(\Phi(a_1), a_2, \ldots, a_n)$$

for all $a_1, a_2, \ldots, a_n \in \mathcal{A}$. Hence, it follows from Proposition 5.6.25 that there exist a $\lambda \in Z(\mathcal{A})$ and a mapping $\gamma : \mathcal{A} \to Z(\mathcal{A})$ vanishing at $p_n(a_1, a_2, \ldots, a_n)$ for all $a_1, a_2, \ldots, a_n \in \mathcal{A}$ such that $\Phi(a) = \lambda a + \gamma(a)$ for all $a \in \mathcal{A}$. Therefore,

$$G_L(a) = \Phi(a) + L(a) = \lambda a + \delta(a) \quad \text{for all} \quad a \in \mathcal{A},$$

where $\lambda \in Z(\mathcal{A})$ and $\delta = \gamma + L$ is a nonlinear Lie n-derivation. This completes the proof of the theorem.

Remark 5.6.27 *The above result reduces the description of a nonlinear generalized Lie n-derivation to that of a nonlinear Lie n-derivation. For example, using the description of nonlinear Lie n-derivations on triangular algebras [121, Theorem 5.9] in the above result, one can obtain the result of Lin [407, Theorem 3.3].*

Let $\mathcal{A}$ be a unital algebra with a nontrivial idempotent e such that $e\mathcal{A}f$ is a faithful $(e\mathcal{A}e, f\mathcal{A}f)$-bimodule and $f\mathcal{A}e = \{0\}$. Then, $\mathcal{A} = e\mathcal{A}e + e\mathcal{A}f + f\mathcal{A}f$ is a triangular algebra. It is easy to see that $\mathcal{A}$ satisfies the property (5.6). Therefore, in view of Theorem 5.6.26 and [121, Theorem 5.9], we obtain the following result, which extends [55, Theorem 5.1] and [407, Theorem 3.3].

Corollary 5.6.1 *Let $\mathcal{A} = e\mathcal{A}e + e\mathcal{A}f + f\mathcal{A}f$ be a $(n-1)$-torsion free triangular algebra. Suppose that*

(i) $Z(e\mathcal{A}e) = Z(\mathcal{A})e$ *and* $Z(f\mathcal{A}f) = Z(\mathcal{A})f$,

(ii) *Either $e\mathcal{A}e$ or $f\mathcal{A}f$ satisfies* (5.5).

Then, every nonlinear generalized Lie n-derivation $G_L : \mathcal{A} \to \mathcal{A}$ can be expressed as $G_L(a) = \lambda a + d(a) + \gamma(a)$ for all $a \in \mathcal{A}$, where $\lambda \in Z(\mathcal{A})$, $d : \mathcal{A} \to \mathcal{A}$ is an additive derivation and $\gamma : \mathcal{A} \to Z(\mathcal{A})$ is a mapping such that $\gamma(p_n(a_1, a_2, \ldots, a_n)) = 0$ for all $a_1, a_2, \ldots, a_n \in \mathcal{A}$.

Let $\mathcal{A}$ be a unital algebra and $M_s(\mathcal{A})$ be the algebra of all $s \times s$ matrices over $\mathcal{A}$, where $s \geq 2$ is an integer. Then, $M_s(\mathcal{A})$ can be written as

$$M_s(\mathcal{A}) = \begin{pmatrix} \mathcal{A} & M_{1\times(s-1)}(\mathcal{A}) \\ M_{(s-1)\times 1}(\mathcal{A}) & M_{s-1}(\mathcal{A}) \end{pmatrix}.$$

Set $e = e_{11}$, a matrix unit and $f = 1 - e_1$. Then, it is easy to see that $M_s(\mathcal{A})$ satisfies (5.6). Moreover, $eM_s(\mathcal{A})e$ is isomorphic to $\mathcal{A}$, $fM_s(\mathcal{A})f$ is isomorphic to $M_{s-1}(\mathcal{A})$, $eM_s(\mathcal{A})f$ is a faithful $(eM_s(\mathcal{A})e, fM_s(\mathcal{A})f)$-bimodule isomorphic to $M_{(s-1)\times 1}(\mathcal{A})$. Note that $Z(M_s(\mathcal{A})) = Z(\mathcal{A}) \cdot I_s$, where I_s is the identity of $M_s(\mathcal{A})$. Thus, assumption (i) of Theorem 5.6.26 is satisfied. Moreover, $M_s(\mathcal{A})(s \geq 2)$ does not contain nonzero central ideals (see [255, Lemma 1]). Hence, assumption (ii) of Theorem 5.6.26 is also satisfied if $s \geq 3$. Therefore, as a direct consequence of Theorem 5.6.26 and [546, Corollary 2], we have the following result:

Corollary 5.6.2 *Let $\mathcal{A}$ be a $(n-1)$-torsion free unital algebra and $M_s(\mathcal{A})$ be a matrix algebra with $s \geq 3$. Then, every nonlinear generalized Lie n-derivation $G_L : M_s(\mathcal{A}) \to M_r(\mathcal{A})$ can be expressed as $G_L(A) = \lambda A + d(A) + \gamma(A)$ for all $A \in M_s(\mathcal{A})$, where $\lambda \in Z(\mathcal{A})$, $d : M_s(\mathcal{A}) \to M_s(\mathcal{A})$ is an additive derivation and $\gamma : M_s(\mathcal{A}) \to Z(\mathcal{A}) \cdot I_s$ is a mapping such that $\gamma(p_n(A_1, A_2, \ldots, A_n)) = 0$ for all $A_1, A_2, \ldots, A_n \in M_s(\mathcal{A})$.*

Recall that an algebra $\mathcal{A}$ is a called a prime algebra if, for any $a, b \in \mathcal{A}$, $a\mathcal{A}b = \{0\}$ implies that either $a = 0$ or $b = 0$. It is easy to see that every prime algebra with a nontrivial idempotent satisfies the property (5.6).

Let $\mathfrak{X}$ be a Banach space over the complex field $\mathbb{C}$ with $dim\mathfrak{X} \geq 2$ and $\mathfrak{B} = \mathcal{B}(\mathfrak{X})$ be the algebra of all bounded linear operators on $\mathfrak{X}$. Then, $\mathfrak{B}$ contains a nontrivial idempotent e and hence can be written as $\mathfrak{B} = e\mathfrak{B}e + e\mathfrak{B}f + f\mathfrak{B}e + f\mathfrak{B}f$. Since $\mathfrak{B}$ is a prime algebra, $\mathfrak{B}$ satisfies (5.6). Moreover, $\mathfrak{B}$, $e\mathfrak{B}e$ and $f\mathfrak{B}f$ are central algebras over $\mathbb{C}$ and hence $eZ(\mathfrak{B})e = Z(e\mathfrak{B}e) = \mathbb{C}e$ and $fZ(\mathfrak{B})f = Z(f\mathfrak{B}f) = \mathbb{C}f$. Thus, all the assumptions of Theorem 5.6.26 are fulfilled. Therefore, in view of [546, Theorem 3.1], we obtain:

Corollary 5.6.3 *Let $\mathfrak{X}$ be a complex Banach space with $dim\mathfrak{X} \geq 2$ and $\mathfrak{B} = \mathcal{B}(\mathfrak{X})$ be the algebra of all bounded linear operators on $\mathfrak{X}$. Then, every nonlinear generalized Lie n-derivation $G_L : \mathfrak{B} \to \mathfrak{B}$ can be expressed as $G_L(A) = \lambda A + d(A) + \gamma(A)$ for all $A \in \mathfrak{B}$, where $\lambda \in \mathbb{C}$, $d : \mathfrak{B} \to \mathfrak{B}$ is an additive derivation and $\gamma : \mathfrak{B} \to \mathbb{C}\mathit{1}$ is a mapping such that $\gamma(p_n(A_1, A_2, \ldots, A_n)) = 0$ for all $A_1, A_2, \ldots, A_n \in \mathfrak{B}$.*

Now, let us introduce the 'higher version' of Lie-type derivations and discuss it on some well-known algebras. Taking into account the definitions of higher derivations and Lie-type derivations, Han [284] introduced a more general concept known as *Lie-type higher derivations.*

Definition 81 *Let $\mathfrak{L} = \{L_r\}_{r\in\mathbb{N}}$ be a family of R-linear mappings $L_r : \mathcal{A} \to \mathcal{A}$ such that $L_0 = id_{\mathcal{A}}$. Then, $\mathfrak{L}$ is called a Lie n-higher derivation if for each $r \in \mathbb{N}$,*

$$L_r(p_n(a_1, a_2, \ldots, a_n)) = \sum_{i_1+i_2+\cdots+i_n=r} p_n(L_{i_1}(a_1), L_{i_2}(a_2), \ldots, L_{i_n}(a_n))$$

holds for all $a_1, a_2, \ldots, a_n \in \mathcal{A}$.

Remark 5.6.28 *In particular, a Lie 2-higher derivation is called a Lie higher derivation and a Lie 3-higher derivation is said to be a Lie triple higher derivation. Lie 2-higher derivations, Lie 3-higher derivations and Lie n-higher derivations are collectively referred to as Lie-type higher derivations. If $\mathfrak{L} = \{L_r\}_{r\in\mathbb{N}}$ is a family of not necessarily linear mappings in the above definitions, then $\mathfrak{L}$ is said to be a nonlinear Lie n-higher derivation. Note that L_1 is always a Lie n-derivation if $\mathfrak{L} = \{L_r\}_{r\in\mathbb{N}}$ is a Lie n-higher derivation. Furthermore, every higher derivation is a Lie n-higher derivation for each $n \geq 2$ and every Lie n-higher derivation is a Lie $(n+k(n-1))$-higher derivation for each $k \in \mathbb{N}$. However, the converse statements are not true in general. Assume that $D = \{d_n\}_{n\in\mathbb{N}}$ is a higher derivation on an algebra $\mathcal{A}$. For any $n \in \mathbb{N}$, let $L_n = d_n + \gamma_n$, where $\{\gamma_n\}_{n\in\mathbb{N}}$ is a sequence of R-linear mappings from $\mathcal{A}$ into its center $Z(\mathcal{A})$ and each γ_n vanishes on all commutators. It is easy to see that $\{L_n\}_{n\in\mathbb{N}}$ is a Lie higher derivation, but not a higher derivation if $\gamma_n \neq 0$ for some n. Likewise, if $\{\gamma_n\}_{n\in\mathbb{N}}$ is a sequence of R-linear mappings from $\mathcal{A}$ into its center $Z(\mathcal{A})$ and each γ_n vanishes on all second commutators, then $\{L_n\}_{n\in\mathbb{N}}$ is a Lie triple higher derivation, but not a Lie higher derivation if $\gamma_n \neq 0$ for some n. A Lie n-higher derivation $\mathfrak{L} = \{L_r\}_{r\in\mathbb{N}}$ is said to be of the standard form if for each $r \in \mathbb{N}$, $L_r = d_r + \gamma_r$, where $\{d_r\}_{r\in\mathbb{N}}$ is a higher derivation and $\{\gamma_r\}_{r\in\mathbb{N}}$ is a family of R-linear mappings $\gamma_r : \mathcal{A} \to Z(\mathcal{A})$ vanishing on all $(n-1)$th commutators of $\mathcal{A}$.*

The fundamental question in the realm of Lie-type higher derivations is under what conditions on an algebra a Lie-type higher derivation is of the standard form. Many authors have studied this problem for various algebras; see for example [251, 284, 286, 416, 490, 494, 554, 559, 560] and references therein. In this context, the following is very interesting in the sense that it enables us to transfer the problems of characterizing Lie-type higher derivations into that of Lie-type derivations.

Theorem 5.6.29 *[251, Theorem 3.1] Let $\mathcal{A}$ be a unital torsion-free algebra over a unital commutative ring R. If every Lie n-derivation of $\mathcal{A}$ is standard, then every Lie n-higher derivation of $\mathcal{A}$ is standard.*

Qi and Hou [494] described the structure of Lie higher derivations on nest algebras. Li and Shen [416] extended the main results of [494] for a triangular algebra by providing some sufficient conditions under which a Lie higher derivation on a triangular algebra is of the standard form.

Theorem 5.6.30 *[416, Theorem 3.1] Let $\mathfrak{A} = \mathrm{Tri}(\mathcal{A}, \mathcal{M}, \mathcal{B})$ be a triangular algebra such that $Z(\mathcal{A}) = \pi_{\mathcal{A}}(Z(\mathfrak{A}))$ and $Z(\mathcal{B}) = \pi_{\mathcal{B}}(Z(\mathfrak{A}))$. Then, every Lie higher derivation on $\mathfrak{A}$ is proper. That is to say, $\mathfrak{L} = \{L_n\}_{n\in\mathbb{N}}$ is a Lie higher derivation on $\mathfrak{A}$ iff for each $n \in \mathbb{N}$, L_n has the form $L_n = d_n + \gamma_n$, where $\{d_n\}_{n\in\mathbb{N}}$ is a higher derivation and $\{\gamma_n\}_{n\in\mathbb{N}}$ is a sequence of R-linear mappings $\gamma_n : \mathfrak{A} \to Z(\mathfrak{A})$ vanishing on all commutators of $\mathfrak{A}$.*

The subsequent result shows that the above theorem is also true for nonlinear Lie higher derivations of triangular algebras.

Theorem 5.6.31 *[561, Theorem 3.1] Let $\mathfrak{A} = \mathrm{Tri}(\mathcal{A}, \mathcal{M}, \mathcal{B})$ be a triangular algebra and $\mathfrak{L} = \{L_n\}_{n\in\mathbb{N}}$ be a nonlinear Lie higher derivation on $\mathfrak{A}$. If $Z(\mathcal{A}) = \pi_{\mathcal{A}}(Z(\mathfrak{A}))$ and $Z(\mathcal{B}) = \pi_{\mathcal{B}}(Z(\mathfrak{A}))$, then $\mathfrak{L} = \{L_n\}_{n\in\mathbb{N}}$ is of standard form. More precisely, each component L_n is of the form $L_n = d_n + \gamma_n$ for all $A \in \mathfrak{A}$, where $\{d_n\}_{n\in\mathbb{N}}$ is an additive higher derivation on $\mathfrak{A}$ and $\{\gamma_n\}_{n\in\mathbb{N}}$ is a sequence of mappings $\gamma_n : \mathfrak{A} \to Z(\mathfrak{A})$ vanishing on all commutators of $\mathfrak{A}$.*

It has been proved in [324, 416] that under certain conditions every (nonlinear) Lie triple derivation on triangular algebras has the standard form. It is natural to ask whether (nonlinear-) Lie triple higher derivations on triangular algebras are of the standard form. This question was addressed by Ashraf and Jabeen in [57], where nonlinear Lie triple higher derivations on triangular algebras were characterized.

Theorem 5.6.32 *[57, Theorem 3.1] Let $\mathcal{A}$ and $\mathcal{B}$ be two unital algebras over a 2-torsion free commutative ring R, and $\mathcal{M}$ be a faithful $(\mathcal{A}, \mathcal{B})$-bimodule. Suppose that $\mathfrak{A} = \mathrm{Tri}(\mathcal{A}, \mathcal{M}, \mathcal{B})$ is a triangular algebra satisfying the following conditions:*

(i) $\pi_{\mathcal{A}}(Z(\mathfrak{A})) = Z(\mathcal{A})$ *and* $\pi_{\mathcal{B}}(Z(\mathfrak{A})) = Z(\mathcal{B})$,

(ii) $\mathcal{A}$ *or* $\mathcal{B}$ *satisfies* (5.4).

Then, every nonlinear Lie triple higher derivation $\mathfrak{L} = \{L_n\}_{n\in\mathbb{N}}$ on $\mathfrak{A}$ is of the standard form. That is, there exist an additive higher derivation $\{d_n\}_{n\in\mathbb{N}}$ on $\mathfrak{A}$

and a sequence $\{\gamma_n\}_{n\in\mathbb{N}}$ *of mappings* $\gamma_n : \mathfrak{A} \to Z(\mathfrak{A})$ *vanishing on all second commutators of* $\mathfrak{A}$ *such that for each* $n \in \mathbb{N}$, $L_n = d_n + \gamma_n$.

Motivated by the intensive and powerful work concerning linear mappings of operator algebras (for examples, [1, 333, 453, 460, 461]), Han introduced the concept of Lie-type higher derivations on operator algebras and proposed some potential research topics in [284]. Han mainly studied Lie (triple) higher derivations on operator algebras and showed that every Lie (triple) higher derivation on some classical operator algebras is of the standard form.

Theorem 5.6.33 *[284, Corollary 3.9] Let* $\mathfrak{M}$ *be a von Neumann algebra. Then, every Lie higher derivation on* $\mathfrak{M}$ *has the standard form. In particular, every Lie higher derivation on a properly infinite von Neumann algebra is a higher derivation.*

Theorem 5.6.34 *[284, Corollary 3.10] Let* $\mathfrak{M}$ *be a von Neumann algebra with no abelian summands. Then, every Lie triple higher derivation on* $\mathfrak{M}$ *is of the standard form.*

We have seen that the problem of describing the form of Lie-type derivations has been studied extensively on different kinds of algebras. One direction in the study of this problem is to study conditions under which Lie-type derivations of an algebra $\mathcal{A}$ can be completely determined by action on some subsets of the algebra $\mathcal{A}$. In the year 2010, Lu and Jing [433] initiated the study of Lie derivations of operator algebras by local actions and proved that if $\mathfrak{B}(\mathfrak{X})$ is the algebra of all bounded linear operators acting on a complex Banach space $\mathfrak{X}$ with $dim\mathfrak{X} \geq 2$ and $L : \mathfrak{B}(\mathfrak{X}) \to \mathfrak{B}(\mathfrak{X})$ a linear mapping satisfying $L([A,B]) = [L(A),B] + [A,L(B)]$ for all $A,B \in \mathfrak{B}(\mathfrak{X})$ with $AB = 0$ (resp. $AB = P$, where P is a fixed nontrivial idempotent), then $L = d + \gamma$, where d is a derivation of $\mathfrak{B}(\mathfrak{X})$ and $\gamma : \mathfrak{B}(\mathfrak{X}) \to \mathbb{C}1$ is a linear mapping vanishing at commutators $[A,B]$ with $AB = 0$ (resp. $AB = P$). Motivated by this result, Ji and Qi [330] investigated the same question on triangular algebras, which was further generalized by Lin [406] in the setting of Lie higher derivations of triangular algebras. Later on, many authors obtained similar results in different settings (for example, see [38, 58, 331, 419, 495, 496] and references therein).

Recently, Ashraf et al. [50] investigated the above problem in the setting of Lie-type higher derivations on triangular algebras and proved the following theorem.

Theorem 5.6.35 *[50, Theorem 3.1] Let* $\mathfrak{A} = \mathrm{Tri}(\mathcal{A},\mathfrak{M},\mathcal{B})$ *be a* $(n-1)$*-torsion free triangular algebra such that*

(*i*) $\pi_{\mathcal{A}}(Z(\mathfrak{A})) = Z(\mathcal{A})$ *and* $\pi_{\mathcal{B}}(Z(\mathfrak{A})) = Z(\mathcal{B})$,

(*ii*) $\mathcal{A}$ *or* $\mathcal{B}$ *satisfies* (5.4).

Suppose that $\{L_r\}_{r\in\mathbb{N}}$ is a family of linear mappings $L_r : \mathfrak{A} \to \mathfrak{A}$ such that for each $r \in \mathbb{N}$,

$$L_r(p_n(A_1,A_2,\ldots,A_n)) = \sum_{i_1+i_2+\cdots+i_n=r} p_n(L_{i_1}(A_1),L_{i_2}(A_2),\ldots,L_{i_n}(A_n))$$

for all $A_1,A_2,\ldots,A_n \in \mathfrak{A}$ with $A_1A_2\cdots A_n = 0$. Then, for each $r \in \mathbb{N}$, $L_r = d_r + \gamma_r$, where $\{d_r\}_{r\in\mathbb{N}}$ is a higher derivation on $\mathfrak{A}$ and $\{\gamma_r\}_{r\in\mathbb{N}}$ is a family of linear mappings $\gamma_r : \mathfrak{A} \to Z(\mathfrak{A})$ vanishing at every $(n-1)$th commutator $p_n(A_1,A_2,\ldots,A_n)$ with $A_1A_2\cdots A_n = 0$.

Xiao and Wei [561] studied nonlinear Lie higher derivations of triangular algebras and proved that under certain restrictions every nonlinear Lie higher derivation of a triangular algebra is of the standard form. Ashraf and Jabeen [57] extended this result to the case of nonlinear Lie triple higher derivations of triangular algebras. Recently, Ashraf et al. [26] further extended these results to nonlinear Lie n-higher derivations of triangular algebras. More precisely, they proved the following result:

Theorem 5.6.36 *[26, Theorem 3.1] Let $\mathfrak{A} = \text{Tri}(\mathcal{A},\mathcal{M},\mathcal{B})$ be a $(n-1)$-torsion free triangular algebra such that*

(i) *$\pi_{\mathcal{A}}(Z(\mathfrak{A})) = Z(\mathcal{A})$ and $\pi_{\mathcal{B}}(Z(\mathfrak{A})) = Z(\mathcal{B})$,*

(ii) *$\mathcal{A}$ or $\mathcal{B}$ satisfies* (5.4).

Suppose that $\{L_r\}_{r\in\mathbb{N}}$ is a family of mappings $L_r : \mathfrak{A} \to \mathfrak{A}$ (not necessarily linear) such that for each $r \in \mathbb{N}$,

$$L_r(p_n(A_1,A_2,\ldots,A_n)) = \sum_{i_1+i_2+\cdots+i_n=r} p_n(L_{i_1}(A_1),L_{i_2}(A_2),\ldots,L_{i_n}(A_n))$$

for all $A_1,A_2,\ldots,A_n \in \mathfrak{A}$. Then, for each $r \in \mathbb{N}$, $L_r = d_r + \gamma_r$, where $\{d_r\}_{r\in\mathbb{N}}$ is an additive higher derivation on $\mathfrak{A}$ and $\{\gamma_r\}_{r\in\mathbb{N}}$ a family of mappings $\gamma_r : \mathfrak{A} \to Z(\mathfrak{A})$ vanishing on all $(n-1)$th commutators of $\mathfrak{A}$.

In an attempt to obtain the higher version of Theorem 5.6.16, Han and Wei [286] studied nonlinear Lie higher derivations on unital algebras with nontrivial idempotents (which generalizes the case with $n = 2$ of Theorem 5.6.16) while Ashraf et al. [41] investigated nonlinear Lie triple higher derivations of generalized matrix algebras (which generalizes the case with $n = 3$ of Theorem 5.6.16).

Theorem 5.6.37 *[286, Theorem 4.1] Let $\mathcal{A}$ be a unital algebra with a nontrivial idempotent e satisfying* (5.6) *and $\{L_k\}_{k\in\mathbb{N}}$ be a nonlinear Lie higher derivation on $\mathcal{A}$. Let us assume that*

(i) *$Z(e\mathcal{A}e) = Z(A)e$ and $Z(f\mathcal{A}f) = Z(A)f$;*

(ii) *Either $e\mathcal{A}e$ or $f\mathcal{A}f$ contains no central ideal.*

(iii) *For each* $m \in e\mathcal{A}f$ *the condition* $m \cdot f\mathcal{A}e = \{0\} = f\mathcal{A}e \cdot m$ *implies* $m = 0$,

(iv) *For each* $t \in f\mathcal{A}e$ *the condition* $e\mathcal{A}f \cdot t = \{0\} = t \cdot e\mathcal{A}f$ *implies* $t = 0$,

(v) *For each* $k \in \mathbb{N}$, $eL_k(e)f = 0$ *and* $fL_k(e)e = 0$.

Then, $\{L_k\}_{k\in\mathbb{N}}$ *is of the standard form, that is, for each* $k \in \mathbb{N}$, $L_k = d_k + \gamma_k$, *where* $\{d_k\}_{k\in\mathbb{N}}$ *is an additive higher derivation on* $\mathfrak{A}$ *and* $\{\gamma_k\}_{k\in\mathbb{N}}$ *is a family of mappings* $\gamma_k : \mathfrak{A} \to Z(\mathfrak{A})$ *such that* $\gamma_k([a,b]) = 0$ *for all* $a, b \in \mathcal{A}$.

Theorem 5.6.38 *[41, Theorem 2.1] Let* $G = G(\mathcal{A}, \mathcal{M}, \mathcal{N}, \mathcal{B})$ *be a 2-torsion free generalized matrix algebra such that:*

(i) $Z(\mathcal{A}) = \pi_{\mathcal{A}}(Z(\mathfrak{A}))$ *and* $Z(\mathcal{B}) = \pi_{\mathcal{B}}(Z(\mathfrak{A}))$, *and*

(ii) *Either* $\mathcal{A}$ *or* $\mathcal{B}$ *does not contain nonzero central ideals,*

(iii) $\mathcal{A}$ *or* $\mathcal{B}$ *satisfies* (5.4),

(iv) *If* n *is odd and* $\mathcal{N} \neq \{0\}$, *then for each* $m \in \mathcal{M}$ *the condition* $m\mathcal{N} = \{0\} = \mathcal{N}m$ *implies* $m = 0$,

(v) *If* n *is odd, then for each* $t \in \mathcal{N}$ *the condition* $\mathcal{M}t = \{0\} = t\mathcal{M}$ *implies* $t = 0$.

Then, every nonlinear Lie triple higher derivation is of the standard form, that is, for each $k \in \mathbb{N}$, $L_k = d_k + \gamma_k$, *where* $\{d_k\}_{k\in\mathbb{N}}$ *is an additive higher derivation on* G *and* $\{\gamma_k\}_{k\in\mathbb{N}}$ *is a family of mappings* $\gamma_k : G \to Z(G)$ *such that* $\gamma_k([[A,B],C]) = 0$ *for all* $A, B, C \in G$.

We have seen that (nonlinear-) Lie (triple) higher derivations have been discussed extensively on some algebras (see, for examples, [85, 86, 284, 286, 416, 490, 494, 549, 561, 569] where further references can be found). However, less attention has been paid to the structure of (nonlinear-) Lie-type higher derivations on algebras. To the best of our knowledge, there are no other articles dealing with nonlinear Lie-type higher derivations except for [26, 50, 85, 251].

It is natural to ask the following:

Open Problems 5.6.39 (i) *Do Theorems 5.6.37 and 5.6.38 hold for nonlinear Lie-type higher derivations on generalized matrix algebras (or unital algebras with nontrivial idempotents)?*

(ii) *Do Theorems 5.6.37 and 5.6.38 hold for nonlinear Lie-type higher derivations on arbitrary algebras?*

Combining the notions of higher derivations and generalized Lie-type derivations, one can introduce the following:

Definition 82 *Let* $\mathcal{A}$ *be an algebra and* $\mathfrak{F} = \{\xi_r\}_{r\in\mathbb{N}}$ *be a family of R-linear mappings on* $\mathcal{A}$ *such that* $\xi_0 = id_{\mathcal{A}}$. *Then,* $\mathfrak{F}$ *is called a generalized Lie n-higher* *derivation if there exists a Lie n-higher derivation* $\mathfrak{L} = \{L_r\}_{r\in\mathbb{N}}$ *on* $\mathcal{A}$ *such that for each* $r \in \mathbb{N}$,

$$\xi_r(p_n(a_1,a_2,\ldots,a_n)) = \sum_{i_1+i_2+\cdots+i_n=r} p_n(\xi_{i_1}(a_1),L_{i_2}(a_2),\ldots,L_{i_n}(a_n))$$

holds for all $a_1,a_2,\ldots,a_n \in \mathcal{A}$.

Remark 5.6.40 *In particular, a generalized Lie 2-higher derivation is called a generalized Lie higher derivation and a generalized Lie 3-higher derivation is said to be a generalized Lie triple higher derivation. Generalized Lie 2-higher derivations, generalized Lie 3-higher derivations, and generalized Lie n-higher derivations can be collectively termed as generalized Lie-type higher derivations. If we drop the linearity condition in the above definition, then* $\mathfrak{F}$ *is called a nonlinear generalized Lie n-higher derivation on* $\mathcal{A}$.

Ashraf and Jabeen also introduced the notions of a nonlinear generalized Lie triple higher derivation in [56] and extended Theorem 5.6.20 to the case of nonlinear generalized Lie triple higher derivations on triangular algebras.

Theorem 5.6.41 *[56, Theorem 3.1] Let* $\mathfrak{A} = \mathrm{Tri}(\mathcal{A},\mathcal{M},\mathcal{B})$ *be a triangular algebra. Suppose that* $\mathfrak{A}$ *satisfies the following conditions:*

(*i*) $\pi_{\mathcal{A}}(Z(\mathfrak{A})) = Z(\mathcal{A})$ *and* $\pi_{\mathcal{B}}(Z(\mathfrak{A})) = Z(\mathcal{B})$,

(*ii*) $\mathcal{A}$ *or* $\mathcal{B}$ *satisfies* (5.4).

Then, every nonlinear generalized Lie triple higher derivation $\mathfrak{F} = \{\xi_n\}_{n\in\mathbb{N}}$ *on* $\mathfrak{A}$ *is of the standard form, that is, there exists an additive generalized higher derivation* $D = \{\delta_n\}_{n\in\mathbb{N}}$ *on* $\mathfrak{A}$ *and a sequence* $\{\gamma_n\}_{n\in\mathbb{N}}$ *of mappings from* $\mathfrak{A}$ *into its center which annihilates all Lie triple products such that for each* $n \in \mathbb{N}$, $\xi_n(A) = \delta_n(A) + \gamma_n(A)$ *for all* $A \in \mathfrak{A}$.

However, we believe the above result may be further extended to nonlinear generalized Lie-type derivations. In this context, we propose the following open problems which can be discussed further.

Open Problems 5.6.42 *Does Theorem 5.6.41 hold for*

(*i*) *Nonlinear generalized Lie-type higher derivations on triangular algebras?*

(*ii*) *Nonlinear generalized Lie-type higher derivations on generalized matrix algebras (or unital algebras with nontrivial idempotents)?*

(*iii*) *Nonlinear generalized Lie-type higher derivations on arbitrary algebras?*

We now focus on the study of Lie-type derivations on $*$-algebras over the complex field $\mathbb{C}$. Let $\mathcal{A}$ be an associative $*$-algebra over the complex field $\mathbb{C}$. Recall that an additive mapping $\delta : \mathcal{A} \to \mathcal{A}$ is said to be an additive $*$-derivations if δ is an additive derivation and $\delta(a^*) = \delta(a)^*$ holds for all $a \in \mathcal{A}$. For any $a, b \in \mathcal{A}$, we denote by $[a,b]_\bullet = ab - ba^*$ the skew Lie product ($*$-Lie product) of $a, b \in \mathcal{A}$. Let us define a sequence of polynomials involving the skew Lie product as follows:

$$\begin{aligned} q_1(a_1) &= a_1, \\ q_2(a_1,a_2) &= [q_1(a_1),a_2]_\bullet = [a_1,a_2]_\bullet, \\ &\vdots \qquad\qquad \vdots \\ q_n(a_1,a_2,\ldots,a_n) &= [q_{n-1}(a_1,a_2,\ldots,a_{n-1}),a_n]_\bullet \ (n \geq 2). \end{aligned}$$

Definition 83 *A linear mapping $\phi : \mathcal{A} \longrightarrow \mathcal{A}$ is called a skew Lie n-derivation ($*$-Lie n-derivation) $(n \geq 2)$ if*

$$\phi(q_n(a_1,a_2,\ldots,a_n)) = \sum_{i=1}^{n} q_n(a_1,\ldots,a_{i-1},\phi(a_i),a_{i+1},\ldots,a_n)$$

holds for all $a_1, a_2, \ldots, a_n \in \mathcal{A}$.

Remark 5.6.43 *If the mapping ϕ is not necessarily linear, then it is said to be a nonlinear skew Lie n-derivation. A skew Lie 2-derivation is called a skew Lie derivation and a skew Lie 3-derivation is said to be a skew Lie triple derivation. Skew Lie 2-derivations, skew Lie 3-derivations, and skew Lie n-derivations are collectively referred to as skew Lie-type derivations.*

Skew Lie-type derivations in different backgrounds are extensively studied by several authors (see [325, 401, 428, 564]). Yu and Zhang [564] considered nonlinear skew Lie derivations on factor von Neumann algebras and obtained the following theorem.

Theorem 5.6.44 *[564, Theorem 2.1] Let $\mathcal{M}$ be a factor von Neumann algebra acting on a complex Hilbert space $\mathcal{H}$ with* dim $\mathcal{H} \geq 2$. *Suppose that $\phi : \mathcal{M} \to \mathcal{M}$ is a nonlinear skew Lie derivation. Then, ϕ is an additive $*$-derivation.*

In [401], Li et al. showed that the above result is also true for a von Neumann algebra without nonzero central abelian projections. The result of Yu and Zhang [564] was extended to the case of nonlinear skew Lie triple derivations by Li et al. [428].

Theorem 5.6.45 *[428, Theorem 2.3] Let $\mathcal{M}$ be a factor von Neumann algebra acting on a complex Hilbert space $\mathcal{H}$ with* dim $\mathcal{H} \geq 2$. *Suppose that $\phi : \mathcal{M} \to \mathcal{M}$ is a nonlinear skew Lie triple derivation. Then, ϕ is an additive $*$-derivation.*

Motivated by these results, Lin [409] gave a description of nonlinear $*$-Lie type derivations on von Neumann algebras.

Theorem 5.6.46 *[409, Theorem 3.1] Let $\mathcal{A}$ be a von Neumann algebra without nonzero central abelian projections. Then, a mapping $\phi : \mathcal{A} \to \mathcal{B}(\mathcal{H})$ is a nonlinear $*$-Lie n-derivations if and only if ϕ is an additive $*$-derivation.*

Recall that a standard operator algebra is an algebra of bounded linear operators acting on a complex Hilbert space containing all finite rank operators. Motivated by the study of Yu and Zhang [564], Jing [325] considered nonlinear $*$-Lie derivations on standard operator algebras. Other than showing that each nonlinear $*$-Lie derivation on factor von Neumann algebras is additive in [564], Jing showed that every $*$-Lie derivation on standard operator algebras is automatically linear. Moreover, it is an inner $*$-derivation. The precise statement is as follows:

Theorem 5.6.47 *[325, Theorem 2.14] Let $\mathcal{H}$ be an infinite dimensional complex Hilbert space and $\mathcal{A}$ be a standard operator algebra on $\mathcal{H}$ containing the identity operator I. If $\mathcal{A}$ is closed under the adjoint operation and $\phi : \mathcal{A} \to \mathcal{B}(\mathcal{H})$ is a nonlinear $*$-Lie derivation, then ϕ is a linear $*$-derivation. Moreover, there exists an operator $T \in \mathcal{B}(\mathcal{H})$ satisfying $T + T^* = 0$ such that $\phi(A) = AT - TA$ for all $A \in \mathcal{A}$, that it, ϕ is inner.*

Generalizing the result of Jing [325], Lin in [408] obtained the above result for $*$-Lie-type derivations.

Theorem 5.6.48 *[408, Theorem 3.1] Let $\mathcal{H}$ be an infinite dimensional complex Hilbert space and $\mathcal{A}$ be a standard operator algebra on $\mathcal{H}$ containing the identity operator I. If $\mathcal{A}$ is closed under the adjoint operation and $\phi : \mathcal{A} \to \mathcal{B}(\mathcal{H})$ is a nonlinear $*$-Lie-type derivation, then ϕ is a linear $*$-derivation. Moreover, there exists an operator $T \in \mathcal{B}(\mathcal{H})$ satisfying $T + T^* = 0$ such that $\phi(A) = AT - TA$ for all $A \in \mathcal{A}$, that it, ϕ is inner.*

The notion of (nonlinear-) $*$-Lie type derivations can be extended to (nonlinear-) $*$-Lie type higher derivations.

Definition 84 *Let $\mathcal{S} = \{\phi_n\}_{n\in\mathbb{N}}$ be a family of linear mappings on a $*$-algebra $\mathcal{A}$ such that $\phi_0 = id_{\mathcal{A}}$. Then, $\mathcal{S}$ is called*

(i) a $*$-higher derivation if for each $n \in \mathbb{N}$, $\phi_n(ab) = \sum\limits_{i+j=n} d_i(a)d_j(b)$ and $\phi_n(a^*) = \phi_n(a)^*$ hold for all $a, b \in \mathcal{A}$;

(ii) a $*$-Lie n-higher derivation if for each $n \in \mathbb{N}$,

$$\phi_r(q_n(a_1,a_2,\ldots,a_n)) = \sum_{i_1+i_2+\cdots+i_n=r} q_n(\phi_{i_1}(a_1),\phi_{i_2}(a_2),\ldots,\phi_{i_n}(a_n))$$

holds for all $a_1, a_2, \ldots, a_n \in \mathcal{A}$.

Remark 5.6.49 *It should be remarked that a $*$-Lie 2-higher derivation is called a $*$-Lie higher derivation while a $*$-Lie 3-higher derivation is said to be a $*$-Lie triple higher derivation. $*$-Lie higher derivations, $*$-Lie triple higher derivations and $*$-Lie n-higher derivations are collectively referred to as $*$-Lie type higher derivations. Note that if the mappings in the above definitions are not necessarily* *linear, then $\mathcal{S}$ is called a nonlinear $*$-higher derivation and a nonlinear $*$-Lie n-higher derivation, respectively.*

Zhang et al. [569] considered nonlinear $*$-Lie higher derivations on factor von Neumann algebras and proved the higher version of Theorem 5.6.44 as follows:

Theorem 5.6.50 *[569, Theorem 2.1] Let $\mathcal{M}$ be a factor von Neumann algebra acting on a complex Hilbert space $\mathcal{H}$ with dim $\mathcal{H} \geq 2$. Suppose that $\mathcal{S} = \{\phi_n\}_{n\in\mathbb{N}}$ is a nonlinear $*$-Lie higher derivation on $\mathcal{M}$. Then, $\mathcal{S}$ is an additive $*$-higher derivation.*

Ashraf et al. [86] introduced the notion of a nonlinear $*$-Lie triple higher derivation on a $*$-algebra and extended Jing's result [325] to the case of nonlinear $*$-Lie triple higher derivations on standard operator algebras.

Theorem 5.6.51 *[86, Theorem 3.1] Let $\mathcal{A}$ be a standard operator algebra on an infinite dimensional complex Hilbert space $\mathcal{H}$. Then, every nonlinear $*$-Lie triple higher derivation $\mathcal{S} = \{d_n\}_{n\in\mathbb{N}}$, $d_n : \mathcal{A} \to \mathcal{B}(\mathcal{H})$, is an additive $*$-higher derivation.*

Wani et al. [549] investigated nonlinear $*$-Lie n-higher derivations on standard operator algebras and proved the following result, which is a generalization of Theorems 5.6.48 & 5.6.51.

Theorem 5.6.52 *[549, Theorem 2.1] Let $\mathcal{A}$ be a standard operator algebra on an infinite dimensional complex Hilbert space $\mathcal{H}$. Then, every nonlinear $*$-Lie n-higher derivation $\mathcal{S} = \{d_n\}_{n\in\mathbb{N}}$, $d_n : \mathcal{A} \to \mathcal{B}(\mathcal{H})$ is an additive $*$-higher derivation.*

Recently, Kong and Zhang [353] introduced a new product, called the *bi-skew Lie product*, on $*$-algebras. Let $\mathcal{A}$ be an associative $*$-algebra over the complex field $\mathbb{C}$. For any $a, b \in \mathcal{A}$, $[a,b]_\diamond = ab^* - ba^*$ is called the bi-skew Lie product of $a, b \in \mathcal{A}$. Further, Kong and Zhang [353] proved that a map D (not necessarily linear) on a factor von Neumann algebra $\mathcal{A}$ satisfies $D([a,b]_\diamond) = [D(a),b]_\diamond + [a,D(b)]_\diamond$ for all $a, b \in \mathcal{A}$ iff D is an additive $*$-derivation. Motivated by this result, Ashraf et al. [40] introduced the concept of a nonlinear bi-skew Lie-type derivation as follows: For any $a_1, a_2, \ldots, a_n \in \mathcal{A}$, define $r_1(a_1) = a_1$ and $r_n(a_1, a_2, \ldots, a_n) = [r_{n-1}(a_1, a_2, \ldots, a_{n-1}), a_n]_\diamond$ for all integers $n \geq 2$.

Definition 85 *A linear mapping $D : \mathcal{A} \to \mathcal{A}$ is called a bi-skew Lie n-derivation $(n \geq 2)$ if*

$$D(r_n(a_1, a_2, \ldots, a_n)) = \sum_{i=1}^{n} r_n(a_1, a_2, \ldots, a_{i-1}, D(a_i), a_{i+1}, \ldots, a_n)$$

holds for all $a_1, a_2, \ldots, a_n \in \mathcal{A}$.

Remark 5.6.53 *If the mapping D is not necessarily linear, then it is called a nonlinear bi-skew Lie-type derivation. In particular, a nonlinear bi-skew Lie 2-derivation is called a nonlinear bi-skew Lie derivation and a nonlinear bi-skew Lie 3-derivation is said to be a nonlinear bi-skew Lie triple derivation. It is easy to verify that every additive $*$-derivation is a bi-skew Lie n-derivation for any integer $n \geq 2$. However, the converse is not true in general. For example, the mapping $D : \mathbb{C} \to \mathbb{C}$ defined by $D(x+\mathrm{i}y) = \mathrm{i}y \quad (x+\mathrm{i}y \in \mathbb{C})$ is a bi-skew Lie 2-derivation but not an additive $*$-derivation.*

Ashraf et al. [40] extended the result of Kong and Zhang [353] to the case of nonlinear bi-skew Lie n-derivation on a factor von Neumann algebra and proved the following theorem.

Theorem 5.6.54 *Let $\mathcal{A}$ be a factor von Neumann algebra acting on a complex Hilbert space $\mathcal{H}$ with $dim(\mathcal{A}) \geq 2$. A map $D : \mathcal{A} \to \mathcal{A}$ satisfies*

$$D(r_n(A_1,A_2,\ldots,A_n)) = \sum_{i=1}^{n} r_n(A_1,A_2,\ldots,A_{i-1},D(A_i),A_{i+1},\ldots,A_n)$$

for all $A_1,A_2,\ldots,A_n \in \mathcal{A}$ iff D is an additive $*$-derivation.

Definition 86 *Let $\mathcal{A}$ be an associative $*$-algebra over the complex field $\mathbb{C}$. An additive map $G : \mathcal{A} \to \mathcal{A}$ is said to be an additive generalized $*$-derivation if there exists an additive $*$-derivation D of $\mathcal{A}$ such that $G(AB) = G(A)B + AD(B)$ and $G(A^*) = G(A)^*$ hold for all $A,B \in \mathcal{A}$.*

Definition 87 *Let $\mathcal{A}$ be an associative $*$-algebra over the complex field $\mathbb{C}$. A linear mapping $G : \mathcal{A} \to \mathcal{A}$ is said to be a generalized bi-skew Lie n-derivation if there exists a bi-skew Lie n-derivation D of $\mathcal{A}$ such that*

$$\begin{aligned} G(r_n(A_1,A_2,\ldots,A_n)) &= r_n(G(A_1),A_2,\ldots,A_n) \\ &\quad + \sum_{i=2}^{n} r_n(A_1,A_2,\ldots,A_{i-1},D(A_i),A_{i+1},\ldots,A_n) \end{aligned} \tag{5.8}$$

holds for all $A_1,A_2,\ldots,A_n \in \mathcal{A}$.

Remark 5.6.55 *If the condition of linearity is dropped from the above definition then we call G a nonlinear generalized bi-skew Lie n-derivation associated with nonlinear bi-skew Lie n-derivation D. Its obvious that every bi-skew Lie n-derivation is an example of a generalized bi-skew Lie n-derivation (set $G = D$ in (5.8)). Moreover, the mapping $x \mapsto \lambda x$, where $\lambda \in Z(A)$ with $\lambda^* = \lambda$ is an example of a generalized bi-skew Lie n-derivation (set $G(x) = \lambda x$ for all $x \in \mathcal{A}$ and $D = 0$ in (5.8)).*

Lets assume that $\mathcal{A}$ be a unital $*$-algebra with the identity element I and a non-trivial projection P (that is, $P \neq 0,I$ and $P^2 = P = P^*$), with $P_1 = P$ and $P_2 =$

$I-P_1$. Then, $\mathcal{A}$ can be represented as $\mathcal{A}=P_1\mathcal{A}P_1+P_1\mathcal{A}P_2+P_2\mathcal{A}P_1+P_2\mathcal{A}P_2$. Furthermore, assume that $\mathcal{A}$ satisfies the property

$$A\mathcal{A}P_i=\{0\}\Longrightarrow A=0 \ \ (i=1,2). \tag{5.9}$$

Some examples of unital $*$-algebras with nontrivial projections satisfying the property (5.9) are unital prime $*$-algebras, factor von Neumann algebras, von Neumann algebras with no central summands of type I_1 (see [203, 401] for details).

Observe that if $G:\mathcal{A}\to\mathcal{A}$ is a nonlinear generalized bi-skew Lie n-derivation with an associated nonlinear bi-skew Lie n-derivation $D:\mathcal{A}\to\mathcal{A}$, then it follows that the map $\mathcal{C}=G-D$ satisfies

$$\mathcal{C}(r_n(A_1,A_2,\ldots,A_n))=r_n(\mathcal{C}(A_1),A_2,\ldots,A_n) \tag{5.10}$$

for all $A_1,A_2,\ldots,A_n\in\mathcal{A}$. Therefore, to obtain a description of the map G it suffices to consider a map $\mathcal{C}$ with property (5.10). Following the Benkovič's approach [115], one can easily prove that the map $\mathcal{C}$ is, in fact, a centralizer on $\mathcal{A}$ (that is, $\mathcal{C}(AB)=\mathcal{C}(A)B=A\mathcal{C}(B)$ for all $A,B\in\mathcal{A}$). Hence, one makes the following proposition:

Proposition 5.6.56 *Let $\mathcal{A}$ be a unital $*$-algebra with a nontrivial projection. If a map $\mathcal{C}:\mathcal{A}\to\mathcal{A}$ satisfies $\mathcal{C}(r_n(A_1,\ldots,A_n))=r_n(\mathcal{C}(A_1),\ldots,A_n)$ for all $A_1,A_2,\ldots,A_n\in\mathcal{A}$, then it is of the form $\mathcal{C}(A)=ZA$, where $Z\in Z(\mathcal{A})$ and $\mathcal{C}(A^*)=\mathcal{C}(A)^*$ for all $A\in\mathcal{A}$.*

Let $\mathcal{A}$ be a unital $*$-algebra with a nontrivial idempotent P_1 satisfying (5.9). It is shown in [567, Theorem 2.1] that every nonlinear bi-skew Lie n-derivation $D:\mathcal{A}\to\mathcal{A}$ is an additive $*$-derivation. Combining this result with Proposition 5.6.56, we obtain the following result which characterizes nonlinear generalized bi-skew Lie n-derivations on unital $*$-algebras.

Theorem 5.6.57 *Let $\mathcal{A}$ be a unital $*$-algebra with a nontrivial projection P satisfying (5.9). Then, every nonlinear generalized bi-skew Lie n-derivation $G:\mathcal{A}\to\mathcal{A}$ is of the form $G(A)=ZA+D(A)$ for all $A\in\mathcal{A}$, where $Z\in Z(\mathcal{A})$ and $D:\mathcal{A}\to\mathcal{A}$ is an additive $*$-derivation. Moreover, $G:\mathcal{A}\to\mathcal{A}$ is an additive generalized $*$-derivation.*

Proof. Let $G:\mathcal{A}\to\mathcal{A}$ be a nonlinear generalized bi-skew Lie n-derivation with associated nonlinear bi-skew Lie n-derivation $D:\mathcal{A}\to\mathcal{A}$. Define a map $\mathcal{C}=G-D$. Then, $\mathcal{C}$ satisfies

$$\mathcal{C}(r_n(A_1,A_2,\ldots,A_n))=r_n(\mathcal{C}(A_1),A_2,\ldots,A_n) \tag{5.11}$$

for all $A_1,A_2,\ldots,A_n\in\mathcal{A}$. Hence, it follow from Proposition 5.6.56 that $\mathcal{C}(A)=ZA$ and $\mathcal{C}(A^*)=\mathcal{C}(A)^*$ for all $A\in\mathcal{A}$, $Z\in Z(\mathcal{A})$. Thus, G is of the form

$G(A) = ZA + D(A)$ for all $A \in \mathcal{A}$, where $Z \in Z(\mathcal{A})$. Since $D : \mathcal{A} \to \mathcal{A}$ is a nonlinear bi-skew Lie n-derivation, in view of [567, Theorem 2.1], D is an additive $*$-derivation. It only remains to be shown that G is an additive generalized $*$-derivation. We assert that G is an additive generalized $*$-derivation with associated additive $*$-derivation D. Clearly, G is additive. For any $A_1, A_2 \in \mathcal{A}$, we have

$$\begin{aligned} G(A_1)A_2 + A_1D(A_2) &= (ZA_1 + D(A_1))A_2 + A_1D(A_2) \\ &= (ZA_1)A_2 + D(A_1)A_2 + A_1D(A_2) \\ &= Z(A_1A_2) + D(A_1A_2) \\ &= G(A_1A_2). \end{aligned}$$

We now show that $G(A^*) = G(A)^*$ for all $A \in \mathcal{A}$. Now, by Proposition 5.6.56, for any $A \in \mathcal{A}$, we get,

$$G(A^*) = \mathcal{C}(A^*) - D(A^*) = \mathcal{C}(A)^* - D(A)^* = (\mathcal{C}(A) - D(A))^* = G(A)^*.$$

Consequently, G is an additive generalized $*$-derivation with associated additive $*$-derivation D. This proves the theorem.

Definition 88 *Let $\mathfrak{D} = \{\phi_n\}_{n\in\mathbb{N}}$ be a family of linear mappings on a $*$-algebra $\mathcal{A}$ such that $\phi_0 = id_{\mathcal{A}}$. Then, $\mathfrak{D}$ is called a bi-skew Lie n-higher derivation if for each $n \in \mathbb{N}$,*

$$\phi_r(r_n(a_1, a_2, \ldots, a_n)) = \sum_{i_1+i_2+\cdots+i_n=r} r_n(\phi_{i_1}(a_1), \phi_{i_2}(a_2), \ldots, \phi_{i_n}(a_n))$$

holds for all $a_1, a_2, \ldots, a_n \in \mathcal{A}$.

Definition 89 *Let $G = \{\xi_n\}_{n\in\mathbb{N}}$ be a family of linear mappings on a $*$-algebra $\mathcal{A}$ such that $\xi_0 = id_{\mathcal{A}}$. Then, G is said to be a generalized bi-skew Lie n-higher derivation if there exists a bi-skew Lie n-higher derivation $\mathfrak{D} = \{\phi_n\}_{n\in\mathbb{N}}$ such that for each $n \in \mathbb{N}$,*

$$\xi_r(r_n(a_1, a_2, \ldots, a_n)) = \sum_{i_1+i_2+\cdots+i_n=r} r_n(\xi_{i_1}(a_1), \phi_{i_2}(a_2), \ldots, \phi_{i_n}(a_n))$$

holds for all $a_1, a_2, \ldots, a_n \in \mathcal{A}$.

If the condition of linearity is dropped from Definitions 88 & 89, then a bi-skew Lie n-higher derivation and a generalized bi-skew Lie n-higher derivation is called a *nonlinear bi-skew Lie n-higher derivation* and a *nonlinear generalized bi-skew Lie n-higher derivation* on $\mathcal{A}$, respectively.

Definition 90 *Let $G = \{\xi_n\}_{n\in\mathbb{N}}$ be a family of linear mappings on a $*$-algebra $\mathcal{A}$ such that $\xi_0 = id_{\mathcal{A}}$. Then, G is said to be a generalized $*$-higher derivation if there exists a $*$-higher derivation $\mathfrak{D} = \{\phi_n\}_{n\in\mathbb{N}}$ such that for each $n \in \mathbb{N}$,*

$$\xi_n(ab) = \sum_{i+j=n} \xi_i(a)\phi_j(b) \ \ and \ \ \xi_n(a^*) = \xi(a)^*$$

hold for all $a, b \in \mathcal{A}$.

Remark 5.6.58 *It can be verified that every $*$-higher derivation is a bi-skew Lie n-higher derivation and every bi-skew Lie n-higher derivation is a generalized bi-skew Lie n-higher derivation. However, the converse statements are not true in general. Note that if $\mathcal{A}$ is a torsion-free $*$-algebra and $\phi : \mathcal{A} \to \mathcal{A}$ is a $*$-derivation (resp. bi-skew Lie n-derivation), then the family of mappings $\mathfrak{D} = \{\phi_n\}_{n\in\mathbb{N}}$ defined by $\phi_0 = id_{\mathcal{A}}$ and $\phi_n = \frac{1}{n!}\phi^n$ is a $*$-higher derivation (resp. bi-skew Lie n-higher derivation) (see [251]). Hence, if we take ϕ to be a bi-skew Lie n-derivation but not necessarily a $*$-derivation (such a map exists see Remark 5.6.53), then $D = \{\phi_n\}_{n\in\mathbb{N}}$ would be a bi-skew Lie n-higher derivation but not necessarily a $*$-higher derivation. Now, consider the two families of mappings $\mathfrak{D} = \{\phi_n\}_{n\in\mathbb{N}}$ and $G = \{\xi_n\}_{n\in\mathbb{N}}$ such that*

$$\phi_n = \begin{cases} id_{\mathcal{A}}, & if\, n = 0 \\ 0, & otherwise. \end{cases} \quad and \quad \xi_n = \begin{cases} id_{\mathcal{A}}, & if\, n = 0 \\ \xi, & otherwise, \end{cases}$$

where $\xi : \mathcal{A} \to \mathcal{A}$ is a mapping defined by $\xi(x) = \lambda x, 0 \neq \lambda \in Z(A)$ with $\lambda^* = \lambda$. Then, $G = \{\xi_n\}_{n\in\mathbb{N}}$ is a generalized bi-skew Lie n-higher derivation with an associated bi-skew Lie n-higher derivation $\mathfrak{D} = \{\phi_n\}_{n\in\mathbb{N}}$ but G is not a bi-skew Lie n-higher derivation.

In view of Theorems 5.6.54 & 5.6.57, it is natural to investigate the following open problems;

Open Problems 5.6.59 (*i*) *Is a nonlinear bi-skew Lie n-higher derivation on a factor von Neumann algebra (more generally, on a unital $*$-algebra) an additive $*$-higher derivation?*

(*ii*) *Is a nonlinear generalized bi-skew Lie n-higher derivation on a unital $*$-algebra an additive generalized $*$-higher derivation?*

5.7 Jordan-type higher derivations on algebras

Analogous to the concept of Lie-type higher derivations, in this section, the notion of Jordan-type higher derivations on some standard algebras can be introduced.

Let $\mathcal{A}$ be an algebra over a commutative unital ring R. Denote by $x \circ y = xy + yx$ the Jordan product of elements $a, b \in \mathcal{A}$.

Definition 91 *An R-linear mapping $\delta : \mathcal{A} \to \mathcal{A}$ is called*

(*i*) *a Jordan derivation if $\delta(a \circ b) = \delta(a) \circ b + a \circ \delta(b)$ holds for all $a, b \in \mathcal{A}$, or equivalently, if $\delta(a^2) = \delta(a)a + a\delta(a)$ holds for all $a \in \mathcal{A}$ in case R is 2-torsion free;*

(*ii*) *a Jordan triple derivation if $\delta((a \circ b) \circ c) = (\delta(a) \circ b) \circ c + (a \circ \delta(b)) \circ c + (a \circ b) \circ \delta(c)$ holds for all $x, y, z \in \mathcal{A}$, or equivalently, if $\delta(aba) =$*

$\delta(a)ba + a\delta(b)a + ab\delta(a)$ holds for all $a, b \in \mathcal{A}$ in case R is 2-torsion free.

Clearly, every derivation is a Jordan derivation. The converse is, in general, not true. If the mapping δ is not necessarily linear in the above definition then δ is called a *nonlinear Jordan derivation* and a *nonlinear Jordan triple derivation*, respectively. In 1957 Herstein [289] proved that every Jordan derivation from a prime ring of characteristic not equal to 2 into itself is a derivation. Herstein's theorem has been extended to different rings and algebras in various directions (see, e.g., [132, 139, 150, 520, 541, 571] and references therein). Motivated by Herstein's result, Zhang and Yu [571] showed that every Jordan derivation of a 2-torsion free triangular algebra is a derivation.

Theorem 5.7.1 *[571, Theorem 2.1] Let $\mathcal{A}, \mathcal{B}$ be unital algebras over a 2-torsion free commutative ring R, and $\mathcal{M}$ be a unital $(\mathcal{A}, \mathcal{B})$-bimodule, which is faithful as a left $\mathcal{A}$-module and also as a right $\mathcal{B}$-module. Then, every Jordan derivation from $\mathfrak{A} = \mathrm{Tri}(\mathcal{A}, \mathcal{M}, \mathcal{B})$ into itself is a derivation.*

Xiao and Wei [554, 559] investigated the above problem in the setting of Jordan (triple) higher derivations of triangular algebras. Let us recall the definition of a Jordan (triple) higher derivation of an algebra.

Definition 92 *Let $\mathcal{A}$ be an algebra over a commutative unital ring R and $D = \{d_n\}_{n\in\mathbb{N}}$ be a sequence of R-linear mappings on $\mathcal{A}$ such that $d_0 = id_{\mathcal{A}}$. Then, D is called*

(i) a Jordan higher derivation on $\mathcal{A}$ if for each $n \in \mathbb{N}$,

$$d_n(a^2) = \sum_{i+j=n} d_i(a)d_j(a) \ for \ all \ a \in \mathcal{A};$$

(ii) a Jordan triple higher derivation on $\mathcal{A}$ if for each $n \in \mathbb{N}$,

$$d_n(aba) = \sum_{i+j+k=n} d_i(a)d_j(b)d_k(a) \quad for \ all \ \ a, b \in \mathcal{A}.$$

If the linear assumption in the above definition is deleted, then D is said to be a *nonlinear Jordan higher derivation* and a *nonlinear Jordan triple higher derivation*, respectively. It is obvious that every higher derivation on an algebra $\mathcal{A}$ is a Jordan higher derivation but the converse need not be true. The converse of this problem was investigated by Xiao and Wei [559] in the setting of triangular algebras.

Theorem 5.7.2 *[559, Theorem 2.5] Let $\mathcal{A}, \mathcal{B}$ be unital algebras over a 2-torsion free commutative ring R, and $\mathcal{M}$ be a unital $(\mathcal{A}, \mathcal{B})$-bimodule, which is faithful as a left $\mathcal{A}$-module and also as a right $\mathcal{B}$-module. Then, every Jordan higher derivation from $\mathfrak{A} = \mathrm{Tri}(\mathcal{A}, \mathcal{M}, \mathcal{B})$ into itself is a higher derivation.*

Further, Wei and Xiao [554] extended this study to Jordan triple higher derivations of triangular algebras and obtained the following theorem.

Theorem 5.7.3 *[554, Theorem 4.4] Let $\mathcal{A},\mathcal{B}$ be unital algebras over a 2-torsion free commutative ring R, and $\mathcal{M}$ be a unital $(\mathcal{A},\mathcal{B})$-bimodule, which is faithful as a left $\mathcal{A}$-module and also as a right $\mathcal{B}$-module. Suppose that $\mathfrak{A} = \mathrm{Tri}(\mathcal{A},\mathcal{M},\mathcal{B})$ is a triangular algebra consisting of $\mathcal{A},\mathcal{B}$ and $\mathcal{M}$ and $D = \{d_n\}_{n\in\mathbb{N}}$ is a sequence of R-linear mappings on $\mathfrak{A}$. Then, the following statements are equivalent:*

(i) *D is a Jordan higher derivation of $\mathfrak{A}$.*

(ii) *D is a Jordan triple higher derivation of $\mathfrak{A}$.*

(iii) *D is a higher derivation of $\mathfrak{A}$.*

Definition 93 *Let $\Delta = \{\delta_n\}_{n\in\mathbb{N}}$ be a sequence of R-linear mappings on an algebra $\mathcal{A}$ such that $\delta_0 = id_{\mathcal{A}}$. Then, Δ is called:*

(i) *a generalized higher derivation on $\mathcal{A}$ if there exists a higher derivation $D = \{d_n\}_{n\in\mathbb{N}}$ on $\mathcal{A}$ such that $\delta_n(ab) = \sum_{i+j=n} \delta_i(a)d_j(b)$ for all $a,b \in \mathcal{A}$ and for each $n \in \mathbb{N}$.*

(ii) *a generalized Jordan higher derivation on $\mathcal{A}$ if there exists a Jordan higher derivation $D = \{d_n\}_{n\in\mathbb{N}}$ on $\mathcal{A}$ such that $\delta_n(a^2) = \sum_{i+j=n} \delta_i(a)d_j(a)$ for all $a \in \mathcal{A}$ and for each $n \in \mathbb{N}$.*

(iii) *a generalized Jordan triple higher derivation on $\mathcal{A}$ if there exists a Jordan triple higher derivation $D = \{d_n\}_{n\in\mathbb{N}}$ on $\mathcal{A}$ such that $\delta_n(aba) = \sum_{i+j+k=n} \delta_i(a)d_j(b)d_k(a)$ for all $a,b \in \mathcal{A}$ and for each $n \in \mathbb{N}$.*

The following theorem is an upgraded version of Theorem 5.7.3.

Theorem 5.7.4 *[554, Theorem 4.7] Let $\mathcal{A},\mathcal{B}$ be unital algebras over a 2-torsion free commutative ring R, and $\mathcal{M}$ be a unital $(\mathcal{A},\mathcal{B})$-bimodule, which is faithful as a left $\mathcal{A}$-module and also as a right $\mathcal{B}$-module. Suppose that $\mathfrak{A} = \mathrm{Tri}(\mathcal{A},\mathcal{M},\mathcal{B})$ is a triangular algebra consisting of $\mathcal{A},\mathcal{B}$ and $\mathcal{M}$ and $\Delta = \{\delta_n\}_{n\in\mathbb{N}}$ is a sequence of R-linear mappings on $\mathfrak{A}$. Then, the following statements are equivalent.*

(i) *Δ is a generalized Jordan higher derivation of $\mathfrak{A}$.*

(ii) *Δ is a generalized Jordan triple higher derivation of $\mathfrak{A}$.*

(iii) *Δ is a generalized higher derivation of $\mathfrak{A}$.*

Ashraf and Jabeen explored the results obtained in [554] and [559] for non-linear Jordan triple (higher) derivations on triangular algebras.

Theorem 5.7.5 *[54, Theorem 1] Let $\mathcal{A},\mathcal{B}$ be unital algebras over a 2-torsion free commutative ring R, and $\mathcal{M}$ be a unital $(\mathcal{A},\mathcal{B})$-bimodule, which is faithful as* *a left $\mathcal{A}$-module and also as a right $\mathcal{B}$-module. Suppose that $\mathfrak{A} = \mathrm{Tri}(\mathcal{A}, \mathcal{M}, \mathcal{B})$ is a triangular algebra consisting of $\mathcal{A}, \mathcal{B}$ and $\mathcal{M}$ and $\delta : \mathfrak{A} \to \mathfrak{A}$ is a mapping (not necessarily linear) satisfying*

$$\delta(ABC + CBA) = \delta(A)BC + A\delta(B)C + AB\delta(C) + \delta(C)BA + C\delta(B)A + CB\delta(A) \tag{5.12}$$

for all $A, B, C \in \mathfrak{A}$. Then, δ is an additive derivation on $\mathfrak{A}$.

Theorem 5.7.6 *[57, Theorem 2.1] Let $\mathcal{A}, \mathcal{B}$ be unital algebras over a 2-torsion free commutative ring R, and $\mathcal{M}$ be a unital $(\mathcal{A}, \mathcal{B})$-bimodule, which is faithful as a left $\mathcal{A}$-module and also as a right $\mathcal{B}$-module. Suppose that $\mathfrak{A} = \mathrm{Tri}(\mathcal{A}, \mathcal{M}, \mathcal{B})$ is a triangular algebra consisting of $\mathcal{A}, \mathcal{B}$ and $\mathcal{M}$ and $D = \{d_n\}_{n \in \mathbb{N}}$ is a sequence of mappings (not necessarily linear) on $\mathfrak{A}$ satisfying*

$$d_n(ABC + CBA) = \sum_{i+j+k=n} (d_i(A)d_j(B)d_k(C) + d_i(C)d_j(B)d_k(A)) \tag{5.13}$$

for all $A, B, C \in \mathfrak{A}$. Then, D is an additive higher derivation on $\mathfrak{A}$.

Remark 5.7.7 *We remark that if δ and d_n (for each n) are linear mappings on a 2-torsion free algebra $\mathfrak{A}$, then the conditions* (5.12) *and* (5.13) *are equivalent to the conditions of a Jordan triple derivation and Jordan triple higher derivation, respectively.*

The notion of derivations on rings was extended to (α, β)-derivations. Many authors have made important contributions to related topics (see [29, 44, 59, 64, 65, 66, 67, 150, 179, 285, 334, 347]). The concept of (α, β)-higher derivations and generalized (α, β)-higher derivations and their various invariants were introduced and studied by the authors in [59, 61, 284] by considering the set of R-linear mappings on any R-algebra $\mathcal{A}$ satisfying the properties given in Definitions 94 and 95.

Definition 94 *Let α and β be two automorphisms on an algebra $\mathcal{A}$. An R-linear mapping $\delta : \mathcal{A} \to \mathcal{A}$ is called*

- (i) *an (α, β)-derivation if $\delta(ab) = \delta(a)\alpha(b) + \beta(a)\delta(b)$ holds for all $a, b \in \mathcal{A}$;*
- (ii) *a Jordan (α, β)-derivation if $\delta(a^2) = \delta(a)\alpha(a) + \beta(a)\delta(a)$ holds for all $a \in \mathcal{A}$;*
- (iii) *a Jordan triple (α, β)-derivation if*

$$\delta(aba) = \delta(a)\alpha(b)\alpha(a) + \beta(a)\delta(b)\alpha(a) + \beta(a)\beta(b)\delta(a)$$

holds for all $a, b \in \mathcal{A}$.

Obviously, any (α,β)-derivation on $\mathcal{A}$ is a Jordan (α,β)-derivation and any Jordan (α,β)-derivation on a 2-torsion free algebra $\mathcal{A}$ is a Jordan triple (α,β)-derivation. However, the converse is not true. Jordan (α,β)-derivations of associative algebras are a subject of research in various areas. An important question in the theory of Jordan derivations is when a Jordan (triple) (α,β)-derivation is an (α,β)-derivation. Liu and Shiue [426] and Lanski [369] independently proved that any Jordan triple (α,β)-derivation (and hence any Jordan (α,β)-derivation) on a 2-torsion free semiprime algebra is an (α,β)-derivation. Motivated by these studies, Han and Wei [285] investigated Jordan (α,β)-derivations on triangular algebras and proved the following theorem.

Theorem 5.7.8 *[285, Theorem 3.9] Let $\mathfrak{A} = \mathrm{Tri}(\mathcal{A},\mathcal{M},\mathcal{B})$ be a 2-torsion free triangular algebra consisting of $\mathcal{A}$, $\mathcal{B}$ and $\mathcal{M}$ and $\delta : \mathfrak{A} \to \mathfrak{A}$ an R-linear mapping on $\mathfrak{A}$. Suppose that $\mathcal{A}$ and $\mathcal{B}$ have only trivial idempotents. Then the following statements are equivalent:*

(i) *δ is a Jordan (α,β)-derivation on $\mathfrak{A}$.*

(ii) *δ is a Jordan triple (α,β)-derivation on $\mathfrak{A}$.*

(iii) *δ is an (α,β)-derivation on $\mathfrak{A}$.*

Han and Wei [285] further extended the concept of (α,β)-derivations to generalized (α,β)-derivations.

Definition 95 *An R-linear mapping $\Delta : \mathcal{A} \to \mathcal{A}$ is called*

(i) *a generalized (α,β)-derivation on $\mathcal{A}$ if there exists an (α,β)-derivation δ on $\mathcal{A}$ such that $\Delta(ab) = \Delta(a)\alpha(b) + \beta(a)\delta(b)$ holds for all $a,b \in \mathcal{A}$;*

(ii) *a generalized Jordan (α,β)-derivation on $\mathcal{A}$ if there exists a Jordan (α,β)-derivation δ on $\mathcal{A}$ such that $\Delta(a^2) = \Delta(a)\alpha(a) + \beta(a)\delta(a)$ holds for all $a \in \mathcal{A}$;*

(iii) *a generalized Jordan triple (α,β)-derivation on $\mathcal{A}$ if there exists a Jordan triple (α,β)-derivation δ on $\mathcal{A}$ such that $\Delta(aba) = \Delta(a)\alpha(b)\alpha(a) + \beta(a)\delta(b)\alpha(a) + \beta(a)\beta(b)\delta(a)$ holds for all $a,b \in \mathcal{A}$.*

The following theorem is a generalized version of Theorem 5.7.8.

Theorem 5.7.9 *[285, Theorem 4.3] Let $\mathfrak{A} = \mathrm{Tri}(\mathcal{A},\mathcal{M},\mathcal{B})$ be a 2-torsion free triangular algebra consisting of $\mathcal{A}$, $\mathcal{B}$ and $\mathcal{M}$ and $\Delta : \mathfrak{A} \to \mathfrak{A}$ an R-linear mapping on $\mathfrak{A}$. Suppose that $\mathcal{A}$ and $\mathcal{B}$ have only trivial idempotents. Then the following statements are equivalent:*

(i) *Δ is a generalized Jordan (α,β)-derivation on $\mathfrak{A}$.*

(ii) *Δ is a generalized Jordan triple (α,β)-derivation on $\mathfrak{A}$.*

(iii) *Δ is a generalized (α,β)-derivation on $\mathfrak{A}$.*

In the year 2010, Ashraf et al. [67] introduced the notion of (σ,τ)-higher derivations in rings and proved that, under certain assumptions, every Jordan (σ,τ)-higher derivation on a prime ring R of characteristic not equal to 2 is a (σ,τ)-higher derivation on R, where σ,τ are endomorphisms of R. Following [67], Ashraf et al. in [61] defined the notion of a (σ,τ)-higher derivation on an algebra $\mathcal{A}$ as follows:

Definition 96 *Let σ, τ be automorphisms on an algebra $\mathcal{A}$ and $D=\{\delta_n\}_{n\in\mathbb{N}}$ be a family of R-linear mappings $\delta_n:\mathcal{A}\to\mathcal{A}$ such that $\delta_0=id_R$. Then, D is said to be*

(i) *a (σ,τ)-higher derivation if for each $n\in\mathbb{N}$*

$$\delta_n(ab)=\sum_{i+j=n}\delta_i(\sigma^{n-i}(a))\delta_j(\tau^{n-j}(b))$$

holds for all $a,b\in\mathcal{A}$;

(ii) *a Jordan (σ,τ)-higher derivation if for each $n\in\mathbb{N}$*

$$\delta_n(a^2)=\sum_{i+j=n}\delta_i(\sigma^{n-i}(a))\delta_j(\tau^{n-j}(a))$$

holds for all $a\in\mathcal{A}$;

(iii) *a Jordan triple (σ,τ)-higher derivation if for each $n\in\mathbb{N}$*

$$\delta_n(aba)=\sum_{i+j+k=n}\delta_i(\sigma^{n-i}(a))\delta_j(\sigma^k\tau^i(b))\delta(\tau^{n-k}(a))$$

holds for all $a,b\in\mathcal{A}$.

Motivated by the papers [67, 285], the (σ,τ)-higher derivation on triangular algebras was studied in [61].

Theorem 5.7.10 *[61, Theorem 3.8] Let $\mathcal{A},\mathcal{B}$ be unital algebras over a 2-torsion free commutative ring R, and $\mathcal{M}$ be a unital $(\mathcal{A},\mathcal{B})$-bimodule, which is faithful as a left $\mathcal{A}$-module and also as a right $\mathcal{B}$-module. Suppose that $\mathfrak{A}=\mathrm{Tri}(\mathcal{A},\mathcal{M},\mathcal{B})$ is a triangular algebra consisting of $\mathcal{A},\mathcal{B}$ and $\mathcal{M}$ and $\Delta=\{\delta_n\}_{n\in\mathbb{N}}$ is a sequence of R-linear mappings on $\mathfrak{A}$. Then the following statements are equivalent:*

(i) *Δ is a Jordan higher (σ,τ)-derivation on $\mathfrak{A}$.*

(ii) *Δ is a Jordan triple higher (σ,τ)-derivation on $\mathfrak{A}$.*

(iii) *Δ is a higher (σ,τ)-derivation on $\mathfrak{A}$.*

Further, the concept of (σ,τ)-higher derivations was extended to generalized (σ,τ)-higher derivations in [59].

Definition 97 *Let σ, τ be automorphisms on an algebra $\mathcal{A}$ and $D = \{\Delta_n\}_{n\in\mathbb{N}}$ be a family of R-linear mappings $\Delta_n : \mathcal{A} \to \mathcal{A}$ such that $\delta_0 = id_R$. Then, D is said to be*

(i) *a generalized (σ,τ)-higher derivation on $\mathcal{A}$ if there exists a (σ,τ)-higher derivation $\{\delta_n\}_{n\in\mathbb{N}}$ on $\mathcal{A}$ such that for each $n \in \mathbb{N}$*

$$\Delta_n(ab) = \sum_{i+j=n} \Delta_i(\sigma^{n-i}(a))\delta_j(\tau^{n-j}(b))$$

holds for all $a,b \in \mathcal{A}$;

(ii) *a generalized Jordan (σ,τ)-higher derivation on $\mathcal{A}$ if there exists a Jordan (σ,τ)-higher derivation $\{\delta_n\}_{n\in\mathbb{N}}$ on $\mathcal{A}$ such that for each $n \in \mathbb{N}$*

$$\Delta_n(a^2) = \sum_{i+j=n} \Delta_i(\sigma^{n-i}(a))\delta_j(\tau^{n-j}(a))$$

holds for all $a \in \mathcal{A}$;

(iii) *a generalized Jordan triple (σ,τ)-higher derivation on $\mathcal{A}$ if there exists a Jordan triple (σ,τ)-higher derivation $\{\delta_n\}_{n\in\mathbb{N}}$ on $\mathcal{A}$ such that for each $n \in \mathbb{N}$*

$$\Delta_n(aba) = \sum_{i+j+k=n} \Delta_i(\sigma^{n-i}(a))\delta_j(\sigma^k\tau^i(b))\delta(\tau^{n-k}(a))$$

holds for all $a,b \in \mathcal{A}$.

The following theorem is a generalized version of Theorem 5.7.10.

Theorem 5.7.11 *[59, Theorem 3.1] Let $\mathfrak{A} = \mathrm{Tri}(\mathcal{A},\mathcal{M},\mathcal{B})$ be a 2-torsion free triangular algebra consisting of $\mathcal{A}$, $\mathcal{B}$ and $\mathcal{M}$ and $\Delta : \mathfrak{A} \to \mathfrak{A}$ an R-linear mapping on $\mathfrak{A}$. Suppose that $\mathcal{A}$ and $\mathcal{B}$ have only trivial idempotents. Then the following statements are equivalent:*

(i) *Δ is a generalized Jordan (σ,τ)-higher derivation on $\mathfrak{A}$.*

(ii) *Δ is a generalized Jordan triple (σ,τ)-higher derivation on $\mathfrak{A}$.*

(iii) *Δ is a generalized (σ,τ)-higher derivation on $\mathfrak{A}$.*

Benkovič and Širovnik [122] considered Jordan derivations of a unital algebra $\mathcal{A}$ having a nontrivial idempotent. It emerges that on unital algebras there exist Jordan derivations that are not derivations. They introduced a term, the singular Jordan derivation, which is a proper Jordan derivation of the form that depends on a Peirce decomposition of the unital algebra $\mathcal{A}$ and proved that under certain conditions every Jordan derivation of $\mathcal{A}$ is the sum of a derivation and a singular Jordan derivation.

Theorem 5.7.12 *[122, Theorem 4.1] Let $\mathcal{A}$ be a unital algebra with a nontrivial idempotent e satisfying* (5.6). *If $\Delta : \mathcal{A} \to \mathcal{A}$ is a Jordan derivation, then there exist a derivation $d : \mathcal{A} \to \mathcal{A}$ and a singular Jordan derivation $\delta : \mathcal{A} \to \mathcal{A}$ such that $\Delta = d + \delta$. Moreover, d and δ are uniquely determined.*

Let $\mathcal{A}$ be a unital algebra with a nontrivial idempotent e such that $e\mathcal{A}f \cdot f\mathcal{A}e = \{0\} = f\mathcal{A}e \cdot e\mathcal{A}f$. It is shown in [122, Remark 3.2] that every singular Jordan derivation of $\mathcal{A}$ is an antiderivarion; that is, a linear mapping $\delta : \mathcal{A} \to \mathcal{A}$ satisfying $\delta(xy) = \delta(y)x + y\delta(x)$ for all $x, y \in \mathcal{A}$. If the bimodule $e\mathcal{A}f$ is faithful as a left $e\mathcal{A}e$-module and also as a right $f\mathcal{A}f$-module, then $\mathcal{A}$ satisfies (5.6). Hence, as an immediate consequence of Theorem 5.7.12, we obtain Corollary 5.7.1.

Corollary 5.7.1 *[122, Corollary 4.2] Let $\mathcal{A}$ be a unital algebra with a nontrivial idempotent e such that $e\mathcal{A}f \cdot f\mathcal{A}e = \{0\} = f\mathcal{A}e \cdot e\mathcal{A}f$ and the bimodule $e\mathcal{A}f$ is faithful as a left $e\mathcal{A}e$-module and also as a right $f\mathcal{A}f$-module. Then, every Jordan derivation of $\mathcal{A}$ can be expressed as the sum of a derivation and an antiderivation.*

Let $\mathcal{A}$ be an algebra. For any $a_1, a_2, \ldots, a_n \in \mathcal{A}$, let us set

$$
\begin{aligned}
s_1(a_1) &= a_1, \\
s_2(a_1, a_2) &= s_1(a_1) \circ a_2 = a_1a_2 + a_2a_1, \\
&\vdots \qquad\qquad \vdots \\
s_n(a_1, a_2, \ldots, a_n) &= s_{n-1}(a_1, a_2, \ldots, a_{n-1}) \circ a_n \quad (n \geq 2).
\end{aligned}
$$

The polynomial $s_n(a_1, a_2, \ldots, a_n)$ $(n \geq 2)$ is called the Jordan n-product of elements $a_1, a_2, \ldots, a_n \in \mathcal{A}$.

Definition 98 *An R-linear mapping $\delta : \mathcal{A} \to \mathcal{A}$ is said to be a Jordan n-derivation if*

$$
\delta(s_n(a_1, a_2, \ldots, a_n)) = \sum_{i=1}^{n} s_n(a_1, \ldots, a_{i-1}, \delta(a_i), a_{i+1}, \ldots, a_n) \tag{5.14}
$$

holds for all $a_1, a_2, \ldots, a_n \in \mathcal{A}$.

It is obvious that a Jordan 2-derivation is a Jordan derivation and a Jordan 3-derivation is a Jordan triple derivation. Generally, if the linearity assumption of δ in the above definition is deleted, then the corresponding mapping is called a *nonlinear Jordan n-derivation.* Such mappings for $n = 2, 3$ on various rings and algebras were studied by many authors. Recently, Qi et al. [492] discussed the structure of nonlinear Jordan n-derivations on unital algebras and generalized many results obtained earlier for Jordan derivations.

Theorem 5.7.13 *[492, Theorem 3.3] Let $\mathcal{A}$ be the unital algebra containing a nontrivial idempotent e satisfying* (5.6). *Assume that the characteristic of $\mathcal{A}$ is* *not equal to 2 and $\delta : \mathcal{A} \to \mathcal{A}$ is a mapping. Then, δ is a nonlinear Jordan n-derivation $(n \geq 2)$ iff δ is an additive Jordan derivation; more concretely, $\delta(a) = d(a) + \phi(a)$ for all $a \in \mathcal{A}$, where $d : \mathcal{A} \to \mathcal{A}$ is an additive derivation and $\phi : \mathcal{A} \to \mathcal{A}$ is an additive singular Jordan derivation.*

As a consequence of the above, we get the following result, which is a generalization of [122, Corollary 4.2].

Corollary 5.7.2 *[492, Corollary 3.4] Let $\mathcal{A}$ be a unital algebra with a nontrivial idempotent e such that $e\mathcal{A}f \cdot f\mathcal{A}e = \{0\} = e\mathcal{A}f \cdot e\mathcal{A}f$ and the bimodule $e\mathcal{A}f$ is faithful as a left $e\mathcal{A}e$-module and also as a right $f\mathcal{A}f$-module. Then, every nonlinear Jordan n-derivation of $\mathcal{A}$ is the sum of an additive derivation and an additive antiderivation.*

Analogous to the notion of a generalized Lie n-derivation, the notion of a generalized Jordan n-derivation can also be introduced.

Definition 99 *The R-linear mapping $\Delta : \mathcal{A} \to \mathcal{A}$ is said to be a generalized Jordan n-derivation if there exists a Jordan n-derivation $\delta : \mathcal{A} \to \mathcal{A}$ such that*

$$\begin{aligned} \Delta(s_n(a_1, a_2, \ldots, a_n)) &= s_n(\Delta(a_1), a_2, \ldots, a_n) \\ &\quad + \sum_{i=2}^{n} s_n(a_1, \ldots, a_{i-1}, \delta(a_i), a_{i+1}, \ldots, a_n) \quad (5.15) \end{aligned}$$

holds for all $a_1, a_2, \ldots, a_n \in \mathcal{A}$.

Remark 5.7.14 *If the condition of linearity is removed from the above definition then Δ is called a nonlinear generalized Jordan n-derivation with associated nonlinear Jordan n-derivation δ. Note that any Jordan n-derivation and the mapping $u \mapsto \lambda u$, where $\lambda \in Z(\mathcal{A})$, are standard examples of a generalized Jordan n-derivation.*

Let $\Delta : \mathcal{A} \to \mathcal{A}$ be a nonlinear generalized Jordan n-derivation with associated nonlinear Jordan n-derivation $\delta : \mathcal{A} \to \mathcal{A}$. Set $\Phi = \Delta - \delta$. Then, it follows from (5.14) and (5.15) that Φ satisfies

$$\Phi(s_n(a_1, a_2, \ldots, a_n)) = s_n(\Phi(a_1), a_2, \ldots, a_n) \quad (5.16)$$

for all $a_1, a_2, \ldots, a_n \in \mathcal{A}$. Thus, the problem of describing a nonlinear generalized Jordan n-derivation can be reduced to a description of a mapping satisfying (5.16). Let us first discuss the structure of mappings satisfying (5.16) on unital algebras.

Proposition 5.7.15 *Let $\mathcal{A}$ be a 2-torsion free unital algebra with a nontrivial idempotent e satisfying (5.6). If a map $\Phi : \mathcal{A} \to \mathcal{A}$ satisfies*

$$\Phi(s_n(a_1, a_2, \ldots, a_n)) = s_n(\Phi(a_1), a_2, \ldots, a_n)$$

for all $a_1, a_2, \ldots, a_n \in \mathcal{A}$, then $\Phi(a) = \lambda a$ for all $a \in \mathcal{A}$, where $\lambda \in Z(\mathcal{A})$.

Proof. Let $\Phi : \mathcal{A} \to \mathcal{A}$ be a mapping satisfying

$$\Phi(s_n(a_1, a_2, \ldots, a_n)) = s_n(\Phi(a_1), a_2, \ldots, a_n) \tag{5.17}$$

for all $a_1, a_2, \ldots, a_n \in \mathcal{A}$. Since the Jordan product is commutative, (5.17) yields

$$\begin{aligned} \Phi(s_n(a_1, a_2, \ldots, a_n)) &= \Phi(s_n(a_2, a_1, \ldots, a_n)) \\ &= s_n(\Phi(a_2), a_1, \ldots, a_n) \\ &= s_n(a_1, \Phi(a_2), \ldots, a_n) \end{aligned} \tag{5.18}$$

for all $a_1, a_2, \ldots, a_n \in \mathcal{A}$. Combining (5.17) and (5.18), we have

$$s_n(\Phi(a_1), a_2, \ldots, a_n)) = s_n(a_1, \Phi(a_2), \ldots, a_n) \tag{5.19}$$

for all $a_1, a_2, \ldots, a_n \in \mathcal{A}$. Let us denote $\Phi(1) = \lambda \in \mathcal{A}$. In particular, putting $a_1 = a$ and $a_2 = \ldots = a_n = 1$ into (5.19), noting that $\mathcal{A}$ is 2-torsion free, we get

$$2\Phi(a) = \lambda \circ a \tag{5.20}$$

for all $a \in \mathcal{A}$. Next, if we take $a_1 = a$, $a_2 = a'$ and $a_3 = \ldots = a_n = 1$ into (5.17), we obtain

$$\Phi(2^{n-2}(a \circ a')) = 2^{n-2}(\Phi(a) \circ a')$$

for all $a, a' \in \mathcal{A}$, which can be further rewritten as

$$2\Phi(2^{n-2}(a \circ a')) = 2^{n-2}(2\Phi(a) \circ a').$$

In view of (5.20), the above equation gives

$$\begin{aligned} \lambda \circ (2^{n-2}(a \circ a')) &= 2^{n-2}((\lambda \circ a) \circ a') \\ \lambda \circ (a \circ a') &= (\lambda \circ a) \circ a' \end{aligned}$$

for all $a, a' \in \mathcal{A}$. Therefore, $[[\lambda, a], a'] = \lambda \circ (a \circ a') - (\lambda \circ a) \circ a' = 0$ for all $a, a' \in \mathcal{A}$. Since $\mathcal{A}$ satisfies the condition (5.5), we conclude that $\lambda \in Z(\mathcal{A})$. Hence, it follows from (5.20) that $\Phi(a) = \lambda a$ for all $a \in \mathcal{A}$, where $\lambda \in Z(\mathcal{A})$.

Applying Proposition 5.7.15, we now obtain the following result which characterizes nonlinear generalized Jordan n-derivations of unital algebras with nontrivial idempotents.

Theorem 5.7.16 *Let $\mathcal{A}$ be a 2-torsion free unital algebra with a nontrivial idempotent e satisfying* (5.6). *Then, every nonlinear generalized Jordan n-derivation $\Delta : \mathcal{A} \to \mathcal{A}$ is additive. More precisely, Δ is of the form $\Delta(a) = \lambda a + \delta(a)$ for all $a \in \mathcal{A}$, where $\lambda \in Z(\mathcal{A})$ and $\delta : \mathcal{A} \to \mathcal{A}$ is an additive Jordan n-derivation.*

Proof. Let us assume that $\Delta : \mathcal{A} \to \mathcal{A}$ be a nonlinear generalized Jordan n-derivation with an associated nonlinear Jordan n-derivation $\delta : \mathcal{A} \to \mathcal{A}$. In view of [492, Theorem 2.1], δ is an additive Jordan n-derivation. Set $\Phi = \Delta - \delta$. By the definitions of Δ, δ, it follows that $\Phi : \mathcal{A} \to \mathcal{A}$ is a mapping satisfying

$$\Phi(s_n(a_1, a_2, \ldots, a_n)) = s_n(\Phi(a_1), a_2, \ldots, a_n)$$

for all $a_1, a_2, \ldots, a_n \in \mathcal{A}$. Hence, by Proposition 5.7.15, $\Phi(a) = \lambda a$ for all $a \in \mathcal{A}$, where $\lambda \in Z(\mathcal{A})$. Therefore, Δ is of the form $\Delta(a) = \Phi(a) + \delta(a) = \lambda a + \delta(a)$ for all $a \in \mathcal{A}$, where $\lambda \in Z(\mathcal{A})$. Note that Φ, δ are additive and hence Δ is also additive.

Let $\mathcal{A}$ be a 2-torsion free unital algebra with a nontrivial idempotent e satisfying (5.6). It is shown in [492, Theorem 3.3] that every additive Jordan n-derivation $\delta : \mathcal{A} \to \mathcal{A}$ is of the form $\delta(a) = d(a) + \phi(a)$ for all $a \in \mathcal{A}$, where $d : \mathcal{A} \to \mathcal{A}$ is an additive derivation and $\phi : \mathcal{A} \to \mathcal{A}$ is an additive singular Jordan derivation. Combining the above result with Theorem 5.7.16, we establish obtain the following corollary.

Corollary 5.7.3 *Let $\mathcal{A}$ be a 2-torsion free unital algebra with a nontrivial idempotent e satisfying* (5.6). *Then, every nonlinear generalized Jordan n-derivation $\Delta : \mathcal{A} \to \mathcal{A}$ is of the form $\Delta(a) = \lambda a + d(a) + \phi(a)$ for all $a \in \mathcal{A}$, where $\lambda \in Z(\mathcal{A})$, $d : \mathcal{A} \to \mathcal{A}$ is an additive derivation and $\phi : \mathcal{A} \to \mathcal{A}$ is an additive singular Jordan derivation.*

Let $\mathcal{A}$ be a unital algebra with a nontrivial idempotent e such that $e\mathcal{A}f \cdot f\mathcal{A}e = \{0\} = e\mathcal{A}f \cdot e\mathcal{A}f$. It is shown in [122, Remark 3.2] that every singular Jordan derivation of $\mathcal{A}$ is an antiderivation; that is, an additive mapping $\phi : \mathcal{A} \to \mathcal{A}$ satisfying $\phi(ab) = \phi(b)a + b\phi(a)$ for all $a, b \in \mathcal{A}$. If the bimodule $e\mathcal{A}f$ is faithful as a left $e\mathcal{A}e$-module and also as a right $f\mathcal{A}f$-module, then $\mathcal{A}$ satisfies (5.6) and we obtain Corollary 5.7.4.

Corollary 5.7.4 *Let $\mathcal{A}$ be a unital algebra with a nontrivial idempotent e such that $e\mathcal{A}f \cdot f\mathcal{A}e = \{0\} = e\mathcal{A}f \cdot e\mathcal{A}f$ and the bimodule $e\mathcal{A}f$ is faithful as a left $e\mathcal{A}e$-module and also as a right $f\mathcal{A}f$-module. Then, every nonlinear generalized Jordan n-derivation $\Delta : \mathcal{A} \to \mathcal{A}$ is of the form $\Delta(a) = \lambda a + d(a) + \phi(a)$ for all $a \in \mathcal{A}$, where $\lambda \in Z(\mathcal{A})$, $d : \mathcal{A} \to \mathcal{A}$ is an additive derivation and $\phi : \mathcal{A} \to \mathcal{A}$ is an additive antiderivation.*

Now, applying Theorem 5.7.16 on triangular algebras together with [492, Corollary 3.5], we get Corollary 5.7.5.

Corollary 5.7.5 *Let $\mathcal{A} = e\mathcal{A}e + e\mathcal{A}f + f\mathcal{A}f$ be a 2-torsion free triangular algebra. Then, every nonlinear generalized Jordan n-derivation $\Delta : \mathcal{A} \to \mathcal{A}$ can be expressed as $\Delta(a) = \lambda a + \delta(a)$ for all $a \in \mathcal{A}$, where $\lambda \in Z(\mathcal{A})$ and $\delta : \mathcal{A} \to \mathcal{A}$ is an additive derivation.*

Applying Theorem 5.7.16 on prime algebras together with [492, Corollary 3.10], we obtain Corollary 5.7.6.

Corollary 5.7.6 *Let $\mathcal{A}$ be a unital 2-torsion free prime algebra with a nontrivial idempotent. Then, every nonlinear generalized Jordan n-derivation $\Delta : \mathcal{A} \to \mathcal{A}$ can be expressed as $\Delta(a) = \lambda a + \delta(a)$ for all $u \in \mathcal{A}$, where $\lambda \in Z(\mathcal{A})$ and $\delta : \mathcal{A} \to \mathcal{A}$ is an additive derivation.*

Recall that a standard operator algebra $\mathcal{A}$ is a subalgebra of $\mathcal{B}(\mathfrak{X})$ containing all bounded finite rank operators including the identity operator. We know that $\mathcal{A}$ is a prime algebra containing many nontrivial idempotents. Therefore, Corollary 5.7.6 yields Corollary 5.7.7.

Corollary 5.7.7 *Let $\mathcal{A}$ be a standard operator algebra. Then, every nonlinear generalized Jordan n-derivation $\Delta : \mathcal{A} \to \mathcal{A}$ can be expressed as $\Delta(a) = \lambda a + \delta(a)$ for all $a \in \mathcal{A}$, where $\lambda \in Z(\mathcal{A})$ and $\delta : \mathcal{A} \to \mathcal{A}$ is an additive derivation.*

Let $\mathcal{B}(\mathbf{H})$ be the algebra of all bounded linear operators of a complex Hilbert space $\mathbf{H}$. A von Neumann algebra $\mathcal{M}$ is a C^*-subalgebra of $\mathcal{B}(\mathbf{H})$ which is closed in strong operator topology and contains the identity operator. If $\mathcal{M}$ is a factor von Neumann algebra, that is, $Z(\mathcal{M}) = \mathbb{C}\mathbf{1}$, then $\mathcal{M}$ is a prime algebra and hence satisfies (5.6) Therefore, by Corollary 5.7.6, we have Corollary 5.7.8.

Corollary 5.7.8 *Let $\mathcal{M}$ be a factor von Neumann algebra. Then, every nonlinear generalized Jordan n-derivation $\Delta : \mathcal{M} \to \mathcal{M}$ can be expressed as $\Delta(a) = \lambda a + \delta(a)$ for all $a \in \mathcal{M}$, where $\lambda \in Z(\mathcal{M})$ and $\delta : \mathcal{M} \to \mathcal{M}$ is an additive derivation.*

Analogous to the notion of (generalized-) a Lie-type higher derivation, one can also introduce and study (generalized-) Jordan-type higher derivations.

Definition 100 *Let $\mathbb{N}$ be the set of non-negative integers. Let $\mathcal{A}$ be an algebra and $\mathfrak{J} = \{J_r\}_{r \in \mathbb{N}}$ be a family of R-linear mappings $J_r : \mathcal{A} \to \mathcal{A}$ such that $J_0 = id_{\mathcal{A}}$. Then, $\mathfrak{J}$ is said to be a Jordan n-higher derivation if for each $r \in \mathbb{N}$,*

$$J_r(s_n(a_1, a_2, \ldots, a_n)) = \sum_{i_1+i_2+\cdots+i_n=r} s_n(J_{i_1}(a_1), J_{i_2}(a_2), \ldots, J_{i_n}(a_n))$$

holds for all $a_1, a_2, \ldots, a_n \in \mathcal{A}$.

Definition 101 *Let $\mathcal{A}$ be an algebra and $G = \{\xi_r\}_{r \in \mathbb{N}}$ be a family of R-linear mappings on $\mathcal{A}$ such that $\xi_0 = id_{\mathcal{A}}$. Then, G is called a generalized Jordan n-higher derivation if there exists a Jordan n-higher derivation $\{J_r\}_{r \in \mathbb{N}}$ on $\mathcal{A}$ such that for each $r \in \mathbb{N}$,*

$$\xi_r(s_n(a_1, a_2, \ldots, a_n)) = \sum_{i_1+i_2+\cdots+i_n=r} s_n(\xi_{i_1}(a_1), J_{i_2}(a_2), \ldots, J_{i_n}(a_n))$$

holds for all $a_1, a_2, \ldots, a_n \in \mathcal{A}$.

Remark 5.7.17 *In particular, a (generalized-) Jordan 2-higher derivation is called a (generalized-) Jordan higher derivation and a (generalized-) Jordan*

3-higher derivation is said to be a (generalized-) Jordan triple higher derivation. (Generalized-) Jordan 2-higher derivations, (generalized-) Jordan 3-higher derivations and (generalized-) Jordan n-higher derivations can be collectively termed as (generalized-) Jordan-type higher derivations. If the condition of linearity is dropped from Definitions 100 & 101, then a Jordan n-higher derivation and a generalized Jordan n-higher derivation is called a nonlinear Jordan n-higher derivation and a nonlinear generalized Jordan n-higher derivation, respectively.

We have seen that Jordan-type derivations in different backgrounds have been extensively discussed (see, for examples, [122, 289, 492, 571], where further references can be found). In the light of the systematic results obtained in [54, 57, 122, 492, 554, 559, 571], it would be interesting to extend these results to the case of nonlinear (generalized-) Jordan-type higher derivations. In this regard, it is investigating Open Problem 5.7.18.

Open Problems 5.7.18 (*i*) *Is a nonlinear Jordan-type higher derivation on an algebra an additive higher derivation?*

(*ii*) *Is a nonlinear generalized Jordan-type higher derivation on an algebra an additive generalized higher derivation?*

(*iii*) *Is a nonlinear generalized Jordan-type higher derivation $\{\xi_r\}_{r\in\mathbb{N}}$ on an algebra $\mathcal{A}$ of the form $\xi_r(x) = \lambda_r x + J_r(x)$ for each $r \in \mathbb{N}$, where $\lambda_r \in Z(A)$ and $\{J_r\}_{r\in\mathbb{N}}$ an additive Jordan type derivation?*

Next, we discuss the notions of Jordan-type derivations and Jordan-type higher derivations in the setting of $*$-algebras. Let $\mathcal{A}$ be an associative $*$-algebra over the complex field $\mathbb{C}$. For any $a, b \in \mathcal{A}$, we denote the *skew Jordan product ($*$-Jordan product)* of a and b by $a \bullet b = ab + ba^*$. Such products based on the Jordan bracket naturally appear in relation to the so-called Jordan $*$-derivations and play an important role in the problem of representability of quadratic functionals by sesquilinear functionals on left-modules over $*$-algebras (see[462, 465, 515]). The product is workable to characterize ideals, see [145, 462]. Particular attention has been paid to understanding mappings which preserve the product $ab + ba^*$ between $*$-algebras refer to [91, 92, 198, 203, 311, 403].

The interrelation between additive and multiplicative structures of an algebra has been investigated by many researchers over the past few decades. In particular, they have studied the conditions under which bijective mappings between algebras preserving the multiplicative structure necessarily preserve the additive structure as well. The most fundamental result in this direction is due to Martindale III [447] who proved that every bijective multiplicative mapping from a prime ring containing a nontrivial idempotent onto an arbitrary ring is necessarily additive. Later, several authors considered the Jordan-type product or Lie-type product and proved that, on certain associative algebras or rings, bijective mappings which preserve any of those products are automatically additive, refer to [25, 70, 72, 86, 91, 145, 198, 200, 203, 311, 312, 325, 402, 403, 404, 411, 464, 514, 564, 572, 574].

Given the consideration of Lie-type and Jordan-type derivations, Lin in [409] introduced the notion of a $*$-Jordan-type derivation on a $*$-algebra. Let us consider a sequence of polynomials with $*$-Jordan product: $t_1(x_1) = x_1$ and

$$t_n(x_1, x_2, \ldots, x_n) = t_{n-1}(x_1, x_2, \ldots, x_{n-1}) \bullet x_n \text{ for all integers } n \geq 2.$$

Definition 102 *Let $\mathcal{A}$ be a $*$-algebra. A linear mapping $d : \mathcal{A} \longrightarrow \mathcal{A}$ is called a $*$-Jordan n-derivation if it satisfies*

$$d(t_n(x_1, x_2, \ldots, x_n)) = \sum_{k=1}^{n} t_n(x_1, \ldots, x_{k-1}, d(x_k), x_{k+1}, \ldots, x_n) \quad (5.21)$$

for all $x_1, x_2, \ldots, x_n \in \mathcal{A}$.

Remark 5.7.19 *If the mapping d is not necessarily linear in the above definition, then it is called a nonlinear $*$-Jordan n-derivation. In particular, a $*$-Jordan 2-derivation is called a $*$-Jordan derivation and a $*$-Jordan 3-derivation is called a $*$-Jordan triple derivation. These mappings are collectively referred to as $*$-Jordan-type derivations. Note that every additive $*$-derivation is a (nonlinear-) $*$-Jordan-type derivation. However, the converse is not true in general. Hence, a natural problem is to study whether every (nonlinear-) $*$-Jordan derivation on a $*$-algebra is an additive $*$-derivation.*

$*$-Jordan-type derivations in different backgrounds are extensively studied by several authors (refer to [401, 410, 528, 573]). Let $\mathcal{H}$ be a complex Hilbert space and $\mathcal{B}(\mathcal{H})$ be the algebra of all bounded linear operators on $\mathcal{H}$. In [401], Li et al. showed that if $\mathfrak{A} \subseteq \mathcal{B}(\mathcal{H})$ is a von Neumann algebra with no central summands of type I_1 (=central abelian projections), then $d : \mathfrak{A} \longrightarrow \mathcal{B}(\mathcal{H})$ is a nonlinear $*$-Jordan derivation iff d is an additive $*$-derivation. This result has been extended to the case of nonlinear $*$-Jordan triple derivations by Zhao and Li [573]. Taghavi et al. [528] and Zhang [568] independently studied nonlinear $*$-Jordan derivations on factor von Neumann algebras and proved that every nonlinear $*$-Jordan derivation between factor von Neumann algebras is an additive $*$-derivation. Inspired by the aforementioned works, Lin [410] concentrated on giving a description of nonlinear $*$-Jordan-type derivations on von Neumann algebras and obtained Theorem 5.7.20.

Theorem 5.7.20 *[410, Theorem 3.1] Let $\mathcal{A}$ be a von Neumann algebra without nonzero central abelian projections. Then, a mapping $\delta : \mathcal{A} \rightarrow \mathcal{B}(\mathcal{H})$ is a nonlinear $*$-Jordan n-derivations iff δ is an additive $*$-derivation.*

Let us define the concept of a $*$-Jordan- type higher derivation of a $*$-algebra.

Definition 103 *Let $t_n(a_1,a_2,\dots,a_n)$ be the polynomial defined by n indeterminates $a_1,a_2\dots,a_n$ and their Jordan multiple $*$-products. Let $\mathbb{N}$ be the set of non-negative integers and $D=\{d_m\}_{m\in\mathbb{N}}$ be a family of linear mappings on an $*$-algebra $\mathcal{A}$, that is, $d_m:\mathcal{A}\to\mathcal{A}$ such that $d_0=id_{\mathcal{A}}$, the identity mapping on $\mathcal{A}$. Then, D is called a $*$-Jordan n-higher derivation if for each $m\in\mathbb{N}$, D satisfies the condition*

$$d_m(t_n(a_1,a_2,\dots,a_n))=\sum_{i_1+i_2+\cdots+i_n=m} t_n(d_{i_1}(a_1),d_{i_2}(a_2),\dots,d_{i_n}(a_n)) \quad (5.22)$$

for all $a_1,a_2,\dots,a_n\in\mathcal{A}$.

Remark 5.7.21 *It is to be noted that $D=\{d_m\}_{m\in\mathbb{N}}$ is called a $*$-Jordan higher derivation whenever $n=2$, and is called a $*$-Jordan triple higher derivation whenever $n=3$. $*$-Jordan higher derivations, $*$-Jordan triple higher derivations and $*$-Jordan n-higher derivations are collectively referred to as $*$-Jordan-type higher derivations. Further, if the condition of linearity is dropped from the above definition we call D a nonlinear $*$-Jordan n-higher derivation.*

Many algebraists have made outstanding contributions for higher derivations on some classical algebras (refer to [86, 267, 459, 494, 554, 561, 569, 573]). Motivated by the aforementioned works, Wani et al. [550] investigated nonlinear $*$-Jordan n-higher derivations on von Neumann algebra without nonzero central abelian projections and proved that every nonlinear $*$-Jordan n-higher derivation $D=\{d_m\}_{m\in\mathbb{N}}$ from a von Neumann algebra $\mathfrak{A}$ to $\mathcal{B}(\mathcal{H})$ is an additive $*$-higher derivation.

Theorem 5.7.22 *[550, Theorem 2.1] Let $\mathfrak{A}$ be a von Neumann algebra without nonzero central abelian projections and $t_n(X_1,X_2,\dots,X_n)$ be a polynomial defined by n indeterminates $X_1,\dots,X_n$ with their Jordan multiple $*$-products. Then, a family $D=\{d_m\}_{m\in\mathbb{N}}$ of mappings $d_m:\mathfrak{A}\to\mathcal{B}(\mathcal{H})$ satisfies the rule*

$$d_m(t_n(A_1,A_2,\dots,A_n))=\sum_{i_1+i_2+\cdots+i_n=m} t_n(d_{i_1}(A_1),d_{i_2}(A_2),\dots,d_{i_n}(A_n))$$

for all $A_1,A_2,\dots,A_n\in\mathfrak{A}$ and for each $m\in\mathbb{N}$ iff $D=\{d_m\}_{m\in\mathbb{N}}$ is an additive $*$-higher derivation.

Recently, significant attention has been paid to the study of mappings involving a 'new product', called the bi-skew Jordan product, on $*$-algebras. Let $\mathcal{A}$ be an associative $*$-algebra over the complex field $\mathbb{C}$. For any $a_1,a_2\in\mathcal{A}$, we denote the bi-skew Jordan product of a_1 and a_2 by $a_1\diamond a_2=a_1a_2^*+a_2a_1^*$. Similar types of products based on the involution operation $(*)$, such as the skew Lie product $([a_1,a_2]_\bullet=a_1a_2-a_2a_1^*)$, the skew Jordan product $(a_1\bullet a_2=a_1a_2+a_2a_1^*)$, the bi-skew Lie product $([a_1,a_2]_\diamond=a_1a_2^*-a_2a_1^*)$, etc., have been extensively studied on various $*$-algebras (refer to [264, 265, 266, 463, 465, 485, 512, 513, 514, 515, 526, 527]). These products play an important role in some research topics, and hence their study has attracted the attention of many authors over the past decades. For example, the skew Lie product was extensively studied because, by the fundamental theorem of Šemrl in [513], maps of the form $T \mapsto TA - AT^*$ naturally arise in the problem of representing quadratic functionals with sesquilinear functionals (see [512, 514]). Particular attention has been paid to investigating maps which preserve the aforementioned products between different kinds of $*$-algebras (refer to [24, 203, 262, 263, 264, 265, 427, 526]).

Let $\mathcal{A}$ be an associative $*$-algebra. For any $a_1, a_2, \ldots, a_n \in \mathcal{A}$, lets define a sequence of polynomials as follows: $u_1(a_1) = a_1$ and $u_n(a_1, a_2, \ldots, a_n) = u_{n-1}(a_1, a_2, \ldots, a_{n-1}) \diamond a_n$ for all integers $n \geq 2$.

Definition 104 *A linear mapping $D : \mathcal{A} \to \mathcal{A}$ is called a bi-skew Jordan n-derivation $(n \geq 2)$ if*

$$D(u_n(a_1, a_2, \ldots, a_n)) = \sum_{i=1}^{n} u_n(a_1, \ldots, a_{i-1}, D(a_i), a_{i+1}, \ldots, a_n) \tag{5.23}$$

holds for all $a_1, a_2, \ldots, a_n \in \mathcal{A}$.

Remark 5.7.23 *If the mapping D is not necessarily linear, then it is called a nonlinear bi-skew Jordan n-derivation. In particular, a bi-skew Jordan 2-derivation is called a bi-skew Jordan derivation and a bi-skew Jordan 3-derivation is said to be a bi-skew Jordan triple derivation. Note that every $*$-derivation is a bi-skew Jordan n-derivation but the converse is generally not true. For example, consider the matrix algebra $\mathcal{A} = M_2(\mathbb{F})$ over any field $\mathbb{F}$ of characteristic 2 with the usual transpose involution '$*$', and define a map $D : \mathcal{A} \to \mathcal{A}$ by $D\begin{pmatrix} a & b \\ c & d \end{pmatrix} = \begin{pmatrix} a+d & b+c \\ b+c & a+d \end{pmatrix}$. Then, D is a bi-skew Jordan 2-derivation but not a $*$-derivation.*

Ashraf et al. in their paper [42] introduced the concept of a nonlinear bi-skew Jordan-type derivation and investigated its structure on a factor von Neumann algebra.

Theorem 5.7.24 *[42, Theorem 2.1] Let $\mathcal{A}$ be a factor von Neumann algebra acting on a complex Hilbert space $\mathcal{H}$ with $dim(\mathcal{A}) \geq 2$. A map $D : \mathcal{A} \to \mathcal{A}$ satisfies*

$$D(u_n(A_1, A_2, \ldots, A_n)) = \sum_{i=1}^{n} u_n(A_1, A_2, \ldots, A_{i-1}, D(A_i), A_{i+1}, \ldots, A_n)$$

for all $A_1, A_2, \ldots, A_n \in \mathcal{A}$ iff D is an additive $*$-derivation.

The following result extended the above result to the case of a nonlinear bi-skew Jordan n-derivation of a unital $*$-algebra.

Theorem 5.7.25 *[43, Theorem 2.1] Let $\mathcal{A}$ be a unital $*$-algebra with a nontrivial projection P_1 satisfying* (5.9)*. Then, a map $D : \mathcal{A} \to \mathcal{A}$ is a nonlinear bi-skew Jordan n-derivation iff D is an additive $*$-derivation.*

Definition 105 *Let $\mathcal{A}$ be an associative $*$-algebra over the complex field $\mathbb{C}$. A linear mapping $G : \mathcal{A} \to \mathcal{A}$ is said to be a generalized bi-skew Jordan n-derivation if there exists a bi-skew Jordan n-derivation D of $\mathcal{A}$ such that*

$$\begin{aligned} G(u_n(a_1,a_2,\ldots,a_n)) &= u_n(G(a_1),a_2,\ldots,a_n) \\ &\quad +\sum_{i=2}^{n} u_n(a_1,a_2,\ldots,a_{i-1},D(a_i),a_{i+1},\ldots,a_n) \qquad (5.24) \end{aligned}$$

holds for all $a_1,a_2,\ldots,a_n \in \mathcal{A}$.

Remark 5.7.26 *If the mapping G is not necessarily linear in the above definition, then G is called a nonlinear generalized bi-skew Jordan n-derivation with associated nonlinear bi-skew Jordan n-derivation D. It is easy to see that any bi-skew Jordan n-derivation is an example of a generalized bi-skew Jordan n-derivation (set $G = D$ in* (5.24)*).*

Observe that if $G : \mathcal{A} \to \mathcal{A}$ is a nonlinear generalized bi-skew Jordan n-derivation with associated nonlinear bi-skew Jordan n-derivation $D : \mathcal{A} \to \mathcal{A}$, then it follows from (5.23) and (5.24) that a map $\delta = G - D$ satisfies

$$\delta(u_n(a_1,a_2,\ldots,a_n)) = u_n(\delta(a_1),a_2,\ldots,a_n) \qquad (5.25)$$

for all $a_1,a_2,\ldots,a_n \in \mathcal{A}$. Therefore, to obtain a description of the map G it suffices to consider a map δ satisfying (5.25). The following proposition provides a description of such maps on unital $*$-algebras containing nontrivial projections.

Proposition 5.7.27 *[43, Proposition 3.1] Let $\mathcal{A}$ be a unital $*$-algebra with a nontrivial projection. Then, a map $\delta : \mathcal{A} \to \mathcal{A}$ satisfying $\delta(u_n(A_1,A_2,\ldots,A_n)) = u_n(\delta(A_1),A_2,\ldots,A_n)$ for all $A_1,A_2,\ldots,A_n \in \mathcal{A}$ is of the form $\delta(A) = ZA$ for all $A \in \mathcal{A}$, where $Z \in Z(\mathcal{A})$ and $\delta(A^*) = \delta(A)^*$ for all $A \in \mathcal{A}$.*

Theorem 5.7.28 *[43, Theorem 3.1] Let $\mathcal{A}$ be a unital $*$-algebra with a nontrivial projection P_1 satisfying* (5.9)*. Then, every nonlinear generalized bi-skew Jordan n-derivation $G : \mathcal{A} \to \mathcal{A}$ is of the form $G(A) = ZA + D(A)$ for all $A \in \mathcal{A}$, where $Z \in Z(\mathcal{A})$ and $D : \mathcal{A} \to \mathcal{A}$ is an additive $*$-derivation. Moreover, $G : \mathcal{A} \to \mathcal{A}$ is an additive generalized $*$-derivation.*

Proof. Let $G : \mathcal{A} \to \mathcal{A}$ be a nonlinear generalized bi-skew Jordan n-derivation with associated nonlinear bi-skew Jordan n-derivation $D : \mathcal{A} \to \mathcal{A}$. Then, the map $\delta = G - D$ satisfies

$$\delta(u_n(A_1,A_2,\ldots,A_n)) = u_n(\delta(A_1),A_2,\ldots,A_n)$$

for all $A_1, A_2, \ldots, A_n \in \mathcal{A}$. Hence, it follows from Proposition 5.7.27 that $\delta(A) = ZA$ and $\delta(A^*) = \delta(A)^*$ for all $A \in \mathcal{A}$, where $Z \in Z(\mathcal{A})$. Thus, $G(A) = ZA + D(A)$ for all $A \in \mathcal{A}$, where $Z \in Z(\mathcal{A})$. Moreover, in view of Theorem 5.7.25, D is an additive $*$-derivation. We only need to show that G is an additive generalized $*$-derivation. We assert that G is an additive generalized $*$-derivation with associated additive $*$-derivation D. Clearly, G is additive. For any $A_1, A_2 \in \mathcal{A}$, we have

$$\begin{aligned} G(A_1)A_2 + A_1D(A_2) &= (ZA_1 + D(A_1))A_2 + A_1D(A_2) \\ &= (ZA_1)A_2 + D(A_1)A_2 + A_1D(A_2) \\ &= Z(A_1A_2) + D(A_1A_2) \\ &= G(A_1A_2). \end{aligned}$$

Now, by Proposition 5.7.27, for any $A \in \mathcal{A}$, we get

$$G(A^*) = \delta(A^*) - D(A^*) = \delta(A)^* - D(A)^* = (\delta(A) - D(A))^* = G(A)^*.$$

Consequently, G is an additive generalized $*$-derivation with associated additive $*$-derivation D.

Definition 106 *Let $D = \{\phi_n\}_{n\in\mathbb{N}}$ be a family of linear mappings on a $*$-algebra $\mathcal{A}$ such that $\phi_0 = id_{\mathcal{A}}$. Then, D is called a bi-skew Jordan n-higher derivation if for each $r \in \mathbb{N}$,*

$$\phi_r(u_n(a_1, a_2, \ldots, a_n)) = \sum_{i_1+i_2+\cdots+i_n=r} u_n(\phi_{i_1}(a_1), \phi_{i_2}(a_2), \ldots, \phi_{i_n}(a_n))$$

holds for all $a_1, a_2, \ldots, a_n \in \mathcal{A}$.

Definition 107 *Let $G = \{\xi_n\}_{n\in\mathbb{N}}$ be a family of linear mappings on a $*$-algebra $\mathcal{A}$ such that $\xi_0 = id_{\mathcal{A}}$. Then, G is said to be a generalized bi-skew Jordan n-higher derivation if there exists a bi-skew Jordan n-higher derivation $D = \{\phi_n\}_{n\in\mathbb{N}}$ such that for each $r \in \mathbb{N}$,*

$$\xi_r(u_n(a_1, a_2, \ldots, a_n)) = \sum_{i_1+i_2+\cdots+i_n=r} u_n(\xi_{i_1}(a_1), \phi_{i_2}(a_2), \ldots, \phi_{i_n}(a_n))$$

holds for all $a_1, a_2, \ldots, a_n \in \mathcal{A}$.

If the condition of linearity is dropped from Definitions 106 & 107, then a bi-skew Jordan n-higher derivation and a generalized bi-skew Jordan n-higher derivation is called a *nonlinear bi-skew Jordan n-higher derivation* and a *nonlinear generalized bi-skew Jordan n-higher derivation*, respectively.

Remark 5.7.29 *It is easy to verify that every $*$-higher derivation is a bi-skew Jordan n-higher derivation and every bi-skew Jordan n-higher derivation is a generalized bi-skew Jordan n-higher derivation. However the converse statements are generally, not true (see Remark 5.6.58).*

We close this chapter with the following;

Open Problems 5.7.30 (i) *Is a nonlinear bi-skew Jordan n-higher derivation on a unital* $*$*-algebra an additive* $*$*-higher derivation?*

(ii) *Is a nonlinear generalized bi-skew Jordan n-higher derivation on a unital* $*$*-algebra an additive generalized* $*$*-higher derivation?*

Chapter 6

Applications of Derivations

6.1 Introduction

This chapter is devoted to presenting some characterizations of derivations and their variants in the cases satisfying some appropriate local properties. From what the literature has proposed in recent years, it is well known that there is a close and precise connection between the behavior of a derivation and the intrinsic structure of the ring in which it is defined. In the first part, we therefore turn our attention to derivations and their variants satisfying certain algebraic properties such as Engel-type conditions (hypercentralizing derivations) and strong commutativity preserving conditions (strong preservers). Subsequently, we develop one of the main themes of the theory of noncommutative rings, that is the study of those identites usually called *quadratic*, whose definition is strictly correlated with the concept of commuting and centralizing maps. Initially, we focus our attention on the differential quadratic identity $[d(x), \delta(x)] = 0$, for d, δ derivations on a prime ring. Starting from this just mentioned aspect, we finish with a more general level of topics, namely those relating to the concept of the generalized differential identity, i.e., the one $[F(x), G(x)] = 0$, for F and G generalized derivations on a prime ring.

The next section is dedicated to the study of generalized derivations satisfying certain properties of nilpotency, i.e., generalized derivations whose evaluations on a suitable subset of R are nilpotent elements in R. Extending the problem to a wider range, the question we shall therefore answer is the following: suppose that $F \neq 0$ is a generalized derivation of R such that, for every $x \in T$, a subset of R, either $F(x)$ is invertible or $F(x)$ is nilpotent in R, then must R have a very special structure? In the final part of the chapter we deal with two aspects closely related with differential identities' theory: Lie-type mappings, whose study is part of Herstein's famous program, and the the analysis of some subrings and subgroups generated by commutators involving derivations, whose size within the ring is sufficiently large, to the point that the informations deduced about their structure allows the study of an entire ring.

In what follows, we continue to use the various notations developed in earlier chapters. In particular, unless otherwise stated, R is a prime (or semiprime) ring with a right Martindale quotient ring Q_r and an extended centroid $C = Z(Q_r)$.

6.2 Hypercentralizing derivations

The first important result on centralizing maps is Posner's second theorem from 1957, which asserts that the existence of a nonzero centralizing derivation on a prime ring R implies that R must be commutative (see Theorem 2.2.13). The assumption on the primeness of the ring is natural in some sense. A trivial example which justifies the presence of the hypothesis on the primeness of the ring is that of the direct product $R = R_1 \times R_2$ of two nonzero rings R_1 and R_2, which is not a prime ring. If R_1 is a commutative ring having a nonzero derivation d_1 and R_2 is any noncommutative ring, then R is a noncommutative ring and $d(x_1,x_2) = (d_1(x_1),0)$ is a nonzero commuting derivation on R. The theorem has been extremely influential, proof of this is the fact that, starting from this result, several authors studied the relationship between the structure of a prime ring R and the behavior of an additive mapping f which satisfies the Engel-type condition $[f(x),x]_k = 0$ on some subsets S of R. We recall that the Engel-condition is defined by $[f(x),x]_k = [[f(x),x]_{k-1},x]$, for all $x \in S \subseteq R$ and $k > 1$. A map satisfying the Engel-type condition $[f(x^n),x^n]_k = 0$ on $S \subseteq R$, for $k,n \geq 1$, is said to be *hypercentralizing* on S.

In [364] Lanski generalizes the Posner's result to higher commutators, extending it from prime rings to Lie ideals in prime rings. In his paper, Lanski first proves Theorem 6.1.1 which is crucial to his arguments:

Theorem 6.2.1 *[364, Theorem 1] Let R be a prime ring, I a nonzero ideal of R, d a nonzero derivation of R and $k \geq 1$ fixed. If $[d(x),x]_k = 0$ for any $x \in I$, then R is commutative.*

The generalization to Lie ideals still makes use of the tools related to the differential identities theory in conjunction with the generalized polynomial identities theory.

Theorem 6.2.2 *[364, Theorem 2] Let R be a prime ring, L a noncommutative Lie ideal of R, d a nonzero derivation of R and $k \geq 1$ fixed. If $[d(x),x]_k = 0$ for any $x \in L$, then $char(R) = 2$ and $R \subseteq M_2(K)$, the 2×2 matrix ring over a filed K.*

Theorem 6.2.3 *[366, Theorem 1] Let R be a prime ring, L a non-zero left (right) ideal of R, d a nonzero derivation of R and $n, k \geq 1$ fixed integers. If $[d(x^n), x^n]_k = 0$ for any $x \in L$, then R is commutative.*

Here, by using Kharchenko's theory of differential identities (see [345] and [346]), we first expound the proof of Theorem 6.1.4 concerning hypercentralizing derivations on a Lie ideal of a prime ring R:

Theorem 6.2.4 *Let R be a prime ring, L a non-central Lie ideal of R, d a nonzero derivation of R and $k, n \geq 1$ fixed integers such that $[d(u^n), u^n]_k = 0$, for all $u \in L$. Then $dim_C(RC) = 4$.*

Proof: In all that follows, we assume $dim_C(RC) > 4$, to obtain a contradiction.

Since L is a non-central Lie ideal of R, and by Theorem 1.2.8, there exists a non-zero ideal I of R such that $0 \neq [I, R] \subseteq L$.

First suppose that d is an inner derivation induced by the element $q \in Q_r$. Then $q \notin C$ since $d \neq 0$ and $d(x) = [q, x]$, for any $x \in R$. By assumption it follows that $[q, [x_1, x_2]^n]_{k+1}$ is a non-trivial generalized polynomial identity for I, as well as for R. Let's set $f(x_1, x_2) = [x_1, x_2]^n$.

By [184, Theorem 2], $[q, f(x_1, x_2)]_{k+1}$ is also an identity for RC as well as for Q_r. In case C is infinite, we obtain

$$[q, f(r_1, r_2)]_{k+1} = 0$$

for all $r_1, r_2 \in Q_r \otimes_C \overline{C}$, where $\overline{C}$ is the algebraic closure of C. Since both Q_r and $Q_r \otimes_C \overline{C}$ are centrally closed, we may replace R by Q_r or $Q_r \otimes_C \overline{C}$ according as C is finite or infinite. Thus we may assume that R is centrally closed over C which is either finite or algebraically closed. By Martindale's theorem [447], R is a primitive ring having a non-zero socle with C as the associated division ring, and R is isomorphic to a dense ring of linear transformations on some vector space V over C. Hence, either R is a finite-dimensional central division algebra over C or R contains nontrivial idempotent elements.

Consider the following subset of R:

$$A = \{a \in R: \text{ such that } \forall r = (r_1, r_2) \in R^2, \ [a, f(r)]_{k+1} = 0\}.$$

It is easy to see that A is C-subspace of R which is invariant under the action of all inner automorphisms of R. If R is a finite-dimensional central division algebra, then $[R, R] \subset A$ follows from [34, Theorem 7]. On the other hand, if R contains nontrivial idempotents, by [182, Theorem], and since $dim_C(RC) \neq 4$, it follows that either $A \subseteq C$ or $[I', R] \subseteq A$, where I' is a noncentral ideal of R. In either case, since $q \in A$ and we suppose that $q \notin C$, we have to study the only case $[I', R] \subseteq A$, which implies that $[[I', R], f(r_1, r_2)]_{k+1} = 0$, for all $r_1, r_2 \in R$. By the well known fact that I and R satisfy the same polynomial identities, we have

$[[R,R],f(r_1,r_2)]_{k+1}=0$, for all $r_1,r_2\in R$. Next suppose that d is not inner. The expansion of

$$\Big[d\big(f(x_1,x_2)\big),f(x_1,x_2)\Big]_k$$

yields that R satisfies

$$\Big[\sum_{i+j=n-1}[x_1,x_2]^i\big([d(x_1),x_2]+[x_1,d(x_2)]\big)[x_1,x_2]^j,f(x_1,x_2)\Big]_k.$$

Since d is an outer derivation, and by Kharchenko's theory, R satisfies the identity

$$\Big[\sum_{i+j=n-1}[x_1,x_2]^i\big([y_1,x_2]+[x_1,y_2]\big)[x_1,x_2]^j,f(x_1,x_2)\Big]_k. \tag{6.1}$$

For $r\in R$, replace y_i by $[r,x_i]$ in (6.1). Thus we see that

$$\Big[r,f(r_1,r_2)\Big]_{k+1}=0\ \ \forall r,r_1,r_2\in R.$$

The previous argument says that, in either case R satisfies the identity

$$\Big[[y_1,y_2],f(x_1,x_2)\Big]_{k+1} \tag{6.2}$$

Since R is a PI-ring, then $Z(R)\neq 0$ and $R=RC$ is a finite-dimensional central simple algebra. Assume firstly $char(R)=0$. Say $dim_C(RC)=m^2$, for some integer $m\geq 2$. Thus we have, for any $r_1,r_2\in R$, $[[RC,RC],f(r_1,r_2)]_k=0$. If $\delta(x)=[x,f(r_1,r_2)]$, the inner derivation induced by the valutation $f(r_1,r_2)$, it follows that $\delta^k([RC,RC])=0$. By [451, Theorem 2 and Corollary 1], there exists $z=z(r_1,r_2)\in C$, depending on the choice of r_1,r_2, and an integer $t\geq 1$, such that $(f(r_1,r_2)-z)^t=0$.

If R is a division ring then, for any $r_1,r_2\in R$ there exists $z=z(r_1,r_2)\in Z(R)$, depending on the choice of r_1,r_2, such that $f(r_1,r_2)=z$, which contradicts the fact that $dim_C(RC)>4$.

Now let $UD(C,m)$ be the generic division algebra of dimension m^2 over its center. It is well known that R, RC and $UD(C,m)$ satisfy the same polynomial identities, because C is infinite. Hence $UD(C,m)$ satisfies $[[x,y],f(x_1,x_2)]_k$ as R does. By the above division ring case, $UD(C,m)$ satisfies $[f(x_1,x_2),x_3]$, hence also R satisfies $[f(x_1,x_2),x_3]$, which is a contradiction again.

Suppose now that $char(R)=p>0$. Fix the integer $t\geq 0$ such that $t=min\{s\geq 0:p^s\geq k+1\}$. Then, by (6.2), for any $s_1,s_2,r_1,r_2\in R$ it follows that

$$0=\Big[\big[[s_1,s_2],f(r_1,r_2)\big]_k,f(r_1,r_2)\Big]_{p^t-k}=\Big[[s_1,s_2],f(r_1,r_2)^{p^t}\Big].$$

Let G be the additive subgroup generated by the polynomial $f(x_1,x_2)^{p^t}$.

By [183], and since $dim_C(RC) > 4$, we have, $f(x_1,x_2)^{p^t}$ cannot be central valued on R, so, there exists a non-central Lie ideal L' of R such that $L' \subseteq A$. In this last case we get $[[s_1,s_2],u] = 0$ for all $s_1,s_2 \in R$ and $u \in L'$, that is L' centralizes $[R,R]$, which easily implies that L' centralizes on R, a contradiction.

♣

Note that the proof of the Lie ideals case is fully developed on the basis that a noncentral Lie ideal of R contains all the commutators $[x,y]$. It is therefore natural to consider the situation when $[d(x),x]_k = 0$ for all $x = f(x_1,\ldots,x_n)$, where $f(X_1,\ldots,X_n)$ is a polynomial over C. From this context the idea of extending the previous Theorem 6.1.4 to Theorem 6.1.5 is founded:

Theorem 6.2.5 *[383, Theorem] Let R be a prime ring, $f(x_1,\ldots,x_n)$ a multilinear polynomial over C, d a nonzero derivation of R and $k \geq 1$ a fixed integer such that*

$$\Big[d\big(f(r_1,\ldots,r_n)\big),f(r_1,\ldots,r_n)\Big]_k = 0, \forall r_1,\ldots,r_n \in R.$$

Then $f(x_1,\ldots,x_n)$ is central valued on R except when $char(R) = 2$ and $R \subseteq M_2(C)$.

We would like to remark that, as a natural generalization, Lee extends Theorems 6.2.5 and 6.2.4 by imposing the condition $\big[d\big(f(r_1,\ldots,r_n)\big),f(r_1,\ldots,r_n)\big]_k = 0$, where $f(x_1,\ldots,x_n)$ is not necesarily multilinear. More precisely, in [377, Theorem 2] Lee proves that, if R is a prime ring, $f(x_1,\ldots,x_n)$ a polynomial over C, d a nonzero derivation of R, I a nonzero left ideal of R and $k \geq 1$ a fixed integer such that

$$\Big[d\big(f(r_1,\ldots,r_n)\big),f(r_1,\ldots,r_n)\Big]_k = 0, \forall r_1,\ldots,r_n \in I$$

then $IC = RCe$ for some idempotent element e in the socle of RC and the following statements hold:

1. $f(x_1,\ldots,x_n)$ is central valued on $eRCe$ except when C is finite or $0 < char(R) \leq k+1$.
2. If $char(R) = p > 0$ then $f(x_1,\ldots,x_n)^{p^s}$ is central valued on $eRCe$ for some integer $s \geq 1$, except when $p = 2$ and $dim_C(eRCe) = 4$.
3. If $f(x_1,\ldots,x_n)$ is multilinear, then it is central valued on $eRCe$ except when $char(R) = 2$ and $dim_C(eRCe) = 4$.

One of the main topics that have been treated in the last two decades is that of proving a generalization of the known results for derivations to the case of generalized derivations, i.e., those maps $g : R \to R$, defined on the ring R, such as to satisfy the rule $g(xy) = g(x)y + xd(y)$, for any $x,y \in R$, and for some derivation

d in R. Now we consider some problem related to the study of hypercentralizing generalized derivations. Motivated by the results above, we shall prove Theorems 6.1.6 and 6.1.7.

Theorem 6.2.6 *[6, Theorem 1] Let R be a noncommutative prime ring and G be a generalized derivation of R such that $[G(r^k), r^k]_n = 0$ for all $r \in R$, where k, n are fixed positive integers. Then there exists $a \in C$ such that $G(x) = ax$ for all $x \in R$.*

Theorem 6.2.7 *[6, Theorem 2] Let R be a noncommutative prime ring with a nonzero left ideal I. Let G be a generalized derivation of R such that $[G(r^k), r^k]_n = 0$ for all $r \in I$, where k, n are fixed positive integers. Then there exists $c \in Q_r$, the Utumi quotient ring of R, such that $g(x) = xc$ and $I(c - \alpha) = 0$ for a suitable $\alpha \in C$. In particular $G(x) = \alpha x$, for all $x \in I$.*

As well known, the generalized derivation G can be uniquely extended to Q_r and assumes the form $G(x) = ax + d(x)$, for some $a \in Q_r$ and a derivation d on Q_r.

We begin with Lemma 6.1.8.

Lemma 6.2.8 *Let $R = M_t(F)$, where F is a field, $t \geq 2$ and $a, b \in R$. Suppose*

$$[ar^k, r^k]_n + [b, r^k]_{n+1} = 0 \quad \text{for all} \quad r \in R \tag{6.3}$$

where k, n are fixed positive integers. Then $a, b \in F$.

Proof: Let e be an idempotent in R. Setting $x = e$ in (6.3) and right multiplying by $(1-e)$, we see that $eb = ebe$ for any idempotent e. Thus b is a diagonal matrix. Note that ubu^{-1} must be diagonal for each invertible element $u \in R$, since $[(uau^{-1})r^k, r^k]_n + [(ubu^{-1}), r^k]_{n+1} = 0$ for all $r \in R$. Write $b = \sum \beta_i e_{ii}$ where, $\beta_i \in F$. Then for each $j > 1$, we see that $\beta_j - \beta_1$, the $(1, j)-$entry of $(1 + e_{1j})b(1 + e_{1j})^{-1}$, equals 0. That is, $\beta_j = \beta_1$ for $j > 1$, hence $b \in F$. Thus by (6.3) we get $[ar^k, r^k]_n = 0$ for all $r \in R$. Taking $r = e$ in the last equation we obtain $[ae, e]_n = 0$. Expansion of $(1-e)[ae, e]_n = 0$ yields $ae = eae$ for any idempotent e. Thus a is a diagonal matrix. Similarly we have $a \in F$. This proves the lemma.

♣

Lemma 6.2.9 *Let R be a prime ring and $a, b \in R$ such that $[ar^k, r^k]_n + [b, r^k]_{n+1} = 0$ for all $r \in R$, where k, n are positive integers. Then $a, b \in Z(R)$.*

Proof: Suppose on the contrary that either $a \notin C$ or $b \notin C$. Say $a \notin C$. Set $f(X) = [aX^k, X^k]_n + [b, X^k]_{n+1} = [a, X^k]_n X^k + [b, X^k]_{n+1}$. Then $f(X)$ is a non-trivial generalized identity (GPI) for R. By [184], $f(X)$ is also a GPI for Q_r. Denote by K either the algebraic closure of C or C according as C is either infinite or finite, respectively. Then, by a standard argument (refer to [392, Proposition]), $f(X)$ is also a *GPI* for $Q_r \otimes_C K$. Since $Q_r \otimes_C K$ is a centrally closed prime

K-algebra [257, Theorems 2.5 and 3.5], by replacing R, C with $Q_r \otimes_C K, K$, respectively we may assume R is centrally closed and C is either finite or algebraic closed. By Martindale's theorem [447] R is a primitive ring having a non-zero socle H with C as its the associated division ring.

Since $a \notin C$, we have $[a,h] \neq 0$ for some $h \in H$. By Litoff's theorem [258], there exists an idempotent $e \in H$ such that $h, ah, ha \in eRe$. Note that $ef(eXe)e$ is a *GPI* for R. Thus, $[(eae)X^k, X^k]_n + [ebe, X^k]_{n+1}$ is a GPI for eRe. Since $eRe \cong M_s(C)$ for some $s \geq 1$ then eae is central in eRe by Lemma 6.2.8. Hence $ah = eahe = eaeh = heae = ha$, contrary to the choice of h. Similarly if $b \notin C$, we get a contradiction.

♣

Corollary 6.2.1 *Let R be a prime ring and $a \in R$ such that $[ar^k, r^k]_n = 0$ for all $r \in R$, where k, n are fixed positive integers. Then $a \in Z(R)$.*

Lemma 6.2.10 *Let R be a prime ring with nonzero left ideal I and $a, b \in R$ such that*

$$[ar^k, r^k]_n + [b, r^k]_{n+1} = 0 \quad \text{for all} \quad r \in I$$

where k, n are positive integers. Then there exist $\alpha, \beta \in C$ such that $I(b-\beta) = 0$, $I(a-\alpha) = 0$ and $a+b = \alpha+\beta$.

Proof: The first aim is to prove that either the lemma holds or R is a GPI-ring.

Note that if $a \in C$, then the condition reduces to $[b, r^k]_{n+1} = 0$ for all $r \in I$. In this case we infer that either $b \in C$ or R is commutative by Theorem 6.2.3. Thus we consider only the case when $a \notin C$.

First suppose that for any $u \in I$, $\{u, ua\}$ are linearly C- dependent. This means that for any $u \in I$ there exist $\alpha_u \in C$ such that $ua = \alpha_u u$, in other words $u(a - \alpha_u) = 0$. Standard arguments show that in this case there exists a fixed $\alpha \in C$ such that $I(a-\alpha) = 0$. Denote $a' = (a-\alpha)$ so that $Ia' = 0$. By our assumption we have that, for all $r \in I$

$$0 = [a'r^k, r^k]_n + [b, r^k]_{n+1} = a'r^{k(n+1)} + [b, r^k]_{n+1}. \tag{6.4}$$

Left multiplying (6.4) by $r \in I$, we get $r[b, r^k]_{n+1} = 0$. In particular, for any $v \in I, r \in R$ we get $rv[b, (rv)^k]_{n+1} = 0$. If $\{v, vb\}$ are linearly C-dependent for any $v \in I$, as above we have $I(b-\beta) = 0$, for a suitable $\beta \in C$. In this case, from (6.4) we get, for all $x \in I$ and $b' = b - \beta$

$$0 = a'r^{k(n+1)} + [b', r^k]_{n+1} = (a' + b')r^{k(n+1)}$$

for all $r \in I$. Hence, by [188, Lemma 3], since R is prime and $I \neq 0$, we get $a' + b' = 0$.

Suppose now that there exists $v \in I$ such that $\{v, vb\}$ are linearly C-independent.

In this case $xv[b,(xv)^k]_{n+1}$ is a non-trivial *GPI* for R. Therefore we have proved that, if for any $u \in I$, $\{u, ua\}$ are linearly C-dependent, then either the lemma holds or R is a GPI-ring. Following the same argument we may prove that, if for any $u \in I$, $\{u, ub\}$ are linearly C-dependent, then either the lemma holds or R is a GPI-ring. We omit it for brevity.

Suppose now $u_0, v_0 \in I$ such that $\{u_0, u_0a\}$ are linearly C-independent and also $\{v_0, v_0b\}$ are linearly C-independent.

Let $r_R(I)$ be the right annihilator of I. If $\{1,a,b\}$ are linearly C-independent modulo $r_R(I)$, by [375, Lemma 3], either R is a P.I.-ring or there exists $x_0 \in I$ such that $\{x_0, x_0a, x_0b\}$ are linearly C-independent. In this last case

$$x_0[a(xx_0)^k,(xx_0)^k]_n + x_0[b,(xx_0)^k]_{n+1}$$

is a non-trivial generalized polynomial identity for R.

Then suppose that $\{1,a,b\}$ are linearly C-dependent modulo $r_R(I)$, that is $I(\alpha a + \beta b + \gamma) = 0$, for suitable $\alpha, \beta, \gamma \in C$. If $\alpha = \beta = 0$ then $\gamma = 0$, we may assume that either $\alpha \neq 0$ or $\beta \neq 0$.

If $\alpha = 0$ and $\beta \neq 0$, we may write $I(b+\lambda) = 0$, for suitable $\lambda \in C$. If we denote $b' = b + \lambda$, we have $Ib' = 0$, moreover,

$$[ar^k, r^k]_n + [b', r^k]_{n+1} = 0$$

for all $r \in I$. Left multiplying by $y \in I$ we get,

$$0 = y[ar^k, r^k]_n + y[b', r^k]_{n+1} = y[ar^k, r^k]_n.$$

In particular for any $s \in R$,

$$u_0[a(su_0)^k, (su_0)^k]_n = 0$$

that is

$$u_0[a(xu_0)^k, (xu_0)^k]_n$$

is a non-trivial generalized polynomial identity for R.

Now let $\alpha \neq 0$ and $\beta = 0$. In this case we may assume that there exists $\lambda \in C$ such that $I(a+\lambda) = 0$ and denote $a' = a + \lambda$, so that $Ia' = 0$. As above, for $y, r \in I$ we get,

$$0 = y[a'r^k, r^k]_n + y[b, r^k]_{n+1} = y[b, r^k]_{n+1}.$$

In particular

$$v_0[b, (xv_0)^k]_{n+1}$$

is a non-trivial generalized polynomial identity for R.

Finally suppose that both $\alpha \neq 0$ and $\beta \neq 0$. We may assume that there exist $0 \neq \beta', 0 \neq \gamma' \in C$ such that, for all $y \in I$,

$$ya = y(\beta' b + \gamma').$$

We start again from

$$x_0[a(rx_0)^k, (rx_0)^k]_n + x_0[b, (rx_0)^k]_{n+1} = 0$$

for any $r \in R$, $x_0 \in I$. By calculation we get,

$$x_0[a(rx_0)^k, (rx_0)^k]_n + x_0[b, (rx_0)^k]_{n+1} =$$

$$\beta' x_0[b(rx_0)^k, (rx_0)^k]_n + x_0[b, (rx_0)^k]_{n+1}.$$

In particular, since $\{v_0, v_0 b\}$ are linearly C-independent,

$$\beta' v_0[b(xv_0)^k, (xv_0)^k]_n + v_0[b, (xv_0)^k]_{n+1}$$

is a non-trivial generalized polynomial identity for R.

By [447] we may assume that Q_r is a primitive ring with a nonzero socle H and we have $I(a-\lambda) = 0$ and $I(b-\lambda') = 0$ iff $HI(a-\lambda) = 0$ and $HI(b-\lambda') = 0$. Replacing R, I by H, HI respectively we may assume that R is simple and equal to its own socle and $RI = I$. Let e be an idempotent in I.

By our assumption, for any $r \in I$, the following holds:

$$[ar^k, r^k]_n + [b, r^k]_{n+1} = 0. \tag{6.5}$$

Choose $r = e$ in (6.5) so that $[ae, e]_n + [b, e]_{n+1} = 0$. Formally multiply this equation by $(1-e)$ from the right-hand side. Thus we get $eb(1-e) = 0$ that is $eb = ebe \in I$.

Now replace r by $ese \in eHe$ in (6.5), for an arbitrary $s \in H$, then

$$[a(ese)^k, (ese)^k]_n + [b, (ese)^k]_{n+1} = 0$$

and left multiplying by e we get,

$$[(eae)(ese)^k, (ese)^k]_n + [(ebe), (ese)^k]_{n+1} = 0.$$

Since eHe is a central simple algebra in which the hypothesis of Lemma 6.2.9 holds, we infer that eae, ebe are both central in eHe.

Obviously $C = Z(Q_r)$, centralizes $H \subseteq Q_r$, so that $Ce \subseteq Z(eHe)$, for any idempotent element $e \in H$. On the other hand, if $h_0 \in Z(eHe)$, then for any $h \in H$ we have $[h_0, ehe] = 0$ that is $h_0ehe = eheh_0$. By [447, Theorem 1], it follows that $h_0e = \omega e$, for some $\omega \in C$. Therefore $Z(eHe) \subseteq Ce$, which means that $Z(eHe) = Ce$.

In light of this we have $eae \in Ce$ and $ebe = eb \in Ce$.

In particular we get $e(b-\beta) = 0$ for a suitable $\beta \in C$. Since this holds for all $e^2 = e \in I$, by the regularity of H it follows that $I(b-\beta) = 0$.

Now put $r = e$ in (6.5), again, then we get $[ae,e]_n + [b,e]_{n+1} = 0$; subsequently, left multiply it by $(1-e)$. It follows that $(1-e)(a+b)e = 0$, i.e., $ae - eae + be - ebe = 0$.

Since we know that $eae \in Ce$ and $ebe \in Ce$, it follows that $ae + be \in Ce$. Also in this case, all that holds for any $e^2 = e \in I$. Thus get $(a+b-\lambda)e = 0$ which implies $(a+b-\lambda)I = 0$ for a suitable $\lambda \in C$. Since I is a left ideal, we get $(a+b-\lambda) = 0$. This means that

$$b = \lambda - a, \qquad I(a-(\lambda-\beta)) = 0.$$

In other words we conclude that there exist $\alpha, \beta, \lambda \in C$ such that $I(b-\beta) = 0, I(a-\alpha) = 0$ where $a+b = \lambda$ and $\alpha = \lambda - \beta$.

♣

From this Corollary 6.1.2 follows.

Corollary 6.2.2 *Let R be a prime ring with nonzero left ideal $I$and $a \in R$ such that $[ar^k, r^k]_n = 0$ for all $r \in I$, where k,n are positive integers. Then $a \in C$.*

The proof of Theorem 6.2.6

Since $G(x) = ax + d(x)$ for some $a \in Q_r$ and a derivation d on Q_r, then R satisfies the differential identity

$$[ax^k + d(x^k), x^k]_n.$$

If $d = 0$, then $G(x) = ax$ and $[ar^k, r^k]_n = 0$ for all $r \in R$. By Corollary 6.2.1 we get $a \in C$. So we may assume that $d \neq 0$.

In light of Kharchenko's theory, we divide the proof into two cases:

Case 1. Let d be the inner derivation induced by the element $b \in Q_r \setminus C$, that is $d(x) = [b,x]$ for all $x \in Q_r$. Thus R satisfies the generalized polynomial identity

$$[ax^k, x^k]_n + [b, x^k]_{n+1}.$$

By [184], it is also a generalized polynomial identity for Q_r. Then by Lemma 6.2.9, it follows that $b \in C$, a contradiction.

Case 2. Now let d be an outer derivation of Q_r. Set $P(Y,X) = \sum_{i=0}^{k-1} X^i Y X^{k-1-i}$. Note that $d(x^k) = P(d(x),x)$. Then R satisfies the differential identity

$$[ax^k + d(x^k), x^k]_n = [ax^k + P(d(x),x), x^k]_n.$$

By Kharchenko's theory again, R satisfies $[ax^k + P(y,x), x^k]_n$.

Taking $y = 0$, it follows that R satisfies $[ax^k, x^k]_n$. Hence $a \in C$ by Corollary 6.2.1. Then, according to the hypothesis, $[d(r^k), r^k]_n = 0$ for all $r \in R$. Therefore $d = 0$ from Theorem 6.2.3, a contradiction.

♣

The proof of Theorem 6.2.7

If we assume $G(x) = ax$ for some $a \in Q_r$ (i.e., $d = 0$), the hypothesis is $[ar^k, r^k]_n = 0$ for all $r \in I$. Then $a \in C$ by Corollary 6.2.2. In this case the proof is complete. So we may assume that $d \neq 0$.

In light of Kharchenko's theory, we divide the proof into two cases:

Case 1. Let d be the inner derivation induced by the element $b \in Q_r$, that is $d(x) = [b,x]$ for all $x \in Q_r$. Thus I satisfies the identity

$$[a(x)^k, (x)^k]_n + [b, (x)^k]_{n+1} \tag{6.6}$$

and a fortiori it follows that RI satisfies the identity (6.6). Since by [184] R and Q_r satisfy the same GPIs, Q_rI satisfies the identity $[ax^k, x^k]_n + [b, x^k]_{n+1}$. Applying Lemma 6.2.10 to Q_rI, there exist $\alpha, \beta, \lambda \in C$ such that $Q_rI(b - \beta) = 0$, $Q_rI(a - \alpha) = 0$ and $a + b = \lambda = \alpha + \beta$. From this we have that $I(b - \beta) = 0$ and $I(a - \alpha) = 0$. Thus we conclude that, for all $x \in R$

$$G(x) = ax + [b,x] = ax + bx - xb = (a+b)x - xb = x(\alpha + \beta - b)$$

and if we put $c = \alpha + \beta - b$, it follows that, for all $x \in R$, $G(x) = xc$ with $I(c - \alpha) = 0$.

In particular, if we put $b' = b - \beta$, we get, for all $x \in I$,

$$G(x) = ax + [b,x] = ax + [b',x] = ax + b'x = \alpha x.$$

Case 2. Now let d be an outer derivation of Q_r. To continue the proof we take $P(Y,X) = \sum_{i=0}^{k-1} X^i Y X^{k-1-i}$, a non-commuting polynomial in variables X and Y. Note that $d(x^k) = P(d(x),x)$. Since

$$[ax^k + P(d(x),x), x^k]_n$$

is an identity for I, then for any $u \in I - C$,

$$[a(xu)^k + P(d(xu),xu), (xu)^k]_n$$

is an identity for R. Thus R satisfies,

$$[a(xu)^k + G(d(x)u + xd(u), xu), (xu)^k]_n.$$

Since d is an outer derivation, by Kharcenko's results, R satisfies the identity

$$[a(xu)^k + P(yu + xd(u), xu), (xu)^k]_n. \tag{6.7}$$

Taking $y = 0$ in (6.7) it follows that R satisfies

$$[a(xu)^k + G(xd(u), xu), (xu)^k]_n. \tag{6.8}$$

By the linearity of $P(Y,X)$ in Y, subtracting equation (6.8) from (6.7) proves that R satisfies

$$[G(yu, xu), (xu)^k]_n$$

which implies that R satisfies

$$[\sum_{i+j=k-1} (xu)^i (yu)(xu)^j, (xu)^k]_n$$

that is

$$\sum_{i+j=k-1} (xu)^i [(yu), (xu)^k]_n (xu)^j. \tag{6.9}$$

Clearly (6.9) is a nontrivial GPI for R, since $u \notin C$. As remarked above, also in this case, by Martindale's theorem, RC is a primitive ring with a non-zero socle H. Note again that H is simple and HI satisfies the same basic conditions as I ([372]). In particular RI, HI and I satisfy the same GPIs. Now replace R by H and I by HI, then, without loss of generality, R is simple and equal to its own socle and $RI = I$. Let $e^2 = e$ some non-trivial idempotent $e \in I$. So for all $r, s \in R$,

$$\sum_{i+j=k-1} (re)^i [(se), (re)^k]_n (re)^j = 0$$

and choosing $s = (1-e)t \in R$, for any $t \in R$, we have

$$(1-e)(te)(re)^{k(n+1)-1} = 0.$$

This leads to the contradiction that either $e = 0$ or $e = 1$.

Thus any idempotent element of I is trivial, that is, $I = R$. Therefore we have to consider the condition

$$\sum_{i+j=k-1} r^i [s, r^k]_n r^j = 0$$

for all $r, s \in R$. Since R is a P.I.-ring, it is well known that there exists F_m, the ring of all $m \times m$ matrices over a field F, such that R and F_m satisfy the same polynomial identities. In particular for all $r, s \in F_m$, we have

$$\sum_{i+j=k-1} r^i [s, r^k]_n r^j = 0. \tag{6.10}$$

For $m \geq 2$, $r = e_{11}$ and $s = e_{21}$ in (6.10). Thus have the contradiction $e_{21} = 0$. Thus $m = 1$ and R is commutative.

♣

Continuing to analyze the behavior of the hypercentralizing maps, we now focus on the action of a generalized skew derivation F, that is an additive map $F : R \longrightarrow R$ such that $F(xy) = F(x)y + \alpha(x)d(y)$, for all $x, y \in R$, where d is a skew derivation d of R with an associated automorphism α. In all that follows $f(x_1, \ldots, x_n)$ will be a multilinear polynomial over C in n non-commuting variables and we denote this by $S = f(R)$ the set of all the evaluations of $f(x_1, \ldots, x_n)$ in R, more precisely $S = \{f(r_1, \ldots, r_n) : r_1, \ldots, r_n \in R\}$. In the last part of this section, we will provide a proof for the following result:

Theorem 6.2.11 *[8, Theorem] Let R be a prime ring of characteristic not equal to 2, Q_r its right Martindale quotient ring, C its extended centroid, $F : R \to R$ a nonzero generalized skew derivation of R and $k \geq 1$ a fixed integer. If $f(x_1, \ldots, x_n)$ is not central-valued on R, such that*

$$[F(f(r_1, \ldots, r_n)), f(r_1, \ldots, r_n)]_k = 0$$

for all $r_1, \ldots, r_n \in R$, then either there exists $\lambda \in C$ such that $F(x) = \lambda x$, or one of the following holds:

1. *$char(R) = p > 0$ and $f(x_1, \ldots, x_n)^{p^s}$ is central-valued on R, for a suitable $s \geq 0$;*
2. *$F(x) = ax + xb$, for all $x \in R$ and $a, b \in Q_r$ such that $a - b \in C$, and $f(x_1, \ldots, x_n)^2$ is central valued on R.*

In the sequel we will always assume that $f(x_1, \ldots, x_n)$ is not central-valued on R.

We firstly consider the case when $F : R \to R$ is a nonzero inner generalized skew derivation on a prime ring R induced by the elements $b, c \in Q_r$ and $\alpha \in Aut(R)$, that is $F(x) = bx - \alpha(x)c$, for all $x \in R$.

Proposition 6.2.12 *Let R be a prime ring of characteristic not equal to 2, $k \geq 1$ a fixed integer, b, c elements of Q_r. If*

$$\Big[bu - \alpha(u)c, u\Big]_k = 0$$

for all $u \in S$, then one of the following holds:

1. *There exists an invertible element $q \in Q_r$ such that $\alpha(x) = qxq^{-1}$, for all $x \in R$, and $q^{-1}c \in C$, $b - c \in C$;*
2. *$u^2 \in C$ for all $u \in S$, $b + c \in C$ and $\alpha = 1$ is the identity map in R;*

3. $char(R) = p > 0$ *and there exists a suitable integer* $s \geq 0$ *such that* $u^{p^s} \in C$, *for all* $u \in S$.

In order to prove Proposition 6.2.12 we also need the following lemmas:

Lemma 6.2.13 *[212, Proposition 1] Let R be a prime ring,* $a, b \in R$, *and* $k \geq 1$ *a fixed integer such that* $[au - ub, u]_k = 0$, *for all* $u \in S = f(R)$. *Then either* $a, b \in C$ *or one of the following conclusions holds:*

1. *if* $char(R) = 0$ *then* $f(x_1, \dots, x_n)$ *is central-valued on R;*
2. *if* $char(R) = p > 0$, *then* $f(x_1, \dots, x_n)^{p^s}$ *is central-valued on R, for a suitable* $s \geq 0$, *unless when* $char(R) = 2$ *and R satisfies the standard identity* s_4;
3. $a + b \in C$ *and* $f(x_1, \dots, x_n)^2$ *is central-valued on R.*

Theorem 6.2.14 *[212, Theorem 1] Let R be a prime ring, G a non-zero generalized derivation of R and* $k \geq 1$ *a fixed integer. If* $[G(u), u]_k = 0$, *for all* $u \in S = f(R)$, *then either there exists* $\lambda \in C$ *such that* $G(x) = \lambda x$, *or one of the following holds:*

1. $char(R) = 0$ *and* $f(x_1, \dots, x_n)$ *is central-valued on R;*
2. $char(R) = p > 0$ *and* $f(x_1, \dots, x_n)^{p^s}$ *is central-valued on R, for a suitable* $s \geq 0$, *unless when* $char(R) = 2$ *and R satisfies the standard identity* s_4;
3. $G(x) = ax + xb$, *where* $a, b \in Q_r$ *such that* $(a - b + \alpha) = 0$, *for* $\alpha \in C$, *and* $f(x_1, \dots, x_n)^2$ *is central-valued on R.*

Lemma 6.2.15 *[219, Lemma 2.2] Let R be a prime ring,* $b, c, q \in R$, *and* $k \geq 1$ *a fixed integer such that q is invertible and* $[bx - qxq^{-1}c, x]_k = 0$, *for all* $x \in R$. *Then* $q^{-1}c \in C$ *and* $b - c \in C$.

We denote

$$\Phi(x_1, \dots, x_n) = \Big[bf(x_1, \dots, x_n) - \alpha(f(x_1, \dots, x_n))c, f(x_1, \dots, x_n)\Big]_k \tag{6.11}$$

so that $\Phi(r_1, \dots, r_n) = 0$ for all $r_1, \dots, r_n \in R$. Notice that, if $c = 0$, then R satisfies the generalized polynomial identity

$$\Big[bf(x_1, \dots, x_n), f(x_1, \dots, x_n)\Big]_k$$

and, by Lemma 6.2.13, one of the following holds:

1. $b \in C$;
2. If $char(R) = 0$ then $f(x_1, \dots, x_n)$ is central-valued on R;

3. If $char(R) = p > 0$, then $f(x_1,\ldots,x_n)^{p^s}$ is central-valued on R, for a suitable $s \geq 0$, unless when $char(R) = 2$ and R satisfies the standard identity s_4.

In any case, the Proposition 6.2.12 is proved. Therefore, in the hypothesis of Proposition 6.2.12, we may always assume $c \neq 0$.

Remark 6.2.16 *Let Y be an invertible matrix, so that Y^2 is central in $M_2(C)$ and let $X \in M_2(C)$ be such that $[X,Y]_k = 0$. Then $[X,Y] = 0$.*

Proof: By the hypothesis, suppose that Y is an invertible matrix, so that Y^2 is central in $M_2(C)$ and $X \in M_2(C)$ satisfies $[X,Y]_k = 0$. An easy computation shows that

$$0 = \Big[X,Y\Big]_k = 2\Big[X,Y\Big]_{k-2}Y^2 - 2Y\Big[X,Y\Big]_{k-2}Y.$$

Multiplying by Y^{-1} from the right, we have

$$2\Big[X,Y\Big]_{k-2}Y - 2Y\Big[X,Y\Big]_{k-2} = 0,$$

and so $[X,Y]_{k-1} = 0$ since $char(R) \neq 2$. Hence, repeating this process $(k-2)$-times, we finally get $[X,Y] = 0$.

♣

We begin with the case when $\alpha \in Aut(R)$ is an X-inner automorphism of R, that is: there exists an invertible element $q \in Q_r$ such that $\alpha(x) = qxq^{-1}$, for all $x \in R$:

Lemma 6.2.17 *Let R be a prime ring of characteristic not equal to 2, $k \geq 1$ a fixed integer, b, c, q elements of Q_r. Assuming $c \neq 0$, q is invertible and*

$$\Big[bu - quq^{-1}c, u\Big]_k = 0$$

for all $u \in S = f(R)$, then one of the following holds:

1. *If $char(R) = 0$ then $f(x_1,\ldots,x_n)$ is central-valued on R;*
2. *If $char(R) = p > 0$, then $f(x_1,\ldots,x_n)^{p^s}$ is central-valued on R, for a suitable $s \geq 0$, unless when $char(R) = 2$ and R satisfies the standard identity s_4;*
3. *$q^{-1}c \in C$ and $b - c \in C$;*
4. *$f(x_1,\ldots,x_n)^2$ is central-valued on R, $q \in C$ and $b + c \in C$.*

Proof: By our assumption, R satisfies (6.11), that is $\Phi(x_1,\ldots,x_n)$ is a generalized polynomial identity for R. Since R and Q_r satisfy the same generalized polynomial identities with automorphisms Q_r, they also satisfy (6.11). Note that if

$\{q^{-1}c, 1\}$ are linearly C-dependent, then $q^{-1}c \in C$ and we are done by Lemma 6.2.13. Hence, consider the case when $\{q^{-1}c, 1\}$ are linearly C-independent, that is $q^{-1}c \notin C$. For the same reason we also may consider $q \notin C$. We will prove that in this situation a number of contradictions follow.

Since both $q^{-1}c \notin C$ and $q \notin C$, we see that (6.11) is a non-trivial generalized polynomial identity for Q_r. As remarked above, without loss of generality, we assume that Q_r is a primitive ring having a non-zero socle H with C its associated division ring, which is isomorphic to a dense ring of linear transformations of a vector space V over C. Of course, we may assume that $dim_C V \geq 2$. If $dim_C V = \infty$, then by Lemma 2 in [556], Q_r satisfies the following generalized identity $[bx - qxq^{-1}c, x]_k$. Therefore, by Lemma 6.2.15 we get $q^{-1}c \in C$ and $b - c \in C$.

Consider now the case when $dim_C V = t$, a finite positive integer. In this condition Q_r is a simple ring which satisfies a non-trivial generalized polynomial identity. By Theorem 2.3.29 in [507], $Q_r \subseteq M_t(C)$, moreover $M_t(C)$ satisfies the same generalized identity of Q_r, hence

$$\Big[bf(r_1,\ldots,r_n) - qf(r_1,\ldots,r_n)q^{-1}c, f(r_1,\ldots,r_n)\Big]_k = 0$$

for any $r_1,\ldots,r_n \in M_t(C)$. Furthermore,$f(x_1,\ldots,x_n)$ is a non-central polynomial for $M_t(C)$. By our assumption that $dim_C V \geq 2, t \geq 2$ and Remark 1.2.11, there exists a sequence of matrices $r_1,\ldots,r_n \in M_t(C)$ such that $f(r_1,\ldots,r_n) = \sum_i \alpha_i e_{ii} = D$ a non-central diagonal matrix, for $\alpha_i \in C$. Suppose $r \neq s$ such that $\alpha_r \neq \alpha_s$. For all $l \neq m$, let $\psi \in Aut_C(M_t(C))$ defined by $\psi(x) = \psi(\sum_{ij} \alpha_{ij} e_{ij}) = \sum_{ij} \alpha_{ij} e_{\sigma(i)\sigma(j)}$, where σ is a permutation in the symmetric group of n elements, such that $\sigma(r) = l$ and $\sigma(s) = m$. Thus $\psi(D)$ is a diagonal matrix in $f(R)$ with (l,l) and (m,m) distinct entries and some diagonal non-zero entries.

Now let φ be the inner automorphism of $M_t(C)$ induced by the element e_{lm}, namely $\varphi(x) = (1 + e_{lm})x(1 - e_{lm})$, for all $x \in M_t(C)$. Since the set $f(R)$ is invariant under the action of φ, we see that $E = \varphi(D) = D + (\alpha_m - \alpha_l)e_{lm}$ belongs to $S = f(R)$. For brevity denote $\beta = \alpha_m - \alpha_l \neq 0$. Analogously, $\varphi'(x) = (1 - e_{lm})x(1 + e_{lm})$, for all $x \in M_t(C)$. It follows that $E' = \varphi'(D) = D + (\alpha_l - \alpha_m)e_{lm} \in S = f(R)$.

We take $b = \sum b_{ij}e_{ij}$, $q = \sum q_{ij}e_{ij}$ and $p = q^{-1}c = \sum p_{ij}e_{ij}$ for $b_{ij}, q_{ij}, p_{ij} \in C$.

In case of $f(x_1,\ldots,x_n) = E$ in (6.11), we multiply e_{lm} from the right and the left to get

$$\beta^k e_{lm}(bD - qDp - \beta q e_{lm} p)e_{lm} = 0. \tag{6.12}$$

On the other hand, for $f(x_1,\ldots,x_n) = E'$ in (6.11), we multiply e_{lm} from the right and the left and get

$$\beta^k e_{lm}(bD - qDp + \beta q e_{lm} p)e_{lm} = 0. \tag{6.13}$$

Hence, by comparing (6.12) with (6.13), and since $\beta \neq 0$ and $char(C) \neq 2$, it follows that

$$q_{ml}p_{ml} = 0, \quad \forall l \neq m. \tag{6.14}$$

In case $t \geq 3$, together with [222, Proposition 1], (6.14) $\Rightarrow$ either $q \in C$ or $p \in C$. Hence the desired conclusion follows from Lemma 6.2.13.

We now assume $t = 2$ and $Q_r = M_2(C)$, with $char(C) \neq 2$. Again by (6.12) and (6.13), we have

$$\alpha_1 b_{21} - \alpha_1 q_{21} p_{11} - \alpha_2 q_{22} p_{21} = 0. \tag{6.15}$$

Our next aim is to show that at least one of p and q is a diagonal matrix in $M_2(C)$. To do this, we assume that p and q are not diagonal matrices and prove that a contradiction follows.

Without loss of generality, suppose $p_{21} \neq 0$, thus by (6.14) we get $q_{21} = 0$. Moreover, $q_{12} \neq 0$ since q is not diagonal, and both $q_{11} \neq 0$ and $q_{22} \neq 0$ since q is invertible. Hence (6.14) implies that $p_{12} = 0$ and (6.15) implies

$$\alpha_1 b_{21} - \alpha_2 q_{22} p_{21} = 0. \tag{6.16}$$

In particular we notice that $\alpha_1 b_{21} \neq 0$, since $\alpha_2 q_{22} p_{21} \neq 0$.

Now let χ be the inner automorphism of $M_2(C)$ defined by $\chi(x) = (\alpha_2 e_{12} + e_{21})x(e_{12} + \alpha_2^{-1} e_{21})$, for all $x \in M_2(C)$. Since the set $f(R)$ is invariant under the action of χ, then $E'' = \chi(D) = \alpha_2 e_{11} + \alpha_1 e_{22}$ belongs to $S = f(R)$. By repeating the same above argument, we get

$$\alpha_2 b_{21} - \alpha_1 q_{22} p_{21} = 0. \tag{6.17}$$

By (6.16) and (6.17) it follows that $b_{21} = \alpha_2^{-1}\alpha_1 q_{22} p_{21}$ and $b_{21} = \alpha_1^{-1}\alpha_2 q_{22} p_{21}$, i.e., $\alpha_2^{-1}\alpha_1 = \alpha_1^{-1}\alpha_2$, which $\Rightarrow$ $\alpha_1^2 = \alpha_2^2$ and so $\alpha_1 + \alpha_2 = 0$, i.e., $\alpha_2 = -\alpha_1$. Hence we have $D = \alpha_1 e_{11} - \alpha_2 e_{22} = \alpha_1(e_{11} - e_{22})$. This $\Rightarrow$ both $0 \neq D^2 \in C$ and

$$b_{21} + q_{22} p_{21} = 0. \tag{6.18}$$

Now using (6.12) and (6.13) and the condition, $char(C) \neq 2$ we also have

$$e_{lm}(bD - qDp)e_{lm} = 0.$$

In particular, since $\alpha_1 \neq 2$ we get $e_{21}(bD - qDp)e_{21} = 0$ implying,

$$b_{12} - q_{12} p_{22} = 0. \tag{6.19}$$

Consider automorphisms of R, $\varphi(x) = (1-e_{21})x(1+e_{21})$ and $\varphi\prime(x) = (1+e_{21})x(1-e_{21})$. Moreover $E = \varphi(D) = D - 2\alpha_1 e_{21} = \alpha_1(e_{11} - e_{22}) - 2\alpha_1 e_{21}$ is invertible such that E belongs to $S = f(R)$ and $0 \neq E^2 \in C$. So using Remark 6.2.16 we get $T = \Big[bE - qEp, E\Big] = 0$. In particular using (6.19) the $(1,2)$-entry of the matrix T is

$$\alpha_1^2(q_{11}p_{12} + q_{11}p_{11}) = 0. \tag{6.20}$$

Since $\alpha_1^2 \neq 0$ and $p_{12} = 0$ we have $q_{11}p_{11} = 0$ and so our assumption $q_{11} \neq 0$ implies $p_{11} = 0$. Calculating the $(2,1)$-entry of the matrix T we also have

$$b_{11} - b_{22} - 2b_{12} + q_{12}p_{21} - q_{22}p_{22} = 0. \tag{6.21}$$

Now we consider the following automorphisms of R: $\psi(x) = (1+e_{12})x(1-e_{12})$ and $\psi'(x) = (1-e_{12})x(1+e_{12})$. Notice that $\varepsilon = \psi(D) = D - 2\alpha_1 e_{12} = \alpha_1(e_{11} - e_{22}) - 2\alpha_1 e_{12}$ is an invertible matrix of $S = f(R)$ and $0 \neq \varepsilon^2 \in C$. So using Remark 6.2.16 we get $T' = \Big[b\varepsilon - q\varepsilon p, \varepsilon\Big] = 0$. In particular using (6.19) the $(1,2)$-entry of the matrix T' is

$$b_{11} - b_{22} - 2b_{21} + q_{22}p_{22} - q_{12}p_{21} - 2q_{11}p_{22} - q_{11}p_{21} = 0. \tag{6.22}$$

Since D is an invertible matrix of $S = f(R)$ and $0 \neq D^2 \in C$, using Remark 6.2.16 we get $\Big[bD - qDp, D\Big] = 0$ implying,

$$bD^2 - qDpD - DbD + DqDp = 0. \tag{6.23}$$

Furthermore, since $\varepsilon = D - 2\alpha_1 e_{12}$ and $\Big[b\varepsilon - q\varepsilon p, \varepsilon\Big] = 0$ we get $bD^2 - q(D - 2\alpha_1 e_{12})p(D - 2\alpha_1 e_{12}) - (D - 2\alpha_1 e_{12})b(D - 2\alpha_1 e_{12}) + (D - 2\alpha_1 e_{12}) q(D - 2\alpha_1 e_{12})p = 0$. Using (6.23) in the preceding relationship we obtain

$$\begin{aligned} &2\alpha_1 qDpe_{12} + 2\alpha_1 qe_{12}pD - 4\alpha_1^2 qe_{12}pe_{12} + 2\alpha_1 Dbe_{12} + 2\alpha_1 e_{12}bD \\ &- 4\alpha_1^2 e_{12}be_{12} - 2\alpha_1 Dqe_{12}p - 2\alpha_1 e_{12}qDp + 4\alpha_1^2 e_{12}qe_{12}p = 0. \end{aligned} \tag{6.24}$$

Similarly since $\varepsilon' = \psi'(D) = D + 2\alpha_1 e_{12} = \alpha_1(e_{11} - e_{22}) + 2e_{12}$ is invertible such that $\varepsilon' \in S = f(R)$ and $0 \neq (\varepsilon')^2 \in C$, using Remark 6.2.16 we get $\Big[b\varepsilon' - q\varepsilon' p, \varepsilon'\Big] = 0$. So according to (6.23) the last relationship $\Rightarrow$

$$\begin{aligned} &- 2\alpha_1 qDpe_{12} - 2\alpha_1 qe_{12}pD - 4\alpha_1^2 qe_{12}pe_{12} - 2\alpha_1 Dbe_{12} - 2\alpha_1 e_{12}bD \\ &- 4\alpha_1^2 e_{12}be_{12} + 2\alpha_1 Dqe_{12}p + 2\alpha_1 e_{12}qDp + 4\alpha_1^2 e_{12}qe_{12}p = 0. \end{aligned} \tag{6.25}$$

Comparing (6.24) and (6.25) we get

$$-8\alpha_1^2(qe_{12}pe_{12}+e_{12}be_{12}-e_{12}qe_{12}p)=0.$$

Since $char(C)\neq 2$ and $\alpha_1^2\neq 0$ we have

$$qe_{12}pe_{12}+e_{12}be_{12}-e_{12}qe_{12}p=0.$$

Since $e_{12}qe_{12}=q_{21}e_{12}$ and $q_{21}=0$ the preceding relationship gives, $qe_{12}pe_{12}+e_{12}be_{12}=0$. Multiplying the last relation from the left by e_{11}, we get

$$b_{21}+q_{11}p_{21}=0. \tag{6.26}$$

Now using the fact that $E=\varphi(D)=D-2\alpha_1e_{21}$ is invertible with E belonging to $S=f(R)$ and $0\neq E^2\in C$, by 6.2.16 we get $T=\Big[bE-qEp,E\Big]=0$. So we have

$$bD^2-q(D-2\alpha_1e_{21})p(D-2\alpha_1e_{21})-(D-2\alpha_1e_{21})b(D-2\alpha_1e_{21})+ \\ (D-2\alpha_1e_{21})q(D-2\alpha_1e_{21})p=0. \tag{6.27}$$

Moreover since $E'=\varphi'(D)=D+2\alpha_1e_{21}=\alpha_1(e_{11}-e_{22})-2e_{21}$ is invertible with $E'\in S=f(R)$ and $0\neq(E')^2\in C$, utilizing Remark 6.2.16 we get $\Big[bE'-qE'p,E'\Big]=0$. So the preceding relationship $\Rightarrow$

$$bD^2-q(D+2\alpha_1e_{21})p(D+2\alpha_1e_{21})-(D+2\alpha_1e_{21})b(D+2\alpha_1e_{21})+ \\ (D+2\alpha_1e_{21})q(D+2\alpha_1e_{21})p=0. \tag{6.28}$$

Comparing (6.27) and (6.28) and using (6.23) we obtain

$$qe_{21}pe_{21}+e_{21}be_{21}-e_{21}qe_{21}p=0.$$

Since $p_{12}=0$ we get $e_{21}pe_{21}=p_{12}e_{21}=0$. So the above relationship reduces to

$$e_{21}be_{21}-e_{21}qe_{21}p=0.$$

Right multiplying the previous relationship with e_{11} we get, $b_{12}-q_{12}p_{11}=0$. Since $p_{11}=0$ we have $b_{12}=0$. Using this fact in (6.19) we get, $q_{12}p_{22}=0$. Our assumption $q_{12}\neq 0$ implies that $p_{22}=0$. Moreover by (6.18) and (6.26) we have

$(q_{11}-q_{22})p_{21}=0$ and using $p_{21}\neq 0$ in the preceding relationship we obtain $q_{11}=q_{22}$. Using these facts and (6.18) in (6.21) and (6.22) respectively, we get,

$$b_{11}-b_{22}+q_{12}p_{21}=0$$

and

$$b_{11}-b_{22}+2q_{22}p_{21}-q_{12}p_{21}-2q_{11}p_{21}=0.$$

Comparing the last two relations and using $char(C)\neq 2$ we have $(q_{11}-q_{22})p_{21}+q_{12}p_{21}=0$. Since $q_{11}=q_{22}$ we get $q_{12}p_{21}=0$, contradicting the assumption, $q_{12}\neq 0$ and $p_{21}\neq 0$.

The consequence $\Rightarrow$ either q or p is a diagonal matrix. Standard argument and easy computations show that either q or p is a central matrix in $M_2(C)$, and we may conclude as above by using Lemma 6.2.13.

♣

The proof of Proposition 6.2.12

Let

$$\Phi(x_1,\dots,x_n,x_1^\alpha,\dots,x_n^\alpha)=\Big[bf(x_1,\dots,x_n)-\alpha(f(x_1,\dots,x_n))c,f(x_1,\dots,x_n)\Big]_k.$$

By this assumption, Q_r satisfies $\Phi(x_1,\dots,x_n,x_1^\alpha,\dots,x_n^\alpha)$. In case that there exists an invertible element $q\in Q_r$ such that $\alpha(x)=qxq^{-1}$, for all $x\in Q_r$, the conclusion follows from Lemma 6.2.17. Thus we may consider the case α is not X-inner and prove that a contradiction follows.

Firstly we note that the x_i^α-word degree of $\Phi(x_i,x_i^\alpha)$ is 1. Applying [187, Theorem 3] yields Q_r satisfying the generalized polynomial identity

$$\Big[bf(x_1,\dots,x_n)-f^\alpha(y_1,\dots,y_n)c,f(x_1,\dots,x_n)\Big]_k$$

where, $f^\alpha(y_1,\dots,y_n)=\sum_{\sigma\in S_n}\alpha(\gamma_\sigma)\alpha(y_{\sigma(1)})\cdot\alpha(y_{\sigma(2)})\cdots\alpha(y_{\sigma(n)})$. In particular

$$\Big[f^\alpha(y_1,\dots,y_n)c,f(x_1,\dots,x_n)\Big]_k$$

is a generalized polynomial identity for Q_r. By Theorem 6.2.5 for inner derivations, we assume $char(R)\neq 2$, it follows that $f^\alpha(y_1,\dots,y_n)c$ is central-valued on R. Notice that, if $c\in C$ then, by the primeness of R, $f^\alpha(y_1,\dots,y_n)$ is central-valued on R, which is a contradiction. Thus we may assume $c\notin C$. Since $f^\alpha(y_1,\dots,y_n)c$ is central-valued, then

$$[f^\alpha(y_1,\dots,y_n)c,f^\alpha(y_1,\dots,y_n)]$$

is a generalized polynomial identity for R. Hence, by applying Lemma 6.2.13, we get $char(R) = 2$ ($\wedge$ R satisfies s_4), which is again a contradiction.

♣

The proof of Theorem 6.2.11

It is known that automorphisms, derivations and skew derivations of R can be extended to Q_r. Moreover there exist $a \in Q_r$ and a skew derivation d of R such that $F(x) = ax + d(x)$, for all $x \in R$. Here we denote the automorphism of R associated with d and with α. Let us denote

$$\begin{aligned}&\Phi(x_1,\dots,x_n,d(x_1),\dots,d(x_n))\\&= \left[af(x_1,\dots,x_n) + d(f(x_1,\dots,x_n)), f(x_1,\dots,x_n)\right]_k.\end{aligned}$$

Thus R satisfies

$$\Phi(x_1,\dots,x_n,d(x_1),\dots,d(x_n)). \tag{6.29}$$

By [190, Theorem 2] we know that R and Q_r satisfy the same generalized polynomial identities with a single skew derivation. Thus Q_r satisfies (6.29).

When α is the identity mapping on R, d is an usual derivation of R. And hence F becomes a generalized derivation of R. In this case, the required conclusions follow from Theorem 6.2.14. Hence in what follows we always assume that $1_R \neq \alpha \in Aut(R)$.

We denote by $f^d(x_1,\dots,x_n)$ the polynomial obtained from $f(x_1,\dots,x_n)$ by replacing each coefficient γ_σ with $d(\gamma_\sigma)$. It should be noted that

$$\begin{aligned}&d\left(\gamma_\sigma \cdot x_{\sigma(1)} \cdot x_{\sigma(2)} \cdots x_{\sigma(n)}\right) = d(\gamma_\sigma)x_{\sigma(1)} \cdot x_{\sigma(2)} \cdots x_{\sigma(n)} +\\&+\alpha(\gamma_\sigma)\sum_{j=0}^{n-1} \alpha(x_{\sigma(1)} \cdot x_{\sigma(2)} \cdots x_{\sigma(j)})d(x_{\sigma(j+1)})x_{\sigma(j+2)} \cdots x_{\sigma(n)}.\end{aligned}$$

So we have

$$\begin{aligned}&d(f(x_1,\dots,x_n)) = f^d(x_1,\dots,x_n) +\\&+\sum_{\sigma\in S_n} \alpha(\gamma_\sigma)\sum_{j=0}^{n-1} \alpha(x_{\sigma(1)} \cdot x_{\sigma(2)} \cdots x_{\sigma(j)})d(x_{\sigma(j+1)})x_{\sigma(j+2)} \cdots x_{\sigma(n)}.\end{aligned}$$

In case d is X-inner, there exists $c \in Q_r$ such that $d(x) = cx - \alpha(x)c$, for all $x \in R$, so that $F(x) = (a+c)x - \alpha(x)c$ and

$$\left[(a+c)u - \alpha(u)c, u\right]_k = 0$$

for all $u \in S = f(R)$. Hence by Proposition 6.2.12, we infer that one of the following holds:

1. There exists an invertible element $q \in Q_r$ such that $\alpha(x) = qxq^{-1}$, for all $x \in R$, and $q^{-1}c \in C$, $a \in C$; therefore $F(x) = ax$.
2. $u^2 \in C$ for all $u \in S$, $b + c \in C$ and $\alpha = 1$ the identity map in R;
3. $char(R) = p > 0$ and there exists a suitable integer $s \geq 0$ such that $u^{p^s} \in C$, for all $u \in S$.

Assume now that d is not X-inner. Since Q_r satisfies $\Phi(x_1,\ldots,x_n,d(x_1),\ldots,d(x_n))$, then it also satisfies

$$\big[af(x_1,\ldots,x_n) + f^d(x_1,\ldots,x_n), f(x_1,\ldots,x_n)\big]_k +$$

$$\big[\sum_{\sigma\in S_n} \alpha(\gamma_\sigma) \sum_{j=0}^{n-1} \alpha(x_{\sigma(1)} \cdot x_{\sigma(2)} \cdots x_{\sigma(j)}) d(x_{\sigma(j+1)}) x_{\sigma(j+2)} \cdots x_{\sigma(n)}, f(x_1,\ldots,x_n)\big]_k.$$

By [190, Theorem 1], it follows that Q_r satisfies $\Phi(x_1,\ldots,x_n,y_1,\ldots,y_n)$, that is

$$\big[af(x_1,\ldots,x_n) + f^d(x_1,\ldots,x_n), f(x_1,\ldots,x_n)\big]_k +$$

$$\big[\sum_{\sigma\in S_n} \alpha(\gamma_\sigma) \sum_{j=0}^{n-1} \alpha(x_{\sigma(1)} \cdot x_{\sigma(2)} \cdots x_{\sigma(j)}) y_{\sigma(j+1)} x_{\sigma(j+2)} \cdots x_{\sigma(n)}, f(x_1,\ldots,x_n)\big]_k.$$

In particular, for any $i = 1,\ldots,n$, Q_r satisfies

$$\big[\sum_{\sigma\in S_n} \alpha(\gamma_\sigma) \alpha(x_{\sigma(1)} \cdot x_{\sigma(2)} \cdots x_{\sigma(i-1)}) y_{\sigma(i)} x_{\sigma(i+1)} \cdots x_{\sigma(n)}, f(x_1,\ldots,x_n)\big]_k. \tag{6.30}$$

Here we divide the argument into two subcases. Let us first consider the case when α is an inner automorphism of R. Then there exists an invertible element $q \in Q_r$ such that $\alpha(x) = qxq^{-1}$ for all $x \in R$. Since $1_R \neq \alpha \in Aut(R)$, we may assume that $q \notin C$. Moreover, it is clear that $\alpha(\gamma_\sigma) = \gamma_\sigma$ for all coefficients involved in $f(x_1,\ldots,x_n)$. Replacing each $y_{\sigma(i)}$ with $qx_{\sigma(i)}$ in (6.30), we observe that for any $i = 1,\ldots,n$, Q_r satisfies

$$\big[q \sum_{\sigma\in S_n} \gamma_\sigma x_{\sigma(1)} \cdot x_{\sigma(2)} \cdots x_{\sigma(i-1)} x_{\sigma(i)} x_{\sigma(i+1)} \cdots x_{\sigma(n)}, f(x_1,\ldots,x_n)\big]_k,$$

that is

$$\big[qf(x_1,\ldots,x_n), f(x_1,\ldots,x_n)\big]_k.$$

Note that $q \notin C$ and $f(x_1,\ldots,x_n)$ is not central-valued on Q_r. Combining these facts with Lemma 6.2.13 yields $char(R) = p > 0$ and there exists $s \geq 1$ such that $f(x_1,\ldots,x_n)^{p^s}$ central-valued on Q_r. We now assume that α is X-outer. Finally, in this case we show that if $f(x_1,\ldots,x_n)$ is not central-valued on R, then we have a contradiction.

We note that any x_i^α-word degree of (6.30) is 1. Applying [187, Theorem 3] to relation (6.30), Q_r satisfies the generalized polynomial identity

$$\Big[\sum_{\sigma\in S_n} \alpha(\gamma_\sigma) z_{\sigma(1)}\cdot z_{\sigma(2)}\cdots z_{\sigma(i-1)} y_{\sigma(i)} x_{\sigma(i+1)}\cdots x_{\sigma(n)}, f(x_1,\ldots,x_n)\Big]_k \tag{6.31}$$

for all $i = 1,\ldots,n$. In particular, we choose:

- for all $i \geq 2$, $y_{\sigma(i)} = 0$;
- for all $i \geq 2$, $z_{\sigma(i)} = 0$.

Therefore by (6.31) we assert that Q_r satisfies the generalized polynomial identity

$$\Big[y_1 \sum_{\sigma\in S_{n-1}} \alpha(\gamma_\sigma) x_{\sigma(2)}\cdots x_{\sigma(n)}, f(x_1,\ldots,x_n)\Big]_k. \tag{6.32}$$

We put, $\sum_{\sigma\in S_{n-1}} \alpha(\gamma_\sigma) x_{\sigma(2)}\cdots x_{\sigma(n)} = t_1(x_2,\ldots,x_n)$. Then Q_r satisfies the generalized polynomial identity

$$\Big[y_1 t_1(x_2,\ldots,x_n), f(x_1,\ldots,x_n)\Big]_k. \tag{6.33}$$

Therefore there exists a suitable field K and an integer $t \geq 1$ such that Q_r and the matrix ring $M_t(K)$ satisfy the same polynomial identities. In particular, $M_t(K)$ satisfies the identity $[y_1 t_1(x_2,\ldots,x_n), f(x_1,\ldots,x_n)]_k$. Since $f(x_1,\ldots,x_n)$ is not central-valued on Q_r, we may assume $t \geq 2$. Thus, there exists a sequence of matrices $r_1,\ldots,r_n \in M_t(C)$ such that $f(r_1,\ldots,r_n) = \sum_i \alpha_i e_{ii} = D$ a non-central diagonal matrix, for $\alpha_i \in C$. Moreover the (l,l) and (m,m) entries are distinct and at least one of them is not equal to zero. Without loss of generality we may consider directly $\alpha_1 \neq \alpha_2$ and $\alpha_1 \neq 0$. Here we also denote $t(r_2,\ldots,r_n) = T_D$. In this situation we have

$$[yT_D, D]_k = 0 \tag{6.34}$$

for all $y \in M_t(K)$. Here we also denote $f^\alpha(x_1,\ldots,x_n)$ the polynomial obtained from $f(x_1,\ldots,x_n)$ by replacing each coefficient γ_σ with $\alpha(\gamma_\sigma)$. Note that the $(1,1)$-entry of the matrix $f^\alpha(r_1,\ldots,r_n) \neq 0$ is not zero. Multiplying both sides of (6.34) by e_{12}, and noting that $\alpha_1 \neq \alpha_2$, we see that the $(2,1)$-entry of the matrix yT_D is zero, for all $y \in M_t(K)$. Analogously the $(1,2)$-entry of the matrix yT_D is zero, for all $y \in M_t(K)$. This $\Rightarrow$ the first and second columns of the matrix $T_D = 0$. Let us start again from (6.31) and fix an index $j \in \{1,\ldots,n\}$. We choose:

- for all $i \neq j$, $y_{\sigma(i)} = 0$;
- for all $i \neq j$, $z_{\sigma(i)} = 0$.

Therefore by (6.31) we arrive at that Q_r that satisfies the generalized polynomial identity

$$\Big[y_j \sum_{\sigma\in S_{n-1}} \alpha(\gamma_\sigma) x_{\sigma(1)}\cdots x_{\sigma(j-1)} x_{\sigma(j+1)}\cdots x_{\sigma(n)}, f(x_1,\ldots,x_n)\Big]_k. \tag{6.35}$$

Let us adopt a new notation for later discussion.

$$\sum_{\sigma\in S_{n-1}} \alpha(\gamma_\sigma)x_{\sigma(1)}\cdots x_{\sigma(j-1)}x_{\sigma(j+1)}\cdots x_{\sigma(n)} = t_j(x_1,\ldots,x_{j-1},x_{j+1},\ldots,x_n).$$

Thus Q_r satisfies the generalized polynomial identity

$$\left[y_j t_j(x_1,\ldots,x_{j-1},x_{j+1},\ldots,x_n), f(x_1,\ldots,x_n)\right]_k.$$

By using the same argument as above, the first and second columns of the matrix $t_j(r_1,\ldots,r_{j-1},r_{j+1},\ldots,r_n)$ are = 0. Finally we note that

$$f^\alpha(x_1,\ldots,x_n) = \sum_j x_j t_j(x_1,\ldots,x_{j-1},x_{j+1},\ldots,x_n),$$

where any t_j is a multilinear polynomial of degree $n-1$ and x_j never appears in any monomial of t_j. This leads to the contradiction that the $(1,1)$-entry of the matrix $f^\alpha(r_1,\ldots,r_n)$ is = 0.

♣

6.3 Strong commutativity preservers

The theory of linear preservers is devoted to analyzing those maps on algebras that preserve some properties of some elements. The target of the theory is to describe such maps and/or to determine the structure of the algebra on which these functions are defined. In this regard, of course the number of publications dealing with automorphisms and antiautomorphisms is huge, as such maps lend themselves perfectly to preserve algebraic properties of elements. Our approach is based on the study of maps preserving the commutativity. Specifically, here we recall two definitions given in Chapter 3. More precisely, if $S \subseteq R$, the map $F : S \to S$ is called *commutativity preserving* on S if $[x,y]=0 \Rightarrow [F(x),F(y)]=0$; it is called *strong commutativity preserving* (for brevity we will always say SCP) on S if $[F(x),F(y)] = [x,y]$, for all $x,y\in S$. The natural possibility when an additive mapping preserves commutativity appears in a paper by Brešar and Miers [143]. They showed that any additive mapping F which is SCP on a semiprime ring R is of the form $F(x) = \lambda x + \mu(x)$, where $\lambda \in C$, $\lambda^2 = 1$ and $\mu : R \to C$ is an additive map of R into C. Recently in [414] Lin and Liu extended this result to Lie ideals, in case the ring R is prime. More precisely they proved that if L is a non-central Lie ideal of R and F is an additive mapping satisfying $[F(x),F(y)] - [x,y] \in C$, for all $x,y\in L$, then $F(x) = \lambda x + \mu(x)$, where $\lambda\in C$, $\lambda^2 = 1$ and $\mu : R\to C$, unless when $char(R) = 2$ and R satisfies the standard identiy s_4 of degree 4.

In [109] Bell and Daif proved that if R is a semiprime ring admitting a derivation d which is SCP on the right ideal I of R, then $I \subseteq Z(R)$. As derivations are widely studied, in recent years many authors began to study generalized derivations which are SCP on some subsets of prime and semiprime rings, in order to obtain appropriate results generalizing the former ones on derivations. In [454] Ma, Xiu and Niu described the structure of a generalized derivation which is SCP on one-sided ideals of a prime ring. More precisely, in case of a prime ring R with right ideal I, it has been proved that any generalized derivation F which is SCP on I is of the form $F(x) = ax + xb$ for all $x \in R$, with $aI = (0)$. Moreover $b = \pm 1$ or else $b \notin C$, $RC \subseteq End(V_K)$, where V_K is a vector space over the field K.

This last cited result is extended for prime rings in [418, Theorem 1.3] by Liu. He studied the case when I is a right ideal of R, F and G are generalized derivations of R, such that $[F(x), G(y)] = [x, y]$ for all $x, y \in I$ and obtained the complete description of F and G, and also the description their action on I. Subsequently, in [425], Liu and Liau generalized the result to the case when L is a non-central Lie ideal of R, F and G are generalized derivations of R, such that $[F(x), G(y)] = [x, y]$ for all $x, y \in L$. They proved that either $R \subseteq M_2(K)$, for a field K, or there exists $0 \neq \alpha \in C$ and a map $\mu : L \to C$ such that $G(x) = \alpha x$ and $F(x) = \alpha^{-1} x$ for all $x \in R$.

A natural generalization is represented by the following:

Theorem 6.3.1 *Let R be a prime ring, Q_r its right Martindale quotient ring, C its extended centroid. Let $F : R \to R$ and $G : R \to R$ be non-zero generalized derivations on R, $f(x_1, \ldots, x_n)$ a multilinear polynomial in n non-commuting variables over C, which is not central valued on R, and $S = \{f(r_1, \ldots, r_n) : r_1, \ldots, r_n \in R\}$, such that $[F(x), G(y)] = [x, y]$, for all $x, y \in S$. Then there exists $\lambda \in C$ such that, for any $x \in R$, $G(x) = \lambda x$ and $F(x) = \lambda^{-1} x$, unless when $R \subseteq M_2(C)$, the 2×2 matrix ring over C.*

Proof: Denote by A the additive subgroup generated by the set S. Assuming that R does not satisfy s_4 and since $f(x_1, \ldots, x_n)$ is not central valued on R, by [183] it follows that there exists a non-central Lie ideal L of R such that $L \subseteq A$. Moreover, for any $x, y \in A$ we have that $[F(x), G(y)] = [x, y]$, so that it holds for all $x, y \in L$. Then the conclusion follows from [425, Theorem 1.1 and Corollary 1.2].

♣

We would like to show that in case R satisfies the standard identity s_4, there exist suitable examples of generalized derivations F and G satisfying the condition $[F(x), G(y)] = [x, y]$, for all $x, y \in S$, but F and G are not two-sided centralizers of R. More precisely, let K be any field, $R = M_2(K)$, $f_1(x_1, \ldots, x_n)$ and $f_2(x_1, \ldots, x_m)$ multilinear polynomials over K and define

$$f(x_1, \ldots, x_{n+m}) = [f_1(x_1, \ldots, x_n), f_2(x_{n+1}, \ldots, x_{n+m})]$$

and

$$S = \{f(r_1, \ldots, r_{n+m}) : r_1, \ldots, r_{n+m} \in R\}.$$

We recall that any commutator $[x,y] \in M_2(K)$ is a matrix with zero trace. Hence any element of S is of the form $\alpha e_{11} + \beta e_{12} + \gamma e_{21} - \alpha e_{22}$, for suitable $\alpha, \beta, \gamma \in K$ depending on the choice of $x_1, \ldots, x_{n+m}$. Choose both $F(x) = e_{11}x + xe_{11}$ and $G(x) = e_{11}x + xe_{11}$, for all $x \in R$. By easy computations one can see that $[F(x), G(y)] = [x,y]$, for all $x, y \in S$ and of course neither F nor G is a two-sided centralizer of R.

The following definition was introduced in [222]:

Definition 108 *If $S \subseteq R$, the map $F : R \to R$ is called* Engel-condition preserving *on S if $[x,y]_k = 0 \Rightarrow [F(x), F(y)]_k = 0$, for all $x, y \in S$ and fixed $k \geq 1$; F is called* strong Engel-condition preserving *(SEP for brevity) on $S \subseteq R$ if $[F(x), F(y)]_k = [x,y]_k$, for all $x, y \in S$ and fixed $k \geq 1$. More generally we say that $F : R \to R$ and $G : R \to R$ are mutually SEP on $S \subseteq R$ if $[G(x), F(y)]_k = [x,y]_k$, for all $x, y \in S$ and fixed $k \geq 1$.*

Connecting these definitions, starting from Theorem 6.3.1, we would like to exhibit the following in this section:

Theorem 6.3.2 *[222, Theorem 1] Let R be a prime ring of characteristic not equal to 2, Q_r its right Martindale quotient ring, C its extended centroid, $f(x_1, \ldots, x_n)$ a multilinear polynomial in n non-commuting variables over C and $S = \{f(r_1, \ldots, r_n) : r_1, \ldots, r_n \in R\}$. Let $F : R \to R$ and $G : R \to R$ be non-zero generalized derivations on R, such that F and G are mutually SEP on S, that is there exists $h \geq 1$ a fixed integer, such that $[F(x), G(y)]_h = [x,y]_h$, for all $x, y \in S$. If $f(x_1, \ldots, x_n)$ is not central-valued on R, then one of the following holds:*

1. *There exists $\lambda \in C$ such that, for any $x \in R$, $G(x) = \lambda x$ and $F(x) = \lambda^{-h} x$;*
2. *$char(R) = p \geq 3$ and there exist $\lambda \in C$ and $s \geq 1$ such that, for any $x \in R$, $G(x) = \lambda x$ and $f(x_1, \ldots, x_n)^{p^s}$ is central-valued on R;*
3. *R satisfies s_4, the standard identity of degree 4.*

For proving Theorem 6.3.2 we need a series of intermediate results, each of which could be of independent interest for readers.

In all that follows (unless stated otherwise) we will always assume that R is a prime ring, F and G non-zero generalized derivations on R, $f(x_1, \ldots, x_n)$ a multilinear polynomial which is not central-valued on R and

$$S = \{f(r_1, \ldots, r_n) : r_1, \ldots, r_n \in R\}$$

the set of all the evaluations of $f(x_1, \ldots, x_n)$ in R, such that F and G are mutually SCP on S. More precisely there is $h \geq 1$ a fixed integer, such that $[F(x), G(y)]_h = [x,y]_h$, for all $x, y \in S$.

Lemma 6.3.3 *Suppose both F and G are not inner generalized derivations of R. Then $char(R) = 2$ and R satisfies s_4.*

Proof: It is known that there exist $a,b \in Q_r$ and d,δ derivations of Q_r such that $F(x) = ax + d(x)$ and $G(x) = bx + \delta(x)$, for all $x \in R$. Firstly we mention that, $d \neq 0$ and $\delta \neq 0$ since both F and G are not inner.

Hence R satisfies the generalized differential identity

$$\begin{aligned}&\Big[af(x_1,\ldots,x_n) + f^d(x_1,\ldots,x_n) + \sum_i f(x_1,\ldots,d(x_i),\ldots,x_n),\\ &bf(y_1,\ldots,y_n) + f^\delta(y_1,\ldots,y_n) + \sum_i f(y_1,\ldots,\delta(y_i),\ldots,y_n)\Big]_h \\ &-[f(x_1,\ldots,x_n),f(y_1,\ldots,y_n)]_h.\end{aligned} \tag{6.36}$$

In case d and δ are C-linearly independent modulo inner derivations of Q_r, then, by Kharchenko's differential identities R satisfies

$$\begin{aligned}&\Big[af(x_1,\ldots,x_n) + f^d(x_1,\ldots,x_n) + \sum_i f(x_1,\ldots,z_i,\ldots,x_n),\\ &bf(y_1,\ldots,y_n) + f^\delta(y_1,\ldots,y_n) + \sum_i f(y_1,\ldots,t_i,\ldots,y_n)\Big]_h \\ &-[f(x_1,\ldots,x_n),f(y_1,\ldots,y_n)]_h.\end{aligned} \tag{6.37}$$

and in particular R satisfies the blended component

$$\Big[f(z_1,x_2,\ldots,x_n),f(t_1,y_2,\ldots,y_n)\Big]_h \tag{6.38}$$

that is, for $c = f(z_1,x_2,\ldots,x_n) \notin C$,

$$\Big[c,f(y_1,\ldots,y_n)\Big]_h \tag{6.39}$$

is a non-trivial generalized polynomial identity for R. Since $c \notin C$, then by Theorem 6.2.5 we conclude that $char(R) = 2$ and R satisfies s_4.

Thus we finally assume that there exist $0 \neq \alpha \in C$ and $q \in Q_r$, such that $\delta(x) = \alpha d(x) + [q,x]$, for all $x \in R$.

By (6.36) we say R satisfies

$$\begin{aligned}&\Big[af(x_1,\ldots,x_n) + f^d(x_1,\ldots,x_n) + \sum_i f(x_1,\ldots,d(x_i),\ldots,x_n),\\ &bf(y_1,\ldots,y_n) + f^{\alpha d}(y_1,\ldots,y_n) + \\ &\alpha\sum_i f(y_1,\ldots,d(y_i),\ldots,y_n) + [q,f(y_1,\ldots,y_n)]\Big]_h \\ &-[f(x_1,\ldots,x_n),f(y_1,\ldots,y_n)]_h\end{aligned} \tag{6.40}$$

and since d is not inner, it follows that

$$\begin{aligned}&\Big[af(x_1,\dots,x_n)+f^d(x_1,\dots,x_n)+\sum_i f(x_1,\dots,z_i,\dots,x_n),\\&bf(y_1,\dots,y_n)+f^{\alpha d}(y_1,\dots,y_n)+\\&\alpha\sum_i f(y_1,\dots,t_i,\dots,y_n)+[q,f(y_1,\dots,y_n)]\Big]_h\\&-[f(x_1,\dots,x_n),f(y_1,\dots,y_n)]_h\end{aligned} \tag{6.41}$$

is a generalized identity for R. As above, for $c=f(z_1,x_2,\dots,x_n)\notin C$,

$$\alpha\Big[c,f(t_1,y_2,\dots,y_n)\Big]_h \tag{6.42}$$

is a non-trivial generalized polynomial identity for R, and we are done by the same argument as above.

♣

Lemma 6.3.4 *Suppose that there exist $a,b\in Q_r$ such that $G(x)=ax+xb$, for all $x\in R$. Then either R satisfies s_4, the standard identity of degree 4, or F is an inner generalized derivation of R.*

Proof: As above assume $F(x)=px+d(x)$, for some $p\in Q_r$ and derivation d on Q_r.

Firstly we consider, if $d=0$, then $F(x)=px$ and we are done. Thus we consider case with $d\neq 0$.

Hence, R satisfies the generalized differential identity

$$\begin{aligned}&[pf(x_1,\dots,x_n)+f^d(x_1,\dots,x_n)+\\&\sum_i f(x_1,\dots,d(x_i),\dots,x_n),af(y_1,\dots,y_n)+f(y_1,\dots,y_n)b]_h\\&-[f(x_1,\dots,x_n),f(y_1,\dots,y_n)]_h.\end{aligned} \tag{6.43}$$

We notice that in case d is an inner derivation of R, that is $d(x)=cx-xc$, for all $x\in R$ and fixed $c\in Q_r$, then $F(x)=(p+c)x-xc$ is an inner generalized derivation of R and we are done. Therefore we now assume that d is not an inner derivation of R and prove that a contradiction occurs unless when R satisfies the standard identity s_4.

Since d is not inner, by (6.43) R satisfies

$$\begin{aligned}&[pf(x_1,\dots,x_n)+f^d(x_1,\dots,x_n)+\\&\sum_i f(x_1,\dots,z_i,\dots,x_n),af(y_1,\dots,y_n)+f(y_1,\dots,y_n)b]_h\\&-[f(x_1,\dots,x_n),f(y_1,\dots,y_n)]_h.\end{aligned} \tag{6.44}$$

and in particular R satisfies the blended component

$$[f(z_1,x_2,\ldots,x_n),af(y_1,\ldots,y_n)+f(y_1,\ldots,y_n)b]_h. \tag{6.45}$$

Assume both a and b are central elements of R, say $0\neq\lambda=a+b$. Therefore by (6.45), R satisfies $\lambda^h[f(x_1,\ldots,x_n),f(y_1,\ldots,y_n)]_h$. Moreover, since $f(x_1,\ldots,x_n)$ is not central-valued on R, there exist $a_1,\ldots,a_n\in R$ such that $u=f(a_1,\ldots,a_n)\notin C$ and $[u,f(r_1,\ldots,r_n)]_h=0$ for all $r_1,\ldots,r_n\in R$. In this case, by Theorem 6.2.5 R satisfies the standard identity s_4. Moreover, in light of the previous argument, in the sequel we may suppose that either $a\notin C$ or $b\notin C$. As already remarked, without loss of generality, we assume that Q_r is a primitive ring having a non-zero socle H with C its associated division ring, which is isomorphic to a dense ring of linear transformations of a vector space V over C.

If $dim_C V=t$ is finite, then $Q_r\cong M_t(C)$, the $t\times t$ matrix ring over C. If $t\leq 2$ we are done, so that we consider $t\geq 3$ and prove that a contradiction follows.

Since $f(x_1,\ldots,x_n)$ is not central-valued on Q_r and as a consequence of Remark 1.2.12, we have that $e_{ij}\in S$, for all $i\neq j$. In particular let i,j,k form different pairs and consider both $e_{ij}\in S$ and $e_{kj}\in S$. By (6.45) it follows

$$[e_{kj},ae_{ij}+e_{ij}b]_h=0$$

and left multiplying by e_{kk} we get $e_{kj}(ae_{ij})^h=0$, that is a is a diagonal matrix in $M_t(C)$. Analogously, for $e_{ji}\in S$ and $e_{jk}\in S$ and by (6.45) we also have

$$[e_{jk},ae_{ji}+e_{ji}b]_h=0$$

and right multiplying by e_{kk} it follows as above that b is a diagonal matrix in $M_t(C)$. Notice that for any automorphism φ of $M_t(C)$ we get,

$$[f(z_1,x_2,\ldots,x_n),\varphi(a)f(y_1,\ldots,y_n)+f(y_1,\ldots,y_n)\varphi(b)]_h$$

therefore $\varphi(a)$ and $\varphi(b)$ satisfy the same condition for a and b, that is both $\varphi(a)$ and $\varphi(b)$ are diagonal matrices. Denote $a=\sum_{r=1}^t a_re_{rr}$, $b=\sum_{r=1}^t b_re_{rr}$, $\varphi(a)=\sum_{r=1}^t a'_re_{rr}$ and $\varphi(b)=\sum_{r=1}^t b'_re_{rr}$. Choose any $i\neq j$ and let $\varphi(x)=(1+e_{ij})x(1-e_{ij})$. Since the (i,j)-entries of $\varphi(a)$ and $\varphi(b)$ are zero, by computations it follows $0=a_i-a_j$ and $0=b_i-b_j$, that is both a and b are central matrices, a contradiction.

Hence we assume that $dim_C V=\infty$. Then, by (6.45) and [556, Lemma 2], we get

$$[x,ay+yb]_h \tag{6.46}$$

is a generalized identity for Q_r. We firstly show that b must be an element of C. To do this, we assume that $b\notin C$ and prove that a contradiction follows. If $b\notin C$, then there exists $v\in V$ such that v and bv are C-linearly independent vectors.

Moreover, since $dim_C V = \infty$, clearly there is $w \in V$ such that $\{v, bv, w\}$ is a C-linearly independent set. If also $\{v, bv, w, bw\}$ is a C-linearly independent set, by the density of Q_r there are $x_0, y_0 \in Q_r$ such that

$$x_0 v = v, x_0 w = 0, y_0 v = 0, y_0 w = 0, y_0(bv) = w \tag{6.47}$$

and $y_0(bw) = w$. Therefore, by (6.46) it follows the contradiction

$$0 = [x_0, ay_0 + y_0 b]_h v = (-1)^h w \neq 0.$$

On the other hand, if $bw = \lambda v + \mu(bv) + \eta w$, for suitable $\lambda, \mu, \eta \in C$, by the same choices made in (6.47), we get $y_0 bw = \mu w$. Hence, again by (6.46), we have,

$$0 = [x_0, ay_0 + y_0 b]_h v = (-1)^h \mu^h w$$

that, i.e., $\mu = 0$. In other words, we have proved that, if $\{v, bv, w\}$ is C-linearly independent then there exist $\lambda, \eta \in C$ (depending on the choice of $w \in V$) such that $bw = \lambda v + \eta w$. It is obvious that $\{v, bv, w - v\}$ is again a C-linearly independent set. Thus, by the above argument, there are $\lambda', \eta' \in C$ such that $b(w - v) = \lambda' v + \eta'(w - v)$. This implies that $\{v, bv, w\}$ must be linearly C-dependent, which is a contradiction, as desired.

We may then assume $b \in C$ and $a \notin C$, that is Q_r satisfies the identity

$$[x, (a+b)y]_h. \tag{6.48}$$

For $e = e^2 \in H$, we have $eQ_re \cong M_m(C)$, with $m = dim_C Ve$. Since Q_r satisfies $e[exe, e(a+b)eye]_h e$, then eQ_re satisfies $[x, e(a+b)e]_h$. As we have seen above, either $e(a+b)e$ is central in eQ_re, or eQ_re satisfies $s_4(x_1, \dots, x_4)$, or $f(x_1, \dots, x_n)$ is central valued on eQ_re. Since $a \notin C$, $a+b$ doesn't centralize the nonzero socle H, so $(a+b)h_0 \neq h_0(a+b)$, for some $h_0 \in H$. Also, $f(x_1, \dots, x_n)$ is not central-valued on H, otherwise Q_r satisfies the polynomial identity $[f(x_1, \dots, x_n), x_{n+1}]$, which is a contradiction. So, $[f(h_1, \dots, h_n), h_{n+1}] \neq 0$, for some $h_1, \dots, h_{n+1} \in H$. For the same reason, $s_4(h_{n+2}, \dots, h_{n+5}) \neq 0$, for some $h_{n+2}, \dots, h_{n+5} \in H$. By Litoff's theorem (see [104, Theorem 4.3.11]), there is an idempotent $e \in H$ so that $h_0 a, ah_0, h_0, \dots, h_{n+5}$ are all in eQ_re. Then the only possible case should be $e(a+b)e \in Ce$, that is

$$(a+b)h_0 = e(a+b)h_0 = e(a+b)eh_0 = h_0 e(a+b)e = h_0(a+b)e = h_0(a+b)$$

which is again a contradiction.

♣

Lemma 6.3.5 *Suppose there exist $a, b \in Q_r$ such that $F(x) = ax + xb$, for all $x \in R$. Then either $char(R) = 2$ and R satisfies s_4, or G is an inner generalized derivation of R.*

Proof: As above we may consider the case $G(x) = px + d(x)$, where $d \neq 0$ is not an inner derivation of R, if not we are done.

Since R satisfies the generalized differential identity

$$
\begin{aligned}
&[af(x_1,\ldots,x_n) + f(x_1,\ldots,x_n)b, pf(y_1,\ldots,y_n) + f^d(y_1,\ldots,y_n) + \\
&\sum_i f(y_1,\ldots,d(y_i),\ldots,y_n)]_h \\
&- [f(x_1,\ldots,x_n), f(y_1,\ldots,y_n)]_h
\end{aligned}
\tag{6.49}
$$

and since d is not inner, R satisfies the generalized identity

$$
\begin{aligned}
&[af(x_1,\ldots,x_n) + f(x_1,\ldots,x_n)b, pf(y_1,\ldots,y_n) + f^d(y_1,\ldots,y_n) + \\
&\sum_i f(y_1,\ldots,z_i,\ldots,y_n)]_h \\
&- [f(x_1,\ldots,x_n), f(y_1,\ldots,y_n)]_h.
\end{aligned}
\tag{6.50}
$$

In particular R satisfies

$$[af(x_1,\ldots,x_n) + f(x_1,\ldots,x_n)b, f(z_1,y_2,\ldots,y_n)]_h. \tag{6.51}$$

Assume there exist $c_1,\ldots,c_n \in R$ such that

$$u = af(c_1,\ldots,c_n) + f(c_1,\ldots,c_n)b \notin C.$$

Since by (6.51) $[u, f(r_1,\ldots,r_n)]_h = 0$ for all $r_1,\ldots,r_n \in R$, by Theorem 6.2.5 it follows that $char(R) = 2$ and $R \subseteq M_2(C)$.

On the other hand, if $af(x_1,\ldots,x_n) + f(x_1,\ldots,x_n)b \in C$, for all $x_1,\ldots,x_n \in R$, then by (6.49) we have $[f(x_1,\ldots,x_n), f(y_1,\ldots,y_n)]_h$ a polynomial identity for R. As in the proof of the previous Lemma, since $f(x_1,\ldots,x_n)$ is not central-valued on R, again by Theorem 6.2.5, it follows that $char(R) = 2$ and $R \subseteq M_2(C)$.

♣

As a consequence of Lemmas 6.3.3, 6.3.4 and 6.3.5, one of the following holds:

1. Either R satisfies s_4 ($R \subseteq M_2(\overline{C})$);
2. or F and G are both inner generalized derivations of R.

In order to prove Theorem 6.3.2, in what follows we may consider just the second case, that is there exist $a,b,c,q \in Q_r$ such that $F(x) = ax + xb$ and $G(x) = cx + xq$, for all $x \in R$. Hence we infer that R satisfies the generalized identity

$$
\begin{aligned}
&\Phi(x_1,\ldots,x_n; y_1,\ldots,y_n) = \\
&[af(x_1,\ldots,x_n) + f(x_1,\ldots,x_n)b, cf(y_1,\ldots,y_n) + f(y_1,\ldots,y_n)q]_h \\
&- [f(x_1,\ldots,x_n), f(y_1,\ldots,y_n)]_h.
\end{aligned}
$$

We restart with the following Lemma, without its proof. This is because it can be proved by using a standard argument related to the theory of generalized polynomial identities:

Lemma 6.3.6 *Let $F : R \to R$ and $G : R \to R$ be non-zero generalized derivations on R, such that F and G are mutually SEP on S. Suppose that R does not satisfy any non-trivial generalized polynomial identity with coefficients in Q_r, then there exists $\lambda \in C$ such that, for any $x \in R$, $G(x) = \lambda x$ and $F(x) = \lambda^{-h}x$.*

We recall that, in case $y \in R$ such that $0 \neq y^2 \in Z(R)$, the center of R, and $char(R) \neq 2$, then thc computation of $[x,y]_h$, for all $x \in R$, is as follows:

1. $[x,y]_h = 2^{h-1}y^{h-1}(yx - xy)$, if h is an even integer;
2. $[x,y]_h = 2^{h-1}y^{h-1}(xy - yx)$, if h is an odd integer.

Lemma 6.3.7 *Let $R \cong M_t(C)$, be the ring of all $t \times t$ matrices over C, with $t \geq 3$. Let $F(x) = ax + xb$, for all $x \in R$, and fixed $a,b \in R$; moreover let $G(x) = \lambda x$, for all $x \in R$ and fixed $0 \neq \lambda \in C$. Then either $a,b \in C$ and $a + b = \lambda^{-h}$, or $char(R) = p > 0$ and there exists $s \geq 1$ such that $f(x_1,\ldots,x_n)^{p^s}$ is central-valued in R.*

Proof: By our assumption, for all $x,y \in S$ we have $\lambda^h[ax+xb,y]_h = [x,y]_h$, in particular for $x = y$, it follows, $\lambda[ax+xb,x]_h = 0$. From the main result in [212, Proposition 1] and the condition, $t \geq 3$, either $char(R) = p > 0$ and there exists $s \geq 1$ such that $f(x_1,\ldots,x_n)^{p^s}$ is central-valued in R, or $b = a+\gamma$ for a suitable $\gamma \in C$ and $f(x_1,\ldots,x_n)^2$ is central-valued on R. In the latter case, we may consider $char(R) \neq 2$, if we are done again. Moreover, without loss of generality, we may assume that h is an odd integer and

$$\lambda^h 2^{h-1}y^{h-1}[ax + xa + \gamma x, y] = 2^{h-1}y^{h-1}[x,y]$$

which means

$$[\lambda^h ax + \lambda^h xa + \lambda^h \gamma x - x, y] = 0 \tag{6.52}$$

for all $x,y \in S$, and $0 \neq y^2 \in C$. Since $f(x_1,\cdots,x_n)$ is not central-valued, from Remark 1.2.11, there exists some even sequence u from R such that $f(u)$ is a non-central diagonal matrix. Therefore, there exists some $D = f(u)$ =diag$(\alpha_1,\alpha_2,\ldots,\alpha_n)$ such that

$$[\lambda^h ax + \lambda^h xa + \lambda^h \gamma x - x, D] = 0 \tag{6.53}$$

for all $x \in S$. Moreover, since $D = f(u)$ is not central, there must exist some distinct $1 \leq s,r \leq t$ such that $\alpha_s \neq \alpha_r$. Thus, since the set $S = f(R)$ is invariant under the action of all C-linear automorphisms of $M_t(C)$, for any given distinct j,k, by taking some suitable conjugation, we may assume that $\alpha_k \neq \alpha_j$. Thus, for $i \neq j,k$ and $x = e_{ij} \in S$, according to (6.53) we get,

$$[\lambda^h ae_{ij} + \lambda^h e_{ij}a + \lambda^h \gamma e_{ij} - e_{ij}, \sum_{r=1}^{t} \alpha_r e_{rr}] = 0. \tag{6.54}$$

Say $a = \sum a_{rs}e_{rs}$, with $a_{rs} \in C$. Multiplying (6.54) on the right by e_{jj} and also on the left by e_{kk}, it follows, $\lambda^h(\alpha_j - \alpha_k)a_{ki}e_{kj} = 0$, that is $a_{ki} = 0$ for all $k \neq i$. Hence a is a diagonal matrix, $a = \sum a_{rr}e_{rr}$. Let $l \neq m$ and φ be the automorphism of $M_t(C)$ defined as $\varphi(x) = (1+e_{lm})x(1-e_{lm})$, for all $x \in M_t(C)$. Note that, by (6.52), $[\lambda^h\varphi(a)x + \lambda^h x\varphi(a) + \lambda^h\gamma x - x, y] = 0$, for all $x, y \in S$, and $0 \neq y^2 \in C$. Therefore $\varphi(a)$ must be a diagonal matrix. By calculations we have, $\varphi(a) = a + e_{lm}a - ae_{lm} - e_{lm}ae_{lm}$, in particular $a_{ll} = a_{mm}$, for all $l \neq m$, i.e., a is a central matrix. Hence both $a \in C$ and $b \in C$ and $F(x) = \mu x$, with $\mu = a + b = 2a + \gamma$.

Thus, by (6.53), $[(\lambda^h\mu - 1)x, D] = 0$, for all $x \in S$. In case $\lambda^h\mu = 1$ we make the desired conclusion. In the other case it follows $[S, D] = (0)$, and by Theorem 6.2.5, since $t \geq 3$ and $f(x_1, \ldots, x_n)$ is not central-valued, we have the contradiction $D \in C$.

♣

Lemma 6.3.8 *Let $R \cong M_t(C)$, the ring of all $t \times t$ matrices over C, with $t \geq 3$. Let $G(x) = ax + xb$, for all $x \in R$, and fixed $a, b \in R$; moreover, let $F(x) = \lambda x$, for all $x \in R$ and fixed $0 \neq \lambda \in C$. Then $a, b \in C$ and $(a+b)^h = \lambda^{-1}$.*

Proof: Let $i, j, k \geq 1$ form different pairs; also, let $x = e_{ij} \in S$, $y = e_{kj} \in S$. By our assumption we have $\lambda[e_{ij}, ae_{kj} + e_{kj}b]_h = 0$, and left multiplying by e_{ii}, we get, $\lambda e_{ij}(ae_{kj})^h = 0$, i.e., a is a diagonal matrix in R. By using the previous argument we infer that a is central in R. Thus $\lambda[x, y(a+b)]_h = [x, y]_h$, for all $x, y \in S$. Once again for $x = y = e_{ij}$, we get $\lambda(e_{ij}(a+b)^h e_{ij})^h = 0$, that is $a+b$ is a diagonal matrix and as above $a + b = \mu \in C$. Finally, since $\lambda\mu^h[x, y]_h = [x, y]_h$, for all $x, y \in S$, by Theorem 6.2.5, we get the contradiction $S \subseteq C$, unless when $\lambda\mu^h = 1$.

♣

Remark 6.3.9 Here we consider the following situation: K is a field of characteristic different from 2, $R = M_t(K)$ is the matrix ring over K and $t \geq 3$. In all that follows let a, b be elements of R, with $a = \sum_{r,s=1}^{t} a_{rs}e_{rs}$ and $b = \sum_{r,s=1}^{t} b_{rs}e_{rs}$, for $a_{rs}, b_{rs} \in K$. Here we always suppose that a, b satisfy the following condition:

$$\begin{gathered}\text{If} \quad i \neq j \quad \text{are fixed integers such that} \quad a_{ij}b_{ij} = 0 \quad \text{then} \\ \text{for any inner automorphism} \quad \varphi \quad \text{of} \quad R \quad \text{it follows} \\ a'_{ij}b'_{ij} = 0 \quad \text{where} \quad \varphi(a) = \sum_{r,s=1}^{t} a'_{rs}e_{rs}, \quad \varphi(b) = \sum_{r,s=1}^{t} b'_{rs}e_{rs}.\end{gathered} \tag{6.55}$$

In [222, Proposition 1] it is proved that, if $a_{ij}b_{ij} = 0$ for all $i \neq j$, then either $a \in K$ or $b \in K$.

Lemma 6.3.10 *Let $R \cong M_t(C)$ be the ring of all $t \times t$ matrices over C, with $t \geq 3$ and char$(C) \neq 2$. Let $G(x) = xa$ and $F(x) = bx$, for all $x \in R$, and fixed $a, b \in R$.*

Then either $a,b \in C$ and $a = b^{-h}$, or $b \in C$, $char(R) = p > 0$ and there exists $s \geq 1$ such that $f(x_1,\dots,x_n)^{p^s}$ is central-valued on R.

Proof: Denote $a = \sum_{r,s=1}^t a_{rs}e_{rs}$ and $b = \sum_{r,s=1}^t b_{rs}e_{rs}$. Since $f(x_1,\dots,x_n)$ is not central valued on R, as above we have that $e_{ij}, e_{ki} \in S$, for any $i,j,k \geq 1$ paired differently. Hence $[e_{ij}a, be_{ki}]_h = 0$, and right multiplying by e_{kk}, we get, $(be_{ki})^h e_{ij}ae_{kk} = 0$. Thus $b_{ik}a_{jk} = 0$, moreover, for any automorphism φ of R we have, $[x\varphi(a), \varphi(b)y]_h = [x,y]_h$, for all $x,y \in S$. Thus the elements, $\varphi(a)$ and $\varphi(b)$ satisfy the same condition on a and b. Let $\varphi(x) = (1+e_{ij})x(1-e_{ij})$, for $i \neq j$ and denote $\varphi(a) = \sum_{r,s=1}^t a'_{rs}e_{rs}$ and $\varphi(b) = \sum_{r,s=1}^t b'_{rs}e_{rs}$. Since $b'_{ik}a'_{jk} = 0$, for $k \neq i,j$, then by computation we have $0 = (b_{ik} + b_{jk})a_{jk} = b_{jk}a_{jk}$.

Therefore $a_{jk}b_{jk} = 0$ for all $j \neq k$, moreover a and b satisfy the condition (6.55). Thus, according to Remark 6.3.9 either $a \in C$ or $b \in C$. In any case we get one of the required conclusions, applying either Lemma 6.3.8 (if $a \in C$), or Lemma 6.3.7 (if $b \in C$).

♣

Lemma 6.3.11 *Let $R \cong M_t(C)$ be the ring of all $t \times t$ matrices over C, with $t \geq 3$ and $char(C) \neq 2$. Let $G(x) = ax$ and $F(x) = xb$, for all $x \in R$, and fixed $a,b \in R$. Then either $a,b \in C$ and $a = b^{-h}$, or $b \in C$, $char(R) = p > 0$ and there exists $s \geq 1$ such that $f(x_1,\dots,x_n)^{p^s}$ is central-valued on R.*

Proof: Denote $a = \sum_{r,s=1}^t a_{rs}e_{rs}$ and $b = \sum_{r,s=1}^t b_{rs}e_{rs}$. As above $e_{kj}, e_{ji} \in S$, for any $i,j,k \geq 1$ paired in different sets. Hence $[ae_{kj}, e_{ji}b]_h = 0$, and left multiplying by e_{ii}, it follows $a_{ik}b_{ij} = 0$, moreover, for any automorphism φ of R we also have, $[\varphi(a)x, y\varphi(b)]_h = [x,y]_h$, for all $x,y \in S$. Thus the elements $\varphi(a)$ and $\varphi(b)$ satisfy the same condition on a and b. By repeating the previous argument, we fix $\varphi(x) = (1+e_{kj})x(1-e_{kj})$ and denote $\varphi(a) = \sum_{r,s=1}^t a'_{rs}e_{rs}$ and $\varphi(b) = \sum_{r,s=1}^t b'_{rs}e_{rs}$. Since $a'_{ik}b'_{ij} = 0$, for $k \neq i,j$, then by computation we have $0 = a_{ik}(b_{ij} - b_{ik}) = -a_{ik}b_{ik}$.

Once again a and b satisfy the condition (6.55), moreover $a_{ij}b_{ij} = 0$ for all $i \neq j$. By Remark 6.3.9 either $a \in C$ or $b \in C$, and as above we arrive at the expected conclusions.

♣

The previous Lemmas allow us to obtain a definitive result in the case of a matrix ring:

Proposition 6.3.12 *Let $R \cong M_t(C)$, the ring of all $t \times t$ matrices over C, with $t \geq 3$ and $char(C) \neq 2$. Then one of the following holds:*

1. *There exists $\lambda \in C$ such that, for any $x \in R$, $G(x) = \lambda x$ and $F(x) = \lambda^{-h}x$;*
2. *$char(R) = p \geq 3$ and there exist $\lambda \in C$ and $s \geq 1$ such that, for any $x \in R$, $G(x) = \lambda x$ and $f(x_1,\dots,x_n)^{p^s}$ is central-valued on R.*

Proof: If $h = 1$ we obtain the required conclusion by Theorem 6.3.1. Thus we assume $h \geq 2$. In light of all previous results, we may consider the only case when R satisfies

$$\begin{aligned}&\Phi(x_1,\dots,x_n;y_1,\dots,y_n) = \\ &[af(x_1,\dots,x_n)+f(x_1,\dots,x_n)b, cf(y_1,\dots,y_n)+f(y_1,\dots,y_n)q]_h \\ &-[f(x_1,\dots,x_n), f(y_1,\dots,y_n)]_h.\end{aligned} \tag{6.56}$$

Let $e_{ij}, e_{kj} \in S$, for any $i, j, k \geq 1$ in different pairs and apply the main condition. Therefore we get $[ae_{ij}+e_{ij}b, ce_{kj}+e_{kj}q]_h = 0$, and left multiplying by e_{ii} and also right multiplying by e_{kk}, it follows $e_{ij}b(e_{kj}q)^h e_{kk} = 0$. If, $b = \sum_{r,s=1}^{t} b_{rs}e_{rs}$ and $q = \sum_{r,s=1}^{t} q_{rs}e_{rs}$, the previous identity means that $b_{jk}q_{jk} = 0$. Moreover it is clear that b, q satisfy the condition (6.55), so that, according to Remark 6.3.9, either b or q is central.

By using the same argument, we may prove that either a or c is central (we omit the proof for brevity). Hence we have 4 different cases:

1. $a, b \in C$. In this case, by Lemma 6.3.8, $c, q \in C$ and $c+q = (a+b)^{-h}$;
2. $c, q \in C$. In this case, by Lemma 6.3.7, either $a, b \in C$ and $a+b = (c+q)^{-h}$, or $char(R) = p > 0$ and there exists $s \geq 1$ such that $f(x_1,\dots,x_n)^{p^s}$ is central valued in R.
3. $a, q \in C$. In this case, by Lemma 6.3.10, either $b, c \in C$ and $(a+b) = (c+q)^{-h}$, or $c \in C$, $char(R) = p > 0$ and there exists $s \geq 1$ such that $f(x_1,\dots,x_n)^{p^s}$ is central-valued on R.
4. $b, c \in C$. In this case, by Lemma 6.3.11, either $a, q \in C$ and $(a+b) = (c+q)^{-h}$, or $q \in C$, $char(R) = p > 0$ and there exists $s \geq 1$ such that $f(x_1,\dots,x_n)^{p^s}$ is central-valued on R.

In any case we are done.

♣

As a particular case of Proposition 6.3.12, we have Corollary 6.2.1.

Corollary 6.3.1 *Let $R \cong M_t(C)$, the ring of all $t \times t$ matrices over C, with $t \geq 3$ and $char(C) \neq 2$. Let F and G be non-zero mutually SEP generalized derivations on R, defined as $F(x) = ax+xb$, $G(x) = cx+xq$, for all $x \in R$ and fixed $a, b, c, q \in R$. Then $a, b, c, q \in C$ and $(a+b) = (c+q)^{-h}$.*

The proof of Theorem 6.3.2

As above we may consider the only case when R satisfies

$$\begin{aligned}&\Phi(x_1,\dots,x_n;y_1,\dots,y_n) = \\ &[af(x_1,\dots,x_n)+f(x_1,\dots,x_n)b, cf(y_1,\dots,y_n)+f(y_1,\dots,y_n)q]_h \\ &-[f(x_1,\dots,x_n), f(y_1,\dots,y_n)]_h.\end{aligned} \tag{6.57}$$

Moreover, by Lemma 6.3.6, we may asume that (6.57) is a non-trivial generalized polynomial identity for R, as well as for Q_r. Here we repeat the previous argument and, without loss of generality, we assume that Q_r is a primitive ring having a non-zero socle H with C its associated division ring. Then Q_r is isomorphic to a dense ring of linear transformations of a vector space V over C. If $dim_C V = t$ is finite, then $Q_r \cong M_t(C)$ and the conclusions follow from Proposition 6.3.12. Hence we assume $dim_C V = \infty$.

In this final case, (6.57) and [556, Lemma 2] $\Rightarrow$

$$\Phi(x,y) = \Big[ax+xb, cy+yq\Big]_h - \Big[x,y\Big]_h \tag{6.58}$$

is a generalized identity for Q_r.

Since H cannot satisfy any polynomial identity, there are $y_1,\dots,y_4 \in H$ such that $s_4(y_1,\dots,y_4) \neq 0$. Let $y_0 \in H$. Since, $ay_0+y_0b \in H$ and $cy_0+y_0q \in H$, by Litoff's theorem [104, Theorem 4.3.11], there exists an idempotent element $e \in H$ such that

$$y_0, y_1, \dots, y_4, \quad ay_0+y_0b, \quad cy_0+y_0q \in eQ_re \cong M_k(C)$$

for some positive integer k. Define $g_e : eQ_re \to eQ_re$ and $f_e : eQ_re \to eQ_re$ by

$$g_e(x) = (ea)x + x(be), \quad \forall x \in eQ_re$$

and

$$f_e(x) = (ec)x + x(qe), \quad \forall x \in eQ_re$$

and note that

$$[g_e(x), f_e(y)]_h - \Big[x,y\Big]_h$$

is satisfied by eQ_re. Application of Corollary 6.3.1, and $s_4(y_1,\dots,y_4) \neq 0$, $\Rightarrow$ $ea, be, ec, qe \in Ce$ and

$$[g_e(s), s] = 0, \quad [f_e(s), s] = 0, \quad \forall s \in eQ_re. \tag{6.59}$$

In particular, from (6.59) and the fact that $y_0, ay_0+y_0b, cy_0+y_0q \in eQ_re$, it follows that

$$0 = [g_e(y_0), y_0] = e[ay_0+y_0b, y_0]e = [ay_0+y_0b, y_0] \tag{6.60}$$

and also

$$0 = [f_e(y_0), y_0] = e[cy_0 + y_0q, y_0]e = [cy_0 + y_0q, y_0]. \tag{6.61}$$

By repeating the above argument for any element of H, H satisfies the generalized polynomial identities

$$[aX + Xb, X] \text{ and } [cX + Xq, X]. \tag{6.62}$$

Since H is an ideal of Q_r, both $[aX + Xb, X]$ and $[cX + Xq, X]$ are generalized polynomial identities for Q_r. Moreover, if we assume that $b \notin C$, then there exists $0 \neq v \in V$, such that $\{v, bv\}$ are linearly C-independent. By $\dim_C V = \infty$ we can fix $w \in V$ such that $\{v, bv, w\}$ are linearly C-independent. By the density of Q_r, there exist $s \in Q_r$ such that

$$sv = 0; \quad s(bv) = w; \quad sw = -w.$$

Hence we have the contradiction

$$0 = [as + sb, s]v = w \neq 0.$$

Therefore, $b \in C$ and, by the same argument, $q \in C$. Thus, both

$$[(a+b)X, X] \text{ and } [(c+q)X, X] \tag{6.63}$$

are satisfied by H. Finally, if we suppose that $a + b \notin C$, then there exists $0 \neq v' \in V$, such that $\{v', (a+b)v'\}$ are linearly C-independent. By the density of Q_r, there exist $s' \in Q_r$ such that

$$s'v' = v'; \quad s'(a+b)v' = 0$$

$\Rightarrow$

$$0 = [(a+b)s', s']v' = (a+b)v'$$

which is a contradiction. Therefore $a + b \in C$ and analogously $c + q \in C$. Namely, $a + b = \lambda$ and $c + q = \mu$, so that (6.58) reduces to

$$\lambda\mu^h \Big[x, y\Big]_h - \Big[x, y\Big]_h \tag{6.64}$$

which means $\lambda\mu^h = 1$, as desired.

♣

6.4 Quadratic differential identities

Let R be an associative ring. A quadratic functional identity is an identity having the form

$$\sum_i \sum_j a_{ij} E_i(x_1, \ldots, x_m) F_j(x_1, \ldots, x_n) + \sum_h \sum_k G_h(x_1, \ldots, x_p) H_k(x_1, \ldots, x_t) b_{hk} = 0$$

where

$$E_i : R^m \to Q_r \quad F_j : R^n \to Q_r \quad G_h : R^p \to Q_r \quad H_k : R^t \to Q_r$$

are additive maps and $a_{ij}, b_{jk} \in Q_r$. A quadratic functional identity is called **differential** if the involved maps are derivations, generalized derivations, skew derivations, generalized skew derivations, products or compositions of the previously cited maps. In particular, several papers in available literature are devoted to the study of additive maps $F, G : R \longrightarrow Q_r$ satisfying the relation $q(x) = F(x)G(x) \pm G(x)F(x) = 0$, for all x in a suitable $S \in R$. In [361] Lanski showed that if d and δ are non-zero derivations on R, such that $[d(x), \delta(x)] \in Z(R)$, for all $x \in R$, then either there exists λ in the extended centroid C of R such that $d = \lambda\delta$ or $char(R) = 2$ and $R \subseteq M_2(C)$, the ring of 2×2 matrices over C. In [363] Lanski extended this last result to the Lie ideals. In [376] Lee studied the case when $[d(x), \delta(x)] \in Z(R)$, for any $x \in \rho$, a non-zero right ideal of R, proving that either there exists $\lambda \in C$ such that $d = \lambda\delta$ or $d(\rho)\rho = \delta(\rho)\rho = 0$, unless $char(R) = 2$ and ρ satisfies the identity $s_4(x_1, \ldots, x_4)x_5$. Later Beidar, Brešar and Chebotar proved that if $[d(x), F(x)] = 0$, for all $x \in R$, where F is an additive map on R and d is a derivation of R, then there exist $\lambda \in C$ and an additive map $\mu : R \longrightarrow C$, such that $F(x) = \lambda d(x) + \mu(x)$, for any $x \in R$ [100]. Following this line of investigation, in [274] M. Fošner and Vukman proved that if F, G are generalized derivations of a prime ring R of characteristic not equal to 2, such that $F(x)G(x) + G(x)F(x) = 0$, for all $x \in R$, then either $F = 0$ or $G = 0$. More recently, in [499] Rania and Scudo considered a quadratic differential identity involving derivations and generalized derivations on the set of values of a multilinear polynomial. They proved that, if R is a prime ring of characteristic not equal to 2, $f(x_1, \ldots, x_n)$ is a multilinear polynomial over C, not central valued on R, d a derivation of R and G a generalized derivation of R such that

$$G\big(f(r_1,\ldots,r_n)\big)d\big(f(r_1,\ldots,r_n)\big) + d\big(f(r_1,\ldots,r_n)\big)G\big(f(r_1,\ldots,r_n)\big) = 0$$

for all $r_1, \ldots, r_n \in R$, then either $d = 0$ or $G = 0$, unless there exists $\lambda \in C$ such that $G(x) = \lambda x$ for all $x \in R$ and $f(x_1, \ldots, x_n)^2$ is central-valued on R. A similar quadratic differential identity involving skew derivations on the set of values of a multilinear polynomial was studied in [96]. More precisely in [96, Theorem 1.3] has been proved as follows: If R is a prime ring of characteristic not equal to 2, $f(x_1, \ldots, x_n)$ is a multilinear polynomial over C, not central-valued on R, F and G skew derivations of R, associated with the automorphism α, such that

$$F\big(f(r_1,\ldots,r_n)\big)G\big(f(r_1,\ldots,r_n)\big) + G\big(f(r_1,\ldots,r_n)\big)F\big(f(r_1,\ldots,r_n)\big) = 0$$

for all $r_1, \ldots, r_n \in R$, then either $F = 0$ or $G = 0$.

Conforming to the described foregoing, in this section we present two results related to the differential identity having the form $[F(x), G(x)] = 0$:

Theorem 6.4.1 *[228, Theorem 1] Let R be a prime K-algebra, where K is a commutative ring with 1, and let $f(x_1,\ldots,x_n)$ be a multilinear polynomial over K, not centrally valued on R. Suppose that d and δ are nonzero derivations of R such that*

$$[d(f(r_1,\ldots,r_n)),\delta(f(r_1,\ldots,r_n))] \in Z(R)$$

for all $r_1,\ldots,r_n \in R$. Then $\delta = \lambda d$ for some $\lambda \in C$ except when $char(R) = 2$ and $\dim_C RC = 4$.

Theorem 6.4.2 *[234, Theorem 1.1] Let R be a prime ring of characteristic not equal to 2, L a non-central Lie ideal of R. Let $F : R \to R$ and $G : R \to R$ be non-zero generalized derivations on R. If $[F(u),G(u)] = 0$, for all $u \in L$, then one of the following holds:*

1. *There exists $\lambda \in C$ such that, for any $x \in R$, $G(x) = \lambda F(x)$;*
2. *R satisfies s_4, the standard identity of degree* 4.

The proof of Theorem 6.4.1

We start by studying the case when $R = M_m(C)$ is the algebra of $m \times m$ matrices over the field C and d and δ are nonzero inner derivations on R. Let I be the identity matrix of R. The first result we prove is as follows:

Proposition 6.4.3 *Let $R = M_m(C)$, where C is a field, and let $f(x_1,\cdots,x_n)$ be a multilinear polynomial over C, not centrally valued on R. Suppose that $A,B \in R$ satisfy*

$$\Big[[A,f(r_1,\ldots,r_n)],[B,f(r_1,\ldots,r_n)]\Big] \in C$$

for all $r_1,\cdots,r_n \in R$. Then A,B and I are C-dependent except when $m = 2$ and $char(C) = 2$.

We prove Proposition 6.4.3 by a sequence of lemmas which give information on the entries of A and B. First, we say, if φ is an inner automorphism of $M_m(C)$, then

$$[[\varphi(A),\varphi(s)],[\varphi(B),\varphi(s)]] = \varphi\left([[A,s],[B,s]]\right) \in C$$

for all $s \in f(R)$. Since $f(R)$ is invariant under the action of all inner automorphisms of R, we have

$$[[\varphi(A),t],[\varphi(B),t]] \in C \quad \text{for all } t \in f(R).$$

Clearly $\{\varphi(A),\varphi(B),I\}$ are linearly dependent iff $\{A,B,I\}$ are linearly dependent. Hence, to prove our result, we may replace A with $\varphi(A)$ and B with $\varphi(B)$. Moreover, if r^t denotes the transpose of the matrix $r \in M_m(C)$ then $f(r_1,\ldots,r_n)^t = f^*(r_1^t,\ldots,r_n^t)$, where, $f^*(x_1,\ldots,x_n)$ is the polynomial obtained from $f(x_1,\ldots,x_n)$ by reversing the order of the indeterminates in any monomial, i.e., $f^*(x_1,\ldots,x_n) = \alpha_{id.}x_n\ldots x_1 + \sum_{\sigma\in S_n,\sigma\neq id.}\alpha_\sigma x_{\sigma(n)}\cdots x_{\sigma(1)}$. Clearly,

$$[[A,f(r_1,\ldots,r_n)],[B,f(r_1,\ldots,r_n)]]\in C \quad \text{for all } r_i\in R$$

iff

$$[[A^t,f^*(r_1,\ldots,r_n)],[B^t,f^*(r_1,\ldots,r_n)]]\in C \quad \text{for all } r_i\in R.$$

Moreover, $\{A,B,I\}$ are linearly dependent iff $\{A^t,B^t,I\}$ are linearly dependent. Clearly, by our assumption, $f^*(x_1,\ldots,x_n)$ is not centrally valued on R, so in order to prove our result we may replace the given matrices A,B by their transposes A^t and B^t.

We write $A=\sum_{s,t}a_{st}e_{st}$ and $B=\sum_{s,t}b_{st}e_{st}$, where $a_{st},b_{st}\in C$, Thus, by Remark 1.2.12, for $i\neq j$ we have:

$$\big[[A,e_{ij}],[B,e_{ij}]\big]\in C \tag{6.65}$$

Remark 6.4.4 *Let V be a m-dimensional vector space over the field C with a fixed basis. Let us denote by T the linear map induced on V by the matrix A. If A is not scalar then there exists a vector v such that $v,T(v)$ are linearly independent. Hence $\{v,T(v)\}$ is contained in an ordered basis L and let A^P be the matrix of T relative to this new basis. Then A^P is conjugate to A by a suitable invertible matrix P; moreover, for some $i\neq j$ the (i,j)-entry of A^P is nonzero. More precisely, if $T(v)$ and v are respectively the i-th vector and the j-th one of the ordered basis L then the j-th column of the matrix A^P has just one non-zero entry, namely we have $(A^P)_{ij}=1$.*

Recalling the previous fact, we begin with:

Lemma 6.4.5 *Let $m\geq 3$ and $i\neq j$, if $a_{ij}\neq 0$ then $b_{kj}=\frac{b_{ij}}{a_{ij}}a_{kj}$ and $b_{ik}=\frac{b_{ij}}{a_{ij}}a_{ik}$ for every $k\neq i,j$.*

Proof: For $i\neq j$, we have $\big[[A,e_{ji}],[B,e_{ji}]\big]\in C$, thus we obtain:

$$\begin{aligned}\big[[A,e_{ji}],[B,e_{ji}]\big] &= Ae_{ji}Be_{ji}-e_{ji}ABe_{ji}+e_{ji}Ae_{ji}B\\ &\quad -Be_{ji}Ae_{ji}+e_{ji}BAe_{ji}-e_{ji}Be_{ji}A\\ &= (b_{ij}A-a_{ij}B)e_{ji}+e_{ji}(a_{ij}B-b_{ij}A)+e_{ji}(BA-AB)e_{ji}\\ &= \big[b_{ij}A-a_{ij}B,e_{ji}\big]+e_{ji}[B,A]e_{ji}\in C\end{aligned} \tag{6.66}$$

Let $k\neq i,j$. We see the (k,i)-entry and (j,k)-entry of (6.66) to get $b_{ij}a_{kj}=a_{ij}b_{kj}$ and $a_{ij}b_{ik}=b_{ij}a_{ik}$, respectively. In particular:

$$\text{if}\quad a_{ij}\neq 0\quad\text{then}\quad b_{kj}=\frac{b_{ij}}{a_{ij}}a_{kj}\quad\text{and}\quad b_{ik}=\frac{b_{ij}}{a_{ij}}a_{ik}\quad\text{for every}\quad k\neq i,j. \tag{6.67}$$

♣

Lemma 6.4.6 *Let $i \neq j$, if $a_{ij} \neq 0$ but $b_{ij} = 0$, then $b_{st} = 0$ for every $s \neq t$ unless $s = j$ or $t = i$.*

Proof: Let s,t be indices such that $j \neq s \neq t \neq i$. If $m = 2$ then $s = i$, $t = j$ we are done. Let $m \geq 3$, then, by Lemma 6.4.5, we have $b_{kj} = 0 = b_{ik} =$ for every $k \neq i, j$. Hence we may assume $t \neq j$ and $s \neq i$, that is $\{s,t\} \cap \{i,j\} = \emptyset$. For $\alpha \in C$ we consider the invertible matrix $P = I + \alpha e_{si}$ and let $X^P = P^{-1}XP$ for every $X \in M_m(C)$. Note that $(A^P)_{sj} = a_{sj} - \alpha a_{ij}$ and $(B^P)_{sj} = b_{sj} - \alpha b_{ij} = 0$. Moreover $(A^P)_{st} = a_{st} - \alpha a_{it}$ and $(B^P)_{st} = b_{st} - \alpha b_{it} = b_{st}$. By choosing α such that $(A^P)_{sj} = a_{sj} - \alpha a_{ij} \neq 0$, by (6.67), we obtain $b_{st} = (B^P)_{st} = \frac{(B^P)_{sj}}{(A^P)_{sj}}(A^P)_{st} = 0$.

♣

Lemma 6.4.7 *If $m \geq 3$, and there exist $i \neq j$ such that $a_{ij} \neq 0$ but $b_{ij} = 0$, then B is a scalar matrix.*

Proof: We first show that all off-diagonal entries of B are zero. By Lemma 6.4.6, it is remained to be shown that $b_{ki} = 0$ for every $k \neq i$ and $b_{jk} = 0$ for every $k \neq j$. Since $f(x_1, \cdots, x_n)$ is not centrally valued, by Lemma 1.2.11 there exists some even sequence u from R such that $f(u)$ is a non-central diagonal matrix. Therefore, there exists some $D = f(u) = \mathrm{diag}(\alpha_1, \alpha_2, \ldots, \alpha_n)$ such that $[[A,D],[B,D]]$ is centrally valued. Moreover, since $D = f(u)$ is not central and $m \geq 3$ there must exist some distinct $1 \leq p, q, r \leq m$ such that $(\alpha_p - \alpha_q)(\alpha_p - \alpha_r) \neq 0$. Thus, since the set $f(R)$ is invariant under the action of all C-linear automorphisms of $M_m(C)$, for any given distinct i, j, k, by taking some suitable conjugation, we may assume that $(\alpha_i - \alpha_j)(\alpha_i - \alpha_k) \neq 0$.

Let $A' = [A,D]$ and $B' = [B,D]$. Then $A'_{st} = a_{st}(\alpha_t - \alpha_s)$ and $B'_{st} = b_{st}(\alpha_t - \alpha_s)$ for $1 \leq s, t \leq m$ and we say, if $B'_{st} \neq 0$ then either $s = j$ or $t = i$. By hypothesis, $E = [A', B']$ is in the center, hence it must be with a zero off-diagonal entry. For every $k \neq i, j$, note that the (k, j)-entry of E is $E_{kj} = -B'_{ki}A'_{ij} = -(\alpha_j - \alpha_i)(\alpha_i - \alpha_k)a_{ij}b_{ki}$. Since $(\alpha_i - \alpha_j)(\alpha_i - \alpha_k) \neq 0$ and $a_{ij} \neq 0$, then we can see, $b_{ki} = 0$.

Repeating the same argument on the transpose of A and B, we get, $b_{jk} = 0$ for every $k \neq i, j$. Thus we have

$$B = \mathrm{diag}(b_{11}, \ldots, b_{nn}) + b_{ji}e_{ji}.$$

Also observe that the (j,k)-entry of E is $-(\alpha_i - \alpha_j)(\alpha_k - \alpha_i)a_{ik}b_{ji}$, so we have $a_{ik}b_{ji} = 0$ for every $k \neq i, j$.

For $\alpha \in C$ let $P = I + \alpha e_{jk}$, then $(A^P)_{ij} = a_{ij} \neq 0, (B^P)_{ij} = b_{ij} = 0$ and $(A^P)_{ik} = a_{ik} + \alpha a_{ij}, (B^P)_{ji} = b_{ji} - \alpha b_{ki} = b_{ji}$. Since $a_{ij} \neq 0$, by choosing α such that $a_{ik} + \alpha a_{ij} \neq 0$ then by the above result, we observe, $0 = (A^P)_{ik}(B^P)_{ji} = (a_{ik} + \alpha a_{ij})b_{ji}$, i.e., $b_{ji} = 0$. Thus B a diagonal matrix.

Finally, for $k \neq i$, let $P = I + e_{ki}$, then $(A^P)_{ij} = a_{ij} \neq 0, (B^P)_{ij} = b_{ij} = 0$. We also have that B^P is a diagonal matrix, hence $0 = (B^P)_{ki} = b_{ki} + b_{kk} - b_{ii} - b_{ik} = b_{kk} - b_{ii}$, i.e., $B = b_{ii}I$ is a scalar matrix, completing the proof.

♣

Lemma 6.4.8 *If $m \geq 3$, then A, B, I are C-dependent.*

Proof: Clearly, if A is not a scalar matrix, then by Remark 6.4.4 some conjugate of A has a non-zero off-diagonal entry. So we may as well assume that A itself has a nonzero off-diagonal entry. Suppose $a_{ij} \neq 0$ for some $i \neq j$, then $B' = B - \frac{b_{ij}}{a_{ij}}A$ also satisfies the original equation and moreover $B'_{ij} = 0$. By Lemma 6.4.7, we must have B' must be a scalar matrix. So $B - \frac{b_{ij}}{a_{ij}}A \in C$, i.e., A, B and I are C-dependent.

♣

To complete the proof of Proposition 6.4.3, the 2×2 matrix ring with characteristic not equal to 2 remains to be computed.

Lemma 6.4.9 *If $m = 2$ and $char(C) \neq 2$, then A, B and I are C-dependent.*

Proof: If A is a scalar matrix, there is nothing to prove. If A is not a scalar matrix, then by Remark 6.4.4 we may assume that A has the form $A = e_{21} + a_{12}e_{12} + a_{22}e_{22}$. Then, by Remark 1.2.12, $\big[[A, e_{12}], [B, e_{12}]\big] \in C$ and we obtain

$$\begin{aligned}&\big[-e_{11} - a_{22}e_{12} + e_{22}, -b_{21}e_{11} + (b_{11} - b_{22})e_{12} + b_{21}e_{22}\big]\\&= 2(-b_{11} + b_{22} - a_{22}b_{21})e_{12} \in C\end{aligned}$$

and so $b_{22} = b_{11} + a_{22}b_{21}$. Since $f(x_1, \cdots, x_n)$ is not centrally valued, as in the proof of Lemma 6.4.7 , there exists some even sequence u from R such that $f(u) = \alpha e_{11} + \beta e_{22}$, where $\alpha \neq \beta$ in C. Hence we have

$$\begin{aligned}&\big[[A, f(u)], [B, f(u)]\big]\\&= \big[(\alpha - \beta)e_{21} - (\alpha - \beta)a_{12}e_{12}, (\alpha - \beta)b_{21}e_{21} - (\alpha - \beta)b_{12}e_{12}\big] \in C\end{aligned}$$

i.e.,

$$(\alpha - \beta)^2(-a_{12}b_{21} + b_{12})(e_{11} - e_{22}) \in C$$

$\Rightarrow b_{12} = b_{21}a_{12}$. So

$$B = \begin{pmatrix} b_{11} & b_{21}a_{12} \\ b_{21} & b_{11} + b_{21}a_{22} \end{pmatrix} = b_{21}A + b_{11}I.$$

So A, B and I are C-dependent.

♣

Hence, the proof of Proposition 6.4.3 is complete.

We now study in detail the case when d and δ are both inner non-zero derivations. More precisely, if d is the inner derivation induced by the element $a \in Q_r$ and δ the one induced by $b \in Q_r$ then let us consider the generalized polynomial $\Phi(x_1,\ldots,x_{n+1}) = [[[a,f(x_1,\ldots,x_n)],[b,f(x_1,\ldots,x_n)]],x_{n+1}]$ in the free product $Q_r *_C C\{x_1,\ldots,x_{n+1}\}$. Our first result is as follows:

Lemma 6.4.10 *Either $\{d,\delta\}$ are linearly C-dependent or $\Phi(x_1,\ldots,x_{n+1})$ is a non trivial generalized polynomial identity for R.*

Proof: By our hypothesis, $\Phi(r_1,\ldots,r_{n+1}) = 0$ for all $r_1,\ldots,r_{n+1} \in R$, i.e., R satisfies the generalized polynomial identity $\Phi(x_1,\ldots,x_{n+1})$. Clearly, we may assume that $\{1,a,b\}$ are linearly C-independent otherwise the derivations d and δ are linearly C-dependent and we are done. If B is a basis of Q_r over C then any element of $T = Q_r *_C C\{x_1,\ldots,x_{n+1}\}$ can be written in the form $g = \sum_i \alpha_i m_i$. In this decomposition the coefficients α_i are in C and the elements m_i are B-monomials, that is $m_i = q_0 y_1 \cdots y_{n+1} q_{n+1}$, with $q_i \in B$ and $y_i \in \{x_1,\ldots,x_{n+1}\}$. As a consequence of [184], since $\{1,a,b\}$ is a subset of some basis of Q_r over C, it easy to see that $\Phi(x_1,\ldots,x_{n+1}) \neq 0 \in T$, that is $\Phi(x_1,\ldots,x_{n+1})$ is a non-trivial generalized polynomial identity for R.

♣

The final result in the inner case is the following:

Proposition 6.4.11 *Let d and δ be both Q_r-inner non-zero derivations, then either $f(x_1,\ldots,x_n)$ is central-valued on R or $\{d,\delta\}$ are linearly C-dependent, except when $char(R) = 2$ and $dim_C RC = 4$.*

Proof: As above, let d be the inner derivation induced by the element $a \in Q_r$, and δ the one induced by $b \in Q_r$. We consider the generalized polynomial $\Phi(x_1,\ldots,x_{n+1}) = [[[a,f(x_1,\ldots,x_n)],[b,f(x_1,\ldots,x_n)]],x_{n+1}]$. By Lemma 6.4.10 we may assume that $\Phi(x_1,\ldots,x_{n+1})$ is a non trivial generalized polynomial identity for R, as well as for Q_r. In case C is infinite, we obtain $\Phi(r_1,\ldots,r_{n+1}) = 0$ for all $r_1,\ldots,r_{n+1} \in Q_r \bigotimes_C \overline{C}$, where $\overline{C}$ is the algebraic closure of C. Since both Q_r and $Q_r \bigotimes_C \overline{C}$ are centrally closed, we may replace R by Q_r or $Q_r \bigotimes_C \overline{C}$ according as C is finite or infinite. Thus we may assume that R is centrally closed over C which is either finite or algebraically closed and $\Phi(r_1,\ldots,r_{n+1}) = 0$, for all $r_1,\ldots,r_{n+1} \in R$. By Martindale's theorem [447], R is a primitive ring having a non-zero socle with C as the associated division ring, and R is isomorphic to a dense ring of linear transformations on some vector space V over C.

Assume first that V is infinite-dimensional over C. Since V is infinite dimensional over C then, as in [556, Lemma 2], the set $f(R)$ is dense on R and so from $[[a,f(r_1,\ldots,r_n)],[b,f(r_1,\ldots,r_n)]] \in Z(R)$, for all $r_1,\ldots,r_n \in R$, we have $[[a,r],[b,r]] \in Z(R)$, for all $r \in R$. As a consequence of [361, Theorem 4], since

a and b don't fall in C, we get the conclusion $d = \gamma\delta$ for some $\gamma \in C$, except when $char(R) = 2$ and $dim_C RC = 4$. Assume next that V is finite-dimensional over C. Then the density of R on V implies that $R \cong M_m(C)$, the ring of all $m \times m$ matrices over C. In this case the conclusion follows by Proposition 6.4.3.

♣

We need a final Lemma:

Lemma 6.4.12 *Let R be a prime K-algebra and $f(x_1,\dots,x_n)$ a multilinear polynomial over K. If, for any $i = 1,\dots,n$,*

$$[f(r_1,\dots,z_i,\dots,r_n), f(r_1,\dots,r_n)] \in Z(R)$$

for all $z_i, r_1,\dots,r_n \in R$ then the polynomial $f(x_1,\dots,x_n)$ is central-valued on R, except when $char(R) = 2$ and $dim_C RC = 4$.

Proof: Let $s \in R$, then by assumption

$$[s, f(r_1,\dots,r_n)]_2 =$$

$$[[s, f(r_1,\dots,r_n)], f(r_1,\dots,r_n)] = [\sum_i f(r_1,\dots,[s,r_i],\dots,r_n), f(r_1,\dots,r_n)] \in Z(R).$$

Hence, $[s, f(r_1,\dots,r_n]_3 = [[s, f(r_1,\dots,r_n)]_2, f(r_1,\dots,r_n)] = 0$ and the result follows by [383, Theorem].

♣

We are now able to provide the final step of the proof for Theorem 6.4.1:

As we said, we can say

$$f(x_1,\dots,x_n) = \alpha_{id}.x_1 \cdot x_2 \cdots x_n + \sum_{\sigma \in S_n, \sigma \neq id} \alpha_\sigma x_{\sigma(1)} \cdots x_{\sigma(n)}$$

where S_n is the permutation group over n elements and all $\alpha_\sigma \in C$.

Moreover $f^d(x_1,\dots,x_n)$ are the polynomials obtained from $f(x_1,\dots,x_n)$ replacing each coefficient α_σ with $d(\alpha_\sigma)$, so that

$$d(f(r_1,\dots,r_n)) = f^d(r_1,\dots,r_n) + \sum_i f(r_1,\dots,d(r_i),\dots,r_n).$$

First suppose that δ and d are C-independent modulo D_{int}.

By assumption, for all $r_1,\dots,r_n \in R$

$$[d(f(r_1,\dots,r_n)), \delta(f(r_1,\dots,r_n))] \in Z(R)$$

that is R satisfies the differential identity

$$\left[[f^d(x_1,\dots,x_n) + \sum_{i \geq 1} f(x_1,\dots,{}^d x_i,\dots,x_n), f^\delta(x_1,\dots,x_n) + \sum_{j \geq 1} f(x_1,\dots,{}^\delta x_j,\dots,x_n)], x_{n+1} \right].$$

By Kharchenko's theorem, R satisfies the polynomial identity

$$\left[[f^d(x_1,\dots,x_n)+\sum_{i\geq 1} f(x_1,\dots,y_i,\dots,x_n), f^{\delta}(x_1,\dots,x_n)+\sum_{j\geq 1} f(x_1,\dots,z_j,\dots,x_n)], x_{n+1}\right].$$

In particular R satisfies any blended component

$$[[f(x_1,\dots,z_i,\dots,x_n), f(x_1,\dots,x_n)], x_{n+1}]$$

in the indeterminates $x_1,\dots,x_{n+1},z_i$, for all $i\geq 1$, which implies that $f(x_1,\dots,x_n)$ is central-valued on R, except when $char(R)=2$ and $dim_C RC=4$, by Lemma 6.4.12.

Let now δ and d be C-dependent modulo D_{int}. Without loss off generality we may suppose $d=\alpha\delta+d_q$, where $\alpha\in C$ and d_q is the inner derivation induced by some element $q\in Q_r$. Note that if $q\in C$, we are done by the conclusion $d=\alpha\delta$. Thus suppose that $q\notin C$. Moreover, by proposition 6.4.11, the derivation δ cannot be inner, otherwise also, we are done.

By the assumptions,

$$[\alpha\delta(f(r_1,\dots,r_n))+[q,f(r_1,\dots,r_n)], \delta(f(r_1,\dots,r_n))]\in Z(R)$$

for all $r_1,\dots,r_n\in R$, i.e.,

$$\left[[q,f(r_1,\dots,r_n)], f^{\delta}(r_1,\dots,r_n)+\sum_{j\geq 1} f(r_1,\dots,\delta(r_j),\dots,r_n)\right]\in Z(R)$$

As above, by Kharchenko's result, R satisfies the generalized polynomial identity

$$\left[[[q,f(x_1,\dots,x_n)], f^{\delta}(x_1,\dots,x_n)+\sum_{j\geq 1} f(x_1,\dots,z_j,\dots,x_n)], x_{n+1}\right].$$

In particular R satisfies the blended component

$$[[[q,f(x_1,x_2,\dots,x_n)], f(z_1,x_2,\dots,x_n)], x_{n+1}].$$

Hence $[q,f(r_1,r_2,\dots,r_n)]_3=0$ for all $r_1,\dots,r_n\in R$. Since $q\notin C$, by [383], we get, $f(x_1,\dots,x_n)$ must be central-valued on R, except when $char(R)=2$ and $dim_C RC=4$.

The proof of Theorem 6.4.2

Here we will study the quadratic functional identity $[F(x),G(x)]=0$, for all $x\in S$ a suitable subset of a prime ring R having characteristic not equal to 2, in case both F and G are non-zero generalized derivations of R. We recall that in [313, Theorem 2], Hvala showed that, for $S=R$, the previous functional identity forces the conclusion $G=\lambda F$, for some $\lambda\in C$.

The aim of the present section is to extend the Hvala's result to the case S is a non-central Lie ideal of R. Under the assumption of Theorem 6.4.2, we prove that one of the following holds:

1. There exists $\lambda\in C$ such that, for any $x\in R$, $G(x)=\lambda F(x)$;
2. R satisfies s_4, the standard identity of degree 4.

Notice that in case R satisfies the standard identity s_4 (that is R embeds in the 2×2 matrix ring $M_2(K)$, for a field K), there exist suitable examples of generalized derivations F and G satisfying the condition $[F(u),G(u)]=0$, for any element $u\in L$, for L a non-central Lie ideal of R, but F and G are linearly independent over C. For example let e_{ij} be the usual matrix unit in $R=M_2(K)$ and consider $L=[R,R]$, $F(x)=e_{11}x+xe_{11}$, $G(x)=e_{22}x+xe_{22}$. Then $[F(u),G(u)]=0$, for all $u\in[R,R]$, but $G\neq\lambda F$, for any $\lambda\in Z(R)$.

In the first part, we suppose that both F and G are inner generalized derivations of R, so that there exist $a,b,c,q\in Q_r$ such that $F(x)=ax+xb$ and $G(x)=cx+xq$, for all $x\in R$. We need the following result (see [218, Lemma 1]):

Lemma 6.4.13 *Let F be an infinite field and $n\geq 2$. If $A_1,\dots,A_k$ are not scalar matrices in $M_n(F)$ then there exists some invertible matrix $P\in M_n(F)$ such that each matrix $PA_1P^{-1},\dots,QA_kQ^{-1}$ has all non-zero entries.*

Moreover:

Remark 6.4.14 *Let R be a prime ring, L a non-central Lie ideal of R, $a,b\in R$ such that $[au+ub,u]=0$, for all $u\in L$. Then $a,b\in Z(R)$, unless when R satisfies the standard identity s_4 (see [33, Theorem 2]).*

We start by studying the following case:

Lemma 6.4.15 *Let K be an infinite field, let $m\geq 2$ and $R=M_m(K)$ be the algebra of $m\times m$ matrices over K, $Z(R)$ the center of R and $L=[R,R]$. Assume that there exist non-zero elements $a,c\in R$ such that $[au,cu]=0$ for all $u\in L$. Then there exists $\eta\in Z(R)$ such that $a=\eta c$.*

Proof: Of course, in case both $a\in Z(R)$ and $c\in Z(R)$, we are done. Assume that either $a\notin Z(R)$ or $c\notin Z(R)$.

If $a \notin Z(R)$ and $0 \neq c \in Z(R)$, then $c[a,u]u = 0$ for all $u \in L$, i.e., $[a,u]u = 0$ for all $u \in L$. Since $char(R) \neq 2$, the last relation leads to the contradiction $a \in Z(R)$ (see [390, Theorem 2]). Similarly, we get a contradiction if we assume $c \notin Z(R)$ and $0 \neq a \in Z(R)$. Hence we suppose that both $a \notin Z(R)$ and $c \notin Z(R)$. By Lemma 6.4.13, we can assume that a and c have all non-zero entries, say $a = \sum_{kl} a_{kl}e_{kl}$ and $c = \sum_{kl} c_{kl}e_{kl}$, for $0 \neq a_{kl}, 0 \neq c_{kl} \in K$.

Fix $i \neq j$ and denote $\eta = \frac{a_{ji}}{c_{ji}} \neq 0$. Let $a' = a - \eta c$ and write $a' = \sum_{kl} a'_{kl}e_{kl}$, then $a'_{ji} = 0$. Since $a = a' + \eta c$, by our assumption it follows $0 = [a'u + \eta cu, cu] = [a'u, cu]$, for all $u \in L$. For $u = e_{ij} \in L$, we get $a'e_{ij}ce_{ij} - ce_{ij}a'_{ij} = 0$ and, since $a'_{ji} = 0$, one has $a'e_{ij}ce_{ij} = 0$. Moreover, since $c_{ji} \neq 0$, it follows that $a'_{ki} = 0$, for any $k \in \{1, \ldots, m\}$. Repeating this process for any $i \in \{1, \ldots, m\}$, it follows that $a' = 0$, that is $a = \eta c$.

♣

Analogously one may prove the following (we have omitted the proof for brevity):

Lemma 6.4.16 *Let K be an infinite field, $m \geq 2$ and $R = M_m(K)$ be the algebra of $m \times m$ matrices over K, $Z(R)$ the center of R and $L = [R,R]$. Assume that there exist non-zero elements $a, c \in R$ such that $[ua, uc] = 0$ for all $u \in L$. Then there exists $\eta \in Z(R)$ such that $a = \eta c$.*

Lemma 6.4.17 *Let K be a field, $m \geq 2$ and $R = M_m(K)$ be the algebra of $m \times m$ matrices over K, $Z(R)$ the center of R and $L = [R,R]$. Assume that there exist $a0$, $c0$, $a, c \in R$ such that $[ua, cu] = 0$ for all $u \in L$. Then $a, c \in Z(R)$, unless when $m = 2$ and $ac = ca \in Z(R)$.*

Proof: For $u = e_{ij} \in L$, we get $e_{ij}ace_{ij} = 0$, then the matrix ac is diagonal. A standard argument shows that $ac \in Z(R)$. By our assumption, $acu^2 - cu^2a = 0$. In case $m = 2$, then $u^2 \in Z(R)$, $(ac - ca)u^2 = 0$ and $ac = ca$. Let $m \geq 3$. By $acu^2 - cu^2a = 0$, it follows $acu^2c - cu^2ac = 0$, i.e., $u^2ac^2 - ac^2u^2 = 0$. Thus $[ac^2, u^2] = 0$, and, since $u^2 \notin Z(R)$, we get $ac^2 \in Z(R)$, that is $c \in Z(R)$, as well as $a \in Z(R)$.

♣

Lemma 6.4.18 *Let K be an infinite field, let $m \geq 3$ and $R = M_m(K)$ be the algebra of $m \times m$ matrices over K, $Z(R)$ the center of R and $L = [R,R]$. Assume that there exist non-zero elements $a, b, p \in R$ such that*

$$[au + ub, pu] = 0 \tag{6.68}$$

for all $u \in L$. Then $b \in Z(R)$ and there exists $\eta \in Z(R)$ such that $(a+b) = \eta p$.

Proof: If $b \in Z(R)$ the conclusion follows from Lemma 6.4.15. On the other hand, if $0 \neq p \in Z(R)$, then by Remark 6.4.14 it follows that $a, b \in Z(R)$ and we are done. Assume that both $p \notin Z(R)$ and $b \notin Z(R)$ and prove that a contradiction follows.

By Lemma 6.4.13, we can assume that p and b have all non-zero entries, say $p = \sum_{kl} p_{kl}e_{kl}$ and $b = \sum_{kl} b_{kl}e_{kl}$, for $0 \neq p_{kl}, 0 \neq b_{kl} \in K$.

For all $i \neq j$ and $u = e_{ii} - e_{jj} \in L$ in (6.68), and both left and right multiplying by e_{kk} $(k \neq i, j)$, we have $e_{kk}p(e_{ii} - e_{jj})be_{kk} = 0$, i.e.,

$$p_{ki}b_{ik} - p_{kj}b_{jk} = 0. \tag{6.69}$$

Let χ be any inner automorphism of R; of course $e_{kk}\chi p(e_{ii} - e_{jj})\chi be_{kk} = 0$, then $\chi(p)$ and $\chi(b)$ satisfy (6.69). In particular for $\chi(p) = (1 + e_{ki})x(1 - e_{ki})$ and $\chi(b) = (1 + e_{ki})x(1 - e_{ki})$, application of (6.69) $\Rightarrow$

$$p_{ii}b_{ik} - p_{kk}b_{ik} - p_{ik}b_{ik} + p_{ij}b_{jk} = 0. \tag{6.70}$$

On the other hand, for $\chi'(p) = (1 - e_{ki})x(1 + e_{ki})$ and $\chi'(b) = (1 - e_{ki})x(1 + e_{ki})$, from (6.69) we also get

$$-p_{ii}b_{ik} + p_{kk}b_{ik} - p_{ik}b_{ik} - p_{ij}b_{jk} = 0. \tag{6.71}$$

Therefore, since $char(K) \neq 2$, comparison of (6.70) with (6.71) leads to the contradiction $p_{ik}b_{ik} = 0$.

♣

Lemma 6.4.19 *Let K be an infinite field, $m \geq 3$ and $R = M_m(K)$ be the algebra of $m \times m$ matrices over K, $Z(R)$ the center of R and $L = [R, R]$. Assume that there exist $a, b, p0$, $a, b, p \in R$ such that*

$$[au + ub, up] = 0 \tag{6.72}$$

for all $u \in L$. Then $a \in Z(R)$ and there exists $\eta \in Z(R)$ such that $(a + b) = \eta p$.

Proof: As above, in case $a \in Z(R)$ the conclusion follows from Lemma 6.4.16. On the other hand, if $0 \neq p \in Z(R)$, then by Remark 6.4.14 it follows that $a, b \in Z(R)$ and we are done. Assume that both $p \notin Z(R)$ and $a \notin Z(R)$ and prove that a contradiction follows.

By Lemma 6.4.13, we can assume that p and a have all non-zero entries, say $p = \sum_{kl} p_{kl}e_{kl}$ and $a = \sum_{kl} b_{kl}e_{kl}$, for $0 \neq p_{kl}, 0 \neq b_{kl} \in K$.

For all $i \neq j$ and $u = e_{ii} - e_{jj} \in L$ in (6.72), and both left and right multiplying by e_{kk} $(k \neq i, j)$, we have $e_{kk}a(e_{ii} - e_{jj})pe_{kk} = 0$, i.e.,

$$a_{ki}p_{ik} - a_{kj}p_{jk} = 0. \tag{6.73}$$

Hence, by using the same argument in Lemma 6.4.18, we have the contradiction $a_{ik}p_{ik} = 0$.

♣

Lemma 6.4.20 *Let K be an infinite field, $m \geq 3$ and $R = M_m(K)$ be the algebra of $m \times m$ matrices over K, $Z(R)$ the center of R and $L = [R,R]$. Assume that there exist $a,b,c,q, \in R$ such that*

$$[au + ub, cu + uq] = 0 \tag{6.74}$$

for all $u \in L$. If $a \notin Z(R)$ and there exist $\alpha, \beta \in Z(R)$ such that $c - \alpha a = \beta$, then $q - \alpha b = -\beta$.

Proof: Firstly we assume that $q - \alpha b$ is not a central matrix in R.

In light of Lemma 6.4.13, we may assume that a and $q - \alpha b$ have all non-zero entries, say $a = \sum_{kl} a_{kl}e_{kl}$ and $q - \alpha b = \sum_{kl} p_{kl}e_{kl}$, for $0 \neq a_{kl}, 0 \neq p_{kl} \in K$. We will prove that in this case we get a contradiction.

Since $c = \alpha a + \beta$, (6.74) reduces to

$$[au + ub, (\alpha a + \beta)u + uq] = 0. \tag{6.75}$$

For all $i \neq j$ and $u = e_{ii} - e_{jj} \in L$ in (6.75), and both left and right multiplying by e_{kk} $(k \neq i, j)$, we have,

$$e_{kk}a(e_{ii} + e_{jj})qe_{kk} - \alpha e_{kk}a(e_{ii} + e_{jj})be_{kk} = 0 \tag{6.76}$$

i.e., $a_{ki}p_{ik} - a_{kj}p_{jk} = 0$. Hence, as in Lemma 6.4.18, we get the contradiction $a_{ik}p_{ik} = 0$.

Therefore $q - \alpha b \in Z(R)$, say $q = \alpha b + \gamma$, for some $\gamma \in Z(R)$. By (6.74), it follows that $(\beta + \gamma)[au + ub, u] = 0$, for all $u \in [R,R]$. Notice that, if $\beta + \gamma \neq 0$, then $[au + ub, u] = 0$ and, by Remark 6.4.14, the contradiction $a \in Z(R)$ follows. Thus $\gamma = -\beta$ and we are done.

♣

At this point, we need to fix the following two auxiliary Lemmas:

Lemma 6.4.21 *Let K be an infinite field, $R = M_m(K)$ be algebra of $m \times m$ matrices over K with $m \geq 3$ and $L = [R,R]$. Suppose there exist $a,b,c,q,v,w \in R$ such that*

$$aucu + vu^2q + uwu + ubuq - cuau - cu^2b - uqvu - uqub = 0 \tag{6.77}$$

for all $u \in L$. Denote $a = \sum_{kl} a_{kl}e_{kl}$, $c = \sum_{kl} c_{kl}e_{kl}$, for suitable a_{kl}, c_{kl} elements of K. If there are $i \neq j$ such that $c_{jk} \neq 0$, for any $k \neq j$, and $a_{ji} = 0$, then $a_{rk} = 0$, for all $k \neq j$ and $r \neq k$.

Proof: For $t \neq s$ and $u = e_{is}$ in (6.77) we have:

$$X = ae_{ts}ce_{ts} + e_{ts}we_{ts} + e_{ts}be_{ts}q - ce_{ts}ae_{ts} - e_{ts}(qv)e_{ts} - e_{ts}qe_{ts}b = 0.$$

From the previous equality it follows that, for all $r \neq t, s$, the (r,s)-entry of the matrix X is zero, so that we have

$$a_{rt}c_{st} - c_{rt}a_{st} = 0, \quad \forall t,s,r \quad \text{different indices.} \tag{6.78}$$

In particular, by (6.78) and for any $r \neq i, j$, we get $a_{ri}c_{ji} - c_{ri}a_{ji} = 0$ and, since $a_{ji} = 0$, it follows $a_{ri}c_{ji} = 0$, that is $a_{ri} = 0$, for any $r \neq i$.

Consider the inner automorphism of R induced by the invertible matrix $P = I - e_{ki}$, for $k \neq i, j$: $\varphi(x) = P^{-1}xP$. Of course

$$\begin{aligned}&\varphi(a)u\varphi(c)u + \varphi(v)u^2\varphi(q) + u\varphi(w)u + u\varphi(b)u\varphi(q)\\ &- \varphi(c)u\varphi(a)u - \varphi(c)u^2\varphi(b) - u\varphi(qv)u - u\varphi(q)u\varphi(b) = 0\end{aligned}$$

for all $u \in [R,R]$. Since $a_{ri} = 0$, for any $r \neq i$, the application of (6.78) to $\varphi(a)$ and $\varphi(c) \Rightarrow$

$$(a_{ii} - a_{kk} - a_{ik})(c_{ji} - c_{jk}) - (c_{ki} + c_{ii} - c_{kk} - c_{ik})(-a_{jk}) = 0. \tag{6.79}$$

Analogously, for $P' = I + e_{ki}$, for $k \neq i, j$: $\varphi'(x) = P'^{-1}xP'$, the application of (6.78) to $\varphi'(a)$ and $\varphi'(c) \Rightarrow$

$$(-a_{ii} + a_{kk} - a_{ik})(c_{ji} + c_{jk}) - (c_{ki} - c_{ii} + c_{kk} - c_{ik})(a_{jk}) = 0. \tag{6.80}$$

Subtracting (6.80) from (6.79), we get

$$2(a_{ii}c_{ji} - a_{kk}c_{ji} + a_{ik}c_{jk} + a_{jk}c_{ki} - a_{jk}c_{ik}) = 0. \tag{6.81}$$

Moreover, by (6.78), $a_{ik}c_{jk} - a_{jk}c_{ik} = 0$ and (6.81) reduces to

$$(a_{ii} - a_{kk})c_{ji} + a_{jk}c_{ki} = 0. \tag{6.82}$$

Let $\chi(x) = T^{-1}xT$ be the inner automorphism of R induced by the invertible matrix $T = I - e_{ki}$, for $k \neq i, j$. By applying (6.82) to $\chi(a)$ and $\chi(c)$, and using relation (6.82) again, it follows that

$$(a_{ii} - a_{kk})(-c_{jk}) - 2a_{ik}c_{ji} + 2a_{ik}c_{jk} + a_{jk}c_{ii} - a_{jk}c_{kk} - a_{jk}c_{ik} = 0. \tag{6.83}$$

By (6.78), $a_{ik}c_{jk} - a_{jk}c_{ik} = 0$, so that (6.83) reduces to

$$(a_{ii} - a_{kk})(-c_{jk}) - 2a_{ik}c_{ji} + a_{ik}c_{jk} + a_{jk}c_{ii} - a_{jk}c_{kk} = 0. \tag{6.84}$$

Consider now the inner automorphism $\omega(x) = TxT^{-1}$ of R induced again by the same invertible matrix T as above. Application of (6.82) to $\omega(a)$ and $\omega(c)$, and using relation (6.82) again, $\Rightarrow$

$$(a_{ii} - a_{kk})c_{jk} + 2a_{ik}c_{ji} + 2a_{ik}c_{jk} - a_{jk}c_{ii} + a_{jk}c_{kk} - a_{jk}c_{ik} = 0. \tag{6.85}$$

Since $a_{ik}c_{jk} - a_{jk}c_{ik} = 0$, then by (6.85) it follows

$$(a_{ii} - a_{kk})c_{jk} + 2a_{ik}c_{ji} + a_{ik}c_{jk} - a_{jk}c_{ii} + a_{jk}c_{kk} = 0. \tag{6.86}$$

Comparing (6.84) with (6.86) we get $2a_{ik}c_{jk} = 0$ and, since $c_{jk} \neq 0$, we have $a_{ik} = 0$, for all $k \neq i, j$.

In other words, we have proved that:

$$a_{ji} = 0, \quad c_{jk} \neq 0 \quad \forall k \neq j \quad \Rightarrow a_{ri} = 0, a_{ik} = 0 \quad \forall r \neq i, \quad \forall k \neq i, j. \tag{6.87}$$

Finally, let $\lambda(x) = H^{-1}xH$ be the inner automorphism of R induced by the invertible matrix $H = I - e_{ij}$, and denote $\lambda(a) = \sum_{hl} a'_{hl}e_{hl}$, $\lambda(c) = \sum_{hl} c'_{hl}e_{hl}$, for suitable a'_{hl}, c'_{hl} elements of K. Notice that $a'_{ji} = a_{ji} = 0$ and, for any $k \neq j$, $c'_{jk} = c_{jk} \neq 0$. By (6.87) it follows that

$$\forall t \neq i, j \rightarrow 0 = a'_{it} = a_{it} + a_{jt} = a_{jt}. \tag{6.88}$$

Since $a_{ji} = 0$ it follows that $a_{jt} = 0$, for all $t \neq j$. Moreover, $c_{jk} \neq 0$, for any $k \neq j$, so that we may apply (6.87) again and obtain

$$a_{ts} = 0, \quad \forall t \neq j, s \neq j, t. \tag{6.89}$$

Putting together (6.88) and (6.89) we get the required result: $a_{rk} = 0$, for all $k \neq j$ and $r \neq k$.

♣

Lemma 6.4.22 *Let K be an infinite field, $R = M_m(K)$ be the algebra of $m \times m$ matrices over K with $m \geq 3$ and $L = [R,R]$. Suppose there exist $a, b, c, q, v, w \in R$ such that*

$$aucu + vu^2q + uwu + ubuq - cuau - cu^2b - uqvu - uqub = 0 \tag{6.90}$$

for all $u \in L$. Denote $a = \sum_{kl} a_{kl}e_{kl}$, $c = \sum_{kl} c_{kl}e_{kl}$, for suitable a_{kl}, c_{kl} elements of K. If $c_{rs} \neq 0$ for all r, s, and there are $i \neq j$ such that $a_{ji} = 0$, then a is central on R.

Proof: The first step is to apply twice Lemma 6.4.21: this forces a to be a diagonal matrix. In fact $a_{ji} = 0$ and $c_{jk} \neq 0$, for any $k \neq j$, $\Rightarrow a_{rk} = 0$ for all $k \neq j$ and $r \neq k$ (in other words, the only non-zero off-diagonal elements of a fall in the

j-th column). In particular, since $m \geq 3$, there exists $t \neq j$ such that $a_{lt} = 0$, for all $l \neq t$. Since $c_{ls} \neq 0$ (for all $s \neq l$), so $a_{nj} = 0$ for all $n \neq j$, as required. Say $a = \sum_k a_{kk}e_{kk}$.

Consider now the inner automorphism of R induced by the invertible matrix $P = I + e_{rj}$, for $r \neq i, j$: $\varphi(x) = P^{-1}xP$. Thus

$$\varphi(a)u\varphi(c)u + \varphi(v)u^2\varphi(q) + u\varphi(w)u + u\varphi(b)u\varphi(q)$$
$$- \varphi(c)u\varphi(a)u - \varphi(c)u^2\varphi(b) - u\varphi(qv)u - u\varphi(q)u\varphi(b) = 0$$

for all $u \in [R,R]$. The (j,i)-entry of $\varphi(a)$ is $a_{ji} = 0$. Moreover, for all $k \neq j$, the (j,k)-entries of $\varphi(c)$ are $c_{jk} \neq 0$. Therefore, again by Lemma 6.4.21 and the above argument, $\varphi(a)$ is a diagonal matrix. By calculations $(\varphi(a))_{rr} = (\varphi(a))_{jj}$, i.e., $a_{jj} = a_{rr}$.

On the other hand, if χ is the inner automorphisms induced by the invertible matrix $Q = I + e_{ri}$ (for $r \neq i, j$), then, as above,

$$\chi(a)u\chi(c)u + \chi(v)u^2\chi(q) + u\chi(w)u + u\chi(b)u\chi(q)$$
$$- \chi(c)u\chi(a)u - \chi(c)u^2\chi(b) - u\chi(qv)u - u\chi(q)u\chi(b) = 0$$

for all $u \in [R,R]$. Since the (j,i)-entry of $\chi(a) = 0$, and for all $k \neq j$, the (j,k)-entries of $\varphi(c)$ are $c_{jk} \neq 0$, again $\chi(a)$ is a diagonal matrix, i.e., $(\chi(a))_{rr} = (\chi(a))_{ii}$ and $a_{ii} = a_{rr} = a_{jj} = \beta$, for all $r \neq i, j$. Thus $a = \beta I$ is a central matrix in R.

♣

Lemma 6.4.23 *Let K be an infinite field, let $R = M_m(K)$ be the algebra of $m \times m$ matrices over K with $m \geq 3$ and $L = [R,R]$. Suppose there exist $a, b, c, q \in R$ such that*

$$[au + ub, cu + uq] = 0$$

for all $u \in L$. Denote $a = \sum_{kl} a_{kl}e_{kl}$, $b = \sum_{kl} b_{kl}e_{kl}$, $c = \sum_{kl} c_{kl}e_{kl}$, $q = \sum_{kl} q_{kl}e_{kl}$ for suitable a_{kl}, b_{kl}, c_{kl}, q_{kl} elements of K. Then one of the following holds:

1. *$c \in C$ and $c + q = 0$;*
2. *$a \in C$ and $a + b = 0$;*
3. *$a, c \in C$ and there exists $\lambda \in C$ such that $a + b = \lambda(c + q)$;*
4. *$b, q \in C$ and there exists $\lambda \in C$ such that $a + b = \lambda(c + q)$;*
5. *There exist $\lambda, \mu \in C$ such that $a = \lambda c + \mu$ and $b = \lambda q - \mu$.*

Proof: Clearly if one of a, b, c or q is a scalar matrix we are done by Lemmas 6.4.18 and 6.4.19. In order to prove the present Lemma, we may assume that a, b, c and q are non-central matrices.

By Lemma 6.4.13, there exists some invertible matrix $P \in M_m(K)$ such that $PaP^{-1} = a'$, $PbP^{-1} = b'$, $PcP^{-1} = c'$ and $PqP^{-1} = q'$ have all non-zero entries. By this conjugation we denote $a' = \sum_{kl} a'_{kl}e_{kl}$, $b' = \sum_{kl} b'_{kl}e_{kl}$, $c' = \sum_{kl} c'_{kl}e_{kl}$, $q' = \sum_{kl} q'_{kl}e_{kl}$, for suitable a'_{kl}, b'_{kl}, c'_{kl} and q'_{kl} elements of K, the conjugates of elements a,b,c,q. Of course

$$[a'u+ub',c'u+uq'] = 0 \tag{6.91}$$

for all $u \in L$. For $i \neq j$ and $u = e_{ij} \in L$, we have

$$a'e_{ij}c'e_{ij} + e_{ij}b'(c'e_{ij} + e_{ij}q') - c'e_{ij}a'e_{ij} - e_{ij}q'(a'e_{ij} + e_{ij}b') = 0.$$

Left multiplying by e_{kk}, for any $k \neq i,j$, we get $a'_{ki}c'_{ji} - c'_{ki}a'_{ji} = 0$. Here we fix $\lambda = \frac{a'_{ki}}{c'_{ki}} \neq 0$ and $\overline{a'} = a' - \alpha c'$. Replacing a' with $\overline{a'} + \lambda c'$ in (6.91) we have

$$\overline{a'}uc'u + a'u^2q' + u(b'c')u + ub'uq' - c'u\overline{a'}u - c'u^2b' - u(q'a')u - uq'ub' = 0$$

for all $u \in L$. Since the (k,i)-entry of the matrix $\overline{a'}$ is zero and $c'_{rs} \neq 0$ for all r,s, then by Lemma 6.4.22 it follows that $\overline{a'}$ is a central matrix. Hence there exists $\mu \in Z(R)$ such that $a' = \lambda c' + \mu$, that is $a = \lambda c + \mu$. Application of Lemma 6.4.20 $\Rightarrow b = \lambda q - \mu$, as required.

♣

Lemma 6.4.24 *Let K be a field, let $R = M_m(K)$ be the algebra of $m \times m$ matrices over K with $m \geq 3$ and $L = [R,R]$. Suppose there exist $a,b,c,q \in R$ such that*

$$[au+ub, cu+uq] = 0 \tag{6.92}$$

for all $u \in L$. Then one of the following holds:

1. *$c \in C$ and $c+q = 0$;*
2. *$a \in C$ and $a+b = 0$;*
3. *$a,c \in C$ and there exists $\lambda \in C$ such that $a+b = \lambda(c+q)$;*
4. *$b,q \in C$ and there exists $\lambda \in C$ such that $a+b = \lambda(c+q)$;*
5. *There exist $\lambda,\mu \in C$ such that $a = \lambda c + \mu$ and $b = \lambda q - \mu$.*

Proof: If one assumes that K is infinite, the conclusion follows from Lemma 6.4.23.

Now let L be an infinite field which is an extension of the field K and let $\overline{R} = M_t(L) \cong R \otimes_K L$. Note that a multilinear polynomial $f(x_1,\ldots,x_n)$ is an identity for on R iff it is an identity on $\overline{R}$. Therefore $\overline{R}$ does not satisfy s_4 and we assume $t \geq 3$. Consider the generalized polynomial

$$P(x_1,x_2) = \Big[a[x_1,x_2] + [x_1,x_2]b, c[x_1,x_2] + [x_1,x_2]q\Big].$$

By relation (6.92), $P(x_1,x_2)$ is a generalized polynomial identity for R. Moreover it is multi-homogeneous of multi-degree $(2,2)$ in the indeterminates x_1,x_2.

Hence the complete linearization of $P(x_1,x_2)$ is a multilinear generalized polynomial $\Theta(x_1,x_2,y_1,y_2)$ in 4 indeterminates. Moreover,

$$\Theta(x_1,x_2,x_1,x_2) = 4P(x_1,x_2).$$

Clearly the multilinear polynomial $\Theta(x_1,x_2,y_1,y_2)$ is a generalized polynomial identity for R and $\overline{R}$ too. Since $char(K) \neq 2$ we obtain $P(r_1,r_2) = 0$, for all $r_1,r_2 \in \overline{R}$, and the conclusion follows from Lemma 6.4.23.

♣

Lemma 6.4.25 *Let R be a prime ring and $a,b,c,q \in R$. Either*

$$P(x_1,x_2) = \Big[a[x_1,x_2] + [x_1,x_2]b, c[x_1,x_2] + [x_1,x_2]q\Big]$$

is a non trivial generalized polynomial identity for R or one of the following conclusions holds:

1. *$c \in C$ and $c+q = 0$;*
2. *$a \in C$ and $a+b = 0$;*
3. *$a,c \in C$ and there exists $\lambda \in C$ such that $a+b = \lambda(c+q)$;*
4. *$b,q \in C$ and there exists $\lambda \in C$ such that $a+b = \lambda(c+q)$;*
5. *There exist $\lambda,\mu \in C$ such that $a = \lambda c + \mu$ and $b = \lambda q - \mu$.*

Proof: If $a \in C$ and $b \in C$ (respectively $c \in C$ and $q \in C$) the conclusion follows from Remark 6.4.14. Hence, in all that follows we may assume that either $a \notin C$ or $b \notin C$, moreover either $c \notin C$ or $q \notin C$.

Now consider the generalized polynomial $P(x_1,x_2) \in Q_r *_C C\{x_1,x_2\}$. By our hypothesis, R satisfies this generalized polynomial identity. If B is a basis of Q_r over C then any element of $T = Q_r *_C C\{x_1,x_2\}$ can be written in the form $g = \sum_i \alpha_i m_i$. In this decomposition the coefficients α_i are in C and the elements m_i are B-monomials, that is $m_i = q_0y_1q_1y_2q_2y_3q_3y_4q_4$, with $q_i \in B$ and $y_i \in \{x_1,x_2\}$. In [184] it is shown that a generalized polynomial $g = \sum_i \alpha_i m_i$ is the zero element of T iff any α_i is zero. As a consequence, let $a_1,\dots,a_k \in Q_r$ be linearly independent over C and $a_1g_1(x_1,x_2) + \dots + a_kg_k(x_1,x_2) = 0 \in T$, for some $g_1,\dots,g_k \in T$. If, for any i, $g_i(x_1,x_2) = \sum_{j=1}^n x_jh_j(x_1,x_2)$ and $h_j(x_1,x_2) \in T$, then $g_1(x_1,x_2),\dots,$ $g_k(x_1,x_2)$ are the zero elements of T. The same conclusion holds if $g_1(x_1,x_2)a_1 + \dots + g_k(x_1,x_2)a_k = 0 \in T$, and $g_i(x_1,x_2) = \sum_{j=1}^n h_j(x_1,x_2)x_j$ for some $h_j(x_1,x_2) \in T$.

We start from

$$P(x_1,x_2)=\Big[a[x_1,x_2]+[x_1,x_2]b,c[x_1,x_2]+[x_1,x_2]q\Big]=0\in T. \tag{6.93}$$

If $\{a,c,1\}$ are linearly independent then $a[x_1,x_2](c[x_1,x_2]+[x_1,x_2]q)=0\in T$. In case $q\notin C$, then $a[x_1,x_2]^2q$ is a non-trivial generalized polynomial identity for R, a contradiction. If $q\in C$ (and so $c\notin C$), it follows $a[x_1,x_2](c+q)[x_1,x_2]=0\in T$, which implies $c+q=0$, is again a contradiction.

Thus we assume there exist $\alpha,\beta,\gamma\in C$ such that $\alpha a+\gamma c+\beta=0$.

If $\alpha=0$ then $c\in C$, so that $q\notin C$, and (6.93) reduces to

$$\Big[a[x_1,x_2]+[x_1,x_2]b,[x_1,x_2](c+q)\Big]=0\in T. \tag{6.94}$$

In case $a\in C$, then $b\notin C$ and

$$\Big[[x_1,x_2](a+b),[x_1,x_2](c+q)\Big]=0\in T. \tag{6.95}$$

Since (6.95) is trivial, $a+b\neq 0$ and $c+q\neq 0$, then there exists $\lambda\in C$ such that $a+b=\lambda(c+q)$. On the other hand, if $a\notin C$, then by (6.94) we get $a[x_1,x_2]^2(c+q)$ is a non-trivial generalized polynomial identity for R, a contradiction.

Notice that the same argument can be used in order to study the case when $\gamma=0$.

Let now $\alpha\neq 0$ and $\gamma\neq 0$. Thus we may write $a=\lambda c+\mu$, for some $\lambda,\mu\in C$ and $\lambda\neq 0$.

Moreover we may assume $c\notin C$, if not also $a\in C$ and we are done by the previous argument. By (6.93) it follows that

$$\Big[\lambda c[x_1,x_2]+\mu[x_1,x_2]+[x_1,x_2]b,c[x_1,x_2]+[x_1,x_2]q\Big]=0\in T. \tag{6.96}$$

Since $c\notin C$ and by (6.96), we have,

$$c[x_1,x_2]\Big(\lambda[x_1,x_2]q-\mu[x_1,x_2]-[x_1,x_2]b\Big)=0\in T. \tag{6.97}$$

We also note that, in case $\{b,q,1\}$ are linearly independent then (6.97) forces $\lambda c[x_1,x_2]^2q=0\in T$, which is a contradiction. Therefore there exist $\alpha',\beta',\gamma'\in C$ such that $\alpha'q+\beta'b+\gamma'=0$.

If $\alpha'=0$, then $b\in C$ (so that $a\notin C$) and (6.97) reduces to

$$c[x_1,x_2]^2(\lambda q-\mu-b)=0\in T \tag{6.98}$$

$\Rightarrow \lambda q - \mu - b = 0$ and $q \in C$ (then $c \notin C$). In this case R satisfies

$$\Big[(a+b)[x_1,x_2],(c+q)[x_1,x_2](c+q)\Big] = 0 \in T$$

and as above we conclude that there exists $\eta \in C$ such that $a+b = \eta(c+q)$.

By using the same argument, we may also prove that if $\beta' = 0$ then $q,b \in C$ $a+b = \eta(c+q)$, for some $\eta \in C$.

Now let $\alpha' \neq 0$ and $\beta' \neq 0$, then $q = \lambda' b + \mu'$, for some $\lambda', \mu' \in C$ and $\lambda' \neq 0$.

Hence we may write (6.97) as follows:

$$c[x_1,x_2]\Big(\lambda\lambda'[x_1,x_2]b + \lambda\mu'[x_1,x_2] - \mu[x_1,x_2] - [x_1,x_2]b\Big) = 0 \in T. \quad (6.99)$$

Clearly, if $b \in C$ then $q \in C$, moreover R satisfies

$$\Big[(a+b)[x_1,x_2],(c+q)[x_1,x_2](c+q)\Big] = 0 \in T$$

and we are done as above. Therefore we assume finally that $b \notin C$, and by (6.99) we have that both

$$(\lambda\lambda' - 1)c[x_1,x_2]^2 b = 0 \in T$$

and

$$(\lambda\mu' - \mu)c[x_1,x_2]^2 = 0 \in T.$$

Thus $\lambda\lambda' = 1$ and $\lambda\mu' = \mu$. Therefore, since $q = \lambda' b + \mu'$ and multiplying by λ, it follows $\lambda q = b + \mu$, as required.

♣

We are now ready to prove Theorem 6.4.2 when both F and G are inner generalized derivations of R:

Proposition 6.4.26 *Let R be a prime ring of characteristic not equal to 2, $F(x) = ax + xb$ and $G(x) = cx + xq$, for all $x \in R$. If $[F(u), G(u)] = 0$, for all $u \in [R,R]$, then one of the following holds:*

1. *$c \in C$ and $c + q = 0$;*
2. *$a \in C$ and $a + b = 0$;*
3. *$a, c \in C$ and there exists $\lambda \in C$ such that $a + b = \lambda(c + q)$;*
4. *$b, q \in C$ and there exists $\lambda \in C$ such that $a + b = \lambda(c + q)$;*
5. *There exist $\lambda, \mu \in C$ such that $a = \lambda c + \mu$ and $b = \lambda q - \mu$;*
6. *R satisfies s_4, the standard identity of degree 4.*

Proof: We may assume that R does not satisfy the standard identity s_4. By Lemma 6.4.25 we may assume that R satisfies the non-trivial generalized polynomial identity

$$P(x_1,x_2) = \Big[a[x_1,x_2]+[x_1,x_2]b, c[x_1,x_2]+[x_1,x_2]q\Big]. \qquad (6.100)$$

We know that this generalized polynomial identity is also satisfied by Q_r. As above, we may replace R by Q_r and assume that R is centrally closed over C which is either finite or algebraically closed. Hence, R is a primitive ring having a non-zero socle H with C as the associated division ring, and eHe is a simple central algebra finite dimensional over C, for any minimal idempotent element $e \in S$. Moreover, R is isomorphic to a dense ring of linear transformations on some vector space V over C.

First assume that V is finite-dimensional over C. Then the density of R on $V \Rightarrow$ that $R \cong M_k(C)$, the ring of all $k \times k$ matrices over C. Since R does not satisfy s_4, we assume $k \geq 3$. In this case the conclusion follows by Lemma 6.4.24.

Assume next that V is infinite-dimensional over C. Since the set $[R,R]$ is dense on R, then from $P(r_1,r_2)=0$, for all $r_1,r_2 \in R$, R satisfies the generalized identity $P(x) = [ax+xb, cx+xq]$, i.e., for all $r \in R$, $[ar+rb, cr+rq]=0$. In this case we get the required conclusion according to [313, Theorem 2].

♣

For proving Theorem 6.4.2, we now assume that there exist $a,b \in Q_r$ and d,δ derivations of Q_r such that $F(x) = ax+d(x)$ and $G(x) = bx+\delta(x)$, for all $x \in R$. Firstly we remark that, in light of Proposition 6.4.26, we may assume that F and G are not simultaneously inner, i.e., d and δ are not simultaneously inner.

In all that follows we suppose that R does not satisfy s_4.

Since L is not central and $char(R) \neq 2$, then there exists a non-zero ideal I of R such that $0 \neq [I,R] \subseteq L$. Therefore, there exists a non-central ideal I of R such that $[F(u),G(u)]=0$, for all $u \in [I,I]$. Since R and I satisfy the same generalized differential identities, we may assume that $[F(u),G(u)]=0$, for all $u \in [R,R]$. Thus R satisfies the generalized differential identity

$$\Big[a[x_1,x_2]+[d(x_1),x_2]+[x_1,d(x_2)], b[x_1,x_2]+[\delta(x_1),x_2]+[x_1,\delta(x_2)]\Big]. \qquad (6.101)$$

If $\delta = 0$ then d is not inner, and by Kharchenko's Theory, R satisfies

$$\Big[a[x_1,x_2]+[y_1,x_2]+[x_1,y_2], b[x_1,x_2]\Big] \qquad (6.102)$$

and in particular R satisfies

$$\Big[[y_1,x_2],b[x_1,x_2]\Big]. \tag{6.103}$$

For $y_1 = x_1$ and applying Remark 6.4.14, it follows $b \in C$. Moreover, if $b = 0$ we get $G = 0$, then we assume $0 \neq b \in C$. By (6.103) it follows that R satisfies the polynomial identity $[[y_1,x_2],[x_1,x_2]]$, then, by Posner's theorem, there exists a suitable field K such that R and $M_t(K)$ satisfy the same polynomial identities. Since R does not satisfy s_4, then $t \geq 3$. On the other hand, for $x_1 = e_{11}$, $x_2 = e_{12}$ and $y_1 = e_{21}$, we get $[[y_1,x_2],[x_1,x_2]] = [e_{22} - e_{11}, e_{12}] = -2e_{12} \neq 0$, a contradiction.

Symmetrically, if we assume $d = 0$ and use the argument as above, a contradiction follows. Hence we may consider both d and δ non-zero derivations of R. In case d and δ are C-linearly independent modulo inner derivations of Q_r, then, by (6.101) and Kharchenko's Theory, R satisfies

$$\Big[a[x_1,x_2]+[y_1,x_2]+[x_1,y_2],b[x_1,x_2]+[z_1,x_2]+[x_1,z_2]\Big]. \tag{6.104}$$

Also in this case, R satisfies the blended component $[[y_1,x_2],[z_1,x_2]]$ and we have a contradiction.

Thus we finally assume that there exist $0 \neq \alpha \in C$ and $q \in Q_r$, such that $\delta(x) = \alpha d(x) + [q,x]$, for all $x \in R$. Note that, in this case d is not an inner derivation, if not both d and δ should be inner, which contradicts our assumption.

By (6.101) R satisfies

$$\Big[a[x_1,x_2]+[d(x_1),x_2]+[x_1,d(x_2)],b[x_1,x_2]+\alpha[d(x_1),x_2]+\alpha[x_1,d(x_2)]+[q,[x_1,x_2]]\Big]. \tag{6.105}$$

Since d is not inner, R satisfies

$$\Big[a[x_1,x_2]+[y_1,x_2]+[x_1,y_2],b[x_1,x_2]+\alpha[y_1,x_2]+\alpha[x_1,y_2]+[q,[x_1,x_2]]\Big]. \tag{6.106}$$

In particular

$$\Big[a[x_1,x_2],b[x_1,x_2]+[q,[x_1,x_2]]\Big] \tag{6.107}$$

is a generalized polynomial identity for R. By Proposition 6.4.26, one of the following holds:

1. $b = 0$ and $q \in C$;

2. $a = 0$;

3. $q \in C$ and there exist $\lambda, \mu \in C$ such that $a = \lambda b + \mu$.

CASE 1: $b = 0$ and $q \in C$.

In this case $F(x) = ax + d(x)$, $G(x) = \alpha d(x)$, for any $x \in R$, and by (6.101) R satisfies

$$\Big[a[x_1,x_2] + [d(x_1),x_2] + [x_1,d(x_2)], \alpha[d(x_1),x_2] + [x_1,d(x_2)]\Big] \tag{6.108}$$

moreover $\alpha \neq 0$, if not $G = 0$. Since d is not inner, then R satisfies

$$\Big[a[x_1,x_2] + [y_1,x_2] + [x_1,y_2], \alpha[y_1,x_2] + \alpha[x_1,y_2]\Big]$$

that is R satisfies

$$\Big[a[x_1,x_2], \alpha[y_1,x_2] + \alpha[x_1,y_2]\Big]$$

and in particular

$$\alpha\Big[a[x_1,x_2], [y_1,x_2]\Big]$$

is a generalized identity for R. By Remark 6.4.14, it follows $a \in C$. Moreover, if $0 \neq a \in C$ then R satisfies the polynomial identity $[[x_1,x_2],[y_1,x_2]]$, and as above we get a contradiction. Thus $a = 0$ and $G(x) = \alpha F(x)$ as required.

CASE 2: $a = 0$.

In this case $F(x) = d(x)$, $G(x) = bx + \alpha d(x) + [q,x]$, for any $x \in R$. Starting from (6.106), R satisfies

$$\Big[[y_1,x_2] + [x_1,y_2], b[x_1,x_2] + \alpha[y_1,x_2] + \alpha[x_1,y_2] + [q,[x_1,x_2]]\Big]$$

that is R satisfies

$$\Big[[y_1,x_2] + [x_1,y_2], b[x_1,x_2] + [q,[x_1,x_2]]\Big]. \tag{6.109}$$

Application of Remark 6.4.14 to (6.109) implies that $q \in C$ and $b \in C$. If $b \neq 0$, then by (6.109) we have that

$$\Big[[y_1,x_2] + [x_1,y_2], [x_1,x_2]\Big]$$

is a polynomial identity for R, in particular R satisfies $[[y_1,x_2],[x_1,x_2]]$ and once again we get a contradiction. Therefore $b = 0$ and $G(x) = \alpha F(x)$.

CASE 3: $q \in C$ and there exist $\lambda, \mu \in C$ such that $a = \lambda b + \mu$.

In this final case $F(x) = (\lambda b + \mu)x + d(x)$, $G(x) = bx + \alpha d(x)$, for any $x \in R$. Moreover, by (6.106), R satisfies

$$\Big[(\lambda b+\mu)[x_1,x_2]+[y_1,x_2]+[x_1,y_2], b[x_1,x_2]+\alpha[y_1,x_2]+\alpha[x_1,y_2]\Big]. \quad (6.110)$$

Firstly we note that

$$\Big[(\lambda b+\mu)[x_1,x_2], b[x_1,x_2]\Big]$$

is satisfied by R and, by Proposition 6.4.26, it follows that $b \in C$. Say $\gamma = \lambda b + \mu \in C$, so that R satisfies

$$\Big[\gamma[x_1,x_2]+[y_1,x_2]+[x_1,y_2], b[x_1,x_2]+\alpha[y_1,x_2]+\alpha[x_1,y_2]\Big]. \quad (6.111)$$

Let $p \in Q_r$ such that $p \notin C$ and replace y_i with $[p, x_i]$ in (6.111), then R satisfies

$$\Big[\gamma[x_1,x_2]+[p,[x_1,x_2]], b[x_1,x_2]+\alpha[p,[x_1,x_2]]\Big]$$

that is

$$\Big[(\gamma+p)[x_1,x_2]-[x_1,x_2]p, (b+\alpha p)[x_1,x_2]-[x_1,x_2](\alpha p)\Big]. \quad (6.112)$$

By Proposition 6.4.26 and since $p \notin C$, we have that:

- either $b = 0$ and $\alpha p \in C$, that is $\alpha = 0$ and $G = 0$, a contradiction;
- or there exist $\eta, \theta \in C$ such that $\gamma + p = \eta(b + \alpha p) + \theta$ and $p = \eta\alpha p + \theta$. Hence, since $p \notin C$, we get $\eta\alpha = 1$ and $\theta = 0$, moreover $\gamma = \eta b$. Therefore $F(x) = \eta bx + d(x)$ and $G(x) = bx + \alpha d(x) = \alpha F(x)$, for any $x \in R$.

Theorem 6.4.2 is then proved.

6.5 Generalized derivations with nilpotent values

In this section we focus our attention on a line of investigation in the literature concerning derivations having nilpotent values. The first such result is due to Herstein:

Theorem 6.5.1 *[298, Theorem 2] If R is a prime ring and d an inner derivation of R satisfying $d(x)^n = 0$, for all $x \in R$ and n a fixed integer, then $d = 0$.*

Many authors extended this result to arbitrary derivations which act either on Lie ideals or on multilinear polynomials in prime and semiprime rings. In 1983 Felzenswalb and Lanski considered derivations satisfying $d(x)^{n(x)} = 0$, for all $x \in I$, an ideal of R, and proved the following theorems.

Theorem 6.5.2 *[261, Theorem 9] If R is a ring with no nil right ideal, I an ideal of R and d a derivation of R satisfying $d(x)^{n(x)} = 0$, for all $x \in R$, then $d(I) = 0$.*

The generalization to Lie ideals is contained in [157]:

Theorem 6.5.3 *[157] If R is a prime ring with no nil right ideal and $char(R) \neq 2$, L a Lie ideal of R and d aderivation of R satisfying $d(x)^{n(x)} = 0$, for all $x \in L$, then $d(L) = 0$.*

The natural extension to multilinear polynomials is:

Theorem 6.5.4 *[556] If $f(x_1,\ldots,x_n)$ is a multilinear polynomial on a prime ring R and $d(f(r_1,\ldots,r_n))^{n(r_1,\ldots,r_n)} = 0$, for all $r_1,\ldots,r_n \in R$ and n depending on the choice of $r_1,\ldots,r_n$, then $f(x_1,\ldots,x_n)$ is central-valued on R provided R contains no non-zero nil right ideal.*

Our purpose here is to provide a proof for a generalized version of the previously cited results, in the case the derivation d is replaced by a generalized derivation F of R. More precisely we prove the following;

Theorem 6.5.5 *[413, Theorem 2] Let R be a prime ring with no non-zero nil right ideal, F a generalized derivation of R, L a noncommutative Lie ideal of R. If, for any $u \in L$, there exists $n = n(u) \geq 1$ such that $F(u)^n = 0$ then $F = 0$.*

Theorem 6.5.6 *Let R be a prime ring with no non-zero nil right ideal, F a generalized derivation of R, L a Lie ideal of R. If, for any $u \in L$, there exists $n = n(u) \geq 1$ such that $F(u)^n = 0$ then one of the following holds:*

1. *$F = 0$;*
2. *$L \subseteq Z(R)$;*
3. *L is commutative, $char(R) = 2$ and R satisfies the standard identity $s_4(x_1,\ldots,x_4)$.*

In order to prove the above cited Theorems, we begin with the case of primitive rings:

Lemma 6.5.7 *Let R be a noncommutative primitive ring, F a generalized derivation of R, with associated derivation d and I a non-zero two-sided ideal of R. If, for any $u \in [I,I]$, there exists $n = n(u) \geq 1$ such that $F(u)^n = 0$, then $F = 0$.*

Proof: Let V_R be a faithful irreducible right R-module with endomorphisms and D, a division ring. It is well known that V can be uniquely extended to a faithful irreducible right Q_r-module, moreover $D = End(V_Q) = End(V_R)$. Since I is a non-zero ideal of R, then V_I is also a faithful irreducible right I-module with $D = End(V_I)$. By the density theorem, I acts densely on V, that is I is a dense subring of D-linear transformations on V. Since every generalized derivation F assumes the form $F(x) = ax + d(x)$, for some $a \in Q_r$ and d derivation on Q_r, we can assert that, for any $u \in [I,I]$, there exists $n = n(u) \geq 1$ such that $\left(au + d(u)\right)^n = 0$.

If $d = 0$, we have $(au)^n = 0$.

Suppose firstly that $dim_D V \geq 3$.

Assume that there exists $v \in V$ such that av, v are linearly D-independent. Therefore there exists $w \in V$ such that av, v, w are linearly D-independent. By the density of I, there exist $r_1, r_2 \in I$ such that

$$r_1 v = 0, r_2 v = w, r_1(av) = 0, r_2(av) = w, r_1 w = v.$$

Moreover, there exists $n = n(r_1, r_2) \geq 1$ such that

$$\left(a[r_1, r_2]\right)^n = 0$$

which in contradiction with

$$0 = \left(a[r_1, r_2]\right)^n v = av \neq 0.$$

Hence, for all $v \in V$, v and av are linearly D-dependent and a starndard argument shows that $a \in Z(Q_r)$, the center of Q_r.

It follows that for any $r_1, r_2 \in I$, there exists $n = n(r_1, r_2) \geq 1$ such that

$$a^n [r_1, r_2]^n = 0.$$

If $a \neq 0$, then $[x_1, x_2]$ is nilpotent in I, implying that R should be commutative, a contradiction.

Let now $dim_D V = 2$, that is $R \subseteq Q_r = M_2(D)$, the 2×2 ring of matrices over D and $(au)^2 = 0$, for any $u \in [R, R]$. Moreover, since R and Q_r satisfy the same generalized polynomial identities, it follows that $(au)^2 = 0$, for any $u \in [Q_r, Q_r]$.

For $i \neq j$, where $i, j \in \{1, 2\}$, $u = e_{ij}$ and $w = e_{ii} - e_{jj}$, we have $(ae_{ij})^2 = 0$, that is a is a diagonal matrix in $M_2(D)$. By using a well known argument, we get $a \in Z(Q_r)$, that is $a^2 u^2 = 0$, for any $u \in [Q_r, Q_r]$. As above, assuming that $0 \neq a \in Z(Q_r)$, we get $[x_1, x_2]^2 = 0$, for any $x_1, x_2 \in Q_r$, which is not possible for $Q_r = M_2(D)$.

Let now $dim_D V = 1$, that is Q_r is a domain. Then $au = 0$, for any $u \in [Q_r, Q_r]$, implying $a = 0$, that is $F = 0$, as required.

Now let $d \neq 0$. Here we will prove in this final case, a number of contradictions follow. Suppose that $dim_D V \geq 2$.

Let $v \in V$, $r \in I$ such that $rv = 0$. Suppose $d(r)v \neq 0$. There exists $w \in V$ such that $d(r)v$ and rw are linearly independent over D. By the density of I, there exist $s_1, s_2 \in I$ such that

$$s_1 d(r)v = w, \quad s_2 rw = v, \quad s_2 d(r)v = 0.$$

Since there exists $n = n(r, s_1, s_2) \geq 1$ such that

$$\left(a[s_1 r, s_2 r] + d([s_1 r, s_2 r])\right)^n = 0$$

then we get the contradiction

$$0 = \left(a[s_1 r, s_2 r] + d([s_1 r, s_2 r])\right)^n v = (-1)^n v \neq 0.$$

Hence, for all $v \in V$ and $r \in I$, if $rv = 0$ then $d(r)v = 0$.

Now we prove that rv and $d(r)v$ are linearly D-dependent, for all $r \in I$ and $v \in V$. In fact, if not, by the density of I, there exists $s \in I$ such that $srv = 0$ and $sd(r)v \neq 0$. On the other hand, since $srv = 0$, and by the above argument, we have $d(sr)v = 0$ and also $d(s)rv = 0$. Then $0 = d(sr)v = d(s)rv + sd(r)v = sd(r)v \neq 0$, a contradiction. This means that, for any $r \in I$ and $v \in V$, there exists $\alpha_{r,v} \in D$, depending on the choice of r and v, such that $d(r)v = rv\alpha_{r,v}$.

Let now $r_1, r_2 \in I$. Then $d(r_1)v = r_1 v \alpha_1$ and $d(r_2)v = r_2 v \alpha_2$. If $r_1 v$ and $r_2 v$ are independent over D, then by $d(r_1)v + d(r_2)v = d(r_1 + r_2)v = (r_1 + r_2)v\beta$, we have $\alpha_1 = \alpha_2 = \beta$. If $r_1 v$ and $r_2 v$ are non-zero and linearly D-dependent, consider the element $r_3 v$, with $r_3 \in I$, such that $r_3 v$ is independent on $r_1 v$ and $r_2 v$. Since $d(r_1)v = r_1 v \alpha_1$, $d(r_2)v = r_2 v \alpha_2$ and $d(r_3)v = r_3 v \alpha_3$, as above we conclude that $\alpha_1 = \alpha_2 = \alpha_3$. Therefore, for a fixed $v \in V$, for all $r \in I$, there exists $\alpha = \alpha_v$, depending on the choice of v, such that $d(r)v = rv\alpha_v$.

Fix now $r \in I$, with $rank(r) > 1$. For all $u, v \in I$, $d(r)u = ru\alpha_u$, $d(r)v = rv\alpha_v$, for suitable α_u and α_v in D. If ru and rv are independent over D, then $ru\alpha_u + rv\alpha_v = d(r)(u+v) = r(u+v)\alpha_{u+v}$ implies $\alpha_u = \alpha_v = \alpha_{u+v}$. If ru and rv are non-zero and linearly dependent over D, consider rw, with $w \in V$, such that rw is independent on ru and rv. Thus, since $d(r)w = rw\alpha_w$, then $\alpha_u = \alpha_v = \alpha_w$. Hence we have proved that there exists $\alpha \in D$ such that $rv\alpha = d(r)v$, for all $r \in I$ and $v \in V$. Moreover $\alpha \neq 0$. In fact, if $\alpha = 0$ then $d(R)V = (0)$, which implies the contradiction $d(R) = 0$.

Now let $r, s \in I$ and $v \in V$. Since $d(rs)v = rsv\alpha$ and also $d(rs)v = d(r)sv + rd(s)v = 2rsv\alpha$, then $rsv\alpha = 0$, i.e., $R^2 V\alpha = (0)$, implying the contradiction $R = 0$.

All the previous arguments say that $dim_D V = 1$, that is Q_r is a division ring. Therefore, by our assumption, $F(u) = 0$, for any $u \in [Q_r, Q_r]$ as well as for any $u \in [R,R]$. Let $x, y \in R$, then

$$0 = F([x,yx]) = F([x,y]x) = [x,y]d(x).$$

In particular, for any $z \in R$,

$$0 = [x,zy]d(x) = [x,z]yd(x)$$

$\Rightarrow$, for any $x \in R$, either $d(x) = 0$ or $x \in Z(R)$. Notice that $G_1 = \{x \in R : d(x) = 0\}$ and $G_2 = \{x \in R : x \in Z(R)\}$ are additive subgroups of R, such that $R = G_1 \cup G_2$. But a group cannot be the union of two its proper subgroups, hence either $R = G_1$ or $R = G_2$. Since R is not commutative, then $R \neq G_2$, that is $d = 0$ in the whole R.

♣

Lemma 6.5.8 *Let R be a noncommutative semiprimitive ring, I a non-zero two-sided ideal of R and $a \in R$. If, for any $u \in [I,I]$, there exists $n = n(u) \geq 1$ such that $(au)^n = 0$, then $a = 0$.*

Proof: Since R is semiprimitive, the Jacobson's radical $J(R)$ is zero. Then R is a subdirect product of primitive rings. For any P primitive ideal of R, let $\overline{R} = \frac{R}{P}$, which is primitive. Moreover $\overline{I} = I + P$ is an ideal of $\overline{R}$ and $\overline{au}$ is nilpotent in $\overline{R}$, for all $u \in [I,I]$. By the primitive case (see the first part of Lemma 6.5.7), for any P primitive ideal of R, one of the following holds:

1. Either $\overline{R} = \frac{R}{P}$ is commutative, that is $[R,R] \subseteq P$;

2. or $\overline{a} = \overline{0}$, that is $a \in P$.

Note that in any case we have $[R,a] \subseteq P$, for any primitive ideal P of R, that is $[R,a] \subseteq \cap P = (0)$. Therefore $a \in Z(R)$ and, for any $u \in [I,I]$, there exists $n = n(u) \geq 1$ such that $a^n u^n = 0$. Suppose $a \neq 0$.

Since $0 \neq a \in Z(R)$ is not a zero-divisor, then $[x_1, x_2]$ is nilpotent in R for all $x_1, x_2 \in I$. In this case it is well knows that R must be commutative, a contradiction. Hence a must be zero, as required.

♣

Lemma 6.5.9 *Let R be a noncommutative semiprimitive ring, F a generalized derivation of R, with associated derivation d, I a non-zero two-sided ideal of R. If, for any $u \in [I,I]$, there exists $n = n(u) \geq 1$ such that $F(u)^n = 0$, then $F = 0$.*

Proof: In light of Lemma 6.5.8, we may consider $d \neq 0$ and show that a number of contradictions follows.

Since R is semiprimitive, the Jacobson's radical $J(R)$ is zero. Then R is a subdirect product of primitive rings. For any P primitive ideal of R, let $\overline{R} = \frac{R}{P}$, which is primitive. Consider the following partition:

$$K_1 = \{P : d(I^2) \subseteq P\}$$

$$K_2 = \{P : d(P) \subseteq P, \quad d(I^2) \nsubseteq P\}$$

$$K_3 = \{P : d(P) \nsubseteq P, \quad d(I^2) \nsubseteq P\}.$$

In addition let $J_i = \cap P$, for $P \in K_i$, $i = 1,2,3$. Moreover $J_1J_2J_3 \subseteq J_1 \cap J_2 \cap J_3 = (0)$ and, by the primeness of R, one J_i must be zero.

If $J_1 = 0$ then $d(I^2) = 0$ and so $d = 0$, a contradiction.

Suppose $J_2 = 0$. Let $\overline{d}$ the derivation of $\overline{R} = \frac{R}{P}$ induced by d be as follows: $\overline{d}(\overline{x}) = \overline{d(x)}$, for all $\overline{x} = x + P$, $x \in R$. Moreover $\overline{I} = I + P$ is an ideal of $\overline{R}$ and

$$\overline{au} + \overline{d}(\overline{u})$$

is nilpotent in $\overline{R}$, for all $u \in [I,I]$. By the primitive case, for all $P \in K_2$, either $\overline{d} = \overline{0}$ or $\overline{R} = \frac{R}{P}$ is commutative. In any case $[d(R),R] \subseteq \cap P = J_2 = (0)$, that is R is commutative or $d = 0$ (and this is a contradiction).

Now suppose $J_3 = 0$. For all $P \in K_3$, $\overline{d(I^2P)}$ is an ideal of $\overline{R}$ and $\overline{d(I^2P)} \neq 0$. For all $x,y \in I^2P$ and $u = [x,d(y)]$ we have that

$$\overline{au + d(u)} = \overline{[d(x),d(y)]} \quad mod.P$$

Hence $[\overline{d(x)},\overline{d(y)}]$ is nilpotent in $\overline{R}$, for all $x,y,z,t \in I^2P$, i.e., $[X,Y]$ is nilpotent in the ideal $\overline{d(I^2P)}$. It is well known that in this case $\overline{d(I^2P)}$ is commutative, that is $\overline{R} = \frac{R}{P}$ is commutative, for all $P \in K_3$, and so R is commutative, which is again a contradiction.

♣

From now on we may assume that R is not a semiprimitive ring, that is, $J(R)$, the Jacobson radical of R, is nonzero.

Lemma 6.5.10 *Let R be a prime ring, with no non-zero nil right ideal, F a generalized derivation of R, d its associated derivation, I a non-zero two-sided ideal of R such that, for any $u \in [I,I]$, there exists $n = n(u) \geq 1$ such that $F(u)^n = 0$. If $d \neq 0$ and $h,c \in I$ such that $hc = 0$, then $hd(c) = d(h)c = 0$.*

Proof: Let $x,y \in R$. Then there exists $n \geq 1$ such that

$$\Big(a[ch,xch] + d([ch,xch])\Big)^n = 0$$

and right multiplying by c, we have: $\left(chxchycd(h)\right)^n c = 0$. In particular, by replacing x with $ycd(h)$ we get $\left(chycd(h)chycd(h)\right)^n c = 0$, that is $\left(cd(h)chy\right)^{n+1} = 0$. Since R does not contain any non-zero nil right ideal, it follows that $cd(h)ch = 0$. Let now $s \in R$ such that $s^2 = 0$. If $q = xs$ and $p = sy$ for $x, y \in I$, then $qp = 0$ and, by the above argument, $0 = syd(xs)syxs = syxd(s)syxs$. Since $d(s)s = -sd(s)$, we have $(d(s)syx)^3 = 0$, and as above it follows $d(s)s = 0$. Moreover, since $hc = 0$, then $(cxh)^2 = 0$, for any $x \in R$. Hence $0 = cxhd(cxh) = cxhd(c)xh$, i.e., $(hd(c)x)^3 = 0$, so $hd(c) = 0$ and clearly $d(h)c = d(hc) - hd(c) = 0$.

♣

At this point, we need a result that will come in handy shortly and whose proof was provided by B. Felzenswalb and C. Lanski in 1983 (more precisely, it is contained in the first part of the proof of [261, Theorem 5]):

Lemma 6.5.11 *[261] Let R be a prime ring with no non-zero nil right ideal, d a non-zero derivation of R, I a non-zero ideal of R, $S = \{s \in R : s^2 = 0\}$ and $T = \{t \in R : atb = 0 \quad \text{if} \quad ab = 0, \quad a, b \in R\}$.*

If $S \neq 0$, then $S \cap T \neq 0$, $V = \{v \in S \cap T : vRv \subseteq T\} \neq 0$ and $vhv = 0$, for all $v \in V$ and for all h nilpotent elements of R.

Proposition 6.5.12 *Let R be a noncommutative prime ring, with no non-zero nil right ideal, F a generalized derivation of R, d its associated derivation, I a non-zero two-sided ideal of R such that, for any $u \in [I, I]$, there exists $n = n(u) \geq 1$ such that $F(u)^n = 0$. Then $F = 0$.*

Proof: Let S, T and V be defined as in Lemma 6.5.11. Our first aim is to prove that $d = 0$.

If $S = 0$, then R is a prime reduced ring and hence it is a domain. Thus $F(u) = 0$, for any $u \in [I, I]$. Since R and I satisfy the same generalized differential identities, then $F(u) = 0$, for any $u \in [I, I]$. Thus $d = 0$ follows by the same argument as in the final part of Lemma 6.5.7.

Now let $S \neq 0$ and assume by contradiction $d \neq 0$.

By Lemma 6.5.11 and since $F(u)$ is nilpotent, for any $u \in [I, I]$, it follows that $vF(u)v = 0$, for any $v \in V$. By the fact that R and I satisfy the same differential identities, we have $vF(u)v = 0$, for any $v \in V$ and $u \in [R, R]$. In other words R satisfies the generalized differential identity

$$v\Big(a[x_1, x_2] + d([x_1, x_2])\Big)v$$

and, in particular, for any $x, y, r \in R$ and $v \in V$,

$$v\Big(a[[x, y], r] + d([[x, y], r])\Big)v = 0.$$

By calculation, we get

$$v[x,y]d(vr)v - vavr[x,y]v = 0. \tag{6.113}$$

On the other hand, we also have

$$v\Big(a[x,vyv] + d([x,vyv])\Big)v = 0$$

leading to the conclusion that $vavyvx = 0$. By the primeness of R, it follows that $vav = 0$. So, by (6.113),

$$v[x,y]d(vr)v = 0 \tag{6.114}$$

that is $d(vr)v = 0$, for any $r \in R$ and $v \in V$. Moreover, from the definition of T and by (6.114),

$$v[x,y]d(vr)tv = 0, \quad \forall t \in T. \tag{6.115}$$

On the other hand, replacing r by rt in (6.114) and using (6.115), we also have

$$0 = v[x,y]d(vrt)v = v[x,y]vrd(t)v$$

which implies $d(t)v = 0$. Hence, for any $z \in J(R)$, $d(t)(1+z)v(1+z)^{-1} = 0$, that is $d(t)zv = 0$, meaning that $d(t) = 0$. Thus $d(T) = 0$ (and of course $d(V) = 0$).

Recall that, for any $v \in V$ and $s \in S$, we have either $sv = 0$ or $vs = 0$ (see the proof of Lemma 6.5.11). Now let $0 \neq v \in V$ and $0 \neq s \in S$. If $vs = 0$, then $vd(s) = 0$ follows from Lemma 6.5.10. In case $sv = 0$, we may write $vs = (1-s)v(1+s) - v \in V$, then $0 = d(vs) = vd(s)$. This means that in any case $vd(s) = 0$. As above, for any $z \in J(R)$, we obtain $(1+z)^{-1}v(1+z)d(s) = 0$, implying $vzd(s) = 0$, that is $VJ(R)d(S) = 0$. Hence $d(S) = 0$ and, for any $s \in S$ and since $sRs \subseteq S$, it follows $0 = d(sRs) = sd(R)s$. Under these conditions, and by using the same argument as in Lemma 6.5.11, it follows that $d = 0$ is a contradiction.

The previous argument shows that, in any case, the derivation d associated with F must be zero, so that $F(x) = ax$, for any $x \in R$. Thus, for any $s \in S$ and $x, y \in I$, there is $n = n(s,x,y) \geq 1$ such that

$$0 = \Big(a[sxsa, sxs]\Big)^n = \Big(asxsasxs\Big)^n$$

which implies $(sasx)^{2n+1} = 0$. Since R does not contain non-zero nil right ideals, we get $sas = 0$, for any $s \in S$. Moreover, for any $x \in I$, there exists $m = m(s,x) \geq 1$ such that

$$0 = \Big(a[x, saxs]\Big)^n = \Big(axsaxs - asaxsx\Big)^m. \tag{6.116}$$

Left multiplying (6.116) by s, it follows $(sax)^{2m+1} = 0$ so that, as above, $sa = 0$, for any $s \in S$. Clearly, for any $z \in J(R)$, $(1+z)^{-1}s(1+z)a = 0$, which means that $sza = 0$, i.e., $sJ(R)a = 0$, for any $s \in S$. Since $S \neq 0$, we finally get $a = 0$ and $F = 0$.

The proof of Theorem 6.5.5

Since L is a noncommutative Lie ideal of R then $I = R[L,L]R$ is a nonzero two-sided ideal of R and $[I,I] \subseteq L$ follows from Theorem 1.2.7. Therefore we may assume that $F(u)$ is nilpotent, for all $u \in [I,I]$, for I a two-sided ideal of R. Hence the conclusion follows from Proposition 6.5.12.

♣

The proof of Theorem 6.5.6

Suppose L is not central. In this case R cannot be commutative. By Theorem 1.2.8, either $[I,I] \subseteq L$, for some ideal I of R, or $char(R) = 2$, L is commutative and R satisfies the standard identity $s_4(x_1,\dots,x_4)$. In the first case, by Proposition 6.5.12, we get the conclusion $F = 0$, as required.

♣

6.6 Generalized derivations with invertible values

Carefully analyzing what was stated in the previous paragraph, it can be deduced that if R is a prime ring with generalized derivation F such that all values of F are nilpotent, then we get very little information on the nature of R. To overcome this lack of information, many authors have followed a different line of investigation in the literature concerning derivations and generalized derivations having invertible values. Bergen, Herstein and Lanski [127], in 1983, classified a ring R, with unity, possessing a nonzero derivation d such that $d(x)$ is either 0 or invertible for all $x \in R$. Later, in 1988, Bergen and Carini [125] extended this result to the case when $d(x)$ is 0 or invertible, for all x in some noncentral Lie ideal L of R (see Chapter 2, Theorems 2.8.1 and 2.8.3). In 1993, Lee [373] studied the more general situation when $d(f(x_1,\dots,x_n))$ is either 0 or invertible for all $x_1,\dots,x_n \in R$, where $f(x_1,\dots,x_n)$ is a multilinear polynomial not central-valued on R, proving that the same conclusion as above holds.

This type of approach seems to have a natural outlet in the study of derivations with invertible or nilpotent values. In 1984 [124], Bergen showed that, if R is a ring with no non-zero nil one-sided ideals and d is a non-zero derivation on R such that $d(x)$ is invertible or nilpotent for all $x \in R$, then R is a division ring or the ring of 2×2 matrices over a division ring. In 2000, a generalization to the case of derivations acting on multilinear polynomials was obtained by Lee and Wong [393]. They proved that, if $f(x_1,\dots,x_n)$ is a multilinear polynomial not central-valued on the prime ring R and d is a non-zero derivation of R, such that $d(f(x_1,\dots,x_n))$ is either invertible or nilpotent, for all $x_1,\dots,x_n \in R$, then R is a division ring or the ring of 2×2 matrices over a division ring, provided that R contains no non-zero nil one-sided ideals. In this sense, it therefore seems natural to ask what we can say about the structure of the ring in the case in which the role played by the derivation, in the aforementioned results, is now played by a generalized derivation. Our next aim is precisely to answer this question, observing in particular what happens in the case in which the generalized derivation is evaluated on a Lie ideal. At the end we will then have demonstrated the following result:

Theorem 6.6.1 *Let R be a prime ring without non-zero nil one-sided ideals, F a non-zero generalized derivation of R, L a noncommutative Lie ideal of R. If $F(u)$ is either invertible or nilpotent, for any $u \in L$, then either R is a division ring or the ring of 2×2 matrices over a division ring.*

The general method for attacking Theorem 6.6.1 can be used for proving the following more general result, whose proof was provided in 2006, by Lin and Liu [412, Theorem 1]:

Theorem 6.6.2 *Let R be a prime ring without non-zero nil one-sided ideals, F a non-zero generalized derivation of R, $f(x_1,\dots,x_n)$ a multilinear polynomial not central-valued on R. If $F(f(x_1,\dots,x_n))$ is either invertible or nilpotent, for all $x_1,\dots,x_n \in R$, then one of the following holds:*

1. *R is a division ring;*
2. *R is the ring of 2×2 matrices over a division ring;*
3. *R is the ring of $m \times m$ matrices over a finite-dimensional division algebra D, where $m \geq 3$, and these exists $\lambda \in Z(R)$ such that $F(x) = \lambda x$, for all $x \in R$. Moreover $(m-1)^2 \cdot dim_{Z(D)} D \leq \left[\frac{deg(f)}{2}\right]^2$.*

The proof of Theorem 6.6.2 as stated above is pretty much the same as that we will provide for Theorem 6.6.1. Looking at the general case, the techniques, related to the study of generalized derivations acting on multilinear polynomials are the same as the ones valid in the discussion on Lie ideals. One should take into account that the action of a generalized derivation on element $f(x_1,\dots,x_n)$, generates more computations. In any case, from a purely theoretical viewpoint, there is no difference between the Lie ideal case and the multilinear polynomial's one.

To prove Theorem 6.6.1, since L is not commutative and by Theorem 1.2.8, we recall that there exists a non-central ideal I of R such that $[I,I] \subseteq L$. Therefore, in all that follows we assume that $F([x,y])$ is either invertible or nilpotent, for any $x,y \in I$.

Lemma 6.6.3 *R is a simple ring.*

Proof: Since R is not commutative and $F \neq 0$, we may assume there are $x_0,y_0 \in I$ such that $F([x_0,y_0])$ is invertible, otherwise, by Proposition 6.5.12, a contradiction follows. On the other hand, $F([x_0,y_0]) = a[x_0,y_0] + d([x_0,y_0]) \in I$, implying that $I=R$. Let M be any non-zero two-sided ideal of R. By our assumption and since $I=R$, $F([x,y])$ is either invertible or nilpotent, for any $x,y \in M$. Using the argument as above, we get $M=R$ and R is simple.

♣

Since R is simple with 1, then it is a dense ring of linear transformations on a vector space V over a division ring D. Moreover $F([x,y])$ is either invertible or nilpotent, for any $x,y \in R$. Our next purpose is to show that R is a simple artinian ring, that is R contains a non-zero transformation of finite rank.

Lemma 6.6.4 *R is an Artinian ring.*

Proof: As we said above we may assume that R is a dense ring of transformations on a (right) vector space V over a division ring D. To prove that R is artinian, we suppose by contradiction that R does not contain non-zero linear transformations of finite rank, so that V is infinite dimensional over D. We also firstly assume $d \neq 0$.

Suppose first that d is V-outer (in the sense of [98]). Pick any D-independent subset $\{v_1,v_2,v_3\}$ of V. By [98, Theorem 4.6], there exist $r_1,r_2 \in R$ such that

$$r_iv_1=0, \quad d(r_i)v_1=0, \quad \text{for all } i=1,2$$

$$r_iv_2=0, \quad \text{for all } i=1,2$$

$$d(r_1)v_2=0, \quad d(r_2)v_2=v_3, \quad r_1v_3=v_2$$

Then $[r_1,r_2]v_1=0$, $[d(r_1),r_2]v_1=0$, $[r_1,d(r_2)]v_1=0$, $[r_1,r_2]v_2=0$, $[d(r_1),r_2]v_2=0$, $[r_1,d(r_2)]v_2=v_2$, so that

$$F([r_1,r_2])v_1=\Big(a[r_1,r_2]+d([r_1,r_2]\Big)v_1=0$$

and

$$F([r_1,r_2])v_2=\Big(a[r_1,r_2]+d([r_1,r_2]\Big)v_2=v_2.$$

Therefore the element $F([r_1,r_2])$ is neither invertible nor nilpotent. This contradiction shows that d is V-inner. Therefore there is an endomorphism T of V such that

$$d(r)u=T(ru)-rT(u), \quad r \in R, u \in V.$$

Of course if T has a finite rank then, for any $r \in R$, $d(r)$ has a finite rank too, and this is a contradiction. Moreover, if u and Tu are linearly dependent for every $u \in V$ then there exists $\lambda \in D$ such that $Tu=u\lambda$, for all $u \in V$ (it is a well known fact, see for example [98, Lemma 7.1]).

In this case, $d(r)u=(ru)\lambda-r(u\lambda)=0$, for all $u \in V, r \in R$, that is $d(R)V=(0)$ and so $d=0$, a contradiction again.

Therefore there is $u_0 \in V$ such that $u_1 = Tu_0$ and u_0 are linearly independent over D. Now let $u_2 \in V$ so that u_0, u_1, u_2 are linearly independent. Since T has an finite rank, there exists $v \in V$ such that v and Tv do not lie in the linear span of u_0, u_1, u_2.

Since R acts densely on V, there are $s_1, s_2 \in R$ such that

$$s_iv = 0, \quad s_iTv = 0, \quad s_iu_0 = 0 \quad i = 1,2$$

$$s_1u_1 = 0, \quad s_2u_1 = -u_2, \quad s_1u_2 = u_0.$$

We have

$$d(s_i)v = T(s_iv) - s_iTv = 0, \quad i = 1,2$$

$$d(s_1)u_0 = T(s_1u_0) - s_1Tu_0 = T(s_1u_0) - s_1u_1 = 0$$

and

$$d(s_2)u_0 = T(s_2u_0) - s_2Tu_0 = T(s_2u_0) - s_2u_1 = u_2$$

Hence $[s_1,s_2]v = 0$, $[d(s_1),s_2]v = 0$, $[s_1,d(s_2)]v = 0$, $[s_1,s_2]u_0 = 0$, $[d(s_1),s_2]u_0 = 0$, $[s_1,d(s_2)]u_0 = u_0$, so that

$$F([s_1,s_2])v = \Big(a[s_1,s_2] + d([s_1,s_2]\Big)v = 0$$

and

$$F([s_1,s_2])u_0 = \Big(a[s_1,s_2] + d([s_1,s_2]\Big)u_0 = u_0.$$

As above, this implies that the element $F([s_1,s_2])$ is neither invertible nor nilpotent, which violates our assumption.

The previous argument shows that d must be zero and $F(x) = ax$, for any $x \in R$.

Here we assume there exists $w_1 \in V$ such that aw_1, w_1 are D-linearly independent and let $w_2, w_3 \in V$ such that $\{aw_1, w_1, w_2, w_3\}$ is a D-linearly independent set of vectors. By the density of R, there are $t_1, t_2 \in R$ such that $t_iw_3 = 0$, for any $i = 1,2$, and

$$t_1w_1 = 0, \quad t_2w_1 = w_2, \quad t_1w_2 = w_1, \quad t_1(aw_1) = 0, \quad t_2(aw_1) = w_2.$$

Thus $a[t_1,t_2]w_3 = 0$, $a[t_1,t_2]w_1 = aw_1$ and $a[t_1,t_2](aw_1) = aw_1$. Once again, $F([t_1,t_2]) = a[t_1,t_2]$ is neither invertible nor nilpotent, a contradiction. Hence, for any $v \in V$, v, av must be linearly D-dependent, that is $a \in Z(R)$, moreover $a \neq 0$ since $F \neq 0$. This means that, for any $x, y \in R$, the commutator $[x,y]$ must be either nilpotent or invertible. To complete the proof, we show that this case cannot occur. We use the computations same as above: let $v_0, v_1, v_2 \in V$ D-linearly independent vectors. Then there exist $z_1, z_2 \in R$ such that $z_1v_0 = z_2v_0 = 0$, $z_1v_1 = 0$, $z_2v_1 = v_2$ and $z_1v_2 = v_1$, implying that $[z_1,z_2]v_0 = 0$ and $[z_1,z_2]v_1 = v_1$, that is $[z_1,z_2]$ cannot be neither invertible nor nilpotent.

All the previous contradictions lead to the conclusion that R must contain some non-zero transformations of finite rank, as required.

♣

Proposition 6.6.5 *Let R be a prime ring without non-zero nil one-sided ideals, F a non-zero generalized derivation of R, I a non-central ideal of R. If $F(u)$ is either invertible or nilpotent, for any $u \in [I,I]$, then either R is a division ring or the ring of 2×2 matrices over a division ring.*

Proof: By previous results we can assume that R is a simple Artinian ring, that is $R = M_k(D)$, for some division ring D, so that $I = R$.

On the other hand, d being a derivation of $M_k(D)$, a well known result due to Polcino Milies (see [486]) says that there exists a derivation $\delta : D \longrightarrow D$ and a matrix $A \in M_k(D)$ such that $d = d_A + \overline{\delta}$, where d_A is the inner derivation induced by A, that is $d_A(x) = Ax - xA$, for all $x \in M_k(D)$, and $\overline{\delta} : M_k(D) \longrightarrow M_k(D)$ is the derivation induced by δ, that is $\overline{\delta}(\sum_{i,j} r_{ij}e_{ij}) = \sum_{i,j}\delta(r_{ij})e_{ij}$ As usual, here e_{ij} are the matrices unit, with 1 in (i,j)-entry and 0 elsewhere. Assume $k \geq 2$. Now fix $i \neq j$ and choose $e_{ij} = [e_{ii}, e_{ij}] \in [R,R]$. Since the matrix $ae_{ij} + d(e_{ij}) = ae_{ij} + [A, e_{ij}]$ has rank ≤ 2, then it is not invertible in $M_n(D)$. Then, by our hypothesis, there is $n \geq 1$ such that $(ae_{ij} + [A,e_{ij}])^n = 0$ and, right mutiplying by e_{ij}, we get $(-1)^n(e_{ij}A)^n e_{ij} = 0$. Hence $A = \sum_i \alpha_i e_{ii}$, where $\alpha_i \in D$, that is A is a diagonal matrix. Moreover we remark that if φ is a $Z(D)$-automorphism of $M_k(D)$, then the derivation $d_\varphi = \varphi d \varphi^{-1}$ satisfies the same condition of d, that is the generalized derivation $F_\varphi(x) = (\varphi a \varphi^{-1})x + (\varphi d \varphi^{-1})(x)$ satisfies the same condition of F. For $k \geq t \geq 3$ and $b \in D$, let $\varphi = \varphi_{t,b}$ be the automorphism of $M_k(D)$ defined by $\varphi(x) = (1 - be_{1t})x(1 + be_{1t})$. Then $\varphi^{-1}(x) = (1 + be_{1t})x(1 - be_{1t})$ and

$$d_\varphi(x) = \varphi d_A(\varphi^{-1}(x)) + \varphi\overline{\delta}(\varphi^{-1}(x)) =$$

$$= \varphi\left(A\varphi^{-1}(x) - \varphi^{-1}(x)A\right) +$$

$$+\varphi\left(\delta(b)e_{1t}x(1 - be_{1t}) + (1 + be_{1t})\overline{\delta}(x)(1 - be_{1t}) + (1 + be_{1t})x(-\delta(b)e_{1t})\right) =$$

$$= \varphi(A)x - x\varphi(A) + (1 - be_{1t})\delta(b)e_{1t}x + \overline{\delta}(x) + x(-\delta(b)e_{1t})(1 + be_{1t}) =$$

$$= \varphi(A)x - x\varphi(A) + \delta(b)e_{1t}x - x\delta(b)e_{1t} + \overline{\delta}(x) =$$

$$d_B(x) + \overline{\delta}(x), \text{ where } B = \varphi(A) + \delta(b)e_{1t}.$$

Therefore, as above, B must be a diagonal matrix. Since $A = \sum_i \alpha_i e_{ii}$, we obtain

$$(-b\alpha_t + \alpha_1 b + \delta(b))\, e_{1t} = 0, \text{ for any } b \in D \text{ and } k \geq t \geq 3.$$

In particular set $b = 1$, then $\alpha_1 = \alpha_t$ for any t, that is A is a scalar matrix, $A = \alpha I_k$, $(\alpha = \alpha_1)$. Therefore $\delta(b) = -(\alpha b - b\alpha)$, which implies $\overline{\delta} = -d_{\alpha I_n}$, $d = d_A + \overline{\delta} = d_{\alpha I_n} - d_{\alpha I_n} = 0$ and $F(x) = ax$, for any $x \in R$. For $r_1 = e_{ii}$ and $r_2 = e_{ij}$ $(i \neq j)$, we have $\left(a([r_1,r_2])\right)^n = \left(ae_{ij}\right)^n$, with a rank ≤ 2, thus it is not invertible in R.

Therefore $\left(ae_{ij}\right)^n = 0$. Left multiplication by e_{ij} yields $e_{ij}(ae_{ij})^n = 0$, that is, a is a diagonal matrix in R, say $a = \sum_i \gamma_i e_{ii}$, $\gamma_i \in D$. Clearly, since $F \neq 0$, there exists $1 \leq j \leq k$ such that $\gamma_h \neq 0$.

Let now $s_1 = e_{ij}$ and $s_2 = e_{ji}$ $(i \neq j)$. Therefore, we have $\left(a([s_1,s_2])\right)^n = \left(ae_{ii} - ae_{jj}\right)^n = \gamma_i^n e_{ii} + (-1)^n \gamma_j^n e_{jj}$, which is neither zero nor invertible since $k \geq 3$.

This contradiction says that $k \leq 2$ and we are done.

♣

The proof of Theorem 6.6.1

Since L is a noncommutative Lie ideal of R then $[I,I] \subseteq L$, where $I = R[L,L]R$ is the nonzero two-sided ideal of R generated by $[L,L]$, follows from Theorem 1.2.7. Therefore we may assume that $F(u)$ is either invertible or nilpotent, for any $u \in [I,I]$, for I a two-sided ideal of R. Hence the conclusion follows from Proposition 6.6.5.

♣

6.7 Lie mappings and related problems

The study of the relationship between the associative and the Lie structure of an associative ring was initiated by Herstein in the beginning of the 1960s (see [290] and [294, Chapters 1, 2]). Herstein focused his attention on the Lie and Jordan ideal structure in the case where the ring is simple. We are specifically referencing the so-called "Herstein program", in which Herstein formulated several conjectures on various "Lie type" maps in associative rings, namely on Lie isomorphisms and Lie derivations. More precisely, let R and R' be associative rings R. A bijective additive map $\varphi : R \to R'$ is a Lie isomorphism if

$$\varphi([x,y]) = [\varphi(x), \varphi(y)] \quad \forall x,y \in R.$$

The standard form of a Lie isomorphism is described in the following example: let $\varphi : R \to R'$ defined by

$$\varphi = \vartheta + \tau \tag{6.117}$$

where,

- $\vartheta : R \to R'$ is either an isomorphism or the negative of an anti-isomorphism;
- $\tau : R \to C'$ is an additive map such that $\tau([x,y]) = 0$, for any $x,y \in R$, where C' is the extended centroid of R'.

Example 6.7.1 *Consider $R = M_k(F)$, for F a field of characteristic not equal to 2 and $k \geq 2$, and let e_{ij} be the usual matrix unit in R. Let φ be the inner automorphism induced by the invertible matrix $g = \sum_{i+j=k} e_{ij}$ and $\vartheta(x) = tr(x) \cdot I_k$, where $tr : R \to F$ is the trace mapping and I_k is the identity in $M_k(F)$. Then the map $\varphi + \vartheta : R \to R$ is a Lie isomorphism of R but it is not an isomorphism.*

We recall that an additive map $\delta : R \to R$ is a Lie derivation of the associative ring R if

$$\delta([x,y]) = [\delta(x),y] + [x,\delta(y)] \quad \forall x,y \in R.$$

As previously, also in the case of a Lie derivation, we propose its standard form:

$$\delta = d + \tau \tag{6.118}$$

where,

- $d : R \to R$ is a derivation
- $\tau : R \to C$ is an additive map such that $\tau([x,y]) = 0$, for any $x,y \in R$, where C is the extended centroid of R.

Example 6.7.2 *Consider $R = M_k(F)$, for F a field and $k \geq 1$, and let e_{ij} be the usual matrix unit in R. Let d be the inner derivation induced by the element e_{11} and $\vartheta(x) = tr(x) \cdot I_k$, where $tr : R \to F$ is the trace mapping and I_k is the identity in $M_k(F)$. Then the map $d + \vartheta : R \to R$ is a Lie derivation of R but it is not a derivation.*

In 1951 (see [306]), Hua proved that any Lie isomorphism $\varphi : M_n(\Delta) \to M_n(\Delta)$, where $n \geq 3$ and Δ is a division ring of characteristic not equal to $2,3$, assumes only the form (6.117). Herstein [290, Problem 2, pages 528-529] conjectured that this should be true for any Lie isomorphism $\varphi : R \to R$ in any simple and perhaps even prime ring R. Martindale (1963, 1969) gives a solution of Herstein's conjecture in the case of simple rings with two orthogonal idempotents whose sum is the identity [446, Main Theorem], primitive rings containing three nonzero orthogonal idempotents whose sum is the identity [442, Theorem 5] and prime rings with two orthogonal idempotents whose sum is the identity [445, Theorem 11], under the assumption that the rings have characteristics not equal to $2,3$. In 1993 Brešar [137, Theorem 3] described Lie isomorphisms of prime rings under assumption that they do not satisfy s_4, the standard identity of degree 4, and the characteristic is not equal to 2. He left open the question of whether the condition that the rings do not satisfy the polynomial identity s_4 could be removed. This assumption was removed by Blau in 2002 [130, Theorem 1.1]:

Theorem 6.7.1 *[137, Theorem 3] and [130, Theorem 1.1] Let R and R' be prime rings such that the characteristic of $\neq 2$. Let $\varphi : R \to R'$ be a Lie isomorphism. Then φ is of the form $\vartheta + \tau$, where ϑ is a homomorphism or a negative of an anti-homomorphism of R into the central closure of R', ϑ is one-to-one, and τ is an additive mapping of R into C sending commutators to zero.*

The analogous question to that for Lie isomorphisms was posed by Herstein in the case of derivations. In [290, Problem 3, page 529] he conjectured that a Lie derivation $\delta : R \to R$ in a prime ring R assumes only the form (6.118). Apparently the first result in this direction was obtained by Kaplansky (unpublished paper) for matrix rings over a field, in the case s_4 is not an identity. But, sticking to what we can find in the literature, for primitive rings of characteristic not equal to 2 having idempotent elements, the conjecture was proved by Martindale in 1964 [443, Theorem 2]. For rings not satisfying s_4, the result was proved by Brešar in 1993 [137, Theorem 5]. Finally, the s_4-case was covered by Brešar and Šemrl in 2003 [147, Theorem 4.3].

Theorem 6.7.2 *[137, Theorem 5] and [147, Theorem 4.3] Let R be a prime ring of characteristic $\neq 2$. Let δ be a Lie derivation of R. Then δ is of the form $d + \tau$, where d is a derivation of R into its central closure and τ is an additive mapping of R into C sending commutators to zero.*

The well-known Posner's first Theorem [487, Theorem 1] states that the composition $d\delta$ of non-zero derivations d and δ of a prime ring R of characteristic not 2, cannot be a derivation of R. Following this line of investigation, in 1998, Hvala [313, Theorem 1] encountered an algebraic study of so-called generalized derivations, taking into consideration the first aspect related to their composition: the product of two generalized derivations. More precisely, the result of Hvala gives conditions equivalent to the product of two generalized derivations being a generalized derivation. It is obvious that Posner's first theorem on products of derivations does not hold for generalized derivations. For instance, the product of two left (right) multipliers is again a left (right) multiplier. Substantially, the usual goal when treating a product of derivations is to describe their forms. In 1998 C. Lanski got the necessary condition when the composition of derivations could be a Lie derivation (see Theorem 2.2.7). A clear special case of Theorem 2.2.7 is the following:

Theorem 6.7.3 *Let R be a prime ring of characteristic not equal to 2, $a, b \in R$ such that $[[a,v],[b,u]] + [[b,v],[a,u]] = 0$, for all $v, u \in [R,R]$. Then either a is a central element of R or b is a central one.*

The method of approaching this problem is to use results on generalized polynomial identities in cases where derivations involved are inner, then concluding by applying the theory of differential identities. Using a similar approach, a one-sided version of Lanski's result for prime rings of characteristic not equal to 2 was proved in 2006:

Theorem 6.7.4 *[209, Theorem 1] Let R be a prime ring of characteristic not equal to 2, I a non-zero right ideal of R, d and δ non-zero derivations of R. If the composition $d\delta$ is a Lie derivation of $[I,I]$ into R, then either $s_4(I,I,I,I)I = 0$ or $\delta(I)I = d(I)I = 0$.*

Analyzing the conclusions of Theorem 6.7.4, we want to give an example that clarifies them.

Example 6.7.3 *Let $R = M_3(F)$, for F a field of characteristic 3, and let e_{ij} be the usual matrix unit in R. Let $I = (e_{11} + e_{22})R$, δ the inner derivation induced by the element e_{13}, d the inner derivation induced by the element e_{12}, that is $\delta(x) = [e_{13}, x] = e_{13}x - xe_{13}$ and $d(x) = [e_{12}, x] = e_{12}x - xe_{12}$, for all $x \in R$. Of course $s_4(x_1,x_2,x_3,x_4)x_5$ is an identity for I and it is easy to verify that $d\delta$ acts as a Lie derivation on $[I,I]$. On the other hand $d(I)I = [e_{12}, I]I \neq 0$.*

The proof of Theorem 6.7.4

Notice that, for all $u, v \in [I,I]$,

$$(d\delta)([u,v]) = [(d\delta)(u), v] + [u, (d\delta)(v)] + [\delta(u), d(v)] + [d(u), \delta(v)].$$

Hence, since we suppose that $(d\delta)$ is a Lie derivation on $[I,I]$, the problem reduces to

$$[\delta(u), d(v)] + [d(u), \delta(v)] = 0 \text{ for all } u, v \in [I,I]. \tag{6.119}$$

For proving Theorem 6.7.4, we assume that the conclusion is false, that is there exist $c_1, c_2, c_3, c_4, c_5, c_6, c_7, c_8, c_9 \in I$ such that $s_4(c_1,c_2,c_3,c_4)c_5 \neq 0$, and either $\delta(c_6)c_7 \neq 0$ or $d(c_8)c_9 \neq 0$.

Lemma 6.7.5 *Let δ and d both be inner derivations such that either $\delta(I)I \neq 0$ or $d(I)I \neq 0$. Then R is a ring satisfying a non-trivial generalized polynomial identity.*

Proof: Let $a, b \in Q_r$ such that $\delta(x) = [a, x]$ and $d(x) = [b, x]$, for all $x \in R$. Without loss of generality we may assume in this context that $\delta(I)I \neq 0$. If $\{y, ay\}$ are linearly C-dependent for all $y \in I$, then there exists $\alpha \in C$, such that $(a - \alpha)I = 0$. If we replace a by $a - \alpha$, since they induce the same inner derivation, it follows that $\delta(I)I = [a - \alpha, I]I = 0$, which is a contradiction. Thus there exists $x_0 \in I$ such that $\{x_0, ax_0\}$ are linearly C-independent and, by (6.119),

$$[[a, [x_0x_1, x_0x_2]], [b, [x_0x_3, x_0x_4]]] + [[b, [x_0x_1, x_0x_2]], [x_0x_3, x_0x_4]]] = 0$$

is a non-trivial generalized polynomial identity for R (we leave the easy proof to the reader). The same argument shows that if $d(I)I \neq 0$ then there exists $x \in I$

such that $\{x,bx\}$ are linearly C-independent and R satisfies a non-trivial GPI in any case.

♣

Lemma 6.7.6 *Let $R = M_n(F)$ be the ring of $n \times n$ matrices over the field F of characteristic not equal to 2 and $n \geq 2$. Let d be a non-zero inner derivation of R, and I a non-zero right ideal of R. If a is a non-zero element of I such that $(d([x_1,x_2])[x_3,x_4] - d([x_3,x_4])[x_1,x_2])a = 0$, for all $x_1,x_2,x_3,x_4 \in I$, then either $s_4(I,I,I,I)I = 0$ or d is induced by an element $b \in R$ such that $(b-\beta)I = 0$, for a suitable $\beta \in Z(R)$.*

Proof: Let $b \in R$ which induces the derivation d, that is $d(x) = [b,x]$, for all $x \in R$. As above, let e_{ij} be the usual matrix unit with 1 in the (i,j)-entry and zero elsewhere and $a = \sum a_{ij}e_{ij}$, $b = \sum b_{ij}e_{ij}$, with a_{ij} and b_{ij} elements of F. We know that I has a number of uniquely determinated simple components: they are minimal right ideals of R and I is their direct sum. Since there exists a set of matrix units that contains the idempotent generator of a given minimal right ideal, we observe that any minimal right ideal is part of a direct sum of minimal right ideals adding to R. Hence, any minimal right ideal of R is a direct sum of minimal right ideals, each of the form $e_{ii}R$. In this light, we may write $I = eR$ for some $e = \sum_{i=1}^{t} e_{ii}$ and $t \in \{1,2,\ldots,n\}$. Since $s_4(I,I,I,I)I = 0$ in case $t \leq 2$, we may suppose that $t \geq 3$.

First of all we want to prove that $b_{rs} = 0$ for all $s \leq t$ and $r \neq s$. To do this, suppose by contradiction that there exist $i \neq j$ such that $b_{ij} \neq 0$ $(j \leq t)$. Without loss of generality we replace b by $b_{ij}^{-1}(b - b_{jj}I_n)$, where I_n is the identity matrix in $M_n(F)$. So we assume $b_{ij} = 1$ and $b_{jj} = 0$. Moreover $a = ex$ for a suitable $x \in R$. Now let $k \leq t$, $k \neq i,j$, $[x_1,x_2] = e_{ki}$, $[x_3,x_4] = e_{ji}$. In this case we have

$$0 = ([b,e_{ki}]e_{ji}a - [b,e_{ji}]e_{ki})a$$

and left multiplying by e_{kk},

$$e_{ki}be_{ji}a = 0$$

that is, since $b_{ij} = 1$, $e_{ii}a = 0$.

On the other hand, setting $[x_1,x_2] = e_{ki}$ and $[x_3,x_4] = e_{jk}$, we have

$$0 = ([b,e_{ki}]e_{jk} - [b,e_{jk}]e_{ki})a = [b,e_{ki}]e_{jk}a = -b_{ij}e_{kk}a.$$

Therefore $e_{rr}a = 0$ for all $r \neq j$, that is $a = e_{jj}a$. Finally, for $[x_1,x_2] = e_{ki}$ and $[x_3,x_4] = e_{kk} - e_{jj}$, it follows

$$0 = ([b,e_{ki}](e_{kk} - e_{jj}) - [b,e_{kk} - e_{jj}]e_{ki})a = e_{ki}be_{jj}a$$

that is $e_{jj}a = 0$. This $\Rightarrow$ $ea = 0$, so that $a = 0$, a contradiction.

This argument says that if $a \neq 0$, then $b_{ij} = 0$ for all $i \neq j$, $j \leq t$. Suppose that $(b-\beta)I \neq 0$, for $\beta \in F$. In this case there exist $1 \leq r,s \leq t$, with $r \neq s$, such that $b_{rr} \neq b_{ss}$.

Let f be the F-automorphism of R defined by $f(x) = (1-e_{rs})x(1+e_{rs})$. Thus we have $f(x) \in I$, for all $x \in I$ and

$$([f(b),[x_1,x_2]][x_3,x_4] - [f(b),[x_3,x_4]][x_1,x_2])f(a) = 0$$

for all $x_1,x_2,x_3,x_4 \in I$. If $a \neq 0$ then $f(a) \neq 0$ and, as above, the (r,s)-entry of $f(b)$ is zero. On the other hand

$$f(b) = (1-e_{rs})b(1+e_{rs}) = b + b_{rr}e_{rs} - b_{ss}e_{rs}$$

i.e., $b_{rr} = b_{ss}$, is a contradiction. This $\Rightarrow$ there exists $\beta \in F$ such that $(b-\beta)I = 0$. Denote $b - \beta = p$. Since b and p induce the same inner derivation d, we have $([p,[x_1,x_2]][x_3,x_4] - [p,[x_3,x_4]][x_1,x_2])a = 0$ with $pI = 0$.

♣

Lemma 6.7.7 *Let R be a prime ring of characteristic not equal to 2, d a non-zero inner derivation of R, I a non-zero right ideal of R. If a is a non-zero element of I such that $(d([x_1,x_2])[x_3,x_4] - d([x_3,x_4])[x_1,x_2])a = 0$, for all $x_1,x_2,x_3,x_4 \in I$, then either $s_4(I,I,I,I)I = 0$ or d is induced by an element $b \in R$ such that $(b-\beta)I = 0$, for a suitable $\beta \in Z(R)$.*

Proof: As a reduction of Lemma 6.7.5, we have that if R is not a GPI ring, then we are done. Thus consider the only case when R satisfies a non-trivial generalized polynomial identity. Thus the Martindale quotient ring Q_r of R is a primitive ring with non-zero socle $H = Soc(Q_r)$. H is a simple ring with minimal right ideals. Let D be the associated division ring of H, then D is a simple central algebra finite dimensional over $C = Z(Q_r)$. Thus $H \otimes_C F$ is a simple ring with minimal right ideals, with F an algebraic closure of C. Let b be an element of R which induces the derivation d. Moreover $([b,[x_1,x_2]][x_3,x_4] - [b,[x_3,x_4]][x_1,x_2])a = 0$, for all $x_1,x_2,x_3,x_4 \in IH \otimes_C F$. Notice that if C is finite, we choose $F = C$.

Now we claim that, for any $c \in IH$, there exists $\beta \in C$ with $(b-\beta)c = 0$. If not then, for some $c \in IH$, $(b-\beta)c \neq 0$ for all $\beta \in C$, so in particular $bc \neq 0$. Since H is regular there exists $g^2 = g \in IH$, such that $c \in gIH$, and $e^2 = e \in H \otimes_C F$, such that

$$g, bg, gb, a, c, bc, cb \in e(H \otimes_C F)e \cong M_n(F) \quad \text{and} \quad n \geq 3.$$

Let $x_1,x_2,x_3,x_4 \in ge(H \otimes_C F)e$ and $a = eae \neq 0$, then

$$0 = e([b,[x_1,x_2]][x_3,x_4] - [b,[x_3,x_4]][x_1,x_2])eae.$$

Applying Lemma 6.7.6, we have that $e(b-\lambda)ec = 0$ for $\lambda \in C$, so $(b-\beta)c = 0$, contradicting the choice of c.

As in Lemma 6.7.5, there exists $\beta \in C$ such that $(b-\beta)I = 0$.

♣

Lemma 6.7.8 *If δ and d are both inner derivations then the Theorem 6.7.4 holds.*

Proof: Let $a, b \in Q_r$ such that $\delta(x) = [a,x]$ and $d(x) = [b,x]$, for all $x \in R$. Since in light of Lemma 2 R satisfies a non-trivial GPI then, without loss of generality R is simple and equal to its own socle and $IR = I$. In fact Q_r has a non-zero socle H with a non-zero right ideal $J = IH$. Note that H is simple, $J = JH$ and J satisfies the same basic conditions as I. Now just replace R by H, I by J and we are done.

Recall that $s_4(c_1,c_2,c_3,c_4)c_5 \neq 0$ and either $\delta(c_6)c_7 \neq 0$ or $d(c_8)c_9 \neq 0$. By the regularity of R, there exists an element $e^2 = e \in IR$ such that $eR = c_1R + c_2R + c_3R + c_4R + c_5R + c_6R + c_7R + c_8R + c_9R$ and $ec_i = c_i$, for $i = 1, \ldots, 9$. We note that $s_4(eRe, eRe, eRe, eRe) \neq 0$ (and $dim_C(eRe) \geq 9$).

Let $x, y, z \in R$, so

$$[[a,[e,ex(1-e)]],[b,[ey,ez]]] + [[b,[e,ex(1-e)]],[a,[ey,ez]]] = 0. \quad (6.120)$$

Denote $A = (1-e)ae$, $B = (1-e)be$. Assume that $A = 0$ but $B \neq 0$. Consider first the case when $\{1-e,(1-e)a\}$ are linearly C-independent. The equation 6.120, multiplied on the left by $(1-e)$, says that

$$-(1-e)b[ey,ez]aex(1-e) + (1-e)b[ey,ez]ex(1-e)a = 0. \quad (6.121)$$

By [445, Theorem 2] and since $\{1-e,(1-e)a\}$ are linearly C-independent, it follows that there exists $\lambda_1 \in C$ such that $-(1-e)b[ey,ez]ae = \lambda_1(1-e)b[ey,ez]e$. Therefore

$$(1-e)b[ey,ez]ex\lambda_1(1-e) + (1-e)b[ey,ez]ex(1-e)a = 0$$

which implies $(1-e)b[ey,ez]e = 0$, again since $\{1-e,(1-e)a\}$ are linearly C-independent. If we denote $T = eR$, $(1-e)b[T,T]T = 0$ forces $(1-e)bT[T,T]T = 0$, so either $(1-e)bT = 0$ or $[T,T]T = 0$. Thus we have that either $B = (1-e)be = 0$ or $[x_1,x_2]x_3$ is an identity for eR. In this last case, a fortiori $s_4(x_1,x_2,x_3,x_4)x_5$ is an identity for eR. In both cases we have a contradiction, since we suppose that $B \neq 0$ and $s_4(c_1,c_2,c_3,c_4)c_5 \neq 0$.

Suppose now that $(1-e)a = \lambda(1-e)$, for some $\lambda \in C$. The equation (6.121) says that

$$-(1-e)b[ey,ez]aex(1-e) + \lambda(1-e)b[ey,ez]ex(1-e) = 0$$

and so

$$-(1-e)b[ey,ez]ae + \lambda(1-e)b[ey,ez]e = 0$$

that is, for $a' = \lambda e - ae$,

$$(1-e)b[ey,ez]a' = 0.$$

Denote $W = [ey,ez]a'$. Since $(1-e)be[Wx_1, ex_2]a' = 0$, for all $x_1, x_2 \in R$, it follows that $(1-e)bex_2Wx_1a' = 0$, and so either $a' = 0$ or $W = 0$. Again denote $T = eR$. If $W = 0$, we have $[T,T]a' = 0$, so that $[T,T]Ta' = 0$, which implies either $a' = 0$ or $[T,T]T = 0$. Since $[eR,eR]e \neq 0$, we have $ae = \lambda e$ in any case.

All the previous arguments say that $(a-\lambda)e = 0$. Replacing a by $a - \lambda = a''$, since they induce the same inner derivation, we may assume that, for all $x,y,z,t \in R$,

$$[[a'',[ex,ey]],[b,[ez,et]]] + [[b,[ex,ey]],[a'',[ez,et]]] = 0. \tag{6.122}$$

Left multiplying (6.122) by $(1-e)$ we have

$$(1-e)be[[ez,et],[ex,ey]]a'' = 0 \tag{6.123}$$

in particular

$$0 = (1-e)be[[ez,et],[ex,ey(1-e)]]a'' = (1-e)be[ez,et]exey(1-e)a''$$

and, by the previous argument, $(1-e)a'' = 0$, that is $a'' = ea''$. In light of this and by (6.123),

$$(eR(1-e)be)[[eze,ete],[exe,eye]](ea''Re) = 0.$$

Let G be the subgroup of eRe generated by the polynomial $[[eze,ete],[exe,eye]]$. It is easy to see that G is a non-central Lie ideal of eRe. In this condition it is well known that $[eRe,eRe] \subseteq G$ and so $eR(1-e)be[eRe,eRe]ea''Re = 0$.

Consider now the simple Artinian ring eRe, then we have

$$eR(1-e)be[ex_1e,ex_2e](ea''Re) = 0 \quad \text{for all} \quad x_1,x_2 \in R.$$

Let $V = [ex_1e,ex_2e](ea''Re)$, so $eR(1-e)beV = 0$. Since

$$(eR(1-e)be)[Vex_1e,ex_2e](ea''Re) = 0$$

then

$$(eR(1-e)be)x_2Vex_1(ea''Re) = 0.$$

It follows that if $(1-e)be \neq 0$ then $a'' = 0$, that is $a = \lambda \in C$, a contradiction. Thus the conclusion is that if $A = (1-e)ae = 0$ then $B = (1-e)be = 0$.

Similarly $A = (1-e)ae = 0$ follows from $B = (1-e)be = 0$.

Now we assume that neither $A = 0$ nor $B = 0$, and proceed to get contradictions, proving $A = B = 0$.

Left multiplying (6.120) by $(1-e)$ and right multiplying by e we get

$$(1-e)aex(1-e)b[ey,ez]e+(1-e)b[ey,ez]ex(1-e)ae+$$

$$(1-e)bex(1-e)a[ey,ez]e+(1-e)a[ey,ez]ex(1-e)be=0.$$

If we denote $A'=A[ye,ze]$, $B'=B[ye,ze]$, it follows that

$$AxB'+B'xA+BxA'+A'xB=0. \qquad (6.124)$$

Consider now the case when A,B are linearly C-independent. As above, it follows that there exist $\alpha_1,\alpha_2,\alpha_3,\alpha_4\in C$ such that $B'=\alpha_1A+\alpha_2B$, $A'=\alpha_3A+\alpha_4B$. So we rewrite (6.124) as follows

$$2\alpha_1AxA+2\alpha_4BxB+(\alpha_2+\alpha_3)AxB+(\alpha_2+\alpha_3)BxA=0$$

that is

$$Ax(2\alpha_1A+(\alpha_2+\alpha_3)B)+Bx(2\alpha_4B+(\alpha_2+\alpha_3)A)=0. \qquad (6.125)$$

Since A,B are C-independent, by (6.125), it follows that $2\alpha_1A+(\alpha_2+\alpha_3)B=0$ and $2\alpha_4B+(\alpha_2+\alpha_3)A=0$, so the independence of A and B forces $\alpha_1=\alpha_4=\alpha_2+\alpha_3=0$.

Therefore we have $B[eRe,eRe]\subseteq CB$. Note that $B[eRe,eRe]\neq 0$. In fact, if $B[eRe,eRe]=0$, since $[eRe,eRe]\neq(0)$ is a non-central Lie ideal of the simple Artinian ring eRe, the contradiction $B=0$ is immediate.

Let $u,v\in[eRe,eRe]$. Hence there exist $\omega_1,\omega_2,0\neq\omega\in C$ such that

$$B[u,v]=\omega B\neq 0, \quad Bu=\omega_1B, \quad Bv=\omega_2B$$

and by calculation we get the contradiction

$$0\neq\omega B=B[u,v]=0.$$

Hence we may assume that A and B are linearly C-dependent, say $A=\alpha B$, for $0\neq\alpha\in C$, so also $A'=\alpha B'$. Equation (6.125) is now $2\alpha BxB'+2\alpha B'xB=0$ and it follows that B and B' must be linearly C-dependent, so that $BxB=0$ and $B=B'=0$. Therefore in any case we have that if $s_4(eR,eR,eR,eR)e\neq 0$ then $(1-e)be=(1-e)ae=0$.

Let $J=eR$, $\overline{J}=\frac{J}{J\cap l_R(J)}$; $\overline{J}$ is a prime C-algebra. Since $d(J)\subseteq J$ and $\delta(J)\subseteq J$, d and δ induce on $\overline{J}$ the following two derivations:

$$\overline{d}:\overline{J}\longrightarrow\overline{J} \quad \text{such that} \quad \overline{d}(\overline{x})=\overline{d(x)}$$

and

$$\overline{\delta} : \overline{J} \longrightarrow \overline{J} \quad \text{such that} \quad \overline{\delta}(\overline{x}) = \overline{\delta(x)}.$$

Therefore we have

$$0 = [\overline{\delta([r_1,r_2])}, \overline{d[r_3,r_4]}] + [\overline{d([r_1,r_2])}, \overline{\delta[r_3,r_4]}]$$

for all $\overline{r_1}, \overline{r_2}, \overline{r_3}, \overline{r_4} \in \overline{J}$. By Theorem 6.7.3 we have that one of the following holds

$$\overline{\delta} = \overline{0}, \quad \overline{d} = \overline{0}, \quad \overline{J} \text{ is commutative.} \tag{6.126}$$

Since $s_4(J,J,J,J)J \neq 0$, the last case cannot occur. On the other hand now we prove that also the other cases lead us to contradictions.

Suppose that the first case occurs, i.e., $\delta(J)J = 0$. In this case there exists an element $q = a - \alpha \in Q_r$, with $\alpha \in C$, such that $(a-\alpha)J = 0$. Moreover a and q induce the same inner derivation δ, so that we have

$$([b,[x_1,x_2]][x_3,x_4] - [b,[x_3,x_4]][x_1,x_2])q = 0 \quad \forall x_1,x_2,x_3,x_4 \in J. \tag{6.127}$$

In particular, for any $r \in R$, choose $[x_1,x_2] = [e, er(1-e)] = er(1-e)$. From (6.127) it follows that $[b,[x_3,x_4]]eR(1-e)q = 0$. If $(1-e)q = 0$, we have $q = eq \in J$. Under this condition, by Lemma 6.7.7 and relation (6.127) it follows that either $q = 0$, which implies the contradiction $a \in C$ and $\delta = 0$, or $(b-\beta)J = 0$ for a suitable $\beta \in C$, that is $d(eR)eR = 0$. So consider the case when $[b,[x_3,x_4]]e = 0$ for all $x_3,x_4 \in J$, and remember that $be = ebe$. This implies that $[ebe,[y_1,y_2]] = 0$ for all $y_1,y_2 \in eRe$, that is either eRe is a commutative central simple algebra or $ebe \in Ce$. In the first case we have the contradiction $0 = s_4(ec_1,ec_2,ec_3,ec_4)ec_5 = s_4(c_1,c_2,c_3,c_4)c_5 \neq 0$. In the second one we get $d(eR)eR = 0$ again. Therefore we conclude that in any case $\delta(eR)eR = d(eR)eR = 0$, which is again a contradiction because of $\delta(c_6)c_7 \neq 0$ or $d(c_8)c_9 \neq 0$.

By a similar argument and assuming $d(J)J = 0$, we conclude.

♣

Before proving Theorem 6.7.4, we premise the following results:

Lemma 6.7.9 *Let R be a prime ring of characteristic different from 2 and I a non-zero right ideal of R. If, for any $x_1,x_2,x_3,x_4 \in I$, $[[x_1,x_2],[x_3,x_4]] = 0$ then $[I,I]I = 0$.*

Proof: First note that if $[y,[I,I]] = 0$, for some $y \in R$, then, for any $s,t \in I$, we have $0 = [y,[st,t]] = [s,t][y,t]$. In particular, for any $x \in IR$, $0 = [sx,t][y,t] = [s,t]x[y,t]$, that is $[s,t]IR[y,t] = 0$. By the primeness of R, we have that either $[s,t]I = 0$, that is $[I,I]I = 0$, or $[y,I] = 0$. In this last case $0 = [y,IR] = I[y,R]$ forcing $y \in Z(R)$.

Therefore, if we assume that $[I,I]I \neq 0$, the assumption $[[I,I],[I,I]] = 0$ forces $0 \neq [I,I] \subseteq Z(R)$. Let $s,t \in I$ be such that $[s,t]I \neq 0$ and $[s,t] \in Z(R)$. Then $2s[s,t] = [s^2,t] \in Z(R)$, so $[s,t] \neq 0$ forces $s \in Z(R)$ and we have the contradiction $[s,I] = 0$.

♣

Lemma 6.7.10 *Let R be a noncommutative prime ring of characteristic $\neq 2$, q a non-central element of R and I a non-zero right ideal of R. If, for any $x_1,x_2,x_3,x_4 \in I$, $[[q,[x_1,x_2]],[x_3,x_4]] = 0$ then $[I,I]I = 0$.*

Proof: Suppose that $[I,I]I \neq 0$. As in Lemma 6.7.9, first we recall that the condition $[y,[I,I]] = 0$ forces $y \in Z(R)$. This means that $[q,[I,I]] \subseteq Z(R)$, since $[[q,[I,I]],[I,I]] = 0$. Moreover we may assume that $[q,[I,I]] \neq 0$, if not then we are finished by Lemma 6.7.9. Note that from $[q,[I,I]] \subseteq Z(R)$ it follows $[q,[[I,I],I]] \subseteq Z(R)$. Expanding this, yields $[[I,I],[q,I]] \subseteq Z(R)$. Since, for all $x \in I$, we have $[[I,I],[q,xq]] \subseteq Z(R)$, then $[[I,I],[q,I]q] \subseteq Z(R)$. Hence

$$0 = [[[I,I],[q,I]q],q] = [q,[q,I]]\,[q,[I,I]]\,.$$

Since the second factor is non-zero and central, we have $[q,[q,I]] = 0$ which $\Rightarrow$, for all $x,y \in I$,

$$0 = [q,[q,xy]] = [q,[q,x]y + x[q,y]] = 2[q,x][q,y].$$

This means that $[q,I][q,I] = 0$ and a fortiori $[q,[I,I]][q,[I,I]] = 0$ giving the contradiction $[q,[I,I]] = 0$.

♣

We therefore have everything we need to prove Theorem 6.7.4.

First suppose that δ and d are C-independent modulo D_{int}.

For $t_1,t_2,t_3,t_4 \in I$ and by relation (6.119), R satisfies

$$[[\delta(t_1x_1),t_2x_2] + [t_1x_1,\delta(t_2x_2)],[d(t_3x_3),t_4x_4] + [t_3x_3,d(t_4x_4)]] +$$
$$[[d(t_1x_1),t_2x_2] + [t_1x_1,d(t_2x_2)],[\delta(t_3x_3),t_4x_4] + [t_3x_3,\delta(t_4x_4)]]$$

that is

$$[[\delta(t_1)x_1 + t_1\delta(x_1),t_2x_2] + [t_1x_1,\delta(t_2)x_2 + t_2\delta(x_2)],[d(t_3)x_3 + t_3d(x_3),t_4x_4] +$$
$$[t_3x_3,d(t_4)x_4 + t_4d(x_4)]] +$$
$$[[d(t_1)x_1 + t_1d(x_1),t_2x_2] + [t_1x_1,d(t_2)x_2 + t_2d(x_2)],[\delta(t_3)x_3 + t_3\delta(x_3),t_4x_4] +$$
$$[t_3x_3,\delta(t_4)x_4 + t_4\delta(x_4)]] = 0$$

By Kharchenko's theory, R satisfies the generalized polynomial identity

$$\begin{aligned}&\Big[[\delta(t_1)x_1+t_1y_1,t_2x_2],[d(t_3)x_3+t_3z_3,t_4x_4]+[t_3x_3,d(t_4)x_4+t_4z_4]\Big]+\\&\Big[[t_1x_1,\delta(t_2)x_2+t_2y_2],[d(t_3)x_3+t_3z_3,t_4x_4]+[t_3x_3,d(t_4)x_4+t_4z_4]\Big]+\\&\Big[[d(t_1)x_1+t_1z_1,t_2x_2],[\delta(t_3)x_3+t_3y_3,t_4x_4]+[t_3x_3,\delta(t_4)x_4+t_4y_4]\Big]+\\&\Big[[t_1x_1,d(t_2)x_2+t_2z_2],[\delta(t_3)x_3+t_3y_3,t_4x_4]+[t_3x_3,\delta(t_4)x_4+t_4y_4]\Big].\end{aligned}$$

In particular R satisfies $[[t_1y_1,t_2x_2],[t_3x_3,t_4z_4]]$, so Q_r satisfies this as well and, for all $y_1=x_2=x_3=z_4=1\in Q_r$, it follows that $[[I,I],[I,I]]=0$. Thus by Lemma 6.7.9 we conclude $[I,I]I=0$, that is $s_4(I,I,I,I)I=0$ which contradicts $s_4(c_1,c_2,c_3,c_4)c_5\neq 0$.

Now let δ and d be C-dependent modulo D_{int}. There exist $\gamma_1,\gamma_2\in C$, such that $\gamma_1\delta+\gamma_2 d\in D_{\text{int}}$, and, by Lemma 6.7.8, it is clear that at most one of the two derivations can be inner. Without loss of generality we may assume $\gamma_1\neq 0$, so that $\delta=\alpha d+ad(q)$, for $\alpha\in C$ and $ad(q)$ the inner derivation induced by the element $q\in Q_r$.

Since if d is inner then also δ is inner, we have that d is an outer derivation of R. As above, for $t_1,t_2,t_3,t_4\in I$, R satisfies

$$\begin{aligned}&\alpha\left[[d(t_1x_1),t_2x_2]+[t_1x_1,d(t_2x_2)],[d(t_3x_3),t_4x_4]+[t_3x_3,d(t_4x_4)]\right]+\\&\qquad[[q,[t_1x_1,t_2x_2]],[d(t_3x_3),t_4x_4]+[t_3x_3,d(t_4x_4)]]+\\&\alpha\left[[d(t_1x_1),t_2x_2]+[t_1x_1,d(t_2x_2)],[d(t_3x_3),t_4x_4]+[t_3x_3,d(t_4x_4)]\right]+\\&\qquad[[d(t_1x_1),t_2x_2]+[t_1x_1,d(t_2x_2)],[q,[t_3x_3,t_4x_4]]]\end{aligned}$$

that is

$$\begin{aligned}&\alpha[[d(t_1)x_1+t_1d(x_1),t_2x_2]+[t_1x_1,d(t_2)x_2+t_2d(x_2)],[d(t_3)x_3+t_3d(x_3),t_4x_4]+\\&\qquad[t_3x_3,d(t_4)x_4+t_4d(x_4)]]+\\&[[q,[t_1x_1,t_2x_2]],[d(t_3)x_3+t_3d(x_3),t_4x_4]+[t_3x_3,d(t_4)x_4+t_4d(x_4)]]+\\&\alpha[[d(t_1)x_1+t_1d(x_1),t_2x_2]+[t_1x_1,d(t_2)x_2+t_2d(x_2)],[d(t_3)x_3+t_3d(x_3),t_4x_4]+\\&\qquad[t_3x_3,d(t_4)x_4+t_4d(x_4)]]+\\&[[d(t_1)x_1+t_1d(x_1),t_2x_2]+[t_1x_1,d(t_2)x_2+t_2d(x_2)],[q,[t_3x_3,t_4x_4]]].\end{aligned}$$

Application of Kharchenko's theory shows that R satisfies

$$\begin{aligned}&\alpha[[d(t_1)x_1+t_1y_1,t_2x_2]+[t_1x_1,d(t_2)x_2+t_2y_2],[d(t_3)x_3+t_3y_3,t_4x_4]+\\&[t_3x_3,d(t_4)x_4+t_4y_4]]+\\&[[q,[t_1x_1,t_2x_2]],[d(t_3)x_3+t_3y_3,t_4x_4]+[t_3x_3,d(t_4)x_4+t_4y_4]]+\\&\alpha[[d(t_1)x_1+t_1y_1,t_2x_2]+[t_1x_1,d(t_2)x_2+t_2y_2],[d(t_3)x_3+t_3y_3,t_4x_4]+\\&[t_3x_3,d(t_4)x_4+t_4y_4]]+\\&[[d(t_1)x_1+t_1y_1,t_2x_2]+[t_1x_1,d(t_2)x_2+t_2y_2],[q,[t_3x_3,t_4x_4]]].\end{aligned}\tag{6.128}$$

In case $\alpha \neq 0$, for $x_1 = x_4 = 0$ in (6.128) we have that R satisfies

$$2\alpha[[t_1y_1, t_2x_2], [t_3x_3, t_4y_4]]$$

so Q_r satisfies this as well and for all $y_1 = x_2 = x_3 = y_4 = 1 \in Q_r$, it follows that $2\alpha[[I,I],[I,I]] = 0$. Hence, if $\alpha \neq 0$, by Lemma 6.7.9 we have the contradiction $[I,I]I = 0$.

Now let $\alpha = 0$. In this case for $x_4 = 0$ in (6.128), we have that R satisfies

$$[[q, [t_1x_1, t_2x_2]], [t_3x_3, t_4y_4]].$$

As above Q_r satisifes this and, taking $x_1, x_2, x_3, y_4 = 1$, it follows that $[[q,[I,I]],[I,I]] = 0$. Then, by Lemma 6.7.10, we conclude again with the contradiction $[I,I]I = 0$.

Similarly, for $\gamma_2 \neq 0$ then $d = \beta\delta + ad(q)$, for some $\beta \in C$. The same above argument gives another contradiction.

6.8 On certain subgroups and subrings with derivations

If d is a derivation of the prime ring R, we shall be concerned about the size of the set $S = \{[d(x),x] \mid x \in R\}$. One way to measure this size is to look at the centralizer $C_S(R) = \{a \in R \mid [a,x] = 0\ \forall x \in S\}$ and the annihilator $A_S(R) = \{a \in R \mid ax = 0\ \forall x \in S\}$ of S in R; the bigger S, the smaller its centralizer and its annihilator should be. Another method for measuring the size of S is to determine how *large* the subring and subgroup generated by S turn out to be. We consider the subgroup (respectively, the subring) generated by S as *large* if it contains a Lie ideal of R (respectively a non-zero ideal of R). In this way we will return to the first mentioned method, given that, in a prime ring, the annihilator of an ideal, as well as that of a Lie ideal, coincides with zero of R. And similarly, the centralizer of an ideal, as well as that of a Lie ideal, is precisely the center of R. In the sequel, we will give the proof of the following theorem:

Theorem 6.8.1 ([558]) *Let R be a noncommutative prime ring of characteristic different from 2, d a non-zero derivation of R, $S = \{[d(x),x] | x \in R\}$. Then the additive subgroups of R generated by S contains a non-central Lie ideal of R.*

We will always be working in the context of prime rings (of characteristic $\neq 2$), however, the result contained in Theorem 6.8.1, in its original form [558], has the analog for 2-torsion free semi-prime rings (not necessarily prime).

The proof of Theorem 6.8.1

Let G be the additive subgroup gnenerated by $S = \{[d(x),x] | x \in R\}$. In 2004, Wong [558] showed that G contains a non-central Lie ideal of R, in the case R

is a 2-torsion free semiprime ring. Here we present a proof based on the original one (main Theorem in [558]), for the case of prime rings of characteristic $\neq 2$.

In what follows we will use the following notations and well known facts:

1. For any additive subgroup G of the ring R, we will denote by $\overline{G}$ the subring of R generated by G.

2. For any A,B additive subgroups of R, the symbol $[A,B]$ will denote the additive subgropus generated by the set $\{[a,b] \mid a \in A,\ b \in B\}$.

Remark 6.8.2 *Let I be an ideal of R and ρ a right ideal of R. Then $[I\rho,I] \subseteq [\rho,I]$.*

Remark 6.8.3 *Let I be an ideal of R and G an additive subgroup of R. Then $[G,I] \subseteq [\overline{G},I]$ ([104, Lemma 9.1.2]).*

Remark 6.8.4 *For any $x,y,z \in R$, the following identity holds: $[xy,z] = [x,yz] + [y,zx]$.*

Since $[d(y+z),y+z] \in G$, for any $y,z \in R$, it follows that

$$[d(z),y] + [d(y),z] = g_0 \in G. \tag{6.129}$$

Starting from $d([z,y]) = [d(z),y] + [z,d(y)]$ and using relation (6.8.2), one has both

$$d([z,y]) = g_0 - [d(y),z] + [z,d(y)] = g_0 - 2[d(y),z]$$

and

$$d([z,y]) = [d(z),y] - g_0 + [d(z),y] = 2[d(z),y] - g_0$$

that is

$$d([z,y]) + 2[d(y),z] = g_0 \in G \tag{6.130}$$

and

$$d([z,y]) - 2[d(z),y] = -g_0 \in G. \tag{6.131}$$

Using Remark 6.8.4 and applying relations (6.130) and (6.131), for any $x,y,z \in R$, it follows that there are $g_1,g_2,g_3 \in G$ such that

$$\begin{aligned}
0 &= d([z,yx]) + d([y,xz]) + d([x,zy]) \\
&= \Big(g_1 + 2[z,d(yx)]\Big) + \Big(g_2 + 2[d(y),xz]\Big) + \Big(g_3 + 2[d(x),zy]\Big) \\
&= g_1 + g_2 + g_3 + 2\Big([z,d(y)x + yd(x)] + [d(y),xz] + [d(x),zy]\Big) \\
&= g_1 + g_2 + g_3 + 2[zd(y),x] + 2[d(x)z,y]
\end{aligned}$$

i.e.,

$$2[zd(y),x] + 2[d(x)z,y] \in G \quad \forall x,y,z \in R. \tag{6.132}$$

In particular, replacing z by $d(z)$ in (6.132), and using (6.132) repeatedly, there exist $g_4, g_5, g_6 \in G$ such that

$$\begin{aligned} 2[d(z)d(y),x] &= g_4 - 2[d(x)d(z),y] \\ &= g_4 - \Big(g_5 + 2[d(y)d(x),z]\Big) \\ &= g_4 - g_5 + \Big(g_6 - 2[d(z)d(y),x]\Big) \end{aligned}$$

implying that

$$4[d(z)d(y),x] = g_4 - g_5 + g_6 \in G. \tag{6.133}$$

Now, in relation (6.133), we replace x by $xd(t)$, for any $t \in R$, so that

$$4[d(z)d(y),xd(t)] \in G. \tag{6.134}$$

On the other hand, by (6.132),

$$2[xd(t),d(z)d(y)] + 2\big[d\big(d(z)d(y)\big)x,t\big] \in G. \tag{6.135}$$

Comparing relations (6.134) and (6.135) it follows

$$4\big[d\big(d(z)d(y)\big)x,t\big] \in G \quad \forall x,y,t,z \in R. \tag{6.136}$$

Firstly we assume there are $y_0, z_0 \in R$ such that $d\big(d(z_0)d(y_0)\big) \neq 0$. In this case, $\rho = d\big(d(z_0)d(y_0)\big)R$ is a non-zero right ideal of R and, by (6.136), we have that $[\rho,R] \subseteq G$. Moreover, by Remark 6.8.2,

$$[R\rho,R] \subseteq [\rho,R] \subseteq G. \tag{6.137}$$

Since R is a noncommutative prime ring, then $R\rho$ is a non-central two-sided ideal of R and by (6.137) we conclude that G contains the Lie ideal $[R\rho,R\rho]$.

We now consider the case when $d\big(d(z)d(y)\big) = 0$, for any $y,z \in R$, that is

$$d^2(z)d(y) + d(z)d^2(y) = 0 \quad \forall y,z \in R. \tag{6.138}$$

Thanks to (6.138), we notice that

$$d^3(xy) = d^3(x)y + xd^3(y) \quad \forall x,y \in R$$

so that $\delta = d^3$ is a derivation of R. Since the nilpotency of a non-zero derivation of a 2-torsion free semiprime ring is always an odd number [192, Theorem], we may divide the rest of the proof into 2 cases: either $\delta^2 \neq 0$ or $\delta = 0$.

Case 1: $\delta^2 \neq 0$.

By $d(d(z)d(y)) = 0$, for any $y, z \in R$, a fortiori we have that $d([d(z), d(y)]) = 0$ and, using relation (6.131), $2[d^2(z), d(y)] \in G$. Then, by (6.129), $2[d^3(z), y] \in G$, for any $y, z \in R$. In other words,

$$2[\delta(R), R] \subseteq G. \tag{6.139}$$

For any $x, y, z \in R$, we also have

$$2[\delta^2(x)y, z] = 2[\delta(\delta(x)y), z] - 2[\delta(x)\delta(y), z]$$

where $[\delta(\delta(x)y), z] \in [\delta(R), R]$ and $[\delta(x)\delta(y), z] \in [\overline{\delta(R)}, R] = [\delta(R), R]$ (by Remark 6.8.3). Therefore $2[\delta^2(R)R, R] \subseteq 2[\delta(R), R]$ and, by Remark 6.8.2 and relation (6.139),

$$2[R\delta^2(R)R, R] \subseteq 2[\delta^2(R)R, R] \subseteq 2[\delta(R), R] \subseteq G.$$

Thus, since $R\delta^2(R)R$ is a non-central ideal of R, G contains the Lie ideal $2[R\delta^2(R)R, R]$ of R.

Case 2: $\delta = 0$.

Since $d^3 = 0$, and using relation (6.131), the following identity holds

$$d[d^2(x)z, y] - 2[d^2(x)d(z), y] \in G \quad \forall x, y, z \in R$$

implying that

$$2d[d^2(x)z, y] - 4[d^2(x)d(z), y] \in G \quad \forall x, y, z \in R. \tag{6.140}$$

On the other hand, by (6.133), we also have

$$4[d^2(x)d(z), y] \in G \quad \forall x, y, z \in R. \tag{6.141}$$

Therefore, by (6.140) and (6.141), it follows that

$$2d[d^2(x)z, y] \in G \quad \forall x, y, z \in R$$

that is

$$2d[d^2(R)R, R] \subseteq G. \tag{6.142}$$

Since $d^2 \neq 0$, $\rho = d^2(R)R$ is a non-zero right ideal of R. Hence, (6.143) and Remark 6.8.2 imply that

$$d[2Rd^2(R)R,R] \subseteq 2d[d^2(R)R,R] \subseteq G \tag{6.143}$$

where $2Rd^2(R)R$ is a non-central ideal of R. Then

$$d[t,y] \in G \quad \forall t \in 2Rd^2(R)R, \;\; y \in R.$$

Since by (6.130) $d[t,y] - 2[d(y),t] \in G$, it follows that

$$2[d(y),t] \in G \quad \forall t \in 2Rd^2(R)R, \;\; y \in R. \tag{6.144}$$

At this point, using the same above argument, we see that

$$2[d^2(R)R,R] \subseteq 2[d(R),R] \subseteq G$$

and

$$2[Rd^2(R)R,R] \subseteq 2[d^2(R)R,R] \subseteq 2[d(R),R] \subseteq G$$

that is G contains the Lie ideal $2[Rd^2(R)R,R]$ of R.

♣

Now, let A be the subring of R generated by $S = \{[d(x),x] | x \in R\}$. In 1993, Brešar and Vukman proved that A contains a non-zero left ideal of R as well as a non-zero right ideal of R [151]. At the end of their manuscript, they proposed an open question: it could be possible to prove that A contains a non-zero two-sided ideal of R. Later, in 1996 Chebotar [174, Theorem 2.1] answered in a positive way on the question of Brešar and Vukman:

Theorem 6.8.5 *[151, Theorem] and [174, Theorem 2.1] Let R be a noncommutative prime ring of characteristic $\neq 2$, d a non-zero derivation of R, $S = \{[d(x),x] | x \in R\}$. Then the subring of R generated by S contains a non-zero two-sided ideal of R.*

Nevertheless, Theorem 6.8.5 can be now considered as a consequence of Theorem 6.8.1, since the subring generated by a non-central Lie ideal of a prime ring contains a non-zero two-sided ideal of the ring (see [294, Lemma 1.3].

6.9 Open problems

We now propose a wide selection of unsolved problems which form an incentive for the development of the area of noncommutative algebra which deals with the study of the structure of rings satisfying algebraic properties involving additive maps, such as derivations, generalized derivation, automorphisms and skew derivations.

Open Problem 6.9.1 *Let R be a prime ring, Q_r its right Martindale quotient ring, C its extended centroid, $F : R \to R$ and $G : R \to R$ generalized skew derivations of R and $k \geq 1$ a fixed integer. Let $f(x_1,\ldots,x_n)$ be a non-central multilinear polynomial over C, such that*

$$[F(f(r_1,\ldots,r_n))f(r_1,\ldots,r_n) - f(r_1,\ldots,r_n)G(f(r_1,\ldots,r_n)), f(r_1,\ldots,r_n)]_k = 0$$

for all $r_1,\ldots,r_n \in R$.

What information can be deduced regarding the structure of R and the form of maps F and G?

Open Problem 6.9.2 *Let R be a prime ring, Q_r its right Martindale quotient ring, C its extended centroid, $F : R \to R$ a non-zero generalized skew derivation of R and $k \geq 1$ a fixed integer. Let $f(x_1,\ldots,x_n)$ be a non-central polynomial over C, such that*

$$[F(f(r_1,\ldots,r_n)), f(r_1,\ldots,r_n)]_k = 0$$

for all $r_1,\ldots,r_n \in R$.

Can we describe the structure of R and the form of F?

Open Problem 6.9.3 *Determine the form of additive maps G_1, G_2, G_3 under the assumption that R is a prime ring, Q_r its right Martindale quotient ring, C its extended centroid, $G_1 : R \to R$, $G_2 : R \to R$, $G_3 : R \to R$ non-zero generalized derivations of R and L a non-central Lie ideal of R, such that*

$$[G_1(x), G_2(y)] = G_3([x,y])$$

for all $x, y \in L$.

Open Problem 6.9.4 *Determine the form of additive maps G_1, G_2 under the assumption that R is a prime ring, Q_r its right Martindale quotient ring, C its extended centroid, $G_1 : R \to R$, $G_2 : R \to R$ non-zero generalized skew derivations of R, L a non-central Lie ideal of R and $k \geq 1$ a fixed integer, such that*

$$[G_1(x), G_1(y)]_k = G_2([x,y]_k)$$

for all $x, y \in L$.

Open Problem 6.9.5 *Determine the form of additive maps G_1, G_2 under the assumption that R is a prime ring, Q_r its right Martindale quotient ring, C its extended centroid, $G_1 : R \to R$, $G_2 : R \to R$ non-zero generalized skew derivations of R, L a non-central Lie ideal of R and $k \geq 1$ a fixed integer, such that*

$$[G_1(x), G_2(y)]_k = [x,y]_k$$

for all $x, y \in L$.

Open Problem 6.9.6 *Determine the structure of a prime ring R and the form of additive maps G_1, G_2, G_3, G_4 under the assumption that $G_1 : R \to R$, $G_2 : R \to R$, $G_3 : R \to R$ and $G_4 : R \to R$ are generalized derivations of R and $f(x_1, \ldots, x_n)$ is a non-central multilinear polynomial, such that*

$$G_1(x)G_2(x) + G_3(x)G_4(x) = 0$$

for all $x \in f(R)$, the set of all evaluations of $f(x_1, \ldots, x_n)$ in R.

Open Problem 6.9.7 *Determine the structure of a prime ring R and the form of additive maps G_1, G_2 under the assumption that $G_1 : R \to R$ and $G_2 : R \to R$ are generalized skew derivations of R and $f(x_1, \ldots, x_n)$ is a non-central multilinear polynomial, such that*

$$G_1(x)G_2(x) + G_2(x)G_1(x) = 0$$

for all $x \in f(R)$, the set of all evaluations of $f(x_1, \ldots, x_n)$ in R.

Open Problem 6.9.8 *Determine the structure of a prime ring R and the form of additive maps G_1, G_2 under the assumption that $G_1 : R \to R$ and $G_2 : R \to R$ are generalized skew derivations of R and $f(x_1, \ldots, x_n)$ is a non-central multilinear polynomial, such that*

$$[G_1(x), G_2(x)] = 0$$

for all $x \in f(R)$, the set of all evaluations of $f(x_1, \ldots, x_n)$ in R.

Open Problem 6.9.9 *Let R be a prime ring, $d : R \to R$ a derivation of R and $f(x_1, \ldots, x_n)$ a non-central multilinear polynomial over C, such that*

$$[d(f(r_1, \ldots, r_n)), f(r_1, \ldots, r_n)]$$

is nilpotent, for all $r_1, \ldots, r_n \in R$.

What information can be deduced regarding the structure of R?

Open Problem 6.9.10 *Let R be a prime ring, $G : R \to R$ a generalized derivation of R and $f(x_1, \ldots, x_n)$ a polynomial over C, such that*

$$G(f(r_1, \ldots, r_n))$$

is nilpotent, for all $r_1, \ldots, r_n \in R$.

What information can be deduced regarding the structure of R?

Open Problem 6.9.11 *Let R be a prime ring, $d : R \to R$ a derivation of R, $f(x_1, \ldots, x_n)$ a non-central multilinear polynomial over C and $m \geq 1$ a fixed integer, such that*

$$[d(f(r_1, \ldots, r_n)), f(r_1, \ldots, r_n)]^m$$

is zero or invertible, for all $r_1, \ldots, r_n \in R$.

What information can be deduced regarding the structure of R?

Open Problem 6.9.12 *Let R be a prime ring, $G: R \to R$ a generalized derivation of R, $f(x_1,\dots,x_n)$ a polynomial over C and $m \geq 1$ a fixed integer, such that*

$$G(f(r_1,\dots,r_n))^m$$

is zero or invertible, for all $r_1,\dots,r_n \in R$.

What information can be deduced regarding the structure of R?

Open Problem 6.9.13 *Let R be a prime ring, $d: R \to R$ and $\delta: R \to R$ non-zero skew derivations of R. It might be possible that $(d\delta)$ acts as a Lie derivation on R.*

Open Problem 6.9.14 *Let R be a prime ring, $G_1: R \to R$ and $G_2: R \to R$ generalized skew derivations of R. It might be possible that (G_1G_2) acts as a Lie derivation on R. In case of a positive answer, what can be said about the form of G_1 and G_2?*

Open Problem 6.9.15 *Let R be a prime ring, $d: R \to R$ a non-zero derivation of R, $f(x_1,\dots,x_n)$ a non-central multilinear polynomial over C. Can the additive subgroup generated by the set*

$$\{[d(x),x] \ : \ x \in f(R)\}$$

contain a Lie ideal or a one-sided ideal of R?

Open Problem 6.9.16 *Let R be a prime ring, $G: R \to R$ a non-zero generalized derivation of R. Can the additive subgroup generated by the set*

$$\{[G(x),x] \ : \ x \in R\}$$

contain a Lie ideal or a one-sided ideal of R?

References

[1] I. Z. Abdullaev, *n-Lie derivations on von Neumann algebras*, Uzbek. Mat. Zh., **5-6**, 3–9 (1992).

[2] M. S. Akhtar and M. Ashraf, *Generalized Lie triple higher derivable maps on rings*, Quaest. Math., **44(7)**, 959–981 (2021).

[3] M. S. Akhtar, M. Ashraf and M. A. Ansari, *Generalized Lie triple derivations on generalized matrix algebras*, Comm. Algebra, **50**(5), 2279–2289 (2022).

[4] A. Alahmadi, S. Ali, A.N. Khan and M. S. Khan, *A characterization of generalized derivations on prime rings*, Comm. Algebra, **44(8)**, 3201–3210 (2016).

[5] E. Albaş and N. Argaç, *Generalized derivations of prime rings*, Algebra Colloq., **11(3)**, 399–410 (2004).

[6] E. Albaş, N. Argaç and V. De Filippis, *Generalized derivations with Engel conditions on one-sided ideals*, Comm. Algebra, **36(6)**, 2063–2071 (2008).

[7] E. Albaş, N. Argaç and V. De Filippis, *Centralizers of generalized skew derivations on multilinear polynomials*, Sib. Math. J., **58(1)**, 1–10 (2017).

[8] E. Albaş, N. Argaç, V. De Filippis and Ç. Demir, *An Engel condition with generalized skew derivations on multilinear polynomials*, Linear Multilinear Algebra, **66(10)**, 1925–1938 (2018).

[9] S. Ali, *Study of derivations and commutativity of rings*, Ph.D. Thesis, Aligarh Muslim University (2002).

[10] S. Ali, *On generalized left derivations in rings and Banach algebras*, Aequat. Math., **81**, 209–226 (2011).

[11] S. Ali, A. S. Alali, S. K. Said Husain and V. Varshney, *Symmetric n-derivations on prime ideals with applications*, AIMS Mathematics, **8(11)**, 27573–27588 (2023).

[12] S. Ali, T. M. Alsuraiheed, N. Parveen and V. Varshney, *Action of n-derivations and n-multipliers on ideals of (semi)-prime rings*, AIMS Mathematics, **8(7)**, 17208–17228 (2023).

[13] S. Ali and N. A. Dar, *On *-centralizing mappings in rings with involution*, Georgian Math. J., **21(1)**, 25–28 (2014).

[14] S. Ali, N. A. Dar and M. Asci, *On derivations and commutativity of prime rings with involution*, Georgian Math. J., **23**, 9–14 (2016).

[15] S. Ali, N. A. Dar and A. N. Khan, *On strong commutativity preserving like maps in rings with involution*, Miskolc Math. Notes, **16**, 17–24 (2015).

[16] A. Ali, S. Ali and V. De Filippis, *Generalized skew derivations with nilpotent values in prime rings*, Comm. Algebra, **42**, 1606–1618 (2014).

[17] A. Ali, S. Ali and V. De Filippis, *Nilpotent and invertible values in semiprime rings with generalized derivations*, Aequat. Math., **82**, 123–134 (2011).

[18] S. Ali, N. N. Rafiquee and V. Varshney, *Certain types of derivations in rings: A survey*, J. Indones. Math. Soc. (2024) (Accepted).

[19] A. Ali and F. Shujat, *On symmetric biderivations of semiprime rings*, Contemporary Ring Theory, 196–208 (2012).

[20] S. Ali and H. Shuliang, *On derivations in semiprime rings*, Algebr Represent Theor., **15**, 1023–1033 (2012).

[21] S. A. Amitsur, *Derivations in simple rings*, Proc. Lond. Math. Soc., **7**, 87–112 (1957).

[22] G. Ancochea, *Le théorème de yon staundt en géometrie projective quaternionienne*, Journal für die reine und angewandte Mathematik, **184**, 192–198 (1942).

[23] G. Ancochea, *On semi-automorphisms of division algebras*, Annals of Mathematics, **48**, 147–153 (1947).

[24] A. J. O. Andrade, E. Barreiro and B. L. M. Ferreira, *$*$-Lie-type maps on alternative $*$-algebra*, J. Algebra Appl., **22(6)**, 2350130 (14 pages) (2023).

[25] R. L. An and J. C. Hou, *A characterization of $*$-automorphism on $\mathcal{B}(H)$*, Acta. Math. Sinica (English Series), **26**, 287–294 (2010).

[26] M. A. Ansari, M. Ashraf and M. S. Akhtar, *Lie triple derivations on trivial extension algebras*, Bull. Iranian Math. Soc., **48**, 1763–1774 (2022).

[27] N. Argaç, *On prime and semiprime rings with derivations*, Algebra Coll., **13(3)**, 371–380 (2006).

[28] N. Argaç and M. S. Yenigül, *Lie ideals and symmetric bi-derivations of prime rings*, Symmetries in Science VI, Edite by B. Gruber, Plenum Press, New York, 41–45 (1993).

[29] N. Argaç and E. Albaş, *On generalized (σ,τ)-derivations*, Sib. Math. J., **43**, 977–984 (2002).

[30] N. Argaç and V. De Filippis, *Actions of generalized derivations on multilinear polynomials in prime rings*, Algebra Colloq., **18(spec01)**, 955–964 (2011).

[31] N. Argaç and C. Demir, *Generalized derivations of prime rings on multilinear polynomials with annihilator conditions*, Turk. J. Math., **37**, 231–243 (2013).

[32] N. Argaç, A. Kaya and A. Kisir, *(σ,τ)-derivations in prime rings*, Math. J. Okayama Univ. (1987).

[33] N. Argaç, L. Carini and V. De Filippis, *An Engel condition with generalized derivations on Lie ideals*, Taiwanese J. Math., **12(2)**, 419–433 (2008).

[34] S. Asano, *On invariant subspace of division algebras*, Kodai Math., **18**, 322–334 (1966).

[35] M. Ashraf, *Structure of certain periodic rings and near-rings*, Rend. Sem. Mat. Univ. Pol. Torino, **53(1)**, 61–67 (1995).

[36] M. Ashraf, *On symmetric bi-derivations in rings*, Rend. Istit. Math. Univ. Trieste, **31(1-2)**, 25–36 (1999).

[37] M. Ashraf and M. S. Akhtar, *Characterizations of Lie triple derivations on generalized matrix algebras*, Comm. Algebra, **48(9)**, 3651–3660 (2020).

[38] M. Ashraf, M. S. Akhtar and A. Jabeen, *Characterizations of Lie triple higher derivations of triangular algebras by local actions*, Kyungpook Math. J., **60(4)**, 683–710 (2020).

[39] M. Ashraf, M. S. Akhtar and M. A., Ansari, *Generalized Lie (Jordan) triple derivations on arbitrary triangular algebras*, Bull. Malaysian Math. Sci. Soc., **44**, 3767–3776 (2021).

[40] M. Ashraf, M. S. Akhter and M.A. Ansari, *Nonlinear bi-skew Lie-type derivations on von Neumann algebras*, Comm. Algebra, **50(11)**, 4766–4780 (2021).

[41] M. Ashraf, M. S. Akhtar and B. A. Wani, *Multiplicative Lie triple higher derivations on generalized matrix algebras*, An. Stiint. Univ. Al. I. Cuza Iasi. Mat. (N.S.), **68(1)**, 23–47 (2022).

[42] M. Ashraf, M. S. Akhter and M. A. Ansari, *Nonlinear bi-skew Jordan-type derivations on factor von Neumann algebras*, Filomat, **37(17)**, 5591–5599 (2023).

[43] M. Ashraf, M. S. Akhter and M. A., Ansari, *Nonlinear generalized bi-skew Jordan n-derivations on* $*$*-algebras*, Bull. Malaysian Math. Sci. Soc., **47**, (18). https://doi.org/10.1007/s40840-023-01611-1 (2024).

[44] M. Ashraf, A. Ali and S. Ali, *On Lie ideals and generalized* (θ, ϕ)*-derivations in prime rings*, Comm. Algebra, **32(8)**, 2977–2985 (2004).

[45] M. Ashraf, A. Ali and S. Ali, *Some commutativity theorems for rings with generalized derivations*, Southeast Asian Bull. Math., **31**, 415–421 (2007).

[46] M. Ashraf, S. Ali and C. Haetinger, *On derivations in rings and their applications*, The Aligarh Bull. Math., **25(2)**, 79–107 (2006).

[47] M. Ashraf, A. Ali and M. R. Mozumder, *On Lie ideals and generalized Jordan* (θ, ψ) *derivations in rings*, Int. J. Pure Appl. Math., **65(4)**, 461–465 (2010).

[48] M. Ashraf and S. Ali, *On generalized Jordan left derivations in rings*, Bull. Korean Math. Soc., **45(2)**, 253–261 (2008).

[49] M. Ashraf and W. S. M. Al-Shammakh, *On generalized* (σ, τ)*-derivations in rings*, Internat. J. Math. Game Theory and Algebra, **12**, 295–300 (2002).

[50] M. Ashraf, M. A. Ansari and M. S. Akhter, *Characterization of Lie-type higher derivations of triangular rings*, Georgian Math. J., https://doi.org/10.1515/gmj-2022-2195 (2022).

[51] M. Ashraf and M. A. Ansari, *Multiplicative generalized Lie n-derivations of unital rings with idempotents*, Rev. Real Acad. Cienc. Exactas Fis. Nat. Ser. A-Mat., **116(92)**, https://doi.org/10.1007/s13398-022-01233-5, (2022).

[52] M. Ashraf and V. De Filippis, *A note on generalized skew derivations on Lie ideals*, Proceedings-Mathematical Sciences, **128**, 1–14 (2018).

[53] M. Ashraf, V. De Filippis and A. Khan, *A result on generalized skew derivations on Lie ideals in prime rings*, Beitr. Algebra Geom., **58**, 341–354 (2017).

[54] M. Ashraf and A. Jabeen, *Nonlinear Jordan triple derivable mappings on triangular algebras*, Pac. J. Appl. Math., **7(4)**, 229–239 (2016).

[55] M. Ashraf and A. Jabeen, *Nonlinear generalized Lie triple derivation on triangular algebras*, Comm. Algebra, **45**, 4380–4395 (2017).

[56] M. Ashraf and A. Jabeen, *Nonlinear generalized Lie triple higher derivation on triangular algebras*, Bull. Iranian Math. Soc., **44(2)**, 513–530 (2018).

[57] M. Ashraf and A. Jabeen, *Nonlinear Lie triple higher derivation on triangular algebras*, Contemporary Mathematics, **715**, https://doi.org/10.1090/conm/715/14403 (2018).

[58] M. Ashraf and A. Jabeen, *Characterization of Lie type derivation on von Neumann algebra with local actions*, Bull. Korean Math. Soc., **58(5)**, 1193–1208 (2021).

[59] M. Ashraf, A. Jabeen and M. S. Akhtar, *Generalized Jordan triple (σ, τ)-higher derivation on triangular algebra*, Filomat, **33(8)**, 2285–2294 (2019).

[60] M. Ashraf, A. Jabeen, M. A. Ansari and M. S. Akhtar, *On multiplicative Lie n-higher derivations of triangular algebras*, Facta Univ. Ser. Math. Inform., **36(5)**, 995–1017 (2021).

[61] M. Ashraf, A. Jabeen and N. Parveen, *On Jordan triple (σ, τ)-higher derivation on triangular algebra*, Spec. Matrices, **6**, 383–393 (2018).

[62] M. Ashraf and M. R. Jamal, *Traces of permuting n-additive maps and permuting n-derivations of rings*, Mediterr. J. Math., **11**, 287–297 (2014).

[63] M. Ashraf, M. R. Jamal and M. R. Mozumder, *On the traces of certain classes of permuting mappings in rings*, Georgian Math. J., **23(1)**, 15–23 (2016).

[64] M. Ashraf and A. Khan, *Generalized (σ, τ)-higher derivations on prime rings*, SpringerPlus, **1**, article-31 (2012).

[65] M. Ashraf and A. Khan, *Generalized Jordan triple (σ, τ)-higher derivations in semiprime rings*, Volume 2014, Article ID 684792, https://doi.org/10.1155/2014/684792 (2014).

[66] M. Ashraf and A. Khan, *On generalized Jordan triple (σ, τ)-higher derivations in prime rings*, ISRN Algebra, **4**, article id-684792, 1–8 (2014).

[67] M. Ashraf, A. Khan and C. Haetinger, *On (σ, τ)-higher derivations in prime rings*, Int. Electron. J. Algebra, **8**, 65–79 (2010).

[68] M. Ashraf, A. Khan and M. R. Jamal, *Traces of permuting generalized n-derivations of rings*, Miskolc Math. Notes, **19**(2), 731–740 (2018).

[69] M. Ashraf and N. Parveen, *Jordan higher derivable mappings on rings*, Algebra, **2014**(672387), 1–9 (2014).

[70] M. Ashraf and N. Parveen, *On Jordan triple higher derivable mappings in rings*, Mediterr. J. Math., **13(4)**, 1465–1477 (2016).

[71] M. Ashraf and N. Parveen, *On Lie higher derivable mappings on prime rings*, Beitr. Algebra Geom., **57**, 137–153 (2016).

[72] M. Ashraf and N. Parveen, *Lie triple higher derivable maps on rings*, Comm. Algebra, **45**, 2256–2275 (2017).

[73] M. Ashraf, N. Parveen and M. R. Jamal, *Traces of permuting n-derivations and commutativity of rings*, Southeast Asian Bull. Math., **38**, 321–332 (2014).

[74] M. Ashraf, M. A. Quadri and N. Rehman, *On Lie ideals and (σ,τ)-derivations on prime rings*, Tamakang J. Math., **32(4)**, 247–252 (2001).

[75] M. Ashraf and N. Rehman, *On derivation and commutativity in prime rings*, EastWest J. Math., **3**, 87–91 (2001).

[76] M. Ashraf and N. Rehman, *On Jordan generalized derivation in rings*, Math. J. Okayama Univ., **42**, 7–9 (2000).

[77] M. Ashraf and N. Rehman, *On Lie ideals and Jordan left derivations of prime rings*, Arch. Math. (Brno), **36**, 201–206 (2000).

[78] M. Ashraf and N. Rehman, *Lie ideals and Jordan triple derivations in rings*, Rend. Sem. Mat. Univ. Padova, **125**, 157–172 (2011).

[79] M. Ashraf and N. Rehman, *On commutativity of rings with derivations*, Results Math., **42**, 3–8 (2002).

[80] M. Ashraf and N. Rehman, *On (σ,τ)-derivations in prime rings*, Archivum Mathematicum, **38**(4), 259–264 (2002).

[81] M. Ashraf, N. Rehman and S. Ali, *On Jordan left derivations of Lie ideals in prime rings*, Southeast Asian Bull. Math., **25**, 379–382 (2001).

[82] M. Ashraf, N. Rehman and S. Ali, *On Lie ideals and Jordan generalized derivation of prime rings*, Indian J. Pure Appl. Math., **34(2)**, 291–294 (2003).

[83] M. Ashraf, N. Rehman and M. A. Quadri, *On (σ,τ)-derivations in certain classes of rings*, Rad. Math., **9**, 187–192 (1999).

[84] M. Ashraf and M. A. Siddeeque, *On *-n-derivations in rings with involution*, Georgian Math. J., **22**(1), 9–18 (2015).

[85] M. Ashraf and B. A. Wani, *Multiplicative Lie triple higher derivations on standard operator algebras*, Math. Reports, **23(73)**, 397–424 (2021).

[86] M. Ashraf, B. A. Wani and F. Wei, *Multiplicative $*$-Lie triple higher derivations of standard operator algebras*, Quaest. Math., **42(7)**, 857–882 (2019).

[87] M. J. Atteya, *Commutativity with derivations of semiprime rings*, Discuss. Math. Gen. Algebra Appl., **40(2)**, 165–175 (2020).

[88] R. Awtar, *Lie and Jordan structure in prime rings with derivations*, Proc. Amer. Math. Soc., **41**, 67–74 (1973).

[89] R. Awtar, *Lie ideals and Jordan Derivations of prime rings*, Proc. Amer. Math. Soc., **90(1)**, 9–14 (1984).

[90] N. Aydin and K. Kaya, *Some generalizations in prime rings with (α,β)-derivation*, Turk. J. Math., **16**, 169–176 (1992).

[91] Z. F. Bai and S. P. Du, *Maps preserving products $XY - YX^*$ on von Neumann algebras*, J. Math. Anal. Appl., **386**, 103–109 (2012).

[92] Z. F. Bai and S. P. Du, *The structure of nonlinear Lie derivation on von Neumann algebras*, Linear Algebra Appl., **436**, 2701–2708 (2012).

[93] I. Bajo, *Lie algebras admitting non-singular prederivations*, Indag. Math. (N.S.), **8(4)**, 433–437 (1997).

[94] R. Banning and M. Mathieu, *Commutativity preserving mappings on semiprime rings*, Comm. Algebra, **25**, 247–265 (1997).

[95] W. E. Baxter and W. S. Martindale, *Jordan homomorphisms of semiprime rings*, J. Algebra, **56**, 457–471 (1979).

[96] N. Baydar Yarbil and V. De Filippis, *A quadratic differential identity with skew derivations*, Comm. Algebra, **46(1)**, 205–216 (2018).

[97] K. I. Beidar, *Rings of quotients of semiprime rings*, Vestnik Moskov. Univ. Ser I Math. Meh. (Engl. Transl. Moscow Univ. Math. Bull.), **33**, 36–42 (1978).

[98] K. I. Beidar and M. Brešar, *Extended Jacobson density Theorem for rings with derivations and automorphisms*, Israel J. Math., **122**, 317–346 (2001).

[99] K. I. Beidar, M. Brešar and M. A. Mikhalev, *Generalized functional identities with (anti)-automorphisms and derivations on prime rings I*, J. Algebra, **215**, 644–665 (1999).

[100] K. I. Beidar, M. Brešar and M. A. Chebotar, *Functional identities with r-independent coefficients*, Comm. Algebra, **30(12)**, 5725–5755 (2002).

[101] K. I. Beidar and M. A. Chebotar, *On Lie derivations of Lie ideals of prime algebras*, Israel J. Math., **123**, 131–148 (2001).

[102] K. I. Beidar and M. A. Chebotar, *On functional identities and d-free subsets of rings I*, Comm. Algebra, **28**, 3925–3951 (2000).

[103] K. I. Beidar and W. S. Martindale III, *On functional identities in prime rings with involution*, J. Algebra, **203**, 491–532 (1998).

[104] K. I. Beidar, W. S. Martindale III and V. Mikhalev, *Rings with generalized identities*, Pure and Applied Math., Dekker, New York (1996).

[105] K. I. Beidar, M. Brešar, M. A. Chebotar and Y. Fong, *Applying functional identities to some linear preserver problems*, Pacific J. Math., **204**, 257–271 (2002).

[106] S. Belkadi, S. Ali and L. Taoufiq, *On nilpotent homoderivations in prime rings*, Comm. Algebra, **51(9)**, 4044–4053 (2023).

[107] S. Belkadi, S. Ali and L. Taoufiq, *On n-Jordan homoderivations in rings*, Georgian Math. J., **31**(1), 1–8 (2023).

[108] H. E. Bell, *On the commutativity of prime rings with derivation*, Quaest. Math., **22**, 329–335 (1999).

[109] H. E. Bell and M. N. Daif, *On commutativity and strong commutativity-preserving maps*, Canadian Math. Bull., **37(4)**, 443–447 (1994).

[110] H. E. Bell and M. N. Daif, *On derivations and commutativity in prime rings*, Acta. Math. Hungar., **66**, 337–343 (1995).

[111] H. E. Bell and M. N. Daif, *Remarks on derivations on semiprime rings*, Int. J. Math. Math. Sci., **15(1)**, 205–206 (1992).

[112] H. E. Bell and L. C. Kappe, *Rings in which derivations satisfy certain algebraic conditions*, Acta Math. Hungar., **53**, 339–346 (1989).

[113] H. E. Bell and W. S. Martindale, *Centralizing mappings of semiprime rings*, Canad. Math. Bull., **30**, 92–101 (1987).

[114] H. E. Bell and N. Rehman, *Generalized derivations with commutativity and anti-commuttaivity conditions*, Math. J. Okayama Univ., **49**, 139–147 (2007).

[115] D. Benkovič, *Generalized Lie derivations of unital algebras with idempotents*, Oper. Matrices, **12(2)**, 357–367 (2018).

[116] D. Benkovič, *Generalized Lie derivations on triangular algebras*, Linear Algebra Appl., **434**, 1532–1544 (2011).

[117] D. Benkovič, *Generalized Lie n-derivations of triangular algebras*, Comm. Algebra, **47(12)**, 5294–5302 (2019).

[118] D. Benkovič, *Jordan derivations and antiderivations on triangular matrices*, Linear Algebra Appl., **397**, 235–244 (2005).

[119] D. Benkovič, *Lie derivations on triangular matrices*, Linear Multilinear Algebra, **55**, 619–629 (2007).

[120] D. Benkovič, *Lie triple derivations of unital algebras with idempotents*, Linear Multilinear Algebra, **63**, 141–165 (2015).

[121] D. Benkovič and D. Eremita, *Multiplicative Lie n-derivations of triangular rings*, Linear Algebra Appl., **436**, 4223–4240 (2012).

[122] D. Benkovič and N. Širovnik, *Jordan derivations of unital algebras with idempotents*, Linear Algebra Appl., **437**, 2271–2284 (2012).

[123] N. Bera, B. Dhara and S. Kar, *A result concerning b-generalized skew derivations on multilinear polynomials in prime rings*, Comm. Algebra, **51(3)**, 887–901 (2023).

[124] J. Bergen, *Lie ideals with regular and nilpotent elements and a result on derivations*, Rend. Circ. Mat. Palermo, **33**, 99–108 (1984).

[125] J. Bergen and L. Carini, *Derivations with invertible values on Lie ideals*, Canad. Math. Bull., **31(1)**, 103–110 (1988).

[126] J. Bergen, I. N. Herstein and J. W. Kerr, *Lie ideals and derivations of prime rings*, J. Algebra, **71**, 259–267 (1981).

[127] J. Bergen, I. N. Herstein and C. Lanski, *Derivations with invertible values*, Canad. Math. Bull., **35(2)**, 300–310 (1983).

[128] R. Berger, Differentiale Höherer Ordnung und Körpererweiterungen bei Primzahlcharakteristik, S.B. Heidelberger Akad. Wiss. Math.-Natur. Kl., 143–202 (1966).

[129] P. S. Blau, *Lie isomorphisms of non-GPI rings with involution*, Comm. Algebra, **27**, 2345–2373 (1999).

[130] P. S. Blau, *Lie isomorphisms of prime rings satisfying St_4*, Southeast Asian Bulletin of Mathematics, **25**, 581–587 (2002).

[131] M. Brešar, *Jordan derivations on semiprime ring*, Proc. Amer. Math. Soc., **104(4)** (1988).

[132] M. Brešar, *Jordan mappings of semiprime rings*, J. Algebra, **127**, 218–228 (1989).

[133] M. Brešar, *A note on derivations*, Math. J. Okayama Univ., **32**, 83–88 (1990).

[134] M. Brešar, *On the distance of the composition of the two derivations to the generalized derivations*, Glasgow Math. J., **33**, 89–93 (1991).

[135] M. Brešar, *On a generalization of the notion of centralizing mappings*, Proc. Amer. Math. Soc., **114**, 641–649 (1992).

[136] M. Brešar, *Centralizing mappings and derivations in prime rings*, J. Algebra, **156**, 385–394 (1993).

[137] M. Brešar, *Commuting traces of biadditive mappings, commutativity-preserving mappings and Lie mappings*, Trans. Amer. Math. Soc., **335**, 525–546 (1993).

[138] M. Brešar, *Functional identities of degree two*, J. Algebra, **175**, 690–720 (1995).

[139] M. Brešar, *Jordan derivations revisited*, Math. Proc. Cambridge Philos. Soc., **139**, 411–425 (2005).

[140] M. Brešar, *Introduction to noncommutative algebra*, Universitext. Springer, Cham (2014).

[141] M. Brešar, M. A. Chebotar and W. S. Martindale III, *Functional identities*, Birkhäuser Verlag, Basel (2007).

[142] M. Brešar, W. S. Martindale and C. R. Miers, *Centralizing maps in prime rings with involution*, J. Algebra, **161**, 342–357 (1993).

[143] M. Brešar and C. R. Miers, *Strong commutativity preserving maps of semiprime rings*, Canad. Math. Bull., **37(4)**, 457–460 (1994).

[144] M. Brešar and B. Hvala, *On additive maps of prime rings*, Bull. Aust. Math. Soc., **51**, 377–381 (1995).

[145] M. Brešar and M. Fošner, *On rings with involution equipped with some new product*, Publ. Math. Debrecen, **57**, 121–134 (2000).

[146] M. Brešar and A. R. Villena, *The noncommutative Singer-Wermer conjecture and Φ-derivations*, J. Lond. Math. Soc., **66(3)**, 710–720 (2002).

[147] M. Brešar and P. Šemrl, *Commuting traces of biadditive mappings revisited*, Comm. Algebra, **31(1)**, 381–388 (2003).

[148] M. Brešar and J. Vukman, *Jordan derivations on prime rings*, Bull. Austral Math. Soc., **37**, 321–322 (1988).

[149] M. Brešar and J. Vukman, *On left derivations and related mappings*, Proc. Amer. Math. Soc., **110(1)**, 7–16 (1990).

[150] M. Brešar and J. Vukman, *Jordan (θ,ϕ)-derivations*, Glas. Mat., **16**, 13–17 (1991).

[151] M. Brešar and J. Vukman, *On certain subrings of prime rings with derivations*, J. Austral. Math. Soc., **54**, 133–141 (1993).

[152] W. Brown and W. Kuan, *Ideals and higher derivations in commutative rings*, Canad. J. Math., Vol. XXIV, No. 3, 400–415 (1972).

[153] B. Brown and N. H. McCoy, *Prime ideals in non-associative rings*, Trans. Amer. Math. Soc., **89**, 245–255 (1958).

[154] D. Burde, *Lie algebra prederivations and strongly nilpotent Lie algebras*, Comm. Algebra, **30(7)**, 3157–3175 (2002).

[155] L. Carini and V. D. De Filippis, *Commutators with power central values on a Lie ideal*, Pac. J. Math., **193**, 269–278 (2000).

[156] L. Carini and V. De Filippis, *Centralizers of generalized derivations on multilinear polynomials in prime rings*, Sib. Math. J., **53(6)**, 1051–1060 (2012).

[157] L. Carini and A. Giambruno, *Lie ideals and nil derivations*, Boll. UMI, **6**, 497–503 (1985).

[158] L. Carini and G. Scudo, *On Posner's theorem with b-generalized skew derivations on Lie ideals*, J. Algebra App., p. 2350057 (2021).

[159] L. Carini, V. De Filippis and G. Scudo, *Power-commuting generalized skew derivations in prime rings*, Mediterr. J. Math., **13**, 53–64 (2016).

[160] L. Carini, V. De Filippis and F. Wei, *Generalized skew derivations co-centralizing multilinear polynomials*, Mediterr. J. Math., **13**, 2397–2424 (2016).

[161] J. Chacron and M. Chacron, *Rings with involution all of whose symmetric elements are nilpotent or regular*, Proc. Amer. Math. Soc., **37**, 397–402 (1973).

[162] J. C. Chang, *(α,β)-derivation with nilpotent values*, Chinese Journal of Mathematics, **22(4)**, 349–55 (1994).

[163] J. C. Chang, *A special identity of (α,β)-derivations and its consequences*, Taiwanese J. Math., **1(1)**, 21–30 (1997).

[164] J. C. Chang, *On the identity $h(x) = af(x) + g(x)b$*, Taiwanese J. Math., **7**, 103–113 (2003).

[165] J. C. Chang, *Generalized skew derivations with annihilating engel conditions*, Taiwanese J. Math., **12**, 1641–1650 (2008).

[166] J. C. Chang, *Annihilators of power values of a right generalized* (α, β)*-derivation*, Bull. Inst. Math. Acad. Sin., **4**, 67–73 (2009).

[167] J. C. Chang, *Generelized skew derivations with nilpotent valves*, Monatsh Math., **161**, 155–160 (2010).

[168] J. C. Chang, *Generalized skew derivations with Engel conditions on Lie ideals*, Bull. Inst. Math. Acad. Sin. (N.S.), **6**, 305–320 (2011).

[169] J. C. Chang, *Generelized skew derivations with power central values on Lie ideals*, Comm. Algebra, **39**, 2241–2248 (2011).

[170] J. C. Chang and J. S. Lin, (α,β)*-derivation with nilpotent values*, Chinese J. Math., **22**, 349–355 (1994).

[171] C. M. Chang and Y. C. Lin, *Derivations on one-sided ideals of prime rings*, Tamsui Oxf. J. Math. Sci., **17**, 139–145 (2001).

[172] S. U. Chase, *A generalization of the ring of triangular matrices*, Nagoya Math. J., **18**, 13–25 (1960).

[173] M. A. Chebotar, *On a composition of derivations of prime rings*, Mascow Univ. Math. Bull., **50**, 22–25 (1995).

[174] M. A. Chebotar, *On certain subrings and ideals of prime rings*, Proceedings of First International Tainan - Moscow Algebra Workshop, Walter de Gruyter, Berlin - New York , 177–180 (1996).

[175] M. A. Chebotar and P. H. Lee, *A note on composition of derivations of prime rings*, Comm. Algebra, **31**(6), 2965–2969 (2003).

[176] M. A. Chebotar, W. F. Ke and P. H. Lee, *Maps characterized by actions on zero products*, Pac. J. Appl. Math., **216**, 217–228 (2004).

[177] H. W. Cheng and F. Wei, *Generalized skew derivations of rings*, Adv. Math. (China), **35(2)**, 237–243 (2006).

[178] L. Chen and J. H. Zhang, *Nonlinear Lie derivations on upper triangular matrices*, Linear Multilinear Algebra, **56(6)**, 725–730 (2008).

[179] W. S. Cheung, *Maps on triangular algebras*, Ph.D. Dissertation, University of Victoria (2000).

[180] W. S. Cheung, *Lie derivation of triangular algebras*, Linear Multilinear Algebra, **51**, 299–310 (2003).

[181] M. C. Chou and K. C. Liu, *An Engel condition with skew derivations*, Monatsh. Math., **158(3)**, 259–270 (2009).

[182] C. L. Chuang, *On invariant additive subgroups*, Israel J. Math., **57**, 116–128 (1987).

[183] C. L. Chuang, *The additive subgroup generated by a polynomial*, Israel J. Math., **59**, 98–106 (1987).

[184] C. L. Chuang, *GPIs' having coefficients in Utumi quotient rings*, Proc. Amer. Math. Soc., **103(3)**, 723–728 (1988).

[185] C. L. Chuang, *On compositions of derivations of prime rings*, Proc. Amer. Math. Soc., **180**, 647–652 (1990).

[186] C. L. Chuang, *Differential identities with automorphisms and antiautomorphisms I*, J. Algebra, **149**, 371–404 (1992).

[187] C. L. Chuang, *Differential identities with automorphisms and antiautomorphisms II*, J. Algebra, **160**, 130–171 (1993).

[188] C. L. Chuang and T. K. Lee, *Rings with annihilator conditions on multilinear polynomials*, Chinese J. Math., **24(2)**, 177–185 (1996).

[189] C. L. Chuang and T. K. Lee, *Finite products of derivations in prime rings*, Comm. Algebra, **30(5)**, 2183–2190 (2002).

[190] C. L. Chuang and T. K. Lee, *Identities with a single skew derivation*, J. Algebra, **288**, 59–77 (2005).

[191] L. O. Chung, *Nil derivations*, J. Algebra, **95(1)**, 20–30 (1985).

[192] L. O. Chung and J. Luh, *Nilpotency of derivations I*, Canad. Math. Bull., **26(3)**, 341–346 (1983).

[193] L. O. Chung and J. Luh, *Nilpotency of derivations II*, Proc. Amer. Math. Soc., **91(3)**, 357–358 (1984).

[194] L. O. Chung and J. Luh, *Nilpotency of derivations on an ideal*, Proc. Amer. Math. Soc., **90(2)**, 211–214 (1984).

[195] L. O. Chung, A. Kovacs and J. Luh, *Algebraic derivations*, preprint.

[196] W. Cortes and C. Haetinger, *On Jordan generalized higher derivations in rings*, Turkish J. Math., **29(1)**, 1–10 (2005).

[197] T. Creedon, *Products of derivations*, Proc. Edinburgh Math. Soc., **41**, 407–410 (1998).

[198] J. L. Cui and C. K. Li, *Maps preserving product $XY - YX^*$ on factor von Neumann algebras*, Linear Algebra Appl., **431**, 833–842 (2009).

[199] J. M. Cusack, *Jordan derivations on rings*, Proc. Amer. Math. Soc., **53(2)**, 321–324 (1975).

[200] M. N. Daif, *When is a multiplicative derivation additive?*, Int. J. Math. Math. Sci., **14(3)**, 615–618 (1991).

[201] M. N. Daif, *Commutativity results for semiprime rings with derivations*, Int. J. Math. Math. Sci., **21(3)**, 471–474 (1998).

[202] M. N. Daif and H. E. Bell, *Remarks on derivations on semiprime rings*, Int. J. Math. Math. Sci., **15**, 205–206 (1992).

[203] L. Dai and F. Lu, *Nonlinear maps preserving Jordan $*$-products*, J. Math. Anal. Appl., **409**, 180–188 (2014).

[204] S. Dakir, H. El Mir and A. Mamouni, *Generalized derivations in ring with involution involving symmetric and skew symmetric elements*, Commun. Korean Math. Soc., **39(1)**, 1–10 (2024).

[205] P. Das, B. Dhara and S. Kar, *Generalized derivations with centralizing conditions in prime rings*, Commun. Korean Math. Soc., **34(1)**, 1–93 (2019).

[206] K. R. Davidson, *Nest algebras*, In: Pitman Research Notes in Mathematics Series, **191**, Longman London, New York (1988).

[207] V. De Filippis, *On the annihilator of commutators with derivation in prime rings*, Rend. Del Circ. Math. Di Palermo, Serie II, **49**, 43–352 (2000).

[208] V. De Filippis, *Generalized derivations and commutators with nilpotent values on Lie ideals*, Tamsui Oxford J. Math. Sci., **22(2)**, 167–175 (2006).

[209] V. De Filippis, *Products of derivations which act as Lie derivations on commutators of right ideals*, Int. J. Math. Math. Sci., **Art. ID 53145**, 1–12 (2006).

[210] V. De Filippis, *Generalized derivations on Lie ideals*, Tamsui Oxford J. Math. Sci., **22(2)**, 167–175 (2006).

[211] V. De Filippis, *Posner's second theorem and an annihilator condition with generalized derivations*, Turkish J. Math., **32**, 197–211 (2008).

[212] V. De Filippis, *Generalized derivations with Engel condition on multilinear polynomials*, Israel J. Math., **171(1)**, 325–348 (2009).

[213] V. De Filippis, *Annihilating and power-commuting generalized skew derivations on lie ideals in prime rings*, Czech. Math. J., **66(2)**, 481–492 (2016).

[214] V. De Filippis, *Annihilators and power values of generalized skew derivations on Lie ideals*, Canad. Math. Bull., **59**, 258–270 (2016).

[215] V. De Filippis, *Product of generalized skew derivations on Lie ideals*, Comm. Algebra, **49(7)**, 2987–3009 (2021).

[216] V. De Filippis and B. Dhara, *A Cocentralizing generalized derivations on multilinear polynomials on right ideals of prime rings*, Demonstratio Math., **XLVI**, 22–36 (2014).

[217] V. De Filippis and O. M. Di Vincenzo, *Posner's second theorem and an annihilator condition*, Math. Pannonica, **12**, 69–81 (2001).

[218] V. De Filippis and O. M. Di Vincenzo, *Vanishing derivations and centralizers of generalized derivations on multilinear polynomials*, Comm. Algebra, **40**, 1918–1932 (2012).

[219] V. De Filippis and O. M. Di Vincenzo, *Hypercentralizing generalized skew derivations on left ideals in prime rings*, Monatsh. Math., **173(3)**, 315–341 (2014).

[220] V. De Filippis and S. Huang, *Generalized derivations on semiprime rings*, Bull. Korean Math Soc., **48(6)**, 1253–1259 (2011).

[221] V. De Filippis and F. Rania, *Multilinear polynomials and cocentralizing conditions in prime rings*, Sib. Math. J., **52(5)**, 813–823 (2011).

[222] V. De Filippis and G. Scudo, *Strong commutativity and Engel condition preserving maps in prime and semiprime rings*, Linear Multilinear Algebra, **61(7)**, 917–938 (2013).

[223] V. De Filippis and F. Wei, *Posner's second theorem and annihilator conditions with generalized skew derivations*, Collect. Math., **129**, 61–74 (2012).

[224] V. De Filippis and M .S. Tammam El-Sayad, *A note on posner's theorem with generalized derivations on lie ideals*, Rendiconti del Seminario Matematico dell'Universit di Padova, **122**, 1–8 (2009).

[225] V. De Filippis and F. Wei, *An Engel condition with X-generalized skew derivations on Lie ideals*, Comm. Algebra, **46(12)**, 5433–5446 (2018).

[226] V. De Filippis and F. Wei, *Centralizers of X-generalized skew derivations on multilinear polynomials in prime rings*, Comm. Math. Stat., **6(1)**, 49–71 (2018).

[227] V. De Filippis and F. Wei, *b-generalized skew derivations on Lie ideals*, Mediterr. J. Math., **15**, 1–24 (2018).

[228] V. De Filippis, O. M. Di Vincenzo and C.Y. Pan, *Quadratic central differential identities on a multilinear polynomial*, Comm. Algebra, **36**, 3671–3681 (2008).

[229] V. De Filippis, N. Rehman and G. Scudo, *Annihilator conditions with generalized skew derivations and Lie ideals of prime rings*, Int. Electron. J. Algebra, **32**, 192–216 (2022).

[230] V. De Filippis, N. Rehman and G. Scudo, *Certain functional identities involving a pair of generalized skew derivations with nilpotent values on Lie ideals*, Comm. Algebra, **51(2)**, 711–723 (2022).

[231] V. De Filippis, G. Scudo and F. Wei, *b-Generalized skew derivations on multilinear polynomials in prime rings*, Polynomial Identities in Algebras, Springer INdAM Series, **44**, 109–138 (2020).

[232] C. Demir and N. Argaç̧, *A result on generalized derivations with Engel conditions on one-sided ideals*, J. Korean Math. Soc., **47(3)**, 483–494 (2010).

[233] C. Demir and N. Argaç, *Prime rings with generalized derivations on right ideals*, Algebra Colloq., **18(Spec.1)**, 987–998 (2011).

[234] Ç. Demir, V. De Filippis and N. Argaç, *A quadratic generalized differential identity on Lie ideals of prime rings*, Linear Multilinear Algebra, **68(9)**, 1835–1847 (2020).

[235] Q. Deng, *On Jordan left derivations*, Math. J. Okayama Univ., **34**, 145–147 (1992).

[236] Q. Deng, *On a conjecture of Vukman*, Int. J. Math. Math. Sci., **20(2)**, 263–266 (1997).

[237] Q. Deng and M. Ashraf, *On strong commutativity preserving mappings*, Results. Math., **30**, 259–263 (1996).

[238] B. Dhara, *Generalized derivations acting as a homomorphism or anti-homomorphism in semiprime rings*, Beitr. Algebra Geom., **53**, 203–209 (2012).

[239] B. Dhara, *A note on generalized skew-derivations on multilinear polynomials in prime rings*, Quaestiones Math., **43(2)**, 251–264 (2020).

[240] B. Dhara, *Generalized skew-derivations acting on Lie ideals with central values in prime rings*, Ann. dell'Universit di Ferrara, **66(1)**, 1–12 (2020).

[241] B. Dhara and S. Ali, *On n-centralizing generalized derivations in semiprime rings with applications to C^*-algebras*, J. Algebra & Appl., **11(6)**, 1250111 (2012).

[242] B. Dhara and N. Argaç, *Generalized derivations acting on multilinear polynomials in prime rings and Banach algebras*, Comm. Math. Stat., **4**, 39–54 (2016).

[243] B. Dhara and V. De Filippis, *Notes on generalized derivations on Lie ideals in prime rings*, Bull. Korean Math. Soc., **46**, 599–605 (2009).

[244] B. Dhara and V. De Filippis, *Co-commutators with generalized derivations in prime and semiprime rings*, Publ. Math. Debrecen, **85(3–4)**, 339–360 (2014).

[245] B. Dhara and V. De Filippis, *Engel conditions of generalized derivations on left ideal and Lie ideals in prime rings*, Comm. Algebra, **48(1)**, 154–167 (2020).

[246] B. Dhara and R. K. Sharma, *Derivations with annihilator conditions in prime rings*, Publ. Math. Debrecen, **71**, 11–20 (2007).

[247] B. Dhara, N. Argaç and K. G. Pradhan, *Annihilator condition of a pair of derivations in prime and semiprime rings*, Indian J. Pure Appl. Math., **47**, 111–124 (2016).

[248] B. Dhara, S. Kar and S. Kuila, *A note on generalized derivations of order 2 and multilinear polynomials in prime rings*, Acta Math. Vietnamica, **47**, 755–773 (2022).

[249] B. Dhara, S. Kar and S. Mondal, *Generalized derivations on Lie ideals in prime rings*, Czech. Math. J., **65**, 179–190 (2015).

[250] B. Dhara, M. A. Raza and N. Rehman, *Commutator identity involving generalized derivations on multilinear polynomials*, Ann. Univ. Ferrara., **62(2)**, 205–216 (2016).

[251] Y. Ding and J. Li, *Lie n-higher derivations and Lie n-higher derivable mappings*, Bull. Aust. Math. Soc., **96**, 223–232 (2017).

[252] O. M. Di Vincenzo, *A result on derivations with algebraic valves*, Canad. Math. Bull., **29**, 432–437 (1986).

[253] O. M. Di Vincenzo, *On the n-th centralizers of a Lie ideal*, Boll. UMI, **7(3-A)**, 77–85 (1989).

[254] J. Dixmier and W. G. Lister, *Derivations of nilpotent Lie algebras*, Proc. Amer. Math. Soc., **8**, 155–157 (1957).

[255] Y. Q. Du and Y. Wang, *Lie derivations of generalized matrix algebras*, Linear Algebra Appl., **437**, 2719–2726 (2012).

[256] M. M. El-Sofy, *Rings with some kinds of Mappings*, Ph.D. Thesis, Cairo University, Branch of Fayoum, Cairo, Egypt (2000).

[257] J. S. Erickson, W. S. Martindale III and J. M. Osborn, *Prime nonassociative algebras*, Pacific J. Math., *60*, 49–63 (1975).

[258] C. Faith and Y. Utumi, *On a new proof of Litoff's theorem*, Acta Math. Acad. Sci. Hung., **14**, 369–371 (1963).

[259] A. K. Faraj, C. Haetinger and A. H. Majeed, *Generalized higher* $(\mathcal{U},\mathcal{R})$*-derivations*, JP J. Algebra Number Theory Appl., **16(2)**, 199–142 (2010).

[260] G. Favre, *Une algèbre de Lie Caractèristiquement nilpotente de dimension 7. CR.*, Acad. Sci. Paris, sér. A **274**, 1338–1339 (1972).

[261] B. Felzenszwalb and C. Lanski, *On the centralizers of ideals and nil derivations*, J. Algebra, **83**, 520–530 (1983).

[262] B. L. M. Ferreira and B. T. Costa, $*$*-Jordan-type maps on* C^**-algebras*, Comm. Algebra, **49(12)**, 5073–5082 (2021).

[263] B. L. M. Ferreira and B. T. Costa, $*$*-Lie-Jordan-type maps on* C^**-algebras*, Bull. Iran. Math. Soc., **48**, 1679–1690 (2022).

[264] R. N. Ferreira, B. L. M. Ferreira, B. T. Costa and A. V. D. Silva, *Reverse* $*$*-Jordan type maps on Jordan* $*$*-algebras*, J. Algebra Appl., https://doi.org/10.1142/S0219498824500154 (2024).

[265] R. N. Ferreira, B. L. M. Ferreira, H. G. Junior and B. T. Costa, $*$*-Lie-type maps on* $\mathcal{C}^*$*-algebras*, Comm. Algebra, **50(12)**, 5145–5154 (2022).

[266] B. L. M. Ferreira and F. Wei, *Mixed* $*$*-Jordan-type derivations on* $*$*-algebras*, J. Algebra Appl., **22(5)**, 2350100, https://doi.org/10.1142/S0219498823501001 (2023).

[267] M. Ferrero and C. Haetinger, *Higher derivations and a theorem by Herstein*, Quaest. Math., **25(2)**, 249–257 (2002).

[268] M. Ferrero and C. Haetinger, *Higher derivations of semiprime rings*, Comm. Algebra, **30**, 2321–2333 (2002).

[269] A. Fošner, N. Baydar and R. Strasek, *Remarks on certain identities with derivations on semiprime rings*, Ukrainian Math. J., **66**, 1609–1614 (2015).

[270] A. Fošner and W. Jing, *Lie centralizers on triangular rings and nest algebras.* Adv. Oper. Theory, **4(2)**, 342–350 (2019).

[271] A. Fošner and W. Jing, *A note on Jordan derivations of triangular rings*, Aequationes Math., **94**, 277–285 (2020).

[272] A. Fošner and N. Peršin, *A result concerning derivations in prime rings*, Glasnik Mat., **48(68)**, 67–79 (2013).

[273] A. Fošner and N. Rehman, *Identities with additive mappings in semiprime rings*, Bull. Korean Math. Soc., **51(1)**, 207–211 (2014).

[274] M. Fošner and J. Vukman, *Identities with generalized derivations in prime rings*, Mediterr. J. Math., **9(4)**, 847–863 (2012).

[275] A. Fošner and J. Vukman, *Some results concerning additive mappings and derivations on semiprime rings*, Publ. Math. Debrecen, **78(3-4)**, 575–581 (2011).

[276] A. Fošner, F. Wei and Z. K. Xiao, *Nonlinear Lie-type derivations of von Neumann algebras and related topics*, Colloq. Math., **132**, 53–71 (2013).

[277] A. Giambruno, J. Z. Goncalves and A. Madel, *Rings with algebraic n-Engel elements*, Comm. Algebra, **22(5)**, 1685–1701 (1994).

[278] A. Giambruno and I. N. Herstein, *Derivations with nilpotent values*, Rend. Circ. Mat. Palermo, **30(2)**, 199–206 (1981).

[279] A. Giambruno, P. Misso and C. P. Milies, *Derivations with invertible values in rings with involution*, Pac. J. Math., **123(1)**, 47–54 (1986).

[280] E. Guven, *On* (σ,τ) *derivations in prime rings*, Int. J. Contemp. Math. Sciences, **3(26)**, 1289–1293 (2008).

[281] C. Haetinger, *Derivações em anéis primos e semiprimos*, M.Phil. Dissertation, MUFRGS, Porto Alegre, Brazil (1994).

[282] C. Haetinger, *Derivações de ordem superior em anéis primos e semiprimos*, Ph.D. Thesis, IMUFRGS, Porto Alegre, Brazil (2000).

[283] C. Haetinger, *Higher derivations of Lie ideals*, Trends in Comp. and Appl. Math., **3(1)**, 141–145 (2002).

[284] D. Han, *Lie type higher derivations on operator algebras*, Bull. Iranian Math. Soc., **40(5)**, 1169–1194 (2014).

[285] D. Han and F. Wei, *Jordan* (α,β)*-derivations on triangular algebras and related mappings*, Linear Algebra Appl., **434**, 259–284 (2011).

[286] D. Han and F. Wei, *Multiplicative Lie higher derivations of unital algebras with idempotents*, Oper. Matrices, **10**, 345–377 (2016).

[287] H. Hasse and F. K. Schmidt, *Noch eine Begründung der theorie der höheren Differrential quotienten in einem algebraischen Funktionenkörper einer Unbestimmten*, J. Reine U. Angew. Math., 215–237 (1937).

[288] I. N. Herstein, *Jordan homomorphisms*, Trans. Amer. Math. Soc., **81**, 331–351 (1956).

[289] I. N. Herstein, *Jordan derivations of prime rings*, Proc. Amer. Math. Soc., **8**, 1104–1110 (1957).

[290] I. N. Herstein, *Lie and Jordan structures in simple, associative rings*, Bull. Amer. Math. Soc., **67**, 517–531 (1961).

[291] I. N. Herstein, *Sui Commutatori Degli Anelli Semplici*, Seminario Mat. e. Fis. di Milano, **33**, 80–86 (1963).

[292] I. N. Herstein, *On a type of Jordan mappings*, An. Acad. Bras. Cienc., **39**, 357–360 (1967).

[293] I. N. Herstein, *Noncommutative rings*, Carus Mathematics Monograph, **15**, 1968.

[294] I. N. Herstein, *Topics in Ring Theory*, Univ. Chicago Press, Chicago (1969).

[295] I. N. Herstein, *On the Lie structure of an associative ring*, J. Algebra, **14**, 561–571 (1970).

[296] I. N. Herstein, *Rings with Involution*, Univ. Chicago Press, Chicago (1976).

[297] I. N. Herstein, *A note on derivations*, Canad. Math. Bull., **21**, 369–370 (1978).

[298] I. N. Herstein, *Center-like elements in prime rings*, J. Algebra, **60**, 567–574 (1979).

[299] I. N. Herstein, *Derivations of prime rings having power central values*, Contemporary Mathematics, **13** AMS, Rhode Island (1982).

[300] Y. Hirano, H. Tominaga and A. Trezepizur, *On a theorem of Posner*, Math. J. Okayama Univ., **27**, 25–32 (1985).

[301] M. Hongan, *A note on semiprime rings with derivation*, Int. J. Math. Math. Sci., **20(2)**, 413–415 (1997).

[302] M. Hongan, *On a theorem of J. Vukman*, Aequationes Math., **52**, 112–115 (1996).

[303] A. Hosseini, *Some conditions under which left derivations are zero*, Filomat, **31**, 3965–3974 (2017).

[304] A. Hosseini and A. Fos̆ner, *The image of Jordan left derivations on algebras*, Bol. Soc. Parana. Mat., **38**, 53–61 (2019).

[305] R. A. Howland, *Lie isomorphisms of derived rings of simple rings*, Trans. Amer. Math. Soc., **145**, 383–396 (1969).

[306] L. Hua, *A theorem on matrices over a field and its application*, J. Chinese Math. Soc. (N.S.), **1**, 110–163 (1951).

[307] S. Huang, *Derivations with Engel conditions in prime and semiprime rings*, Czech. Math. J., **61(4)**, 1135–1140 (2011).

[308] S. Huang, *b-generalized derivations on prime rings*, J. Math. Res. Appl., **41**, 349–354 (2021).

[309] S. Huang and B. Davvaz, *Generalized derivations of rings and Banach algebras*, Comm. Algebra, **41(3)**, 1188–1194 (2013).

[310] L. K. Hua, *On the automorphisms of a field*, Proc. Natl. Acad. Sci. USA, **35**, 386–389 (1949).

[311] D. H. Huo, B. D. Zheng and H. Y. Liu, *Nonlinear maps preserving Jordan triple τ-$*$-products*, J. Math. Anal. Appl., **430**, 830–844 (2015).

[312] D. H. Huo, B. D. Zheng, J. L. Xu and H. Y. Liu, *Nonlinear mappings preserving Jordan multiple $*$-product on factor von Neumann algebras*, Linear Multilinear Algebra, **63**, 1026–1036 (2015).

[313] B. Hvala, *Generalized derivations in rings*, Comm. Algebra, **26**, 1147–1166 (1998).

[314] M. A. Idrissi and L. Oukhtite *Structure of a quotient ring R/P with generalized derivations acting on the prime ideal P and some applications*, *Indian J. Pure Appl. Math.*, **53**, 792–800 (2022).

[315] I. Martin, Isaacs, Algebra, a Graduate Course (1st ed.) (1993) Brooks/Cole.

[316] A. Jabeen, *Lie (Jordan) centralizers on generalized matrix algebras*, Comm. Algebra, **49(1)**, 278–291 (2021).

[317] D. R. Jacobs, *Lie derivations on skew elements of simple rings with involution*, Ph.D. Dissertation, University of Massachusetts (1973).

[318] N. Jacobson, *A note on automorphisms and derivations of Lie algebras*, Proc. Amer. Math. Soc., p. 6 (1955).

[319] N. Jacobson, *Structure of rings*, Amer. Math. Soc. Coll. Pub., **37**, (1956).

[320] N. Jacobson, *PI-algebras. An Introduction, Lecture Notes in Mathematics*, **441**, Springer-Verlag (1975).

[321] N. Jacobson and C. Rickart, *Jordan homomorphisms of rings*, Trans. Amer. Math. Soc., **69**, 479–502 (1950).

[322] A. Jaeger, *Eine algebraische Theorie vertauschbarer Differentiationen für Körper beliebiger Charakteristik"*, J. Reine Angew. Math. (Journal für Mathematik), **190**, 1–21 (1952).

[323] D. W. Jensen, *Nilpotency of derivations in prime rings*, Proc. Amer. Math. Soc., **123(9)**, 2633–2636 (1995).

[324] P. Ji, R. Liu and Y. Zhao, *Nonlinear Lie triple derivations of triangular algebras*, Linear Multilinear Algebra, **60(10)**, 1155–1164 (2012).

[325] W. Jing, *Nonlinear $*$-Lie derivations of standard operator algebras*, Quaest. Math., **38(8)**, 1037–1046 (2016).

[326] W. Jing and S. Lu, *Generalized Jordan derivations on prime rings and standard operator algebras*, Taiwanese J. Math., **7(4)**, 605–613 (2003).

[327] W. Jing and F. Lu, *Additivity of Jordan (Triple) derivations on rings*, Comm. Algebra, **40**, 2700–2719 (2012).

[328] W. Jing and F. Lu, *Lie derivable mappings on prime rings*, Linear Multilinear Algebra, **60**, 167–180 (2012).

[329] W. Jing, S. Lu and P. Li, *Characterisations of derivations on some operator algebras*, Bull. Austral. Math. Soc., **66**, 227–232 (2002).

[330] P. Ji and W. Qi, *Characterizations of Lie derivations of triangular algebras*, Linear Algebra Appl., **435**, 1137–1146 (2011).

[331] P. Ji, W. Qi and X. Sun, *Characterizations of Lie derivations of factor von Neumann algebras*, Linear Multilinear Algebra, **61(3)**, 417–428 (2013).

[332] P. S. Ji and L. Wang, *Lie triple derivations of TUHF algebras*, Linear Algebra Appl., **403**, 399–408 (2005).

[333] B. E. Johnson, *Symmetric amenability and the nonexistence of Lie and Jordan derivations*, Math. Proc. Cambridge Philos. Soc., **120**, 455–473 (1996).

[334] S. Jondrup, *Automorphisms and derivations of upper triangular matrix rings*, Linear Algebra Appl., **221**, 205–218 (1995).

[335] C. R. Jordan and D. A. Jordan, *Lie rings of derivations of associative rings*, London Math. Soc., **2**, 33–41 (1978).

[336] Y. S. Jung, *On left derivations and derivations of Banach algebras*, Bull. Korean Math. Soc., **35**, 659–667 (1998).

[337] Y. S. Jung, *Generalized Jordan triple higher derivations on prime rings*, Indian J. Pure Appl. Math., **36(9)**, 513–524 (2005).

[338] I. Kaplansky, *Semi-automorphisms of rings*, Duke Math. J., **14**, 521–527 (1947).

[339] N. Kawamoto, *On prime ideals of Lie algebras*, Hiroshima Math. J., **4**, 679–684 (1974).

[340] K. Kaya, *On a (σ,τ)-derivations of prime rings*, Doga Tr. J. Math. D. C., **12(2)**, 46–51 (1988).

[341] K. Kaya, E. Güven and M. Soyturk, *On (σ, τ)-derivations of prime rings*, J. Korean Soc. Math. Educ. Ser. B Pure Appl. Math., 13 (2006).

[342] I. B. Kaygorodov and Y. S. Popov, *Alternative algebras admitting derivations with invertible values and invertible derivations*, Izv. Math., **78(5)**, 922–936 (2014).

[343] M. S. Khan, S. Ali and M. Ayedh, *Herstein's theorem for prime ideals in rings with involution involving pair of derivations*, Comm. Algebra, **50(6)**, 2592–2603 (2022).

[344] M. S. Khan and M. A. Khan, *Derivations in prime rings and Banach algebra*, Int. J. Algebra, **4(27)**, 1317–1328 (2010).

[345] V. K. Kharchenko, *Differential identities of prime rings*, Algebra and Logic, **17(2)**, 155–68 (1978).

[346] V. K. Kharchenko, *Differential identites of semiprime rings*, Algebra and Logic, **18(1)**, 58–80 (1979).

[347] R. Khazal, S. Dăscălescu and L. V. Wyk, *Isomorphism of generalized triangular matrix rings and recovery of tiles*, Int. J. Math. Math. Sci., **9**, 533–538 (2003).

[348] E. Killiam, *Lie derivations on skew elements in prime rings with involution*, Canad. Math. Bull., **30**, 344–350 (1987).

[349] G. H. Kim, *A result concerning derivations in noncommutative Banach algebras*, Sci. Math. Jpn., **53** (2001).

[350] K. H. Kim, *On symmetric bi-derivations and commutativity of prime rings*, Gulf J. Math., **7(2)**, 22–30 (2019).

[351] M. A. Knus, A. Merkurjev, M. Rost and J. P. Tignol, *The Book of Involutions*, Colloquium Publications, **44** (1998).

[352] E. Koc Sögütcü, *A characterization of semiprime rings with homoderivations*, Journal of New Theory, **42**, 14–28 (2023).

[353] L. Kong and J. Zhang, *Nonlinear bi-skew Lie derivation on factor von Neumann algebras*, Bull. Iranian Math. Soc., **47(4)**, 1097–1106 (2021).

[354] M. T. Kosan and T. K. Lee, *b-Generalized derivations of semiprime rings having nilpotent values*, J. Aust. Math. Soc., **96**, 326–337 (2014).

[355] M. T. Kosan, T. K. Lee and Y. Zhou, *Identities with Engel conditions on derivations*, Monatsh. Math., **165**, 543–556 (2012).

[356] A. Kovacs, *Nilpotent derivations*, Technion Preprint Series, Mt-453 November (1979).

[357] J. Krempa and J. Matczuk, *On the composition of derivations*, Rend. Circ. Mat. Palermo, **33**, 441–455 (1984).

[358] S. Kurepa, *The Cauchy functional equation and scalar product in vector spaces*, Glas. Mat. Fiz. Astr., **19**, 23–64 (1964).

[359] C. Lanski, *Rings with involution whose symmetric elements are regular*, Proc. Amer. Math. Soc., **33/2**, 264–270 (1972).

[360] C. Lanski, *Differential identities, Lie ideals and Posners theorems*, Pacific J. Math., **134(2)**, 275–297 (1988).

[361] C. Lanski, *Differential identities of prime rings, Kharchenko's theorem and applications*, Contemporary Mathematics, **124**, 118–128 (1992).

[362] C. Lanski, *Derivations nilpotent on subsets of prime rings*, Comm. Algebra, **20(5)**, 1427–1446 (1992).

[363] C. Lanski, *Quadratic central differential identities of prime rings*, Nova J. Algebra Geom., **1(2)**, 185–206 (1992).

[364] C. Lanski, *An Engel condition with derivation*, Proc. Amer. Math. Soc., **118(3)**, 731–734 (1993).

[365] C. Lanski, *Derivations with nilpotent values on left ideals*, Comm. Algebra, **22(4)**, 1305–1320 (1994).

[366] C. Lanski, *An Engel condition with derivation for left ideals*, Proc. Amer. Math. Soc., **125(2)**, 339–345 (1997).

[367] C. Lanski, *Lie ideals and derivations in rings with involution*, Pacific. J. Math, **69(2)** (1997).

[368] C. Lanski, *Left ideals and derivations in semiprime rings*, J. Algebra, **277(2)**, 658–667 (2004).

[369] C. Lanski, *Generalized derivations and nth power maps in rings*, Comm. Algebra, **35**, 3660–3672 (2007).

[370] C. Lanski, *Skew derivations and Engel conditions*, Comm. Algebra, **42(1)**, 139–152 (2014).

[371] C. Lanski and S. Montgomery, *Lie structure of prime rings of characteristic* 2, Pac. J. Math., **42(1)**, 117–136 (1972).

[372] T. K. Lee, *Semiprime rings with differential identities*, Bull. Inst. Math. Acad. Sinica, **20(1)**, 27–38 (1992).

[373] T. K. Lee, *Derivations with invertible values on a multilinear polynomial*, Proc. Amer. Math. Soc., **119**, 1077–1083 (1993).

[374] T. K. Lee, *Semiprime rings with hypercentral derivations*, Canada. Math. Bull., **38**, 445–449 (1995).

[375] T. K. Lee, *Left annihilators characterized by GPIs*, Trans. Amer. Math. Soc., **347**, 3159–3165 (1995).

[376] T. K. Lee, *Derivations and centralizing mappings in prime rings*, Taiwanese J. Math., **1(3)**, 333–342 (1997).

[377] T. K. Lee, *Deivations with Engle conditions on polynomials*, Algebra Colloq., **5(1)**, 13–24 (1998).

[378] T. K. Lee, *Generalized derivation of left faithful rings*, Comm. Algebra, **27(8)**, 4057–4073 (1999).

[379] T. K. Lee, *Generalized skew derivations characterized by acting on zero product*, Pacif. J. Appl. Math. **216(2)**, 293–301 (2004).

[380] T. K. Lee and T. C. Lee, *Derivations with algebraic valves of bounded degree*, Comm. Algebra, **27(2)**, 1911–1920 (1999).

[381] T. K. Lee and C. K. Liu, *Spectrally bounded ϕ-derivations on Banach algebras*, Proc. Amer. Math. Soc., **133(5)**, 1427–1435 (2005).

[382] T. K. Lee and K. S. Liu, *Generalized skew derivations with algebraic valves of bounded degree*, Houst. J. Math., **39(3)**, 733–740 (2013).

[383] P. H. Lee and T. K. Lee, *Derivations with Engel conditions on multilinear polynomials*, Proc. Amer. Math. Soc., **124**, 2625–2629 (1996).

[384] P. H. Lee and T. K. Lee, *Lie ideals of prime rings with derivations*, Bull. Inst. Math. Acad. Sinica, **11**, 75–80 (1983).

[385] P. H. Lee and T. K. Lee, *Note on nilpotent derivations*, Proc. Amer. Math. Soc., **98(1)** (1996).

[386] T. K. Lee and J. S. Lin, *A result on derivations*, Proc. Amer. Math. Soc., **124**, 1687–1691 (1996).

[387] T. K. Lee and J. H. Lin, *Jordan derivations of prime rings with characteristic two*, **462**, 1–15 (2014).

[388] U. Leerawat and S. Khun-in, *On trace of symmetric bi-derivations on rings*, Int. J. Math. Comput. Sci., **16**, 743–752 (2021).

[389] T. K. Lee and W. K. Shiue, *A result on derivations with Engel condition in prime rings*, Southeast Asian Bull. Math., **23**, 437–446 (1999).

[390] T. K. Lee and W. K. Shiue, *Derivations cocentralizing polynomials*, Taiwanese J. Math., **2(4)**, 457–467 (1998).

[391] T. K. Lee and W. K. Shiue, *Identities with generalized derivations*, Comm. Algebra, **29(10)**, 4437–4450 (2001).

[392] P. H. Lee and T. L. Wong, *Derivations cocentralizing Lie ideals*, Bull. Ins. Math. Acad. Sinica, **23**, 1–5 (1995).

[393] P. H. Lee and T. L. Wong, *Derivations with invertible or nilpotent values on multilinear polynomials*, Alg. Coll., **7(1)**, 93–98 (2000).

[394] T. K. Lee and Y. Zhou, *An identity with generalized derivations*, J. Algebra & Appl., **8(3)**, 307–317 (2009).

[395] G. F. Leger and S. Tôgô, *Characteristically nilpotent Lie algebras*, Duke Math. J., **26**, 623–628 (1959).

[396] U. Leron, *Nil and power central polynomials in rings*, Trans. Amer. Math. Soc., **202**, 97–103 (1975).

[397] A. Leroy and J. Matczuk, *Derivations et automorphismes algebriques d'anneux premiers*, Comm. Algebra, **13(6)**, 1245–1266 (1985).

[398] P. K. Liau and C. K. Liu, *An Engel condition with b-generalized derivations for Lie ideals*, J. Algebra & Appl., **17(3)**, 1850046 (2018).

[399] C. Li, X. Fang, F. Lu and T. Wang, *Lie triple derivable mappings on rings*, Comm. Algebra, **42**, 2510–2527 (2014).

[400] P. Li and W. Jing, *Jordan elementary maps on rings*, Linear Algebra Appl., **382**, 237–245 (2004).

[401] C. Li, F. Lu and X. C. Fang, *Non-linear ξ-Jordan $*$-derivations on von Neumann algebras*, Linear Multilinear Algebra, **62(4)**, 466–473 (2014).

[402] C. J. Li, F. Y. Lu and X. C. Fang, *Nonlinear mappings preserving product $XY+YX^*$ on factor von Neumann algebras*, Linear Algebra Appl., **438**, 2339–2345 (2013).

[403] C. J. Li and F. Y. Lu, *Nonlinear maps preserving the Jordan triple 1-$*$-product on von Neumann algebras*, Complex Anal. Oper. Theory, **11**, 109–117 (2017).

[404] C. J. Li, F. Y. Lu and T. Wang, *Nonlinear maps preserving the Jordan triple $*$-product on von Neumann algebras*, Ann. Funct. Anal., **7**, 496–507 (2016).

[405] J. S. Lin, *Derivations with invertible values in semiprime rings with involution*, Chinese Journal of Mathematics, **18(2)**, 175–184 (1990).

[406] W. H. Lin, *Characterizations of Lie higher derivations on triangular algebras*, Indian J. Pure Appl. Math., **51(1)**, 77–104 (2020).

[407] W. H. Lin, *Nonlinear generalized Lie n-derivations on triangular algebras*, Comm. Algebra, **46(6)**, 2368–2383 (2018).

[408] W. H. Lin, *Nonlinear $*$-Lie-type derivations on standard operator algebras*, Acta Math. Hungar., **154(2)**, 480–500 (2018).

[409] W. H. Lin, *Nonlinear $*$-Lie-type derivations on von Neumann algebras*, Acta Math. Hungar., **156**, 112–131 (2018).

[410] W. H. Lin, *Nonlinear $*$-Jordan-type derivations on von Neumann algebras*, arXiv: 1805.16027v1 [math.OA] (2024).

[411] W. H. Lin, *Nonlinear mappings preserving Jordan-type η-$*$-products on von Neumann algebras*, arXiv: 1805.96815v1 [math.OA] (2024).

[412] J. S. Lin and C. K. Liu, *Generalized derivations with invertible or nilpotent values on multilinear polynomials*, Comm. Algebra, **34**, 633–640 (2006).

[413] J. S. Lin and C. K. Liu, *Generalized derivations with nilpotent values on multilinear polynomials*, Taiwanese J. Math., **10(5)**, 1183–1192 (2006).

[414] J. S. Lin and C. K. Liu, *Strong commutativity preserving maps on Lie ideals*, Linear Algebra Appl., **428**, 1601–1609 (2008).

[415] J. S. Lin and C. K. Liu, *Strong commutativity preserving maps in prime rings with involution*, Linear Algebra Appl., **432(1)**, 14–23 (2010).

[416] J. Li and Q. Shen, *Characterization of Lie higher and Lie triple derivation on triangular algebra*, J. Korean Math. Soc., **49(2)**, 419–433 (2012).

[417] C. K. Liu, *Derivations cocentralizing multilinear polynomials on left ideals*, Monatsh. Math., **162**, 297–311 (2011).

[418] C. K. Liu, *Strong commutativity preserving generalized derivations on right ideals*, Monatsh. Math., **166**, 453–465 (2012).

[419] L. Liu, *Lie triple derivations on factor von Neumann algebras*, Bull. Korean Math. Soc., **52**, 581–591 (2015).

[420] C. K. Liu, *An Engel condition with skew derivations for left ideals*, Monatsh. Math., **180(4)**, 833–852 (2016).

[421] C. K. Liu, *An Engel condition with b-generalized derivations*, Linear Multilinear Algebra, **65(2)**, 300–312 (2017).

[422] C. K. Liu, *Additive n-commuting maps on semiprime rings*, Proc. Edinb. Math. Soc., **63(1)**, 193–216 (2020).

[423] C. K. Liu, *An Engel condition with two generalized derivations in prime rings*, Comm. Algebra, **49(2)**, 836–849 (2021).

[424] C. K. Liu, H. Y. Chen and P. K. Liau, *Generalized skew derivations with nilpotent values on left ideals of rings and Banach algebras*, Linear Multilinear Algebra, **62**, 453–465 (2014).

[425] C. K. Liu and P. H. Liau, *Strong commutativity preserving generalized derivations on Lie ideals*, Linear Multilinear Algebra, **59(8)**, 905–915 (2011).

[426] C. K. Liu and W. K. Shiue, *Generalized Jordan triple (θ,ϕ)-derivations on semiprime rings*, Taiwanese J. Math., **11**, 1397–1406 (2007).

[427] C. Li, F. Zhao and Q. Chen, *Nonlinear maps preserving the product X^*Y+Y^*X on von Neumann algebras*, Bull. Iranian Math. Soc., **44**, 729–738 (2018).

[428] C. J. Li, F. F. Zhao and Q. Y. Chen, *Nonlinear skew Lie triple derivations between factors*, Acta Math. Sinica (English Series), **32(7)**, 821–830 (2016).

[429] F. Lu, *Additivity of Jordan Maps on standard operator algebras*, Linear Algebra Appl., **357**, 123–131 (2002).

[430] F. Lu, *Jordan Triple Maps*, Linear Algebra Appl., **357**, 311–317 (2003).

[431] F. Lu, *Jordan Derivable Maps of prime rings*, Comm. Algebra, **38**, 4430–4440 (2010).

[432] J. Luh, *A note on commuting automorphism of prime rings*, Amer. Math. Monthly, **77**, 61–62 (1970).

[433] F. Lu and W. Jing, *Characterizations of Lie derivations of $\mathcal{B}(\mathcal{X})$*, Linear Algebra Appl., **432**, 89–99 (2010).

[434] F. Lu and B. Liu, *Lie Derivable Maps on $B(X)$*, J. Math. Anal. Appl., **372**, 369–376 (2010).

[435] L. Magnin, *Adjoint and trivial cohomology tables for indecomposable Lie algebras of dimension ≤ 7 over $\mathbb{C}$*, Lecture Notes (1995).

[436] G. Maksa, *A remark on symmetric biadditive functions having nonnegative diagonalization*, Glasnik Mat., **15(35)**, 279–282 (1980).

[437] G. Maksa, *On the trace of symmetric bi-derivations*, C.R. Math. Rep. Acad. Sci. Canada., **9**, 303–307 (1987).

[438] A. Mamouni and L. Oukhtite, *Derivations satisfying certain algebraic identities on Jordan ideals*, Arab. J. Math., **1**, 341–346 (2012).

[439] A. Mamouni, L. Oukhtite and M. Samman, *Commutativity theorems for $*$-prime rings with differential identities on Jordan ideals*, ISRN Algebra, Art. ID 729356, 11 pp (2012).

[440] A. Mamouni and L. Oukhtite, *Generalized derivations centralizing on Jordan ideals of rings with involution*, Turkish J. Math., **38(2)**, 225–232 (2014).

[441] A. Mamouni, L. Oukhtite and M. Zerra, *Certain algebraic identities on prime rings with involution*, Comm. Algebra, **49/7**, 2976–2986 (2021).

[442] W. S. Martindale III, *Lie isomorphisms of primitive rings*, Proc. Amer. Math. Soc., **14**, 909–916 (1963).

[443] W. S. Martindale III, *Lie derivations of primitive rings*, Michigan Math. J., **11**, 183–187 (1964).

[444] W. S. Martindale III, *When are multiplicative mappings additive?*, Proc. Amer. Math. Soc., **21**, 695–698 (1969).

[445] W. S. Martindale III, *Lie isomorphisms of prime rings*, Trans. Amer. Math. Soc., **142**, 437–55 (1969).

[446] W. S. Martindale III, *Lie isomorphisms of simple rings*, J. Lond. Math. Soc., **44**, 213–221 (1969).

[447] W. S. Martindale III, *Prime rings satisfying a generalized polynomial identity*, J. Algebra, **12**, 576–584 (1969).

[448] W. S. Martindale III, *Lie isomorphisms of the skew elements of a simple ring with involution*, J. Algebra, **36**, 408–415 (1975).

[449] W. S. Martindale III, *Lie isomorphisms of the skew elements of a prime ring with involution*, Comm. Algebra, **4**, 927–977 (1976).

[450] W. S. Martindale III, *Lie and Jordan mappings*, Contemporary Mathematics, **13**, 173–177 (1982).

[451] W. S. Martindale III and C. R. Miers, *On the iterates of derivations of prime rings*, Pacific J. Math., **104**, 179–190 (1983).

[452] M. Mathieu, *Posner's second theorem deduced from the first*, Proc. Amer. Math. Soc., **114**, 601–602 (1992).

[453] M. Mathieu and A. R. Villena, *The structure of Lie derivations on C^*-algebras*, J. Funct. Anal., **202**, 504–525 (2003).

[454] J. Ma, X. W. Xu and F. W. Niu, *Strong commutativity-preserving generalized derivations on semiprime rings*, Acta Math. Sinica, **24(11)**, 1835–1842 (2008).

[455] J. H. Mayne, *Centralizing automorphisms of prime rings*, Canad. Math. Bull., **19**, 113–115 (1976).

[456] J. H. Mayne, *Ideals and centralizing mappings in prime rings*, Proc. Amer. Math. Soc., **86**, 211–212 (1982).

[457] J. H. Mayne, *Centralizing mappings of prime rings*, Canad. Math. Bull., **27**, 122–126 (1984).

[458] A. Melaibari, N. Muthana and A. Al-Kenani, *Homoderivations on Rings*, General Mathematics Notes, **35(1)**, 1–8 (2016).

[459] C. R. Miers, *Lie homomorphisms of operator algebras*, Pacific J. Math., **38**, 717–735 (1971).

[460] C. R. Miers, *Lie derivations of von Neumann algebras*, Duke Math. J., **40**, 403–409 (1973).

[461] C. R. Miers, *Lie triple derivations of von Neumann algebras*, Proc. Amer. Math. Soc., **71**, 57–61 (1978).

[462] L. Molnár, *A condition for a subspace of $\mathcal{B}(H)$ to be an ideal*, Linear Algebra Appl., **235**, 229–234 (1996).

[463] L. Molnár, *Jordan $*$-derivation pairs on a complex $*$-algebra*, Aequationes Math., **54**, 44–55 (1997).

[464] L. Molnár, *Jordan maps on standard operator algebras*, In Z. Daróczy and Zs. Páles (Ed.), Functional Equations–Results and Advances, Kluwer Academic Publishers, 305–320 (2001).

[465] L. Molnár and P. Šemrl, *Local Jordan $*$-derivations of standard operator algebras*, Proc. Amer. Math. Soc., **125**, 447–454 (1997).

[466] D. Müller, *Isometries of bi-invariant pseudo-Riemannian metrics on Lie groups*, Geom Dedicata, **29**, 65–96 (1989).

[467] Y. Nakai, *On locally finite iterative higher derivations*, Osaka J. Math., **15(3)**, 655–662 (1978).

[468] A. Nakajima, *Categorical properties of generalized derivations*, Sci. Math., **2**, 345–352 (1999).

[469] A. Nakajima, *On generalized higher derivations*, Turkish J. Math., **24**, 295–311 (2000).

[470] B. Nejjar, A. Kacha, A. Mamouni and L. Oukhtite, *Commutativity theorems in rings with involution*, Comm. Algebra, **45**, 698–708 (2017).

[471] L. Oukhtite, *On Jordan ideals and derivations in rings with involution*, Comment. Math. Univ. Carolin., **51**, 389–395 (2010).

[472] L. Oukhtite, *Posner's second theorem for Jordan ideals in rings with involution*, Expo. Math., **29**, 415–419 (2011).

[473] L. Oukhtite and S. Salhi, *On commutativity of σ-prime rings*. Glas. Mat. Ser. III, **41**, 57–64 (2006).

[474] L. Oukhtite and S. Salhi, *On derivations in σ-prime rings*, Int. J. Algebra, **1(5)**, 241–246 (2007).

[475] L. Oukhtite and S. Salhi, *Lie ideals and derivations of σ-prime rings*, Int. J. Algebra, **1(1)**, 25–30 (2007).

[476] L. Oukhtite and S. Salhi, *σ-Lie ideals with derivations as homomorphisms and anti-homomorphisms*, Int. J. Algebra, **1(5)**, 235–239 (2007).

[477] L. Oukhtite and S. Salhi, *On derivations in σ-prime rings*, Int. J. Algebra, **1/5**, 241–246 (2007).

[478] L. Oukhtite, S. Salhi and L. Taoufiq, *Commutativity conditions on derivations and Lie ideals in σ-prime rings*, Beiträge Algebra Geom., **51/1**, 275–282 (2010).

[479] L. Oukhtite, M. Mamouni and M. Ashraf, *Commutativity theorems for rings with differential identities on Jordan ideals*, Comment. Math. Univ. Carolin., **54/4**, 447–457 (2013).

[480] S. K. Ou, D. Y. Wang and R. P. Yao, *Derivations of the Lie algebra of strictly upper triangular matrices over a commutative ring*, Linear Algebra Appl., **424**, 378–383 (2007).

[481] M. A. Öztürk, *Permuting tri-derivations in prime and semi-prime rings*, East Asian Math. J., **15**, 177–190 (1999).

[482] K. H. Park, *On prime and semiprime rings with symmetric n-derivations*, J. Chungcheong Math. Soc., **22**, 451–458 (2009).

[483] T. Pehlivan and E. Albaş, *b-Generalized derivations on prime rings*, Ukr. Math. J., **74**, 953–966 (2022).

[484] B. Peirce, *Linear Associative Algebras*, Van Nostrand, ISBN 978-0-548-94787-6 (1882).

[485] T. C. Pierin, R. N. Ferreira, F. Borges and B. L. M. Ferreira, *Nonlinear mixed $*$-Jordan type derivations on alternative $*$-algebras*, J. Algebra Appl., 2550028 (19 pages) https://doi.org/10.1142/S0219498825500288 (2025).

[486] C. Polcino Milies, *Derivations of full matrix rings*, Atas de XI Eschola de Algebra, IME-USP, S. Paolo, 92–103 (1990).

[487] E. C. Posner, *Derivations in prime rings*, Proc. Amer. Math. Soc. **8**, 1093–1100 (1957).

[488] B. Prajapati and C. Gupta, *On b-generalized skew derivations having centralizing commutator in prime rings*, Comm. Algebra, **50(7)**, 2997–3006 (2022).

[489] C. Procesi, *Rings with Polynomial Identities*, Marcel Dekker (1973).

[490] X. F. Qi, *Characterization of Lie higher derivations on triangular algebras*, Acta Math. Sinica, English Series, **29(5)**, 1007–1018 (2013).

[491] X. F. Qi, *Characterization of Lie n-derivations for reflexive algebras*, Linear Multilinear Algebra, **63**, 1693–1706 (2015).

[492] X. F. Qi, Z. Guo and T. Zhang, *Characterizing Jordan n-derivations of unital rings containing idempotents*, Bull. Iranian Math. Soc., **46**, 1639–1658 (2020).

[493] X. Qi and J. Hou, *Additive Lie (ξ-Lie) derivations and generalized Lie (ξ-Lie) derivations on nest algebras*, Lin. Alg. Appl., **431**, 843–854 (2009).

[494] X. F. Qi and J. Hou, *Lie higher derivations on nest algebras*, Commun. Math. Res., **26**, 131–143 (2010).

[495] X. F. Qi and J. C. Hou, *Characterization of Lie derivations on prime rings*, Comm. Algebra, **39**, 3824–3835 (2011).

[496] X. F. Qi and J. C. Hou, *Characterization of Lie derivations on von Neumann algebras*, Linear Algebra Appl., **438**, 533–548 (2013).

[497] M. A. Quadri, M. S. Khan and N. Rehman, *Generalized deriovations and commutativity of prime rings*, Indian J. Pure & Appl. Math., **34(9)**, 1393–1396 (1998).

[498] H. Radjavi, *Commutativity-preserving operators on symmetric matrices*, Linear Algebra Appl., *461*, 1601–1609 (1984).

[499] F. Rania and G. Scudo, *A quadratic differential identity with generalized derivations on multilinear polynomials in prime rings*, Mediterr. J. Math., **11**, 273–285 (2014).

[500] M. A. Raza, N. Rehman and T. Bano, *A note on an Engel condition with generalized derivations in rings*, J. Math. Ext., **10(2)**, 75–86 (2016).

[501] N. Rehman, *On commutativity of prime rings with generalized derivations*, Math. J. Okayama Univ., **49**, 43–49 (2002).

[502] N. Rehman and M. A. Raza, *On ideals with skew derivations of prime rings*, Miskolc Math. Notes, **15(2)**, 717–724 (2014).

[503] P. Ribenboim, *Higher derivations of rings I*, Revue Roumaine de Math. Pures et Appl., **16**, 77–110 (1971).

[504] P. Ribenboim, *Higher derivations of rings II*, Revue Roumaine de Math. Pures et Appl., **16**, 245–272 (1971).

[505] P. Ribenboim, *Higher order derivations of modules*, Portugaliae Mathematica, **39(1-4)**, 381–397 (1980).

[506] M. P. Rosen, *Lie isomorphisms of a certain class of prime rings*, J. Algebra, **89**, 291–317 (1984).

[507] L. Rowen, *Polynomial identities in ring theory*, Pure and Applied Mathematics (1980).

[508] S. M. Salih and C. Haetinger, *Jordan triple higher derivations on prime rings*, Mathematical Thoery and Modeling, **5(10)** (2015).

[509] A. D. Sands, *Radicals and Morita contexts*, J. Algebra, **24**, 335–345 (1973).

[510] G. S. Sandhu, *On Lie ideals and generalized skew derivation in prime rings*, The Aligarh Bull. Math., **39**, 89–102 (2020).

[511] M. Sapanci, M. A. Ozturk and Y. B. Jun, *Symmetric bi-derivations on prime rings*, East Asian Math. J., **15(1)**, 105–109 (1999).

[512] P. Šemrl, *On Jordan $*$-derivations and an application*, Colloq. Math., **59**, 241–251 (1990).

[513] P. Šemrl, *Quadratic functionals and Jordan $*$-derivations*, Studia Math., **97**, 157–165 (1991).

[514] P. Šemrl, *Quadratic and quasi-quadratic functionals*, Proc. Amer. Math. Soc., **119**, 1105–1113 (1993).

[515] P. Šemrl, *Jordan $*$-derivations of standard operator algebras*, Proc. Amer. Math. Soc., **120**, 515–518 (1994).

[516] P. Šemrl, *Commutativity preserving maps*, Linear Algebra Appl., **429**, 1051–1070 (2008).

[517] R. K. Sharma, B. Dhara, V. De Filippis and C. Garg, *A result concerning nilpotent values with generalized skew derivations on Lie ideals*, Comm. Algebra, **46**, 5330–5341 (2018).

[518] W. K. Shiue, *Annihilators of derivations with Engel conditions*, Rend. Del Circ. Math. Di Palermo, Serie II, **52**, 505–509 (2003).

[519] F. Shutaj and S. Khan, *Left annihilator of generalized derivations on Lie ideals in prime rings*, Rend. Circ. Mat. Palermo, **64**, 77–81 (2015).

[520] A. M. Sinclair, *Jordan homomorphisms and derivations on semisimple Banach algebras*, Proc. Amer. Math. Soc., **24**, 209–214 (1970).

[521] I. M. Singer and J. Wermer, *Derivations on commutative normed algebras*, Math. Ann., **129**(1), 260–264 (1955).

[522] B. Singh, *Higher derivations, normal flatness and analytic products*, J. Algebra, **95**, 236–244 (1985).

[523] M. F. Smiley, *Jordan homomorphisms onto prime rings*, Trans. Amer. Math. Soc., **84**, 426–429 (1957).

[524] G. A. Swain and P. S. Blau, *Lie derivations in prime rings with involution*, Canad. Math. Bull., **42**, 401–411 (1999).

[525] G. A. Swain, *Lie derivations of the skew elements of prime rings with involution*, J. Algebra, **184**, 679–704 (1996).

[526] A. Taghavi and S. Gholampoor, *Maps preserving product A^*B+B^*A on $\mathcal{C}^*$-algebras*, Bull. Iranian Math. Soc., **48**, 757–767 (2022).

[527] A. Taghavi and M. Razeghi, *Nonlinear new product A^*B-B^*A derivations on $*$-algebras*, Proyecciones, **39(2)**, 467–479 (2020).

[528] A. Taghavi, H. Rohi and V. Darvish, *Non-linear $*$-Jordan derivations on von Neumann algebras*, Linear Multilinear Algebra, **64**, 426–439 (2016).

[529] M. P. Thomas, *The image of a derivation is contained in the radical*, Ann. of Math., **128(3)**, 435–60 (1988).

[530] A. R. Villena, *Lie derivations on Banach algebras*, J. Algebra, **226**, 390–409 (2000).

[531] J. Vukman, *Symmetric hi-derivations on prime and semi-prime rings*, Aequationes Math., **38**, 245–254 (1989).

[532] J. Vukman, *Two results concerning symmetric bi-derivations on prime rings*, Aequationes. Math., **40**, 181–189 (1990).

[533] J. Vukman, *Commuting and centralizing mappings in prime rings*, Proc. Am. Math. Soc., **109**, 47–52 (1990).

[534] J. Vukman, *Jordan left derivations on semiprime rings*, Math. J. Okayama Univ., **39** (1997).

[535] J. Vukman, *Identities with derivations on rings and Banach algebras*, Glas. Mat. Ser. III, **40(2)**, 189–199 (2005).

[536] J. Vukman, *On a derivations of prime and semiprime rings*, Demonstr. Math., **XXXVIII**(2), 283–290 (2005).

[537] J. Vukman, *Identities with products of* (α,β)*-derivations on prime rings*, Demonstr. Math. (2006).

[538] J. Vukman, *A note on generalized derivation on semiprime rings*, Taiwanese J. Math., **11(2)**, 367–370 (2007).

[539] J. Vukman, *On left Jordan derivations of rings and Banach algebras*, Aequationes. Math., **75**, 260–266 (2008).

[540] J. Vukman, *Some remarks on derivations in semiprime rings and standard operator algebras*, Glasnik Mat., **46**(66), 43–48 (2011).

[541] J. Vukman and I. Kosi Ulbl, *A note on derivations in semiprime rings*, Int. J. Math. Math. Sci., **17**, 3347–3350 (2005).

[542] J. Vukman and I. Kosi Ulbl, *On some equations related to derivations in rings*, Int. J. Math. Math. Sci., **17**, 2703–2710 (2005).

[543] Y. Wang, *Power-centralizing automorphisms of Lie ideals in prime rings*, Comm. Algebra, **34**, 609–615 (2006).

[544] Y. Wang, *Lie n-derivations of unital algebras with idempotents*, Linear Algebra Appl., **458**, 512–525 (2014).

[545] Y. Wang, *Lie (Jordan) derivations of arbitrary triangular algebras*, Aequationes Math., **93**(6), 1221–1229 (2019).

[546] Y. Wang and Y. Wang, *Multiplicative Lie n-derivations of generalized matrix algebras*, Linear Algebra Appl., **438**, 2599–2616 (2013).

[547] D. Y. Wang and Q. Yu, *Derivations of the parabolic subalgebras of the general linear Lie algebra over a commutative ring*, Linear Algebra Appl., **418**, 763–774 (2006).

[548] B. A. Wani, (ϕ,ψ)*-derivations on semiprime rings and Banach algebras*, Commun. Math., **29**(3), 371–383 (2021).

[549] B. A. Wani, M. Ashraf and M. S. Akhtar, *Multiplicative* $*$*-Lie type higher derivations of standard operator algebras*, Comm. Algebra, https://doi.org/10.1080/00927872.2021.1906266 (2021).

[550] B. A. Wani, M. Ashraf and W. Lin, *Multiplicative* $*$*-Jordan type higher derivations on von Neumann algebras*, Quaestiones Mathematicae, **43(12)**, 1689–1711 (2020).

[551] W. Watkins, *Linear maps that preserve commuting pairs of matrices*, Linear Algebra Appl., **14**, 29–35 (1976).

[552] F. Wei, *Generalized derivations with nilpotent values on semiprime rings*, Acta Math. Sinica, **20(3)**, 453–462 (2004).

[553] F. Wei and Z. Xiao, *Generalized Jordan triple higher derivations on semiprime rings*, Bull. Korean Math. Soc., **46(3)**, 553–565 (2009).

[554] F. Wei and Z. K. Xiao, *Higher derivations of triangular algebras and its generalizations*, Linear Algebra Appl., **435**, 1034–1054 (2011).

[555] A. M. Wolfgang, *A characterisation of Nilpotent Lie algebras by invertible Leibniz-derivations*, Comm. Algebra, **41**(7), 2427–2440 (2013).

[556] T. L. Wong, *Derivations with power central values on multilinear polynomials*, Algebra Colloq., **3(4)**, 369–378 (1996).

[557] T. L. Wong, *Derivations cocentralizing multilinear polynomials*, Taiwanese J. Math., **1**, 31–37 (1997).

[558] T. L. Wong, *On certain subgroups of semiprime rings with derivations*, Comm. Algebra, **32(5)**, 1961–1968 (2004).

[559] Z. K. Xiao and F. Wei, *Jordan higher derivations on triangular algebras*, Linear Algebra Appl., **432**, 2615–2622 (2010).

[560] Z. K. Xiao and F. Wei, *Lie triple derivations of triangular algebras*, Linear Algebra Appl., **437**, 1234–1249 (2012).

[561] Z. K. Xiao and F. Wei, *Nonlinear Lie higher derivations on triangular algebras*, Linear Multilinear Algebra, **60**, 979–994 (2012).

[562] N. B. Yarbil, *Annihilators of power values of b-generalized derivations in prime rings*, Commun. Fac. Sci. Univ. Ank. Ser. A1 Math. Stat., **69(2)**, 1278–1284 (2020).

[563] W. Y. Yu and J. H. Zhang, *Nonlinear Lie derivations of triangular algebras*, Linear Algebra Appl., **432**, 2953–2960 (2010).

[564] W. Y. Yu and J. H. Zhang, *Nonlinear $*$-Lie derivations on factor von Neumann algebras*, Linear Algebra Appl., **437**, 1979–1991 (2012).

[565] H. Yazarli, M. A. Öztürk and Y. B. Jun, *Tri-additive maps and permuting tri-derivations*, Commun. Fac. Sci. Univ. Ank. Ser. A1. Math. Stat., **54**, 1–8 (2005).

[566] S. M. A. Zaidi, M. Ashraf and S. Ali, *On Jordan ideals and left (θ,θ)-derivations in prime rings*, Int. J. Math. Math. Sci., **37-40**, 1957–1964 (2004).

[567] J. Zhang and C. J. Li, *Nonlinear bi-skew Lie-type derivations on $*$-algebras*, Rocky Mountain J. Math. (2023) (to appear).

[568] F. J. Zhang, *Nonlinear skew Jordan derivable maps on factor von Neumann algebras*, Linear Multilinear Algebra, **64**, 2090–2103 (2016).

[569] F. Zhang, X. F. Qi and J. Zhang, *Nonlinear $*$-Lie higher derivations on factor von Neumann algebras*, Bull. Iranian Math. Soc., **42(3)**, 659–678 (2016).

[570] J. H. Zhang, B. W. Wu and H. X. Cao, *Lie triple derivations of nest algebras*, Linear Algebra Appl., **416**, 559–567 (2006).

[571] J. H. Zhang and W. Y. Yu, *Jordan derivations of triangular algebras*, Linear Algebra Appl., **419**, 251–255 (2006).

[572] F. J. Zhang and J. H. Zhang, *Nonlinear Lie derivations on factor von Neumann algebras*, Acta Mathematica Sinica. (Chin. Ser), **54(5)**, 791–802 (2011).

[573] F. F. Zhao and C. J. Li, *Nonlinear $*$-Jordan triple derivations on von Neumann algebras*, Math. Slovaca, **68**, 163–170 (2018).

[574] F. F. Zhao and C. J. Li, *N*onlinear maps preserving the Jordan triple $*$-product between factors, Indag. Math., **29**, 619–627 (2018).

Index

About the Monograph

The fundamental premise of this monograph is the theory of differential identities in associative rings and algebras. Informally, a differential identity in a ring (or algebra) is an identical relation involving arbitrary elements in the underlying rings (or algebras) together with the unknown differential function. This abstract theory has proven to be a powerful tool for resolving a variety of complex problems that arise not only in the theory of rings and algebras but also in areas such as invariant theory, non-commutative geometry, and mathematical physics.

This volume comprises six chapters that present a comprehensive overview of the results and methods that have greatly enriched the subject during the past three decades. The first chapter presents a brief introduction of the subject, which is aimed at making this monograph as self contained as possible, while the next three chapters discuss the diverse types of derivations in rings and algebras, and the interplay between the structure of certain classes of rings and the behavior of the underlying derivations, n-derivations, Lie derivations, pre-derivations, generalized derivations, skew derivations, b-generalized derivations and their respective properties. The fifth chapter delves into the study of various types of higher derivable mappings, their additivity and characterization of Lie and Jordan type higher derivations as the sum of a derivation and a central mapping with certain annihilating properties. Though applications of the results presented in these chapters can be found throughout the book, the final chapter mainly consists of applications of derivations. Besides characterizing derivations and their invariants with local properties, the results of this chapter provide a deeper understanding of the behavior of derivations and the intrinsic structure of the ring over which they are defined.

With over more than 500 references on the subject will be of interest to advanced graduate students, researchers particularly those specializing in algebras and ring theory. The inclusion of the latest results with their historical background in the book makes it a desirable reading for beginners in research as well.

Each chapter is enriched with examples and exercises which make this monograph useful for researchers and throughout the book several open problems are provided to stimulate further research in this area.

FSC
www.fsc.org
MIX
Papier | Fördert gute Waldnutzung
FSC® C083411

Kolibri 360 GmbH
Ferdinand-Jühlke-Straße 7
99095 Erfurt, Deutschland
produktsicherheit@kolibri360.de